W9-CUK-143

Büchner, Schliebs, Winter, Büchel

Industrial Inorganic Chemistry

Distribution

VCH Verlagsgesellschaft, P.O. Box 1260/1280, D-6940 Weinheim (Federal Republic of Germany)

Switzerland: VCH Verlags-AG, P.O. Box, CH-4020 Basel (Switzerland)

Great Britain and Ireland: VCH Publishers (UK) Ltd., 8 Wellington Court, Wellington Street, Cambridge CB1 1HW (Great Britain)

USA and Canada: VCH Publishers, Suite 909, 220 East 23rd Street, New York, NY 10010-4606 (USA)

ISBN 3-527-26629-1 (VCH Verlagsgesellschaft) ISBN 0-89573-610-1 (VCH Publishers)

Industrial Inorganic Chemistry

W. Büchner
R. Schliebs
G. Winter
K. H. Büchel

Translated by David R. Terrell

Prof. Dr. Werner Büchner
Prof. Dr. Reinhard Schliebs †
Prof. Dr. Gerhard Winter
Prof. Dr. Dr. h.c. Karl Heinz Büchel
Bayer AG
D-5090 Leverkusen
Federal Republic of Germany

Published jointly by
VCH Verlagsgesellschaft, Weinheim (Federal Republic of Germany)
VCH Publishers, New York, NY (USA)

Editorial Director: Dr. Hans F. Ebel and Dr. Christina Dyllick-Brenzinger
Production Manager: Dipl.-Ing. (FH) Hans Jörg Maier

Library of Congress Card No. 88-37860

British Library Cataloguing in Publication Data
Industrial inorganic chemistry.
1. Industrial chemicals, —— Encyclopaedias
I. Buchner, W (Werner)
II. [Industrielle anorganische Chemie].
English
611′.003′21
ISBN 3-527-26629-1

Deutsche Bibliothek Cataloguing-in-Publication Data
Industrial inorganic chemistry / W. Büchner ... Transl. by
David R. Terrell. – Weinheim ; Basel (Switzerland) ; Cambridge ;
New York, NY : VCH, 1989
Einheitssacht.: Industrielle anorganische Chemie <engl.>
ISBN 3-527-26629-1 (Weinheim) ...) Pp.
ISBN 0-89573-610-1 (Cambridge ...) Pp.
NE: Büchner, Werner [Mitverf.]; EST

Composition and Printing: Wiesbadener Graphische Betriebe GmbH, D-6200 Wiesbaden 1
Bookbinding: Josef Spinner, Großbuchbinderei GmbH, D-7583 Ottersweier
Printed in the Federal Republic of Germany

Preface

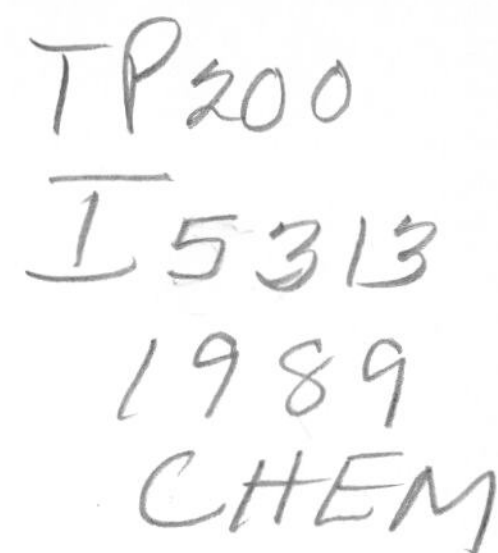

"Industrial Inorganic Chemistry" was first published in German in 1984. The book was well received by students and teachers alike, leading to the publication of a second German edition in 1986. The publishers, VCH Verlagsgesellschaft, were convinced that a wide circle of readers would welcome the appearance of our book in the English language, and their encouragement has led to the preparation of the present, up-dated and revised edition in English.

The basic structure of the German edition has been retained. Changes in the industrial importance of some compounds and processes since the appearance of the German edition have been taken into account and data relating to the US market have been emphasized. Thus the chapter on potassium permanganate has been considerably abridged and that on the membrane process for the manufacture of chlorine and sodium hydroxide expanded.

We are indebted to Dr Podesta and Dr Heine from Bayer AG for their assistance in the revision of the German edition in addition to the institutions and colleagues mentioned in the preface to the German edition.

The book was translated by Dr D. R. Terrell from Agfa-Gevaert NV, to whom we are particularly grateful for the patience and care he devoted to this difficult task. We also wish to acknowledge the contribution of VCH Verlagsgesellschaft in producing this edition.

Leverkusen, Autumn 1988 K. H. Büchel

Preface to the First German Edition

The book "Industrielle Anorganische Chemie" will fill a long term need, which has become even more apparent since the appearance of "Industrielle Organische Chemie" by Weissermel and Arpe*. Although there are comprehensive chapters on this branch of chemistry in a number of encyclopedias and handbooks, a single volume text is lacking that describes concisely the current state of industrial inorganic chemistry.

The authors have been made aware of this need in discussions with students, young chemists, colleagues in neighboring fields, teachers and university lecturers and willingly accepted the suggestion of the publishers to write this text. Changes in the supply of raw materials and their markets and economic and ecological requirements are responsible for the continual reshaping of the inorganic chemical industry. As a result the treatment of industrial processes in the available textbooks seldom keeps pace with these developments.

The inorganic chemical industry is an important branch of industry and its structure is particularly diverse: including a large number of finished products (mineral fertilizers, construction materials, glass, enamels and pigments to name but a few) and basic products for the organic chemical industry such as mineral acids, alkalis, oxidizing agents and halogens. Modern developments in other branches of industry such as chips for microelectronics, video cassettes and optical fibers have only been possible due to the continuous development of the inorganic chemical industry.

This book emphasises the manufacturing processes, economic importance and applications of products. In the sections on production the pros and cons are considered in the context of the raw material situation, economic and ecological considerations and energy consumption, the different situations in different countries also being taken into account. Processes which are no longer operated are at most briefly mentioned. The properties of the products are only considered to the extent that they are relevant for production or applications.

It was necessary to restrict the material to avoid overextending the brief. Metallurgical processes have not been included, except for the manufacture of "chemical" metals (e.g. alkali metals) which is briefly described. Several borderline areas with organic chemistry are considered (e.g. organophosphorus, -silicon and -fluoro products), others are deliberately excluded. A whole chapter is devoted to the nuclear fuel cycle, since it involves so much industrial scale inorganic chemistry and is currently so important.

---

* An English translation of this book is available: K. Weissermel, H. J. Arpe: *Industrial Organic Chemistry,* VCH Verlagsgesellschaft, 1978.

The layout follows that of its sister book "Industrielle Organische Chemie" with the main text being supplemented by marginal notes. These are essentially summaries of the main text and enable the reader to obtain a rapid grasp of the most important facts. The equations are printed on a gray background for the same reason.

At the end of each main section a generally subtitled list of references is provided. This should enable the reader to obtain more detailed information on particular matters with the minimum of effort. In addition to references to original papers and reviews, readers are referred to the important handbooks: Ullmann, Winnacker-Küchler and Kirk-Othmer. The Chemical Economics Handbook of the Stanford Research Institute has frequently been used for economic data.

The documentation system at Bayer AG was invaluable in gathering the important facts for this book. Numerous colleagues have also assisted us:

Outside Bayer AG our thanks are due to Prof. P. Eyerer from Stuttgart University, Dr H. Grewe from Krupp AG, Essen, Dr Ch. Hahn from Hutschenreuther AG, Selb, Dr G. Heymer from Hoechst AG, Knapsack Works, Dr P. Kleinschmit from Degussa, Dr G. König from Martin & Pagenstecher GmbH, Krefeld, Dr R. Kroebel from the Kernforschungszentrum Karlsruhe, Dr G. Kühner from Degussa AG, Prof. F. W. Locher from the Forschungsinstitut der Zementindustrie, Düsseldorf, H. Schmidt from the Ziegeleiforschungsinstitut, Essen, Dr M. Schwarzmann and his colleagues from BASF AG and Dr E. Wege from Sigri Elektrographit GmbH, Meitingen, for technical advice and critical perusal of sections of the manuscript.

Inside Bayer AG our thanks are due to Dr H.-P. Biermann, Dr G. Franz, Dr P. Kiemle, Dr M. Mansmann, Dr H. H. Moretto and Dr H. Niederprüm, who with many other colleagues have helped with the technical realization of the text. In particular we would like to thank Dr Hanna Söll, who with her many years of experience has substantially contributed to the editing of this book.
We also thank Verlag Chemie, which has assimilated the suggestions of the authors with much understanding and has produced this book in such an excellent form.

Leverkusen, Spring 1984 K. H. Büchel

Contents

1 Primary Inorganic Materials . . . 1

1.1 Water . . . 1

1.1.1 Economic Importance . . . 1

1.1.2 Production of Potable Water . . . 2
1.1.2.1 Break-Point Chlorination and Ozonization . . . 3
1.1.2.2 Flocculation and Sedimentation . . . 4
1.1.2.3 Filtration . . . 5
1.1.2.4 Removal of Dissolved Inorganic Impurities . . . 5
1.1.2.5 Removal of Dissolved Organic Impurities . . . 6
1.1.2.6 Safety Chlorination . . . 7
1.1.2.7 Removal of Nitrates . . . 7
1.1.2.8 Removal of Traces of Chlorohydrocarbons from Groundwater . . . 8
1.1.2.9 Production of Soft or Deionized Water . . . 8
1.1.3 Production of Freshwater from Seawater and Brackish Water . . . 10
1.1.3.1 Production by Multistage Flash Evaporation . . . 10
1.1.3.2 Production using Reverse Osmosis . . . 12
References for Chapter 1.1: Water . . . 13

1.2 Hydrogen . . . 14

1.2.1 Economic Importance . . . 14

1.2.2 Hydrogen Manufacture . . . 15
1.2.2.1 Petrochemical Processes and Coal Gasification . . . 15
1.2.2.2 Water Electrolysis . . . 16
1.2.2.3 Other Manufacturing Processes for Hydrogen . . . 17
1.2.2.4 Production of Hydrogen as a Byproduct . . . 18
1.2.3 Hydrogen Applications . . . 19
References for Chapter 1.2: Hydrogen . . . 20

1.3 Hydrogen Peroxide and Inorganic Peroxo Compounds . . . 21

1.3.1 Economic Importance . . . 21

1.3.1.1 Hydrogen Peroxide . . . 21
1.3.1.2 Sodium Perborate and Sodium Carbonate Perhydrate . . . 22
1.3.1.3 Alkali Peroxodisulfates and Sodium Peroxide . . . 22

1.3.2 Production . . . 23
1.3.2.1 Hydrogen Peroxide . . . 23
1.3.2.2 Sodium Perborate . . . 26

1.3.2.3 Sodium Carbonate Perhydrate (Sodium Percarbonate) 27
1.3.2.4 Alkali Peroxodisulfates 27
1.3.2.5 Sodium Peroxide 28

1.3.3 Applications 29
1.3.3.1 Hydrogen Peroxide, Sodium Perborate and Sodium Carbonate Perhydrate 29
1.3.3.2 Alkali Peroxodisulfates and Sodium Peroxide 30
References for Chapter 1.3: Hydrogen Peroxide and Inorganic Peroxo Compounds 30

1.4 Nitrogen and Nitrogen Compounds 30

1.4.1 Ammonia 30
1.4.1.1 Economic Importance 31
1.4.1.2 Synthetic Ammonia Manufacture 31
1.4.1.2.1 General Information 31
1.4.1.2.2 Synthesis Gas Production 33
1.4.1.2.3 Ammonia Synthesis Catalysts 39
1.4.1.2.4 Reaction of Synthesis Gas to Ammonia 41
1.4.1.2.5 Integrated Ammonia Synthesis Plants 43
1.4.1.3 Ammonia Applications 44
References for Chapter 1.4.1: Ammonia 45

1.4.2 Hydrazine 45
1.4.2.1 Economic Importance 45
1.4.2.2 Manufacture of Hydrazine 45
1.4.2.2.1 Raschig Process 46
1.4.2.2.2 Urea Process 48
1.4.2.2.3 Bayer Process 48
1.4.2.2.4 H_2O_2 Process 50
1.4.2.3 Applications for Hydrazine 51
References for Chapter 1.4.2: Hydrazine 52

1.4.3 Hydroxylamine 52
1.4.3.1 Economic Importance and Applications 52
1.4.3.2 Manufacture 53
1.4.3.2.1 Raschig Process 53
1.4.3.2.2 Nitrogen(II) Oxide Reduction Process 53
1.4.3.2.3 Nitrate Reduction Process (DSM/HPO-Stamicarbon) 54
References for Chapter 1.4.3: Hydroxylamine 55

1.4.4 Nitric Acid 55
1.4.4.1 Economic Importance 55
1.4.4.2 Manufacture 56
1.4.4.2.1 Fundamentals of Nitric Acid Manufacture 56
1.4.4.2.2 Plant Types 59
1.4.4.2.3 Process Description 60
1.4.4.2.4 Manufacture of Highly Concentrated Nitric Acid 62

1.4.4.2.5 Tail Gases from Nitric Acid Manufacture 65
1.4.4.3 Nitric Acid Applications 67
References for Chapter 1.4.4: Nitric Acid 68

1.5 Phosphorus and Its Compounds 68

1.5.1 Phosphorus and Inorganic Phosphorus Compounds 68
1.5.1.1 Raw Materials 68
1.5.1.2 Products 70
1.5.1.2.1 Phosphoric Acid 70
1.5.1.2.2 Phosphoric Acid Salts 79
1.5.1.2.3 Phosphorus 85
1.5.1.2.4 Products Manufactured from Phosphorus 90
References for Chapter 1.5.1: Phosphorus and Inorganic Phosphorus Compounds 95

1.5.2 Organic Phosphorus Compounds 96
1.5.2.1 Economic Importance 96
1.5.2.2 Manufacture of Organic Phosphorus Compounds 96
1.5.2.2.1 Phosphoric Acid Triesters 96
1.5.2.2.2 Phosphorus(V) Ester Acids 98
1.5.2.2.3 Thiophosphoric Acid Derivatives 99
1.5.2.2.4 Phosphorous Acid Esters 99
1.5.2.2.5 Phosphonic Acids 100
1.5.2.3 Applications of Organic Phosphorus Compounds 102
1.5.2.3.1 Phosphoric Acid Triesters 102
1.5.2.3.2 Phosphorus(V) Ester Acids 103
1.5.2.3.3 Thiophosphoric Acid Derivatives 103
1.5.2.3.4 Phosphorous Acid Esters 104
1.5.2.3.5 Phosphonic Acids 105
References for Chapter 1.5.2: Organic Phosphorus Compounds. 105

1.6 Sulfur and Sulfur Compounds 105

1.6.1 Sulfur 105
1.6.1.1 Occurrence 105
1.6.1.2 Economic Importance 106
1.6.1.3 Production 106
1.6.1.3.1 From Elemental Sulfur Deposits 106
1.6.1.3.2 From Hydrogen Sulfide and Sulfur Dioxide 107
1.6.1.3.3 From Pyrites 108
1.6.1.4 Applications 108

1.6.2 Sulfuric Acid 108
1.6.2.1 Economic Importance 108
1.6.2.2 Starting Materials for Sulfuric Acid Manufacture 109
1.6.2.2.1 Sulfuric Acid from Sulfur Dioxide 109

1.6.2.2.2 Sulfuric Acid from Waste Sulfuric Acid and Metal Sulfates 115
1.6.2.3 Applications of Sulfuric Acid 120
1.6.3 100% Sulfur Dioxide 120
1.6.4 100% Sulfur Trioxide 122
1.6.5 Disulfur Dichloride 123
1.6.6 Sulfur Dichloride 123
1.6.7 Thionyl Chloride 123
1.6.8 Sulfuryl Chloride 124
1.6.9 Chlorosulfonic Acid 125
1.6.10 Fluorosulfonic Acid 125
1.6.11 Sulfurous Acid Salts 125
1.6.12 Sodium Thiosulfate, Ammonium Thiosulfate 126
1.6.13 Sodium Dithionite and Sodium Hydroxymethanesulfinate 127
1.6.14 Hydrogen Sulfide 129
1.6.15 Sodium Sulfide 130
1.6.16 Sodium Hydrogen Sulfide 130
1.6.17 Carbon Disulfide 131
References for Chapter 1.6: Sulfur and Sulfur Compounds 132

1.7 Halogens and Halogen Compounds 133
1.7.1 Fluorine and Fluorine Compounds 133
1.7.1.1 Economic Importance and Applications 133
1.7.1.2 Raw Materials 133
1.7.1.2.1 Fluorspar 134
1.7.1.2.2 Fluorapatite 135
1.7.1.3 Fluorine and Inorganic Fluorides 135
1.7.1.3.1 Fluorine 135
1.7.1.3.2 Hydrogen Fluoride 137
1.7.1.3.3 Aluminum Fluoride 140
1.7.1.3.4 Sodium Aluminum Hexafluoride, Cryolite 142
1.7.1.3.5 Alkali Fluorides 143
1.7.1.3.6 Hexafluorosilicates 144
1.7.1.3.7 Uranium Hexafluoride 144
1.7.1.3.8 Boron Trifluoride and Tetrafluoroboric Acid 145
1.7.1.3.9 Sulfur Hexafluoride 145
1.7.1.4 Organofluoro Compounds by Electrochemical Fluorination 146
References for Chapter 1.7.1: Fluorine and Fluorine Compounds. . . . 148

1.7.2 Chloralkali Electrolysis, Chlorine and Sodium Hydroxide 149
1.7.2.1 Economic Importance 149
1.7.2.2 Starting Materials 150
1.7.2.3 Manufacturing Processes 152
1.7.2.3.1 Mercury Process 152
1.7.2.3.2 Diaphragm Process 155
1.7.2.3.3 Membrane Process 157
1.7.2.3.4 Evaluation of the Mercury, Diaphragm and Membrane Processes . . . 161
1.7.2.4 Applications for Chlorine and Sodium Hydroxide 162
1.7.2.4.1 Chlorine 162
1.7.2.4.2 Sodium Hydroxide 162
References for Chapter 1.7.2: Chloralkali Electrolysis, Chlorine, Sodium Hydroxide 163

1.7.3 Hydrochloric Acid – Hydrogen Chloride 164
1.7.3.1 Manufacture of Hydrogen Chloride 164
1.7.3.2 Economic Importance of Hydrogen Chloride and Hydrochloric Acid . . 165
1.7.3.3 Electrolysis of Aqueous Hydrochloric Acid Solutions. 166
1.7.3.4 Nonelectrolytic Processes for the Manufacture of Chlorine from Hydrogen Chloride 167
References for Chapter 1.7.3: Hydrochloric Acid – Hydrogen Chloride 168

1.7.4 Chlorine-Oxygen Compounds 168
1.7.4.1 Economic Importance 168
1.7.4.2 Manufacture of Chlorine-Oxygen Compounds 169
1.7.4.2.1 Hypochlorite 169
1.7.4.2.2 Chlorites. 172
1.7.4.2.3 Chlorates 173
1.7.4.2.4 Perchlorates and Perchloric Acid 175
1.7.4.2.5 Chlorine Dioxide 175
1.7.4.3 Applications of Chlorine-Oxygen Compounds 177
References for Chapter 1.7.4: Chlorine-Oxygen Compounds 178

1.7.5 Bromine and Bromine Compounds 178
1.7.5.1 Natural Deposits and Economic Importance 178
1.7.5.2 Manufacture of Bromine and Bromine Compounds 179
1.7.5.2.1 Bromine 179
1.7.5.2.2 Hydrogen Bromide 181
1.7.5.2.3 Alkali Bromides, Calcium Bromide, Zinc Bromide 181
1.7.5.2.4 Alkali Bromates 182
1.7.5.3 Applications for Bromine and Bromine Compounds 182
References for Chapter 1.7.5: Bromine and Bromine Compounds. . . . 184

1.7.6 Iodine and Iodine Compounds. 184
1.7.6.1 Economic Importance 184
1.7.6.2 Manufacture of Iodine and Iodine Compounds 185
1.7.6.2.1 Iodine 185
1.7.6.2.2 Hydrogen Iodide 186

1.7.6.2.3 Alkali Iodides 187
1.7.6.2.4 Alkali Iodates 187
1.7.6.3 Applications for Iodine and Iodine Compounds 187
References for Chapter 1.7.6: Iodine and Iodine Compounds 188

2 Mineral Fertilizers 189

2.1 Phosphorus-Containing Fertilizers 189

2.1.1 Economic Importance 189
2.1.1.1 General Information 189
2.1.1.2 Importance of Superphosphate 190
2.1.1.3 Importance of Triple Superphosphate 190
2.1.1.4 Importance of Ammonium Phosphates 191
2.1.1.5 Importance of Nitrophosphates 191
2.1.1.6 Importance and Manufacture of Thermal (Sinter, Melt) and Basic Slag (Thomas) Phosphates 191

2.1.2 Manufacture of Phosphorus-Containing Fertilizers 192
2.1.2.1 Superphosphate 192
2.1.2.2 Triple Superphosphate 194
2.1.2.3 Ammonium Phosphates 194
2.1.2.4 Nitrophosphates 197

2.2 Nitrogen-Containing Fertilizers 199

2.2.1 Economic Importance 199
2.2.1.1 Importance of Ammonium Sulfate 200
2.2.1.2 Importance of Ammonium Nitrate 200
2.2.1.3 Importance of Urea 201

2.2.2 Manufacture of Nitrogen-Containing Fertilizers 201
2.2.2.1 Ammonium Sulfate 201
2.2.2.2 Ammonium Nitrate 203
2.2.2.3 Urea 204

2.3 Potassium-Containing Fertilizers 208

2.3.1 Occurrence of Potassium Salts 208

2.3.2 Economic Importance of Potassium-Containing Fertilizers 209

2.3.3 Production of Potassium-Containing Fertilizers 210
2.3.3.1 Potassium Chloride 210
2.3.3.2 Potassium Sulfate 212
2.3.3.3 Potassium Nitrate 213
References for Chapter 2: Mineral Fertilizers 213

3 Metals and their Compounds . . . 215

3.1 Alkali and Alkaline Earth Metals and their Compounds . . . 215

3.1.1 Alkali Metals and their Compounds . . . 215
3.1.1.1 General Information . . . 215
3.1.1.2 Lithium and its Compounds . . . 215
3.1.1.2.1 Natural Deposits and Commercial Importance . . . 215
3.1.1.2.2 Metallic Lithium . . . 216
3.1.1.2.3 Lithium Compounds . . . 217
3.1.1.3 Sodium and its Compounds . . . 218
3.1.1.3.1 General Information . . . 218
3.1.1.3.2 Metallic Sodium . . . 218
3.1.1.3.3 Sodium Carbonate . . . 219
3.1.1.3.4 Sodium Hydrogen Carbonate . . . 222
3.1.1.3.5 Sodium Sulfate . . . 223
3.1.1.3.6 Sodium Hydrogen Sulfate . . . 225
3.1.1.3.7 Sodium Borates . . . 226
3.1.1.4 Potassium and its Compounds . . . 228
3.1.1.4.1 General Information . . . 228
3.1.1.4.2 Metallic Potassium . . . 228
3.1.1.4.3 Potassium Hydroxide . . . 229
3.1.1.4.4 Potassium Carbonate . . . 229
References for Chapter 3.1.1: Alkali Metals and their Compounds . . . 230

3.1.2 Alkaline Earth Metals and their Compounds . . . 231
3.1.2.1 General Information . . . 231
3.1.2.2 Beryllium and its Compounds . . . 232
3.1.2.3 Magnesium and its Compounds . . . 233
3.1.2.3.1 Natural Deposits . . . 233
3.1.2.3.2 Metallic Magnesium . . . 233
3.1.2.3.3 Magnesium Carbonate . . . 235
3.1.2.3.4 Magnesium Oxide . . . 236
3.1.2.3.5 Magnesium Chloride . . . 237
3.1.2.3.6 Magnesium Sulfate . . . 238
3.1.2.4 Calcium and its Compounds . . . 239
3.1.2.4.1 Natural Deposits . . . 239
3.1.2.4.2 Metallic Calcium . . . 239
3.1.2.4.3 Calcium Carbonate . . . 240
3.1.2.4.4 Calcium Oxide and Calcium Hydroxide . . . 240
3.1.2.4.5 Calcium Chloride . . . 241
3.1.2.4.6 Calcium Carbide . . . 242
3.1.2.5 Strontium and its Compounds . . . 243
3.1.2.6 Barium and its Compounds . . . 244
3.1.2.6.1 Natural Deposits and Economic Importance . . . 244
3.1.2.6.2 Barium Carbonate . . . 244

3.1.2.6.3 Barium Sulfide 245
3.1.2.6.4 Barium Sulfate 245
References for Chapter 3.1.2: Alkaline Earth Metals and their Compounds 246

3.2 Aluminum and its Compounds 247

3.2.1 General Information 247
3.2.2 Occurrence 247
3.2.3 Aluminum Metal 248
3.2.3.1 Economic Importance 248
3.2.3.2 Manufacture 248
3.2.3.3 Aluminum Applications 249

3.2.4 Aluminum Oxide and Aluminum Hydroxide 250
3.2.4.1 Economic Importance 250
3.2.4.2 Manufacture 251
3.2.4.3 Applications 252
3.2.5 Aluminum Sulfate 253
3.2.6 Aluminum Chloride 253
3.2.7 Sodium Aluminate 254
References for Chapter 3.2: Aluminum and its Compounds 254

3.3 Chromium Compounds and Chromium 255

3.3.1 Chromium Compounds 255
3.3.1.1 Economic Importance 255
3.3.1.2 Raw Material: Chromite 257
3.3.1.3 Manufacture of Chromium Compounds 258
3.3.1.3.1 Chromite Roasting to Alkali Chromates 258
3.3.1.3.2 Alkali Dichromates 261
3.3.1.3.3 Chromium(VI) Oxide ("Chromic Acid") 263
3.3.1.3.4 Chromium(III) Oxide 263
3.3.1.3.5 Basic Chromium(III) Salts (Chrome-Tanning Agents) 264
3.3.1.4 Applications for Chromium Compounds 265

3.3.2 Chromium Metal 266
3.3.2.1 Economic Importance 266
3.3.2.2 Manufacture of Chromium Metal 266
3.3.2.2.1 Chemical Reduction 266
3.3.2.2.2 Electrochemical Reduction of Chrome Alum 267
3.3.2.2.3 Electrochemical Reduction of Chromium(VI) Oxide 268
References for Chapter 3.3: Chromium Compounds and Chromium . . 268

3.4 Silicon and its Inorganic Compounds 269

3.4.1 Elemental Silicon 269
3.4.1.1 General Information and Economic Importance 269
3.4.1.2 Manufacture 270

3.4.1.2.1 Metallurgical Grade Silicon and Ferrosilicon . . . 270
3.4.1.2.2 Electronic Grade (EG) Silicon . . . 271
3.4.1.3 Silicon Applications . . . 273

3.4.2 Inorganic Silicon Compounds . . . 274
References for Chapter 3.4: Silicon and its Inorganic Compounds . . . 276

3.5 Manganese Compounds and Manganese . . . 277

3.5.1 Manganese Compounds . . . 277
3.5.1.1 Economic Importance . . . 277
3.5.1.2 Raw Materials . . . 277
3.5.1.3 Manufacture of Manganese Compounds . . . 279
3.5.1.3.1 Manganese(II) Compounds . . . 279
3.5.1.3.2 Manganese(II,III) Oxide (Mn_3O_4) and Manganese(III) Oxide (Mn_2O_3) 281
3.5.1.3.3 Manganese(IV) Oxide . . . 281
3.5.1.3.4 Potassium Permanganate . . . 284
3.5.1.4 Applications for Manganese Compounds . . . 286

3.5.2 Manganese – Electrochemical Manufacture, Importance and Applications . . . 287
References for Chapter 3.5: Manganese Compounds and Manganese . . . 288

4 Silicones . . . 289

4.1 Structure and Properties . . . 289

4.2 Economic Importance . . . 290

4.3 Silicone Manufacture . . . 291

4.3.1 Synthesis of the Starting Materials . . . 291
4.3.1.1 (Chloro)methylsilanes . . . 291
4.3.1.2 (Chloro)phenyl- and (Chloro)methylphenylsilanes . . . 292
4.3.1.3 Other Industrially Important Silanes . . . 293

4.3.2 Linear Poly(organosiloxanes) . . . 294
4.3.2.1 Synthesis of Oligomeric Precursors . . . 294
4.3.2.2 Synthesis of High Molecular Weight Linear Poly(dimethylsiloxanes) . . 295

4.3.3 Manufacture of Branched Poly(organosiloxanes) . . . 297

4.4 Industrial Silicone Products . . . 298

4.4.1 Silicone Oils . . . 298

4.4.2 Products Manufactured from Silicone Oils . . . 299

4.4.3 Silicone Rubbers . . . 299
4.4.3.1 Room Temperature Vulcanizable Single Component Silicone Rubbers . 299
4.4.3.2 Room Temperature Vulcanizable Two Component Silicone Rubbers . . 301
4.4.3.3 Hot Vulcanizable Peroxide Crosslinked Silicone Rubbers . . . 302

4.4.3.4 Hot Vulcanizable Addition Crosslinkable Silicone Rubbers . . . 302
4.4.3.5 Hot Vulcanizable Liquid Rubbers . . . 303
4.4.3.6 Silicone Rubber Properties . . . 303

4.4.4 Silicone Resins . . . 304

4.4.5 Silicone Copolymers, Block Copolymers and Graft Copolymers . . . 305
References for Chapter 4: Silicones . . . 306

5 Inorganic Solids . . . 307

5.1 Silicate Products . . . 307

5.1.1 Glass . . . 307
5.1.1.1 Economic Importance . . . 307
5.1.1.2 General Information . . . 307
5.1.1.3 Glass Compositions . . . 308
5.1.1.4 Glass Manufacture . . . 310
5.1.1.4.1 Glass Raw Materials . . . 310
5.1.1.4.2 Melting Process . . . 311
5.1.1.4.3 Melting Furnaces . . . 312
5.1.1.5 Molding . . . 314
5.1.1.6 Glass Properties and Applications . . . 316
5.1.1.7 Sol-Gel Process . . . 317
References for Section 5.1.1: Glass . . . 318

5.1.2 Alkali Silicates . . . 319
5.1.2.1 General Information and Economic Importance . . . 319
5.1.2.2 Manufacture of Alkali Silicates . . . 319
5.1.2.3 Applications . . . 321
References for Section 5.1.2: Alkali Silicates . . . 321

5.1.3 Zeolites . . . 321
5.1.3.1 Economic Importance . . . 321
5.1.3.2 Zeolite Types . . . 322
5.1.3.3 Natural Zeolites . . . 325
5.1.3.4 Manufacture of Synthetic Zeolites . . . 325
5.1.3.4.1 From Natural Raw Materials . . . 325
5.1.3.4.2 From Synthetic Raw Materials . . . 325
5.1.3.4.3 By Cation Exchange of Synthetic Zeolites . . . 327
5.1.3.5 Pelletization of Zeolites . . . 327
5.1.3.6 Dehydration of Zeolites . . . 327
5.1.3.7 Applications for Zeolites . . . 328
5.1.3.7.1 As Ion Exchangers . . . 328
5.1.3.7.2 As an Adsorption Agent . . . 328
5.1.3.7.3 For Separation Processes . . . 329
5.1.3.7.4 As Catalysts . . . 329
5.1.3.7.5 Miscellaneous Applications . . . 330
References for Section 5.1.3: Zeolites . . . 330

5.2 Inorganic Fibers . . . 330

5.2.1 General Information . . . 330

5.2.2 Naturally Occurring Inorganic Fibers — Asbestos Fibers . . . 333
5.2.2.1 Deposits, Extraction and Economic Importance . . . 333
5.2.2.2 Properties of Asbestos Fibers . . . 335
5.2.2.3 Applications of Asbestos Fibers . . . 337
References for Section 5.2.2: Asbestos Fibers . . . 342

5.2.3 Synthetic Inorganic Fibers . . . 342
5.2.3.1 Textile Glass Fibers . . . 342
5.2.3.1.1 General Information . . . 342
5.2.3.1.2 Economic Importance . . . 342
5.2.3.1.3 Classification of Textile Glass Fibers . . . 343
5.2.3.1.4 Manufacture . . . 343
5.2.3.1.5 Applications . . . 347
5.2.3.2 Optical Fibers . . . 348
5.2.3.3 Mineral Fiber Insulating Materials . . . 349
5.2.3.3.1 General Information and Economic Importance . . . 349
5.2.3.3.2 Manufacture . . . 350
5.2.3.3.3 Applications . . . 354
References for Sections 5.2.3.1 to 5.2.3.3: Mineral Fiber Insulation Materials . . . 355
5.2.3.4 Carbon Fibers . . . 356
5.2.3.4.1 Properties . . . 356
5.2.3.4.2 Manufacture and Applications . . . 358
5.2.3.4.3 Economic Importance . . . 362
5.2.3.5 Aluminum Oxide Fibers . . . 362
5.2.3.6 Boron Fibers . . . 364
5.2.3.7 SiC-Fibers and SiC-Coated Carbon Fibers . . . 365
5.2.3.8 Metal Fibers . . . 367
5.2.3.8.1 Manufacture . . . 367
5.2.3.8.2 Properties, Applications and Economic Importance . . . 368
5.2.3.9 Whiskers and Miscellaneous Synthetic Short Fibers . . . 369
References for Sections 5.2.3.4 to 5.2.3.9: Other Inorganic Fibers . . . 370

5.3 Construction Materials . . . 371

5.3.1 General Information . . . 371

5.3.2 Lime . . . 372
5.3.2.1 Economic Importance . . . 372
5.3.2.2 Raw Materials . . . 372
5.3.2.3 Quicklime . . . 373
5.3.2.4 Slaked Lime . . . 374
5.3.2.4.1 Wet Slaking of Quicklime . . . 375
5.3.2.4.2 Dry Slaking of Quicklime . . . 375

5.3.2.4.3 Lime Hydrate from Calcium Carbide 376
5.3.2.5 Steam-Hardened Construction Materials 376
5.3.2.6 Applications for Lime 377

5.3.3 Cement 378
5.3.3.1 Economic Importance 378
5.3.3.2 Cement Compositions 378
5.3.3.3 Portland Cement 378
5.3.3.3.1 Raw Materials 378
5.3.3.3.2 Composition of Portland Cement Clinkers 379
5.3.3.3.3 Manufacture of Portland Cement 379
5.3.3.3.4 Applications for Portland Cement 383
5.3.3.4 Slag Cement 383
5.3.3.5 Pozzolan Cements 384
5.3.3.6 Alumina Cement 384
5.3.3.7 Asbestos Cement 385
5.3.3.8 Miscellaneous Cement Types 385
5.3.3.9 Processes in the Solidification of Cement 386

5.3.4 Gypsum 389
5.3.4.1 Economic Importance 389
5.3.4.2 Modifications of Calcium Sulfate 389
5.3.4.3 Natural Gypsum 392
5.3.4.4 Natural Anhydrite 393
5.3.4.5 Synthetic Anhydrite from Hydrofluoric Acid Manufacture 393
5.3.4.6 Synthetic Gypsum 394
5.3.4.6.1 From the Manufacture and Purification of Organic Acids 394
5.3.4.6.2 Synthetic Gypsum from Flue Gas Desulfurization 394
5.3.4.6.3 Synthetic Gypsum from Phosphoric Acid Production. 395
5.3.4.7 Processes in the Setting of Plaster 397

5.3.5 Coarse Ceramic Products for the Construction Industry 398

5.3.6 Expanded Products 399
5.3.6.1 General Information 399
5.3.6.2 Expanded Products from Clays and Shales 399
5.3.6.2.1 Raw Materials 399
5.3.6.2.2 Gas-Forming Reactions in the Manufacture of Expanded Products. . . 401
5.3.6.2.3 Manufacture of Expanded Products 402
5.3.6.3 Expanded Products from Glasses (Foam Glass) 404
5.3.6.4 Applications for Expanded Products 404
References for Chapter 5.3: Construction Materials 405

5.4 Enamels 406

5.4.1 General Information 406

5.4.2 Classification of Enamels 406

5.4.3 Frit Manufacture And Enameling Procedure 409
5.4.3.1 Survey 409

5.4.3.2 Manufacture of Enamel Frits . . . 410
5.4.3.2.1 Raw Materials . . . 410
5.4.3.2.2 Smelting of Frits . . . 410
5.4.3.3 Conversion of the Frits into Coatable Forms . . . 411
5.4.3.4 Enamel Application on Sheet Steel . . . 412
5.4.3.4.1 Wet Application Processes . . . 412
5.4.3.4.2 Dry Application Processes . . . 412
5.4.3.5 Firing of Enamels . . . 413

5.4.4 Applications for Enamel . . . 413
References for Chapter 5.4: Enamels . . . 414

5.5 Ceramics . . . 414

5.5.1 General Information . . . 414

5.5.2 Classification of Ceramic Products . . . 415

5.5.3 General Process Steps in the Manufacture of Ceramics . . . 417

5.5.4 Silicate Ceramic Materials . . . 417
5.4.4.1 Composition and Raw Materials of Clay Ceramic Products . . . 417
5.5.4.2 Extraction and Treatment of Raw Kaolin . . . 419
5.5.4.3 Manufacture of Clay Ceramic Batches . . . 420
5.5.4.4 Forming Processes . . . 421
5.5.4.4.1 Casting Processes . . . 422
5.5.4.4.2 Plastic Forming . . . 423
5.5.4.4.3 Forming by Powder Pressing . . . 423
5.5.4.5 Drying Processes . . . 425
5.5.4.6 Firing of Ceramics . . . 425
5.5.4.6.1 Physical-Chemical Processes . . . 425
5.5.4.6.2 Firing Conditions . . . 426
5.5.4.6.3 Glazes . . . 427

5.5.4.7 Properties and Applications of Clay Ceramic Products . . . 428
5.5.4.7.1 Fine Earthenware . . . 428
5.5.4.7.2 Stoneware . . . 429
5.5.4.7.3 Porcelain . . . 429
5.5.4.7.4 Rapidly Fired Porcelain . . . 430
5.5.4.8 Economic Importance of Clay Ceramic Products . . . 431

5.5.5 Refractory Ceramics . . . 431
5.5.5.1 Definition and Classification . . . 431
5.5.5.2 Silica Products . . . 434
5.5.5.3 Fireclay Products . . . 434
5.5.5.4 Alumina-Rich Products . . . 435
5.5.5.5 Basic Products . . . 436
5.5.5.6 Speciality Refractory Products . . . 437
5.5.5.7 Refractory Silicon Carbide Products . . . 437
5.5.5.8 Economic Importance . . . 438

5.5.6 Oxide Ceramic Products . . . 439
5.5.6.1 General Information . . . 439
5.5.6.2 Aluminum Oxide . . . 440
5.5.6.3 Zirconium(IV) Oxide . . . 442
5.5.6.4 Beryllium Oxide . . . 443
5.5.6.5 Uranium Oxide and Thorium Oxide . . . 443
5.5.6.6 Other Oxidic Ceramics . . . 444
5.5.6.7 Electro- and Magneto-Ceramics . . . 445
5.5.6.7.1 Titanates . . . 445
5.5.6.7.2 Ferrites . . . 446

5.5.7 Nonoxide Ceramics . . . 449
5.5.7.1 Economic Importance . . . 450
5.5.7.2 Manufacturing Processes for Silicon Carbide . . . 450
5.5.7.3 Fine Ceramic Silicon Carbide Products . . . 451
5.5.7.4 Fine Silicon Nitride Ceramic Products . . . 452
5.5.7.5 Manufacture and Properties of Boron Carbide . . . 453
5.5.7.6 Manufacture and Properties of Boron Nitride . . . 454
5.5.7.7 Aluminum Nitride . . . 455
References for Chapter 5.5: Ceramics . . . 455

5.6 Metallic Hard Materials . . . 457

5.6.1 General Information . . . 457

5.6.2 General Manufacturing Processes and Properties of Metal Carbides . . . 458

5.6.3 Carbides of the Subgroup of the IVth Group . . . 460
5.6.3.1 Titanium Carbide . . . 460
5.6.3.2 Zirconium Carbide and Hafnium Carbide . . . 460

5.6.4 Carbides of the Subgroup of the Vth Group . . . 461
5.6.4.1 Vanadium Carbide . . . 461
5.6.4.2 Niobium Carbide and Tantalum Carbide . . . 461

5.6.5 Carbides of the Subgroup of the VIth Group . . . 461
5.6.5.1 Chromium Carbide . . . 461
5.6.5.2 Molybdenum Carbide . . . 462
5.6.5.3 Tungsten Carbide . . . 462
5.6.5.4 Cemented Carbides Based on Tungsten Carbide . . . 463

5.6.6 Thorium Carbide and Uranium Carbide . . . 464

5.6.7 Metal Nitrides . . . 464

5.6.8 Metal Borides . . . 466

5.6.9 Metal Silicides . . . 467
References for Chapter 5.6: Metallic Hard Materials . . . 468

5.7 Carbon Modifications . . . 468

5.7.1 Introduction . . . 468

5.7.2 Diamond . . . 468
5.7.2.1 Economic Importance . . . 469
5.7.2.2 Mining of Natural Diamonds . . . 470
5.7.2.3 Manufacture of Synthetic Diamonds . . . 470
5.7.2.4 Properties and Applications . . . 472

5.7.3 Natural Graphite . . . 472
5.7.3.1 Economic Importance . . . 472
5.7.3.2 Deposits and Mining . . . 473
5.7.3.3 Properties and Applications . . . 474

5.7.4 Mass-Produced Synthetic Carbon and Synthetic Graphite . . . 476
5.7.4.1 Economic Importance . . . 476
5.7.4.2 General Information about Manufacture . . . 476
5.7.4.3 Manufacture of Synthetic Carbon . . . 477
5.7.4.3.1 Raw Materials . . . 477
5.7.4.3.2 Processing . . . 478
5.7.4.3.3 Densification and Molding . . . 479
5.7.4.3.4 Carbonization . . . 479
5.7.4.4 Graphitization of Synthetic Carbon . . . 480
5.7.4.4.1 General Information . . . 480
5.7.4.4.2 Acheson Process . . . 480
5.7.4.4.3 Castner Process . . . 481
5.7.4.4.4 Other Graphitization Processes . . . 481
5.7.4.4.5 Purification Graphitization . . . 482
5.7.4.5 Impregnation and Processing of Carbon and Graphite Articles . . . 482
5.7.4.6 Properties and Applications . . . 483

5.7.5 Special Types of Carbon and Graphite . . . 484
5.7.5.1 Pyrolytic Carbon and Pyrolytic Graphite . . . 484
5.7.5.2 Glassy Carbon and Foamed Carbon . . . 486
5.7.5.3 Graphite Foils and Membranes . . . 487

5.7.6 Carbon Black . . . 488
5.7.6.1 Economic Importance . . . 488
5.7.6.2 Manufacture . . . 489
5.7.6.2.1 General Information . . . 489
5.7.6.2.2 Pyrolysis Processes in the Presence of Oxygen . . . 490
5.7.6.2.3 Pyrolysis Processes in the Absence of Oxygen . . . 493
5.7.6.2.4 Posttreatment . . . 494
5.7.6.3 Properties and Applications . . . 495

5.7.7 Activated Carbon . . . 498
5.7.7.1 Economic Importance . . . 498
5.7.7.2 Manufacture . . . 498
5.7.7.2.1 General Information . . . 498
5.7.7.2.2 Chemical Activation Processes . . . 499
5.7.7.2.3 Gas Activation Processes . . . 500

5.7.7.3 Reactivation and Regeneration of Used Activated Carbons 502
5.7.7.4 Applications for Activated Carbon 503
References for Chapter 5.7: Carbon Modifications 504

5.8 Fillers 505

5.8.1 General Information 505
5.8.2 Economic Importance 506

5.8.3 Natural Fillers 507
5.8.3.1 Silicon-Based Fillers 507
5.8.3.2 Other Natural Fillers 508
5.8.3.3 Beneficiation of Natural Fillers 508

5.8.4 Synthetic Fillers 509
5.8.4.1 Silicas and Silicates 509
5.8.4.1.1 Pyrogenic Silicas 509
5.8.4.1.2 Wet Chemically Manufactured Silicas and Silicates 510
5.8.4.1.3 Posttreatment of Silicas 511
5.8.4.1.4 Glass 512
5.8.4.1.5 Cristobalite 512
5.8.4.2 Aluminum Hydroxide 512
5.8.4.3 Carbonates 513
5.8.4.4 Sulfates 514
5.8.4.5 Other Synthetic Fillers 515

5.8.5 Properties and Applications 515
References for Chapter 5.8: Fillers 518

5.9 Inorganic Pigments 518

5.9.1 General Information and Economic Importance 518

5.9.2 White Pigments 522
5.9.2.1 General Information 522
5.9.2.2 Titanium Dioxide Pigments 523
5.9.2.2.1 Economic Importance 523
5.9.2.2.2 Raw Materials for TiO_2 Pigments 523
5.9.2.2.3 Manufacturing Processes for TiO_2 Pigments 525
5.9.2.2.4 Applications for TiO_2 Pigments 529
5.9.2.3 Lithopone and Zinc Sulfide Pigments 530
5.9.2.4 Zinc Oxide White Pigments 531
5.9.2.4.1 Manufacture 531
5.9.2.4.2 Applications 532

5.9.3 Colored Pigments 532
5.9.3.1 Iron Oxide Pigments 532
5.9.3.1.1 Natural Iron Oxide Pigments 533
5.9.3.1.2 Synthetic Iron Oxide Pigments 534

5.9.3.2 Chromium(III) Oxide Pigments . . . 539
5.9.3.2.1 Manufacture . . . 539
5.9.3.2.2 Properties and Applications of Chromium(III) Oxide . . . 541
5.9.3.3 Chromate and Molybdate Pigments . . . 541
5.9.3.4 Mixed Metal Oxide Pigments, Stains and Oxides . . . 543
5.9.3.5 Cadmium Pigments . . . 546
5.9.3.6 Ferrocyanate Pigments . . . 547
5.9.3.7 Ultramarine Pigments . . . 548

5.9.4 Corrosion Protection Pigments . . . 549

5.9.5 Lustrous Pigments . . . 551
5.9.5.1 Metallic Pigments . . . 552
5.9.5.2 Nacreous Pigments . . . 552
5.9.5.3 Interference Pigments . . . 552

5.9.6 Luminescent Pigments . . . 553

5.9.7 Magnetic Pigments . . . 555
5.9.7.1 General Information and Properties . . . 555
5.9.7.2 Manufacture of Magnetic Pigments . . . 556
References for Chapter 5.9: Inorganic Pigments . . . 558

6 Nuclear Fuel Cycle . . . 559

6.1 Economic Importance of Nuclear Energy . . . 559

6.2 General Information about the Nuclear Fuel Cycle . . . 562

6.3 Availability of Uranium and Thorium . . . 563

6.4 Nuclear Reactor Types . . . 565

6.4.1 General Information . . . 565

6.4.2 Light Water Reactors . . . 566
6.4.2.1 Boiling Water Reactors . . . 567
6.4.2.2 Pressurized Water Reactors . . . 567

6.4.3 Graphite-Moderated Reactors . . . 567
6.4.3.1 Gas-Cooled Reactors . . . 568
6.4.3.2 Light Water-Cooled Reactors . . . 569

6.4.4 Heavy Water Reactors . . . 569

6.4.5 Fast Breeder Reactors . . . 570

6.5 Nuclear Fuel Production . . . 571

6.5.1 Production of Uranium Concentrates . . . 571
6.5.1.1 Uranium from Uranium Ores . . . 571
6.5.1.1.1 Leaching Processes . . . 572
6.5.1.1.2 Separation of Uranium from the Leaching Solutions . . . 573

6.5.1.1.3 Manufacture of Marketable Uranium Compounds ("Yellow Cake") . . 574
6.5.1.2 Uranium from Phosphates 575
6.5.1.3 Uranium from Seawater 576

6.5.2 Conversion of Uranium Concentrates to Uranium Hexafluoride 576
6.5.2.1 General Information 576
6.5.2.2 Wet Process for UF_6 Manufacture 577
6.5.2.3 Dry Process for UF_6 Manufacture 578

6.5.3 ^{235}U-Enrichment 579

6.5.4 Reconversion of UF_6 into Nuclear Fuel 580
6.5.4.1 Conversion into Uranium(IV) Oxide 580
6.5.4.1.1 Wet Processes 581
6.5.4.1.2 Dry Process 582
6.5.4.1.3 Manufacture of UO_2 Pellets 582
6.5.4.2 Other Uranium Nuclear Fuels 582

6.5.5 Fuel Element Manufacture. 584

6.6 Disposal of Waste from Nuclear Power Stations 584

6.6.1 General Information 584

6.6.2 Stages in Nuclear Waste Disposal 585
6.6.2.1 Interim Storage of Irradiated Fuel Elements 585
6.6.2.2 Reprocessing of Spent Fuel Elements 586
6.6.2.3 Further Processing of Uranium and Plutonium Solutions. 590
6.6.2.4 Conditioning of Radioactive Waste 590
6.6.2.5 Permanent Storage of Radioactive Waste. 593

6.6.3 Direct Permanent Storage of Irradiated Fuel Elements 594
References for Chapter 6: Nuclear Fuel Cycle 594

Company Abbreviations Index 597

Subject Index 603

1 Primary Inorganic Materials

1.1 Water

1.1.1 Economic Importance

Water is a raw material which is available on Earth in unlimited quantities. Water is not consumed as, after use, it is fed back sooner or later into the Earth's water circulation. The local availability of water (e.g. in arid regions), especially with the purity necessary for the particular application, is another matter. Cheap high purity water is required for many applications. Average annual statistics for West Germany serve to illustrate the origin and production of water in an industrialized country:

Water: a raw material in principle available in unlimited quantities, since used water is fed back into the Earth's water circulation

- 210×10^9 m^3/a from rainfall, which also supplies the groundwater;
- 80×10^9 m^3/a from rivers flowing into the country.

FRG water supply:

- 210×10^9 m^3/a from rainfall
- 80×10^9 m^3/a from rivers flowing into the country

West German water consumption:

- 26×10^9 m^3/a in power stations
- 12.5×10^9 m^3/a in industry
- 5×10^9 m^3/a for domestic and light industrial use.

FRG water consumption:

• Power stations	26×10^9 m^3/a
• Industry	12.5×10^9 m^3/a
• Domestic	5×10^9 m^3/a

Water usage in industry is much higher, with a recycling rate of ca. 3, at 37×10^9 m^3/a. 50×10^9 m^3 of groundwater is available.

Industrial groundwater usage is small, for example only 13.5% in the province of North Rhine Westphalia (FRG). Personal water consumption is ca. 140 L/day of which ca. 2 L is drunk.

Groundwater availability in West Germany:

- 50×10^9 m^3/a

industrial share in usage small

In the USA in 1980 622×10^9 m^3 of fresh and salt water was processed, which was used for

• Power stations	290×10^9 m^3
• Irrigation	207×10^9 m^3
• Industry	62×10^9 m^3

USA, consumption:

• Power stations	290×10^9 m^3
• Irrigation	207×10^9 m^3
• Industry	62×10^9 m^3

Capacity of desalination plants worldwide in 1980 (under construction or in operation):

- 2.8×10^9 m^3/a

Relative capacity

- Distillation plants 76%
- Reverse osmosis plants 24%

Distribution of capacity:

- Near East and North Africa 65%
- USA and possessions 15%
- Europe and South-East Asia 12%
- USSR 3.5%
- Central and South America 3.5%

In 1980, the world capacity for the production of potable water from sea and brackish water (in operation or under construction) was about 2.8×10^9 m^3. More than 2200 installations exist on land. 76% of their capacity is provided by distillation plants and 24% by reverse osmosis (RO) plants. The number of RO-installations is growing rapidly.

This capacity is distributed as follows:

- Near East and North Africa 65%
- USA and possessions 15%
- Europe and South-East Asia 12%
- USSR 3.5%
- Central and South America 3.5%

In 1985 Saudi-Arabia produced 64×10^6 m^3/a of freshwater using distillation and 35×10^6 m^3/a using reverse osmosis. The distillation plant at Ez-Zuetina in Libya is an example of a modern distillation plant (constructor: Krupp), with a capacity of 3×10^4 $m^3/$day ($\alpha = 11 \times 10^6$ m^3/a) at a cost of 87.5×10^6 US $. In 1986 the largest RO-installation to date was under construction at Ad Dur in Bahrein. This will have a capacity of 4.53×10^4 m^3/day ($\alpha = 16 \times 10^6$ m^3/a).

1.1.2 Production of Potable Water

To obtain potable water the following steps must, depending upon the quality of the untreated water, be partly or completely carried out:

- Break-point chlorination or ozonization
- Flocculation
- Sedimentation
- Filtration
- Treatment with activated carbon
- Safety chlorination
- pH adjustment

The untreated water is more or less contaminated depending upon the source. The treatment for potable water consists of the following steps:

- Break-point chlorination (or ozone treatment)
- Flocculation
- Sedimentation
- Filtration
- Treatment with activated carbon
- Safety chlorination
- pH adjustment

The number of steps carried out in practice depends entirely upon the quality of the untreated water. In the case of spring water only safety chlorination needs to be carried out. In the case of strongly polluted water (e.g. filtered water from wells near the banks of the Rhine or Ruhr) almost all the steps are required. In this way potable water can be obtained even from strongly contaminated water. On the other hand, the production of process water, e.g. for cooling purposes, requires far fewer purification steps.

Additional purification steps are necessary if the water contains large amounts of hardeners (calcium, magnesium ions), free carbon dioxide or iron and manganese. Nitrates in excess of the maximum of 50 mg/L allowed by the WHO guidelines may be present, particularly in rural areas, and this excess nitrate concentration must also be removed from the water. Chlorinated hydrocarbons can, as a result of leaky effluent pipes, also find their way into the groundwater. In the USA traces of chlorinated hydrocarbons have been found in 45% of all waterworks which serve more than 10000 inhabitants.

Further purification steps may also be necessary for:

- Reduction in concentration of water hardeners (calcium and magnesium ions)
- Removal of free carbon dioxide, iron and manganese
- Reduction of nitrate ion concentration
- Removal of chlorinated hydrocarbons

Certain applications require deionized water. This can be obtained by ion exchange.

1.1.2.1 Break-Point Chlorination and Ozonization

In the case of strongly polluted surface water chlorination is the first purification step, after prior removal of any coarse foreign matter. Sufficient chlorine is added to ensure a free chlorine concentration of ca. 0.2 to 0.5 mg/L in the water after treatment (break-point chlorination). Chlorine reacts with water forming hydrochloric acid and the hypochlorite anion, depending upon the pH:

Break-point chlorination:

- Addition of sufficient chlorine to ensure 0.2 to 0.5 mg/L chlorine in the water after treatment.

$$Cl_2 + H_2O \longrightarrow HCl + HClO$$

$$HClO \rightleftharpoons H^+ + ClO^-$$

Chlorination results in:

- Elimination of pathogenic germs, deactivation of viruses
- Oxidation of cations such as iron(II) or manganese(II) to higher valency states
- Chlorination of ammonia to chloramines or nitrogen trichloride
- Chlorination of phenols to chlorophenols
- Chlorination of organic impurities, e.g. to aliphatic chlorohydrocarbons.

Chlorination results in:

- Elimination of pathogenic organisms
- Chlorination of ammonia
- Formation of undesirable organic chloro compounds!

The last two processes mentioned are both undesirable: chlorophenols have a very strong taste and some of the aliphatic chlorohydrocarbons (e.g. chloroform) are also suspected of being carcinogenic. For this reason it has been suggested that the chlorination only be carried out to the stage of chloramines and that the other pollutants be eliminated by using, for example, microbiological degradation processes on activated carbon.

Ozonization is an alternative to chlorination:

- Advantages:
  - no formation of organic chloro compounds
  - subsequent flocculation aided
- Disadvantages:
  - high costs

Ozonization in over 1000 towns on earth. Consumption of ozone in water treatment in West Germany in 1984: ca. 3500 t.

An important alternative to chlorination is the ozonization of water. This is practiced in more than 1000 towns in the world. In West Germany in 1984 3500 t of ozone were used in the treatment of water. Here the above-mentioned problems can not arise. The considerably higher costs and the relatively low stability of ozone in water are, however, problematical as is the limited knowledge regarding the toxicity of substances which are formed by the ozonization of organic compounds. Ozonization aids the subsequent flocculation process. About 0.2 to 1.0 g ozone per m^3 water is required, in exceptional cases up to 3 g/m^3. Another alternative is treatment with chlorine dioxide with which the formation of chlorine-containing organic compounds is less than with chlorination.

1.1.2.2 Flocculation and Sedimentation

Preliminary purification by flocculation is necessary, if the untreated water has a high turbidity and contains colloidal organic impurities. Iron or aluminum salts are added to the water, so that ferric or aluminum hydroxide is precipitated:

$$Al_2(SO_4)_3 + 6H_2O \longrightarrow 2Al(OH)_3\downarrow + 3H_2SO_4$$

$$FeSO_4Cl + 3H_2O \longrightarrow Fe(OH)_3\downarrow + H_2SO_4 + HCl$$

$$Fe_2(SO_4)_3 + 6H_2O \longrightarrow 2Fe(OH)_3\downarrow + 3H_2SO_4$$

Flocculation:

removal of inorganic and organic colloids by adsorption on (in situ produced) aluminum and ferric hydroxide flakes. If necessary addition of flocculation aids

The optimum pH value for flocculation is about 6.5 to 7.5 for aluminum salts and about 8.5 for iron salts. If the natural alkali content of the untreated water is insufficient to neutralize the acid formed, alkali has to be added (e.g. lime or soda ash). In addition flocculation aids such as derivatives of poly(acrylamide) or starch can be added. When aluminum sulfate

$Al_2(SO_4)_3 \cdot 18H_2O$ is used ca. 10 to 30 g/m^3 is added. The very fine hydroxide flakes which precipitate are positively charged and adsorb the negatively charged colloidal organic materials and clay particles.

There is a range of technically established units (e.g. Precipitator, Pulsator, Accentrifloc, Reactivator) for carrying out the flocculation and the separation of the flocculated material in which a well-defined sludge suspension layer is formed, which can be separated off. Some plants operate with sludge recirculation to bring about a better adsorption. The sludge flakes can also be separated off using flotation.

Sedimentation:

separation of the finely distributed sludge resulting from flocculation, optionally with sludge recirculation

1.1.2.3 Filtration

Water having undergone flocculation then has to be filtered. (If the untreated water is of such quality that flocculation and break-point chlorination are not necessary, then filtration is the first treatment step.) The water is generally filtered downwards through a 1 to 2 m high sand filter at a rate is 3 to 5 mm/s. The sand particles are 0.2 to 2 mm in size. When the filter is covered with impurities this increases the filter resistance and it is then cleaned by flushing upwards, if necessary together with air. Alternatively, a multiple-layer filter can be used, optionally combined with a 0.5 m thick anthracite layer. A precoated filter can be used for fine filtration.

Filtration:

separation of undissolved solids over a sand filter, optionally combined with an anthracite filter. Flushed with water or water/air when filter is covered

1.1.2.4 Removal of Dissolved Inorganic Impurities

Untreated water containing a lot of dissolved hydrogen carbonate forms, upon heating, a precipitate consisting mainly of calcium carbonate (carbonate hardness, boiler scale):

$$Ca(HCO_3)_2 \longrightarrow CaCO_3 + CO_2 + H_2O\,.$$

The carbonate hardness can be removed by adding acid, whereupon the more soluble calcium sulfate is formed:

$$Ca(HCO_3)_2 + H_2SO_4 \longrightarrow CaSO_4 + 2CO_2 + 2H_2O\,.$$

Hardeners, especially calcium and magnesium hydrogen carbonates, rendered untroublesome by addition of:

- Sulfuric acid and expulsion of carbon dioxide,
- Calcium hydroxide and separation of the carbonates formed

The resulting carbon dioxide has to be expelled, as carbon dioxide-containing water is corrosive. The hydrogen carbonate can be removed by the addition of calcium hydroxide:

$$Ca(HCO_3)_2 + Ca(OH)_2 \longrightarrow 2CaCO_3 + 2H_2O\,.$$

In a technical variant of this process the calcium hydroxide, as a solution or a suspension, is added to hydrogen carbonate-containing water and the mixture passed over calcium carbonate beads, upon which the freshly formed calcium carbonate is deposited. Fresh beads form on the crystal nuclei added and those beads which become too large are separated off.

Removal of iron(II) and manganese(II) by oxidation of the bivalent ions with air, or if necessary with chlorine, and separation of the oxide hydrates formed. Dissolved carbon dioxide also expelled during air oxidation

Iron and manganese are present as bivalent ions in many waters. They are removed by oxidation to the oxide hydrates, preferably with air, and if necessary after increasing the pH. Treatment with air expels the dissolved carbon dioxide at the same time. If air is an insufficiently powerful oxidation agent, e.g. when much humic acid (which acts as a complexing agent) is present, stronger oxidizing agents such as chlorine are used. These ions are also removed during treatment with activated carbon.

1.1.2.5 Removal of Dissolved Organic Impurities

A large number of different compounds are involved. Apart from the naturally occurring substances such as humic acid, there are substances which come from household, industrial or agricultural effluent or result from their reaction with ozone or chlorine. Typical secondary products from chlorination are e.g. chlorophenols and chloroalkanes (see Sec. 1.1.2.1).

Removal of dissolved organic impurities by treatment with activated carbon:

- In West Germany mostly granulated
- In the USA preferably powdered

Treatment with activated carbon is particularly suitable for the removal of dissolved organic impurities. Two processes are operated:

- Percolation through granular activated carbon, particularly in West Germany
- Stirring in of carbon powder, particularly in the USA.

Powdered activated carbon has the advantages that the quantity can be well matched to the degree of contamination and that the investment costs for this process are low. Powdered activated carbon can not, however, be easily regenerated, unlike granular activated carbon. As the pollutants vary from one water to another, the activated carbon type and the kind of treatment (e.g. number of filters, contact time) have, with granular activated carbon, to be empirically established. During purification the displacement of compounds already adsorbed, such as chloroalkanes into the water, by more easily adsorbable compounds must be avoided.

About 50 to 150 g TOC/m^3 (TOC = total organic carbon) of organic carbon are on average removed from water per day. This value is higher, if the water is not break-point chlorinated (see Sec. 1.1.2.1) or is pretreated with ozone.

Between 50 and 150 g TOC/m^3 water removed by activated carbon per day.

Back flushing is used to remove the sludge from the activated carbon. At longer intervals the filter has to be thermally regenerated under similar conditions to those of its manufacture. This can be carried out either at the waterworks or by the manufacturer of the activated carbon.

Granular activated carbon regenerated by back flushing and from time to time by thermal reactivation

The activated carbon treatment also has effects other than the elimination of dissolved organic impurities:

- Excess chlorine is decomposed
- Ammonia and some of the organic compounds are biologically oxidized
- Iron and manganese oxide hydrates are removed.

Activated carbon treatment also leads to

- Decomposition of excess chlorine
- Biological oxidation of ammonia and organic compounds by microbiological processes on the activated carbon surface
- Removal of iron and manganese

1.1.2.6 Safety Chlorination

After the treatment is finished a safety chlorination is carried out to prevent reinfection of the water in the distribution network. This is also necessary after prior ozonization. Potable water contains about 0.1 to 0.2 mg/L chlorine. Chlorine dioxide can also be used instead of chlorine (e.g. Düsseldorf waterworks).

Safety chlorination:

avoidance of reinfection of potable water in the distribution network by adding 0.1 to 0.2 mg/l chlorine

1.1.2.7 Removal of Nitrates

This is in an early stage of development. In a pilot plant at the Gatzweiler waterworks (Mönchengladbach, West Germany) the nitrate concentration present in the potable water is reduced using a RO-mod-

Nitrate removal at an early stage of development. Possible methods:

- Reverse osmosis
- Ion exchange
- Biological denitrification

ule. The nitrate-enriched effluent is worked up to solid nitrates by a multistep process. Other options are ion exchange or biological denitrification. All these processes are very expensive.

1.1.2.8 Removal of Traces of Chlorohydrocarbons from Groundwater

Removal of chlorohydrocarbons:

- Blowing out with air (while removing the carbon dioxide)
- Removal with activated carbon

These can be flushed out by aerating the water (e.g. together with dissolved carbon dioxide, see Sec. 1.1.2.4). The air used can be recirculated over an activated carbon filter. Further purification of the water can be achieved by additional activated carbon filtration.

1.1.2.9 Production of Soft or Deionized Water

For a number of technical processes, water is required with a lower hardener content than can be achieved by the processes described in Sec. 1.1.2.4. This can be accomplished by ion exchange with solid polymeric organic acids, the "ion exchangers".

Treatment of water with cation exchangers:

exchange of Ca^{2+} and Mg^{2+} for Na^+ or H^+.

Water with less than 0.02 mg/L salt obtained by stepwise treatment over cation and anion exchange beds or "mixed beds". Last residues of organic impurities eliminated with reverse osmosis

When the sodium salt of sulfonated poly(styrene) is used as the cation exchanger, calcium and magnesium ions are exchanged for sodium ions:

$$PS-SO_3^-Na^+ + 0.5\,Ca^{2+} \rightarrow PS-SO_3^-Ca_{0.5}^{2+} + Na^+$$

[PS Poly(styrene)]

Regeneration of ion exchangers charged with calcium and magnesium ions (1 L of ion exchange material can be charged with ca. 40 g of CaO) can be accomplished by reversing the above equation by (countercurrent) elution with 5 to 10% sodium chloride solution. If the hardeners are present as hydrogen carbonate, the eluant becomes alkaline after heating:

$$2NaHCO_3 \longrightarrow Na_2CO_3 + CO_2\uparrow + H_2O$$

If ion exchangers are used in the acid form, then the eluant will be acidic:

$$PS-SO_3^-H^+ + M^+ \rightarrow PS-SO_3^-M^+ + H^+$$

(M^+: monovalent metal ion or equivalent of a multivalent ion)

If (weakly acidic) resins containing carboxy-groups are used, only those hardeners present as hydrogen carbonates are removed:

$$PS-(COOH)_2 + Ca(HCO_3)_2 \longrightarrow PS-(COO^-)_2Ca^{2+} + 2CO_2\uparrow + 2H_2O$$

Subsequent heating expels the carbon dioxide. The exchanger is regenerated with dilute acid.

For very high purity water (for applications such as high performance boilers or in the electronics industry) virtually ion-free water is required. This is achieved in alternate layers of cation and anion exchangers or in so-called "mixed bed exchangers". In these, both strongly acid cationic exchangers in the proton form and basic ion exchangers based on poly(styrene) modified with amino- or ammonium-groups are present, e.g.

$$PS-N(CH_3)_2 \text{ or } PS-N(CH_3)_3^{2+}OH^-$$

Basic ion exchangers remove anions and are regenerated with sodium hydroxide, e.g.:

$$PS-N(CH_3)_3^+OH^- + Cl^- \longrightarrow PS-N(CH_3)_3^+Cl^- + OH^-$$

Upon passing salt-containing water through a mixed bed, the cations are replaced by protons and the anions by hydroxide ions. Protons and hydroxide ions together form water, making the resulting water virtually ion free with an ion residue of 0.02 mg/L. The higher density of anion exchangers (than cationic exchangers) makes the regeneration of mixed beds possible. The mixed bed ion-exchange columns are flushed from the bottom upwards with such a strong current of water that the resins are transported into separate zones, in which they can be regenerated independently of one another.

[For the electronics industry etc. a further purification using reverse osmosis (see also Section 1.1.3.2) is necessary to remove dissolved nonionic organic compounds].

1.1.3 Production of Freshwater from Seawater and Brackish Water

1.1.3.1 Production by Multistage Flash Evaporation

Seawater contains on average 3.5 % by weight of dissolved salts, for the most part sodium chloride. Calcium, magnesium and hydrogen carbonate ions are also present. Potable water should not contain more than 0.05% of sodium chloride and less than 0.1% of dissolved salts.

The removal of such quantities of salt from seawater using ion exchangers would be totally uneconomic. Distillation processes are currently mainly used in the production of potable and irrigation water from seawater.

Most important process for the production of freshwater from seawater:

multistage(vacuum) flash evaporation

Pretreatment: removal of solid and biological contaminants

Multistage (vacuum) flash evaporation (MSF) is particularly favored, Fig. 1.1-1.

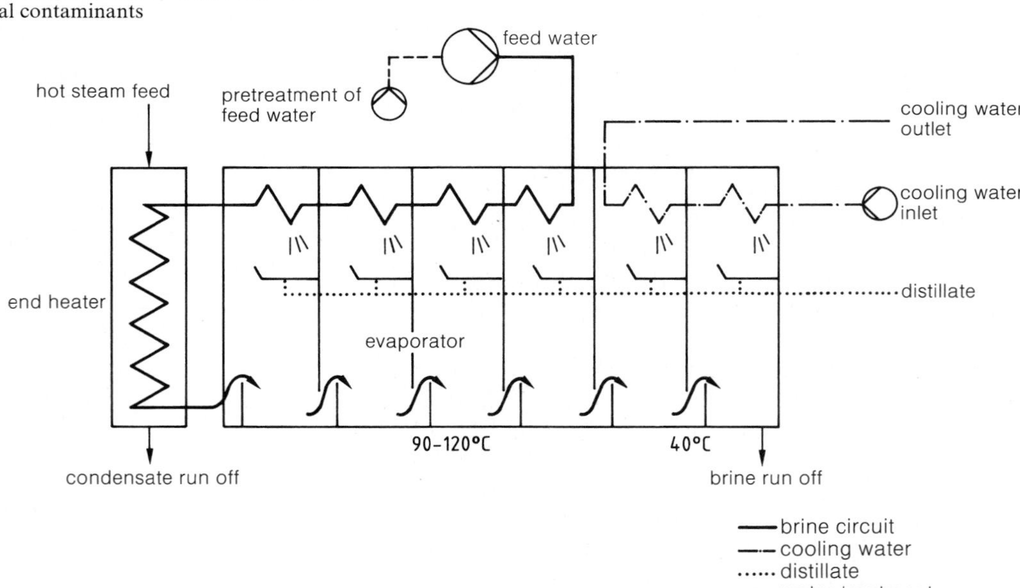

Fig. 1.1-1. Schematic lay-out of an MSF plant.

Seawater is evaporated at temperatures between 90 °C and a maximum of 120 °C in each of up to 45 consecutive stages. The seawater feed is also the coolant for condensing the steam produced and in so doing is heated up as it proceeds from stage to stage. In the first (hottest) stage the energy required for the complete system is supplied by steam using a heat exchanger. Additional seawater is necessary in a supplementary circuit for cooling the steam produced in the last (coolest) stages. This is returned directly to the sea, which represents a considerable energy loss. The rest of the prewarmed water is mixed with some of the water already concentrated through evaporation and being recycled. The rest of the concentrate is blown down. In this way a predetermined "concentration factor" (e.g. 1.5 to 2.0) is obtained. Disposal of this blown-down concentrate also represents an energy loss.

The quality of seawater feed has to fulfill certain requirements: in addition to the removal of coarse foreign matter and biological impurities, hardener elimination or stabilization is necessary. Calcium carbonate and magnesium hydroxide (Brucite) are deposited from untreated seawater onto the heat exchanger surfaces with loss of carbon dioxide, resulting in a strong decrease in the distillation performance of the plant.

Hardener precipitation can be prevented by adding sulfuric acid, whereupon the fairly soluble calcium and magnesium sulfates are formed. However, the quantities of acid required are considerable and the poor accessibility of many of the desalination plant sites makes delivery difficult. In addition, the dosing must be very exact: too little acid leads to incrustation and too much leads to corrosion. Polyphosphates are currently often used for hardener stabilization in the first (hottest) stage at temperatures of up to ca. 90 °C in under-stoichiometric quantities. Poly(maleic acid) derivatives are employed for hardener stabilization in plants operating at temperatures above 90 °C. Above 120 °C calcium sulfate precipitates out as anhydrite, for which there is currently no technically useful remedy.

Precipitation of hardeners is prevented by adding:

- Molar quantities of sulfuric acid
- Polyphosphate or poly(maleic acid) derivatives in under-stoichiometric quantities

The cost of potable water production from seawater is mainly dependent upon the cost of the energy consumed. It is, however, considerably higher than that for potable water produced from freshwater.

1.1.3.2 Production using Reverse Osmosis

Production of potable water from brackish water or seawater by reverse osmosis:

permeation of water with a low salt content through a semipermeable membrane by applying pressure to the side containing saltwater

Another process for the production of potable water from seawater is beginning to become established: reverse osmosis. In osmosis, water permeates through a semipermeable membrane from a dilute solution to a concentrated solution resulting in a hydrostatic pressure increase in the concentrated solution. This process proceeds spontaneously. In reverse osmosis, water with a low salt content is produced by forcing a salt-containing solution through a semipermeable membrane under pressure. To produce a usable quantity of water, the pressure applied must be substantially higher than the equilibrium osmotic pressure. This is 3.5 bar for a 0.5 % by weight salt solution. Pressures of 40 to 70 bar are necessary for water production, the higher the pressure on the feed water side the higher the permeation of water. However, the salt concentration in the water thus produced increases with increasing pressure, as the membrane is unable to retain the salt completely. A multistep process has sometimes to be used.

Membranes mostly made of acetylcellulose or more preferably polyamide. Large pressure differences mean complicated desalination plant construction (in some cases multistage). Pretreatment of water necessary as for distillation plants

The membranes are manufactured from acetylcellulose or, more preferably, polyamide. The technical construction is complicated and made expensive by the large pressure differences and the need for thin membranes. Bundles of coiled thin hollow capillaries (external diameter 0.1 mm, internal diameter 0.04 mm) are, for example, placed in a pressure cylinder (Fig. 1.1-2). These capillaries protrude from the

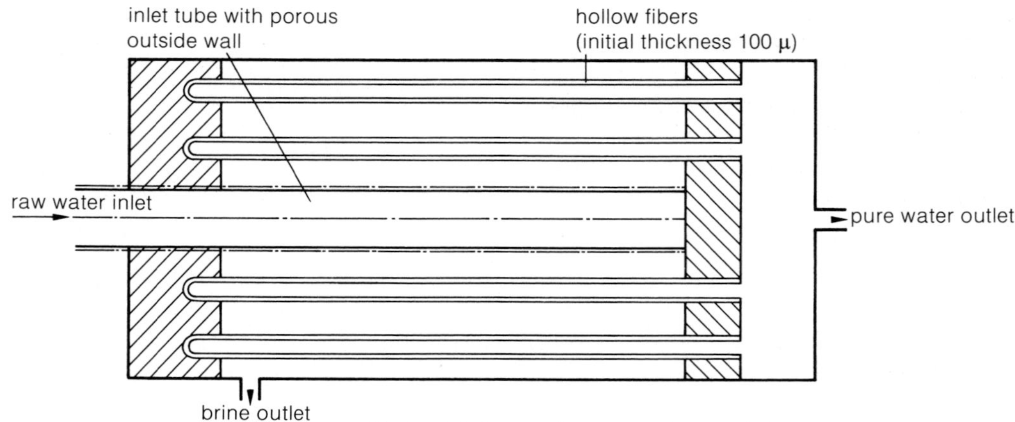

Fig.1.1-2. Schematic lay-out of a RO-module.

ends of the cylinder through plastic sealing layers. Of the (high salt content)-water fed into the cylinder from the other side, 30% passes through the capillary walls into the capillaries and the rest is run off as concentrate and disposed of. An intensive and expensive pretreatment of the feed water is also necessary. In addition to the removal of all colloidal and biological impurities, treatment of the hardeners is necessary for example by acid addition. The use of feed water from wells in the neighborhood of beaches is particularly favored.

Reverse osmosis requires less than 50% of the energy required by multistage flash destillation (8 to 10.6 kWh/m^3/d for the production of water in a plant with a capacity of 19×10^3 m^3/d).

Freshwater production by reverse osmosis is considerably cheaper than flash distillation.

References for Chapter 1.1: Water

Water supply in West Germany:

"Wasser für die Industrie" (Water for Industry), published by the Bundesverband der Deutschen Industrie e.V. (July 1983).

Water supply in the USA:

Chem. Week, pp 34–39 (Feb. 22, 1984).

Reviews:

T. V. Arden,"Water Purification and Recycling", in: *The Modern Inorganic Chemicals Industry*. R.Thompson, (ed), The Chemical Society, Burlington House, London 1977, pp 69–104.

Chem. Week, pp 34–39 (Feb. 22, 1984).

J. H. Lehr et al.: *Domestic Water Treatment*. McGraw-Hill, New York (1980).

J. J. Fried, "Groundwater Resources in the European Community", *Endeavour, New Series*, **9**, pp 112–116 (1985).

"Special Report: Water Treatment", *Chem. Week*, pp 37–51 (Feb. 12, 1986).

Water Desalination:

Gmelin Handbuch der Anorganischen Chemie. 8th Ed., 0 Water Desalting, Supplement Volume 1. Springer-Verlag, Berlin-Heidelberg-New York (1979).

H. Ludwig, "Reverse Osmosis in the Desalination of Brackish Water and Seawater", *Desalination* **36,** pp 153–178 (1981).

H. W. Pohland,"Seawater Desalination by Reverse Osmosis",*Endeavour, New Series* **4,** pp 141–147 (1980).

W. M. Wagner and D. R. Finnegan, "Select a seawater-desalting process", *Chem. Eng.*, pp 71–75 (Feb. 1983).

New Developments in Water Matters:

Schriftenreihe *Vom Wasser* (Publication series about water), published by the Subject Group Wasserchemie (Water chemistry) in der Gesellschaft Deutscher Chemiker. Verlag Chemie, Weinheim-Deerfield Beach, Florida–Basel. (Vol. 56, 1981).

Publications in the field of desalination: *"Desalination"*, Elsevier Scientific Publishing Company, Amsterdam.

L.Awerbuch, "Future Developments in Desalination Processes", *Desalination* **50,** pp 61–70 (1984).

1.2 Hydrogen

1.2.1 Economic Importance

Hydrogen is the most widespread element in the Universe, but only the ninth most common element in the Earth's crust

Hydrogen is by far the most widespread element in the universe, but on Earth (Earth's crust together with its waters and atmosphere) it is only the ninth most common element with 1% by weight (or 15 atomic %). Hydrogen is almost exclusively present as its compounds in the biomass and fossilized raw materials (water, hydrates, hydrocarbons etc.).

Commercially hydrogen has only been utilized as a chemical raw material and industrial chemical. However, particularly since the 1973/74 oil crisis there has been increasing interest in hydrogen as an almost inexhaustible (secondary) energy source instead of, or in addition to, electricity, due to its high energy density per unit mass (121 kJ/g compared with 50.3 kJ/g for methane), its high environmental compatibility, its being nonpoisonous and the ease of its transport and storage. However, the current dependence upon fossil and nuclear fuels as primary energy sources together with the comparatively high price of hydrogen stand in the way of its utilization as an energy source.

Further development of hydrogen technology requires cheap primary energy sources

The future development of hydrogen demand is extremely difficult to predict as most of the hydrogen produced directly or as a byproduct is either further utilized by the producer or vented into the atmosphere (e.g. in coking plants and the chemical and petrochemical industries). The world production in 1974 was estimated to be ca. 270×10^9 m^3 at STP and to have risen annually by 2.5% up to 1986 to ca. 350×10^9 m^3 at STP (31×10^6 t).

1985 Hydrogen production:

- World: ca. 350×10^9 m^3
- USA: ca. 80 to 90×10^9 m^3
- FRG: ca. 17×10^9 m^3

The statistically recorded volume of hydrogen produced in the USA has increased 4% annually in the last 10 years to 3.1×10^9 m^3 in 1985. The USA production in 1985 probably amounted to 80 to 90×10^9 m^3, which was mainly utilized directly by the producer in ammonia synthesis, hydrogenation and hydrocracking. Hydrogen production in West Germany was ca. 17×10^9 m^3 in 1985.

Liquid hydrogen has a small but important market e.g. for rocket fuels and industrial applications. The USA production capacity was 59 000 t/a in 1985.

Only a small part of the hydrogen produced is marketed, most is directly utilized by the producer

1.2.2 Hydrogen Manufacture

90% of the hydrogen is currently produced from fossil raw materials (crude oil, natural gas and coal), which at the same time serve as primary energy sources. The dissociation of water is much more energy intensive. The most important large-scale processes for hydrogen production are currently:

- Steam-reforming of natural gas
- Partial oxidation of heavy oil
- Coal gasification
- Water electrolysis

Raw materials sources for H_2:

- Fossil raw materials (natural gas, oil, coal) account for $>90\%$ of H_2 production
- Water

Hydrogen is also formed in large quantities as a byproduct in petrochemical processes, refineries, coking plants (coke oven gas) and in chemical and electrochemical processes e.g. the chloralkali electrolysis. Other processes such as the photochemical production of hydrogen or thermal dissociation of water are only used in special applications and are currently industrially unimportant.

H_2 byproduct in:

- Refineries
- Petrochemical plants
- Coking plants
- Chemical industry

1.2.2.1 Petrochemical Processes and Coal Gasification

The industrially most important and currently cheapest hydrogen production process is the catalytic steam reforming process in which steam is reacted with natural gas (methane) or light crude oil fractions (propane, butane, naphtha with b.p.'s $\leq 200\,°C$):

77% of H_2 is petrochemically produced. Most important process is currently the steam reforming of methane and low b.p. hydrocarbons

$$CH_4 + H_2O \rightarrow 3\,H_2 + CO, \Delta H = 205\ \text{kJ/mol}$$

Ca. 50% of the hydrogen produced comes from the steam.

In addition to the steam reforming of low b.p. hydrocarbons (accounting for 80% of the USA production), the partial oxidation of heavy fuel oil and crude oil residues is also industrially important. It can be represented by the following equation:

$$2\,C_nH_{2n+2} + n\,O_2 \longrightarrow 2(n+1)H_2 + 2n\,CO$$

This self-sustaining noncatalytic thermal reaction is followed, as in the case of steam reforming, by the conversion of carbon monoxide to carbon dioxide and hydrogen:

$$CO + H_2O \rightarrow CO_2 + H_2, \Delta H = -41 \text{ kJ/mol}$$

Coal gasification is currently too expensive, but can become very attractive as crude oil and natural gas become scarcer

In countries with cheap coal (e.g. South Africa) hydrogen is being increasingly produced by coal and coke gasification (before World War II 90% of hydrogen was produced in this way). This reaction proceeds as follows:

$$3\,C + O_2 + H_2O \rightarrow H_2 + 3\,CO, \Delta H = -286 \text{ kJ/mol}$$

A great future is predicted for coal gasification in the "hydrogen age", particularly when the heat from nuclear processes is utilized.

Since more than half of the hydrogen is utilized in the production of ammonia (fertilizer production) mainly in integrated plants, all three processes are described in the section on the manufacture of synthetic ammonia, Sec. 1.4.1.2.

1.2.2.2 Water Electrolysis

Hydrogen production by the electrolysis of water is currently less important due to the overall low efficiency of electrolysis (ca. 25% including that of the electricity production itself) as well as the high investment costs and electricity prices. Water electrolysis accounts for only ca. 4% of the hydrogen production, this mainly in small units (100 to 1000 m^3 H_2/h) with a purity of $> 99\%$. Hydrogen for ammonia synthesis is also produced in several large plants utilizing cheap hydroelectric power (e.g. in Norway, Egypt, India, Canada) with outputs of up to 33 000 m^3/h.

H_2 production by water electrolysis (ca. 4%) only economic when cheap electricity is available. Nuclear and solar energy possible future primary energy sources

At present, electrolysis cells basically consist of two electrodes separated by an asbestos diaphragm impermeable to gases. 20 to 30% potassium hydroxide is dissolved in the electrolyte to increase its conductivity. The electrolysis is carried out at 80 to 85 °C. The theoretical decomposition potential is 1.24 V with 1.8 to 2.2 V being used in practice due to overvoltage effects etc. The specific energy consumption is about 4.5 to 5.3 kWh/m^3 hydrogen (and 0.5 m^3 oxygen) of which 0.8 to 1.8 kWh is dissipated as heat. Oxygen is produced at the anode and hydrogen at the cathode:

$2\,OH^-$	$\rightarrow H_2O$	$+ 1/2\,O_2 + 2e^-$	anode
$2\,H_2O + 2e^-$	$\rightarrow H_2$	$+ 2\,OH^-$	cathode
H_2O	$\rightarrow H_2$	$+ 1/2\,O_2$	

Heavy water (D_2O) can be produced as a byproduct during electrolysis.

Industrial cells are mainly bipolar consisting of a large number of individual plate cells connected back to back and coupled in blocks according to the filter press principle. The energy consumption can be reduced by 20% by carrying out the electrolysis at ca. 30 bar. Further recent developments are the use of porous electrodes, high temperature steam electrolysis with doped zirconium(IV) oxide as the electrolyte and the SPE-process (*S*olid *P*olymer *E*lectrolyte) with proton-conducting ion exchange membranes.

New electrolysis processes with lower energy consumption such as

- Pressure electrolysis
- Electrolysis with porous electrodes
- High temperature steam electrolysis
- SPE-process

are currently being developed

The recent advances in the field of water electrolysis cannot obscure the fact that the long term aim of hydrogen technology requires cheap nonfossil primary energy sources such as nuclear and solar energy.

1.2.2.3 Other Manufacturing Processes for Hydrogen

The direct thermal dissociation of water:

$$H_2O \rightarrow H_2 + 1/2\,O_2, \qquad \Delta H = -285\ kJ/mol$$

is industrially and commercially impractical due to the > 2000 °C temperatures required. Thermochemical scission is an alternative in which a process aid is added and recycled.

Other processes:

- Thermal scission of water
- Thermochemical cyclic processes
- Iron-steam process
- Photo(electro)chemical processes
- Catalytic scission of ammonia or methanol

An example of such processes which is operated at temperatures below 1000 °C, is the four-stage iron chloride process:

$$3FeCl_2 + 4H_2O \xrightarrow{650\,^\circ C} Fe_3O_4 + 6HCl + H_2$$
$$Fe_3O_4 + 8HCl \xrightarrow{100\text{ to }600\,^\circ C} FeCl_2 + 2FeCl_3 + 4H_2O$$
$$2FeCl_3 \xrightarrow{400\,^\circ C} 2FeCl_2 + Cl_2$$
$$Cl_2 + H_2O \xrightarrow{1000\,^\circ C} 2HCl + 1/2O_2$$

overall: $$H_2 \longrightarrow H_2 + 1/2O_2$$

Of the large number of thermodynamically feasible cycles not one has been developed industrially. The choice of materials, corrosion and the attainment of the high temperatures required (possibly with nuclear or solar energy) are major problems.

Hydrogen production units using solar energy either by direct photochemical processes or by indirect photoelectrochemical processes with the help of solar cells (Sec. 3.4.1.3) are currently only being operated in a few small experimental plants with capacities of up to 20 MW.

For particular purposes hydrogen is produced by the catalytic decomposition of ammonia (over nickel catalysts at 900 °C for hydration or metallurgical purposes) or methanol in dissociation plants.

1.2.2.4 Production of Hydrogen as a Byproduct

H_2 produced as a byproduct in refineries and petrochemical companies particularly from:

- Reforming
- Aromatization
- Production of ethenes from saturated hydrocarbons

Hydrogen-containing gases are formed in large quantities as a byproduct in a multitude of industrial processes:

- Catalytic reforming and cracking processes
- Various petrochemical processes (alkene, ethyne and styrene production)
- Coking plants and blast furnaces (as coke oven or furnace gas)
- Fischer-Tropsch synthesis
- Ammonia synthesis
- Chloralkali- and hydrochloric acid-electrolysis.

The hydrogen-rich gases are mostly utilized directly (e.g. for hydrogenation purposes in the petrochemical industry). Gases with low hydrogen contents on the other hand are mainly used as combustion gases (e.g. in coking plants).

The process used for purifying hydrogen-containing gases depends upon the purity requirements:

- Condensation, fractionation or scrubbing at low temperatures
- Absorption in organic solvents (alcohols, amines, oils)
- Pressure swing adsorption (PSA) on molecular sieves
- Hydrogen penetration through hollow fiber membranes

The latter are particularly adaptable and cost effective.

Hydrogen purification using:

- Low temperature processes
- Absorption processes
- Pressure swing adsorption processes
- Diffusion processes

1.2.3 Hydrogen Applications

Hydrogen is mainly utilized directly and its consumption matches its production. Only small quantities of mostly very pure hydrogen are marketed. These are increasing steadily as the hydrogen logistics as regards transport and storage systems improve.

According to a 1982 study carried out by the International Energy Agency the consumption spectra of hydrogen in eight selected OECD countries in 1978 and 1985 were estimated to be as follows:

Utilization of H_2 in refineries:

- Hydrocracking for quality improvement of oil products
- Hydrotreating i.e. removal of sulfur, nitrogen and oxygen from crude oil and distillates
- Heating purposes

Table 1.2-1. Consumption Spectra of Hydrogen in 1978 and 1985 for Belgium, Canada, West Germany, Japan, the Netherlands, Sweden, Switzerland and the USA in %.

	Chemical industry	Petrochemical industry	Process heating	Synthetic fuels	Metallurgy
1978	63	29	6	1.2	0.6
1985	59	31	7	2.0	1.0

Its utilization as a fuel in combustion engines and fuel cells is currently technically feasible, but economically impractical except for space vehicles.

A considerably increased utilization of hydrogen for hydrogenation of coal, oil shale and tar sands is predicted for the future in connection with the manufacture of synthetic fuels.

H_2 demand is currently strongly increasing in refineries for the treatment of heavy crude and in the future for oil shale, tar sands and oils from coal (to increase the H/C ratio)

Huge future market for H_2 for:

- Coal hydrogenation (liquification) for the manufacture of motor and combustion fuels
- Synthetic natural gas manufacture by the hydrogenation of CO

In the USA in the mid-1980's ca. 85×10^9 m^3/a was consumed. Of the 54×10^9 m^3 of hydrogen produced in the USA in 1983, 49% was utilized in ammonia synthesis, 39% in refinery processes (hydrocracking, hydrotreating, desulfurization by hydrogenation) and 7% in the manufacture of methanol. The remaining 5% was utilized in the hydrogenation of organic chemicals (e.g. aniline and cyclohexane synthesis), in the hydrogenation of fats, in metallurgy (e.g. as a protective gas in nonoxidizing annealing furnaces and in the direct reduction of iron ore) and in the electronics industry (protective gas in semiconductor manufacture). Small quantities are required in the glass industry, for the manufacture of hydrogen chloride, for autogenic welding and cutting (oxyhydrogen blowpipe) and in protective gas welding technology (e.g. with argon-hydrogen).

1983 USA consumption spectrum for H_2:

- Ammonia synthesis 49%
- Refinery processes 39%
- Methanol manufacture 7%
- Hydrogenation of organic chemicals and fats, in the metallurgy, electronics and glass industries and in hydrogen chloride synthesis etc. 5%

Hydrogen gas is transported in steel cylinders or bundles of cylinders at pressures of up to 200 bar or by pipeline. For decades Chemische Werke Hüls has operated a 210 km pressurized pipeline network with a maximum transport capacity of 312×10^6 m^3/a to 18 companies. Increasingly hydrogen is being transported as a liquid under pressure in highly insulated cryogenic containers at $-253\,^\circ C$. In the USA 65% was transported in this way as early as 1982.

H_2 transport:

- As a gas in gas cylinders or pipelines
- As a liquid in pressurized cryogenic containers
- Possibly as a solid in the form of hydrides

Apart from gas cylinders, natural cavities such as porous reservoirs, salt caverns and aquafers can also be used for the storage of gaseous hydrogen with capacities of many 10^8 m^3 of hydrogen at STP. In addition mobile hydride reservoirs are currently being evaluated, in which gaseous hydrogen is reversibly adsorbed as metallic hydrides e.g. titanium iron hydride $TiFeH_{1.95}$ or magnesium nickel hydride $MgNiH_{4.2}$.

Storage and release of H_2 from hydrides:

$$y\,Me + x\,H_2 \rightarrow Me_yH_{2x} + \Delta H$$

References for Chapter 1.2: Hydrogen

General Information:

A. F. Trotman-Dickenson: *Comprehensive Inorganic Chemistry.* 1st Ed., Vol. 1, pp 1–22, Pergamon Press, Oxford (1973).

Ullmanns Encyklopädie der technischen Chemie, 3rd Ed., Vol. 18, pp 501–539, Urban Schwarzenberg, Munich-Berlin-Vienna (1967).

Kirk-Othmer: *Encyclopedia of Chemical Technology*, 3rd Ed., Vol. 12, pp 938–982, John Wiley & Sons Inc., New York-Chichester-Brisbane-Toronto (1981).

W. Balthasar: "Hydrogen production and technology: today, tomorrow and beyond", *Int. J. Hydrogen Energy* **9**, pp 649–668 (1984).

K. Weissermel and H.-J. Arpe: *Industrial Organic Chemistry*, pp 17–30, Verlag Chemie, Weinheim-Deerfield Beach-Basel (1978).

W. N. Smith and J. C. Santangelo: *Hydrogen: Production and Marketing,* ACS Symposium Series 116, Washington (1980).
T. Ohta: *Solar – Hydrogen Systems.* Pergamon Press, Oxford (1979).

Specific Information:

Commercial Information:

"Hydrogen – Product Review", *Chemical Economics Handbook,* Stanford Research Institute, Menlo Park, California, USA (July 1985 and June 1986).
E. Wicke: "Wasserstoff-Energiewirtschaft aus kritischer Sicht" (Hydrogen energy economics critically examined), *Chem.-Ing.-Tech.* **54,** pp 41–52 (1982).
Statistische Bundesamt Wiesbaden, Fachserie 4, Reihe 3.1, p. 112, W. Kohlhammer Verlag, Stuttgart (1985).
International Energy Agency, *World Energy Outlook,* OECD/IEA, Paris (1982).
K. J. Mundo, "Gaserzeugungsverfahren für große Wasserstoffmengen" (Processes for the large scale production of hydrogen), *Brennst.-Wärme-Kraft* **34,** pp 15–21 (1982).

Manufacture:

S. Stucki and A. Menth, "Physikalisch-chemische Probleme bei der Erzeugung und Speicherung chemischer sekundärer Energieträger" (Physical-chemical problems connected with the production and storage of chemical secondary energy carriers), *Ber. Bunsenges. Phys. Chem.* **84,** pp 1008–1013 (1980).
E. Supp and H. Jockel, "Verfahren zur Herstellung von Wasserstoff" (Processes for the manufacture of hydrogen), *Erdöl Kohle* **29,** pp 117–122 (1976).
H. Hofmann, "Die Erzeugung von Wasserstoff mit Hilfe thermochemischer Reaktionen" (Production of hydrogen with the aid of thermochemical reactions), *Chem.-Ing.-Tech.* **48,** pp 87–91 (1976).
R. Lohmüller: "Wasserstoff-Produktion aus Kohlenwasserstoffen" (Hydrogen production from hydrocarbons), *Chem.-Ing.-Tech.* **56,** pp 203–213 (1984).
Adv. Hydrogen Energy, 3(Hydrogen Energy Prog. 4, Vol. 1), pp 47–68, Vol. 4, pp 1417–1430 (1982).
H. Vandenborre: "New developments in alkaline water electrolysis", *Dechema-Monographien,* Vol. 98, pp 313–328, Verlag Chemie, Weinheim-Deerfield Beach-Basel (1985).
T. Yamamoto, *Chem. Econ. Eng. Rev.* **13,** Nr. 1–2, pp 11–16 (1981).

Applications:

Adv. Hydrogen Energy, 4 (Hydrogen Energy Prog. 5, Vol. 3), pp 1171–1186 (1984).
S. I. Wang, D. M. Nicholas and S. P. DiMartino: "Analysis dictates H_2 purification process", *Oil & Gas Journal,* pp 111–117 (Feb. 1984).
J. Schulze and H. Gaensslen: "Wasserstoffmärkte: Wenig Gegenwartsbewegung – faszinierende Zukunft" (Hydrogen markets: little current movement – a fascinating future), *Chem. Ind.* **36,** pp 202–208 (1984).
K. Griesbaum and D. Hönicke, "Kraftstoffe der Zukunft" (Motor fuel of the future), *Chemie in unserer Zeit* **14,** pp 90–101 (1980).
Chem. Mark. Rep. **219,** p. 3 and 36 (1981).

1.3 Hydrogen Peroxide and Inorganic Peroxo Compounds

1.3.1 Economic Importance

1.3.1.1 Hydrogen Peroxide

Commercially hydrogen peroxide is available in a variety of concentrations. The most important are aqueous solutions with 35, 50 and 70% by weight H_2O_2. The following statistics refer to "100% hydrogen peroxide".

H_2O_2 commercially mainly available in concentrations of 35, 50 and 70% by weight

Table 1.3-1. 1987 Hydrogen Peroxide Capacities in 10^3 t/a.

World*	Europe	USA	Canada	South America	Asia
882	430	226	56	30	140

* (excluding communist countries)

1987 World capacity for H_2O_2:
882×10^3 t/a

The consumption in the USA was 114×10^3 t in 1984 and is expected to increase at a rate of 7% per annum.

Interox with a capacity of 350×10^3 t/a is the largest producer worldwide, followed by Degussa with 160×10^3 t/a (of which 100×10^3 t/a is in Europe).

1.3.1.2 Sodium Perborate and Sodium Carbonate Perhydrate

World capacity for sodium perborate in 1986:

700×10^3 t/a

The following capacity and production figures for 1986 are estimates:

Table 1.3-2. 1986 Sodium Perborate Capacities (estimated) in 10^3 t/a.

World	Europe	USA (1980)
700	600	36

Table 1.3-3. Sodium Perborate Production 1986 (estimated) in 10^3 t/a.

World	Europe
500	400

Sodium perborate is produced in almost all western industrialized countries, but particularly in Europe (main application: detergents).

USA capacity for sodium carbonate perhydrate:

16×10^3 t/a

The world capacity for sodium carbonate perhydrate is much smaller than that for sodium perborate. In some plants the two products can be produced alternately. The USA capacity is ca. 16×10^3 t/a.

1.3.1.3 Alkali Peroxodisulfates and Sodium Peroxide

World capacity for alkali peroxodisulfates:

60×10^3 t/a

Details over the capacities and consumption of peroxodisulfates are given in Table 1.3-4.

Table 1.3-4. Capacities and Consumption of Peroxodisulfates in 10^3 t/a for 1979 (more recent figures not available).

	World	Europe	West Germany	Japan
Capacity	60	25 to 30	20	10
Consumption	30 to 35			

The most important compound is the ammonium salt, followed by the potassium and sodium salts. In Canada, there is a pilot plant for the production of Na_2O_2 with a capacity of 300 t/a. In West Germany, Degussa has opened a small plant at Hanau, three years after closing their production plant.

1.3.2 Production

1.3.2.1 Hydrogen Peroxide

Three processes are industrially operated for the production of hydrogen peroxide:

- Oxidation of isopropanol
- Electrochemical oxidation of sulfuric acid or ammonium sulfate
- Anthraquinone process

Production of H_2O_2:

- Oxidation of isopropanol,only in the USSR.
- Electrochemical oxidation of sulfuric acid or ammonium sulfate, currently only of limited interest.
- Anthraquinone process: used for production of more than 95% of H_2O_2.

The first process is only utilized in two plants in the USSR. The electrochemical processes are currently of little importance, 95% of the hydrogen peroxide produced being manufactured using the anthraquinone process. Processes practiced in the past, such as production from barium peroxide, have not been used for a long time.

Isopropanol Oxidation Process

Acetone and hydrogen peroxide are produced with a selectivity of 80% upon multistage liquid phase oxidation of isopropanol with air at 15 to 20 bar and 90 to 140 °C:

Isopropanol oxidation process:

isopropanol oxidized with air to acetone and H_2O_2 at 90 to 140 °C and 15 to 20 bar and the reaction mixture worked up by distillation. Process has disadvantage that weight of acetone produced is double that of H_2O_2

$$CH_3CH(OH)CH_3 + O_2 \longrightarrow CH_3COCH_3 + H_2O_2$$

The degree of conversion is limited to about 30% to suppress side reactions. After the oxidation and diluting the reaction mixture with water, the acetone, unreacted isopropanol and water are distilled off. A ca. 20% hydrogen peroxide solution is taken off from the sump (sump temperature ca. 120 °C). Acetone is separated from the head product and the isopropanol-water solution fed back into the process. The 20% hydrogen peroxide solution is then purified over ion exchangers and concentrated by distillation. The process has one disadvantage: the weight of acetone produced is double that of the hydrogen peroxide.

Electrochemical Processes

Electrochemical processes: anodic oxidation

- From sulfuric acid to peroxodisulfuric acid
- From ammonium sulfate to ammonium peroxodisulfate

subsequent hydrolysis to sulfuric acid or ammonium sulfate respectively and H_2O_2. Separation of H_2O_2 by distillation

In the electrochemical processes, an aqueous solution of sulfuric acid (550 to 570 g/L) (Degussa-Weißenstein Process) or of sulfuric acid (260 g/L) and ammonium sulfate (210 to 220 g/L) (Löwenstein-Riedel Process) is electrochemically oxidized at the anode to peroxodisulfuric acid or ammonium peroxodisulfate respectively and reduced at the cathode producing hydrogen:

$$2\,H_2SO_4 \xrightarrow{\text{electr. energy}} H_2S_2O_8 + H_2$$

or

$$(NH_4)_2SO_4 + H_2SO_4 \xrightarrow{\text{electr. energy}} (NH_4)_2S_2O_8 + H_2$$

Small quantities of ammonium thiocyanate or hydrochloric acid are added to increase the anode potential. The peroxo compound obtained is subsequently hydrolyzed, the hydrolysis proceeding by way of the peroxomonosulfate:

$$H_2S_2O_8 + H_2O \longrightarrow H_2SO_5 + H_2SO_4$$

$$H_2SO_5 + H_2O \longrightarrow H_2SO_4 + H_2O_2$$

The hydrogen peroxide formed is distilled off, the sulfuric acid or sulfuric acid-ammonium sulfate solutions respectively being recycled. The total yield for both processes relative to the electricity consumed is about 70%.

The disadvantages of the electrochemical processes are the high plant and high production costs, due to the high cost of the electricity used.

Anthraquinone Process

Anthraquinone process:

oxidation of a 2-alkyl-anthrahydroquinone with air to a 2-alkyl-anthraquinone and H_2O_2 at 30 to 80 °C and 5 bar. H_2O_2 extracted with water and the quinone reduced with H_2/Pd to the hydroquinone. H_2O_2 produced from H_2 and O_2

The anthraquinone process is based on the following reactions: (1) the oxidation of a 2-alkyl-anthrahydroquinone with air to the corresponding 2-alkyl-anthraquinone and hydrogen peroxide and (2) the catalytic (back)-reduction of the 2-alkyl-anthraquinone to the

2-alkyl-anthrahydroquinone. In this cyclic process hydrogen peroxide is formed from hydrogen and oxygen:

$$\underset{\text{OH}}{\overset{\text{OH}}{\text{R}}} \underset{+H_2[\text{cat.}]}{\overset{+O_2}{\rightleftharpoons}} \underset{\text{O}}{\overset{\text{O}}{\text{R}}} + H_2O_2$$

The alkyl-group substituent R on the anthraquinone differs from manufacturer to manufacturer. In addition to 2-ethyl- (mainly used), 2-*tert*-butyl-, 2-*tert*-amyl- and 2-*sec*-amyl-anthraquinone are also employed. Mixtures of different alkyl anthraquinones can also be used.

The solvent mixture, in which both the quinone- and hydroquinone-compounds must dissolve, is complex. The "working solution" contains, as a solvent for the quinone, mainly a mixture of aromatic compounds such as naphthalene or trimethylbenzene. Polar compounds such as tris-(2-ethylhexyl)-phosphate, diisobutylcarbinol or methylcyclohexanol-acetate are suitable solvents for the hydroquinone.

The anthraquinone process requires a complex solvent mixture:

- "Quinone dissolver": mainly mixtures of aromatic solvents
- "Hydroquinone dissolver": polar solvents, especially esters

The solvent mixture has to fulfill a number of requirements: low solubility in water, low volatility, good dissolving properties, chemical stability under the reaction conditions used, low viscosity etc.

In the first step of the process the anthraquinone is hydrogenated to the hydroquinone with palladium as the preferred catalyst: on carriers, as gauze or in suspension. The reaction is carried out at about 40 °C and pressures up to ca. 5 bar with cooling and only to ca. 50% hydrogenation to suppress side reactions (see below).

The subsequent oxidation proceeds with air at 30 to 80 °C and pressures up to 5 bar, if necessary after catalyst separation and a precautionary filtration. It can be carried out in co- or countercurrent mode, in a single step or a multistep process.

The hydrogen peroxide formed during the oxidation is extracted from the reaction mixture with water e.g. in pulsating packed towers. The extraction yield is ca. 98%. The hydrogen peroxide solutions obtained are 15 to 35% by weight and must be freed from residual organic compounds before they can be concentrated by distillation. Commercial hydrogen peroxide solutions always contain stabilizers, such as diphosphates, organic complexing agents or tin compounds, to prevent their decomposition.

After separating off the hydrogen peroxide, the working solution has to be dried and freed of byproducts e.g. with active aluminum oxide. This occurs in a bypass.

Formation of byproducts – particularly during hydrogenation – complicates the anthraquinone process

In practice, the anthraquinone process is much more complicated than has been described above, in that byproducts such as 1,2,3,4-tetrahydroanthraquinone are formed, particularly in the hydrogenation step. These behave similarly to anthrahydroquinones, but their further hydrogenation leads to the octahydroanthrahydroquinones which are unusable in this process. Other byproducts such as oxanthrones and anthrones can only be partially regenerated. These unusable byproducts have to be removed from the process.

1.3.2.2 Sodium Perborate

Sodium perborate manufacture

1st step: Formation of metaborate from borax and sodium hydroxide

2nd step: Reaction of metaborate with H_2O_2 forming perborate, which is then filtered off from the cooled solution and dried

Sodium perborate (more correctly sodium peroxoborate)

$$Na_2\left[(HO)_2B(O{-}O)_2B(OH)_2\right]$$

exists in hexahydrate, tetrahydrate and anhydrous forms, of which the most important is the hexahydrate (commercially known as sodium perborate tetrahydrate). It is manufactured from borax in a two-step process:

$$1.\ Na_2B_4O_7 + 2\,NaOH \rightarrow 4\,NaBO_2 + H_2O$$

$$2.\ 2\,NaBO_2 + 2\,H_2O_2 + 6\,H_2O \rightarrow Na_2B_2O_4(OH)_4 \cdot 6\,H_2O$$

The first step, the formation of sodium metaborate from borax and sodium hydroxide, is carried out at temperatures up to 90 °C. When impure borax is used the solution is filtered. The second step is carried out at 25 °C and the mixture subsequently cooled to 15 °C and the precipitated sodium peroxoborate hexahydrate filtered off. Stabilizers for the perborate, such as silicates or magnesium salts, may be added to the reaction mixture. Residual moisture (3 to 10%) is removed in a hot air drier. The mother liquor from the second step can be returned to the first step.

The end product contains ca. 10.1 to 10.4% "active oxygen" (theoretically 10.38%). The bulk density of the perborate is adjusted to that of the other components in detergents ("light perborate") by special steps in the process. Older manufacturing processes, which started from sodium peroxide or use electrochemical processes, are no longer used.

Active oxygen content: 10.1 to 10.4%

1.3.2.3 Sodium Carbonate Perhydrate (Sodium Percarbonate)

In contrast with sodium perborate, which is a genuine peroxo compound, sodium percarbonate is only a perhydrate. It has the composition $Na_2CO_3 \cdot 1.5H_2O_2$. It can be manufactured using "dry" and "wet" processes. In a modern dry process, hydrogen peroxide and a sodium carbonate solution are sprayed onto a fluidized bed of sodium percarbonate which is fluidized with warm air. The fines are returned to the process and the oversized particles are ground.

Sodium percarbonate: from sodium carbonate and H_2O_2

- "Dry process" in a fluidized bed
- "Wet process" in solution. End product separated by cooling

In the wet process, sodium carbonate solution and hydrogen peroxide are reacted together in stoichiometric proportions. The percarbonate precipitates out upon cooling after vacuum concentration, if required.

Since percarbonate is much less stable than perborate, stabilizers such as alkali silicates or phosphates are used in both processes. For its use in detergents it can also be coated with an organic or inorganic material to increase its stability. Its active oxygen content is ca. 13.5% (theoretically 15.28%).

Stabilizers have to be added to sodium percarbonate, because it decomposes easily

Active oxygen content ca. 13.5%

1.3.2.4 Alkali Peroxodisulfates

Diammonium peroxodisulfate is obtained by electrolyzing solutions of ammonium sulfate and sulfuric acid, in cells with or without diaphragms, using the Löwenstein-Riedel Process (see Sec. 1.3.2.1):

Diammonium peroxodisulfate produced electrochemically from solutions of sulfuric acid and ammonium sulfate on platinum electrodes

$$(NH_4)_2SO_4 + H_2SO_4 \xrightarrow{\text{electr. energy}} (NH_4)_2S_2O_8 + H_2$$

The voltage used is between 5 and 7 V (theor.: 2.1 V) and the current density between 0.5 and 1 A/cm^2. Graphite or lead cathodes and platinum anodes are used. During electrolysis the solution becomes enriched with peroxodisulfate up to a concentration of 1 to 1.5 mol./L. Pure diammonium peroxodisulfate (purity > 99%) precipitates out upon cooling. After adding ammonium sulfate and sulfuric acid to the mother liquor, it is returned to the electrolysis cell.

Disodium and dipotassium peroxodisulfate manufacture:

electrochemically as with diammonium peroxodisulfate, or from diammonium peroxodisulfate with $KHSO_4$, NaOH or Na_2CO_3 respectively

Disodium and dipotassium peroxodisulfates can also be produced in cells without diaphragms by the electrolysis of the corresponding disulfates. Alkali peroxodisulfates can also be obtained by a metathesis reaction with diammonium peroxodisulfate:

$$(NH_4)_2S_2O_8 + 2KHSO_4 \longrightarrow K_2S_2O_8 + 2NH_4HSO_4$$

$$(NH_4)_2S_2O_8 + 2NaOH \longrightarrow Na_2S_2O_8 + 2NH_3 + H_2O$$

1.3.2.5 Sodium Peroxide

Sodium peroxide manufacture:

1st step : Formation of Na_2O

2nd step: Further oxidation to Na_2O_2

Its manufacture from sodium metal and oxygen from air takes place in two steps:

$$2Na + 0.5O_2 \longrightarrow Na_2O$$

$$Na_2O + 0.5O_2 \longrightarrow Na_2O_2$$

Reaction temperature of 1st step: 200 to 700 °C

The reaction is carried out by adding sodium metal intermittently to sodium monoxide in a rotary tube reactor while passing in air countercurrently. The sodium metal rapidly distributes itself over the surface of the monoxide and in doing so is oxidized. The reaction temperature is 200 to 700 °C and the heat of reaction is sufficient to maintain this temperature. In this way a noncaking material is obtained. The monoxide formed, containing a few per cent of peroxide and less than 1% sodium, is withdrawn from the reactor intermittently.

Reaction temperature of 2nd step: 350 °C

The oxidation of the monoxide to sodium peroxide occurs in a similar reactor at 350 °C. As this step is only slightly exothermic (– 79.5 kJ/mol) heating is needed. The end product is formed as beads with a diameter of 0.5 to 1 mm and a purity of 97 to 98%.

Purity of Na_2O_2: 97 to 98%

1.3.3 Applications

1.3.3.1 Hydrogen Peroxide, Sodium Perborate and Sodium Carbonate Perhydrate

The fields in which hydrogen peroxide is utilized vary considerably from region to region. This is due to different washing temperatures. In Europe, the household wash is carried out at relatively high temperatures. Therefore, the detergents contain perborate or percarbonate, whose manufacture accounts, in Europe and West Germany, for ca. 40% of the hydrogen peroxide produced. Detergents in Europe contain about 15 to 30% perborate with ca. 23% in West Germany.

H_2O_2 applications: vary considerably from region to region. Used little in the USA for the manufacture of per-compounds

In the USA, where the washing temperature is much lower, hypochlorites are used instead. In 1985 the consumption of hydrogen peroxide in the USA was:

- > 20% for cellulose bleaching
- 25% for water purification
- 20% for the production of chemicals (of which 6% was for peroxo compounds)
- 20% for textile bleaching

Uses of H_2O_2:

in Europe:
- 40% Perborate and percarbonate manufacture (for detergents)
- 25% Paper bleaching
- 25% Textile bleaching

in USA:
- 20% Cellulose bleaching
- 25% Water purification
- 20% Production of chemicals
- 20% Textile bleaching

In Europe ca. 25% is used for both paper and textile bleaching.

In the chemical industry, hydrogen peroxide is used in the manufacture of glycerol from allyl alcohol, of epoxy-compounds such as epoxidized soya oil (plasticizer for PVC) and of organic peroxides (e.g. methyl ethyl ketone peroxide, dibenzoyl peroxide) which are used as free radical initiators in polymerization processes. The manufacture of amine oxides, such as lauryl dimethyl amine oxide, is also carried out with hydrogen peroxide (used in rinsing agents for dishwashers).

The consumption of hydrogen peroxide for the manufacture of high quality paper is expected to increase particularly strongly, at least in the USA. In the textile industry, hydrogen peroxide is mainly used for the bleaching of cotton, although it is also used for the bleaching of wool.

The consumption of hydrogen peroxide for the treatment of effluents, especially for the removal of phenols, cyanides and sulfur compounds (hydrogen sulfide), is also expected to increase strongly.

1.3.3.2 Alkali Peroxodisulfates and Sodium Peroxide

Peroxodisulfates:

- Over 65% used as a polymerization initiator

Most of the peroxodisulfate manufactured (> 65%) is used as a polymerization initiator in the production of poly(acrylonitrile), emulsion-polymerized PVC etc. The rest is utilized in numerous application areas (etching of printed circuit boards, bleaching processes etc.).

Sodium peroxide is used in the mining industry and as a laboratory chemical.

References for Chapter 1.3 Hydrogen Peroxide and Inorganic Peroxo Compounds

Reviews:

Kirk-Othmer: *Encyclopedia of Chemical Technology.* 3rd Ed., Vol. 13, pp 12–38, John Wiley & Sons, Inc., New York, Chichester, Brisbane, Toronto (1981).

C. A. Crampton et al.: "The Manufacture, Properties and Uses of Hydrogen Peroxide and Inorganic Peroxy Compounds" in: R.Thompson (ed.) *The Modern Inorganic Chemicals Industry.* The Chemical Society, Burlington House, London 1977, pp 232–272.

Winnacker-Küchler: *Chemische Technologie.* Vol. 2, Anorganische Technologie I (Inorganic technology I), pp 563–606, Carl Hauser Verlag, München, Wien (1982).

Wasserstoffperoxid und seine Derivate (Hydrogen Peroxide and its Derivatives), Ed. W. Weigert, Hüthig Verlag, Heidelberg (1978).

Commercial Information:

Chem. Ind. **XXXIII,** p. 806 (Dec. 1981).

Chem. Week, p. 49 (Nov. 4, 1981).

Chem. Week, p. 19 (Sept. 16, 1981).

Chem. Week, p. 35 (May 6, 1981).

Chem. Ind., **XXXII,** p. 698 (Oct. 1980).

Chem. Week, pp 29–30 (Nov. 17, 1982).

Eur. Chem. News, p. 11 (April 25, 1983).

Chemical Profile, Hydrogen Peroxide, Schnell Publishing Company, Inc.(April 1, 1984).

Chem. Week, pp 19–20 (April 10, 1985).

R. S. Kouk, M. G. Dodson and J. R. Presley, " Pulp Bleaching Applications Lead Way to Increased Peroxide Demand", *Pulp & Paper,* pp 54–56 (Oct. 1985).

Hydrogen Peroxide, *Chemical Economics Handbook.* Stanford Research Institute, Menlo Park, California, USA (April 1980).

1.4 Nitrogen and Nitrogen Compounds

1.4.1 Ammonia

Ammonia is the starting material for almost all industrially produced nitrogen compounds

Ammonia is the starting material for almost all the industrially produced nitrogen compounds and is therefore one of the most important inorganic raw materials. The synthesis of ammonia is very closely coupled with the manufacture of synthetic fertilizers such as ammonium sulfate, ammonium nitrate, urea (Sec. 2.2) and the various ammonium phosphates (Sec. 2.1.2.3) and is thus an important indicator of the state of development of the agricultural industry.

1.4.1.1 Economic Importance

The production capacity for synthetic ammonia has increased strongly in recent decades despite the stagnating consumption in the western industrialized countries since the end of the 1970's. New plants have been installed in the USSR, India, the People's Republic of China and Mexico in particular, increasing the worldwide capacity from 72.2×10^6 t/a in 1975 to 128×10^6 t/a ammonia in 1985 (see Table 1.4.1-1).

Table 1.4.1-1. Ammonia Capacities in the World Regions in 10^6 t NH_3.

	World	North America	USA	Central & South America	Western Europe	Eastern Europe	USSR	Asia & Oceania	Africa
1975	77.2	18.3	14.3	2.6	17.2	8.5	10.9	18.5	1.1
1980	109.7	21.4	18.6	4.5	17.8	11.2	20.0	32.4	2.3
1985*	128.3	20.7	16.4	7.3	17.3	13.9	25.3	40.6	3.4

* estimated

The countries producing the largest quantities are the USSR, the People's Republic of China and the USA. The 1985 world demand was 108×10^6 t and this is expected to rise to ca. 125×10^6 t ammonia by 1990.

In addition to the dedicated synthesis of ammonia, small quantities (< 1%) are produced as a byproduct in coking plants and crude oil refining.

1.4.1.2 Synthetic Ammonia Manufacture

1.4.1.2.1 General Information

Large scale manufacture of synthetic ammonia is exclusively carried out with "synthesis gas", a 1:3 mixture of nitrogen and hydrogen, according to the reaction:

$$1/2\,N_2 + 3/2\,H_2 \rightleftharpoons NH_3,\ \Delta H = -45.93\ \text{kJ/mol}$$

and is based on the investigations of Haber into the equilibrium between nitrogen, hydrogen and ammonia in 1904. The industrial manufacture of ammonia resulted from a later cooperation with Bosch and Mittsch (both from BASF). The first Haber-Bosch production plant started up at BASF in 1913.

Formation of ammonia from nitrogen and hydrogen in the Haber-Bosch process is favored by:

- High pressure
- Low temperature
- Active catalysts
- Removal of the NH_3 from the equilibrated mixture
- Pure gas (as little inert gas as possible)

The exothermic reaction between nitrogen and hydrogen occurs in the presence of suitable catalysts and results in volume reduction, the highest ammonia concentrations in the reaction gas being obtained at the highest possible pressure and the lowest possible temperature. The upper limit for the applied pressure is determined by the cost of compression of the gas mixture and the cost of high pressure plant. The reaction temperature is determined by the type and activity of the catalyst. The removal of ammonia from the reaction gas should be as complete as possible to favor the fresh formation of ammonia. Other important parameters are the contents of inert gas and oxygen compounds in the unreacted synthesis gas.

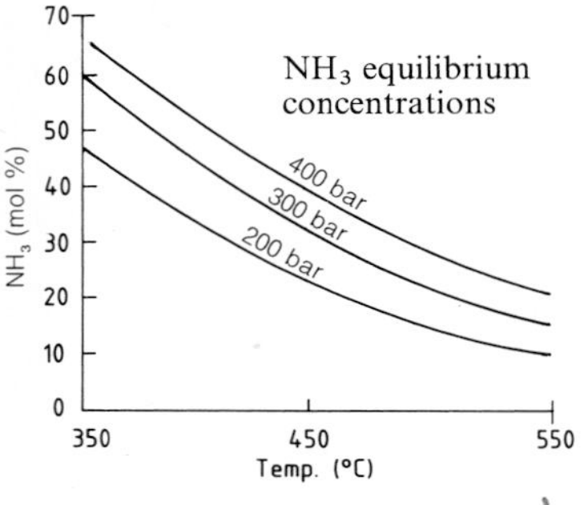

All the ammonia production plants in the world operate according to these basic principles i.e. reaction of nitrogen and hydrogen in a catalyst-filled pressure reactor at temperatures between 400 and 500 °C, pressures between 100 and 1000 bar (depending upon the plant) and removal of the ammonia formed from the reaction gas. The plants differ in their plant design, catalyst composition and production and purification of the synthesis gas. The important process steps in modern ammonia plants are:

Actual synthesis is carried out at temperatures of 400 to 500 °C and pressures of 100 to 1000 bar

NH_3 synthesis consists of the following process steps:

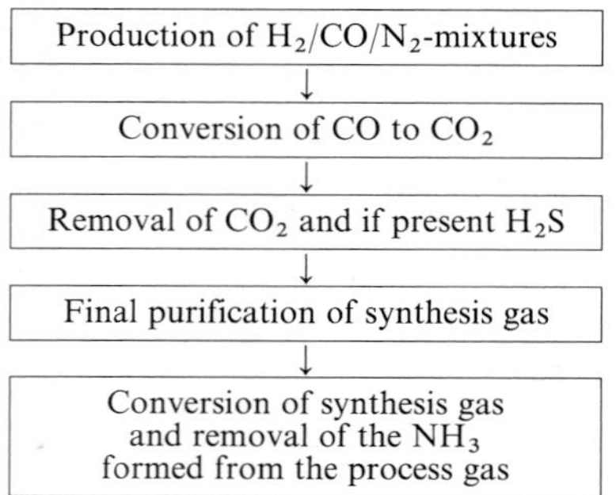

- Production of the nitrogen-hydrogen mixtures (CO and CO_2 are produced as byproducts)
- Conversion of carbon monoxide into carbon dioxide
- Removal of carbon dioxide and (if necessary) hydrogen sulfide
- Final purification of the synthesis gas
- Reaction of the synthesis gases, removal of the ammonia formed from the cyclic process.

1.4.1.2.2 Synthesis Gas Production

Raw Materials

The production of one ton of ammonia requires a mixture of 2400 m^3 of highly purified hydrogen and 800 m^3 of highly purified nitrogen (at 0 °C and 1000 mbar). It is produced using different processes depending upon the raw materials utilized.

Nitrogen from natural gas is frequently used. Alternatively it is extracted from air by low temperation fractionation. The latter is preferred when pure oxygen is required as an oxidizing agent in the production of synthesis gas. Alternatively air is directly employed in the production of synthesis gas and the oxygen is removed during oxidation of the reactants.

Synthesis gas:

- N_2 – from air or natural gas
- H_2 – from reaction of natural gas or naphtha with H_2O (by steam reforming)
  - from heavy fuel oil and H_2O (by partial oxidation)
  - from coal and water (by coal gasification)

Hydrogen is manufactured from hydrocarbons (natural gas, naphtha) or coal and water (Sec. 1.2.2).

The choice of raw material depends upon availability, cost and the investment in the particular plant. Before World War II ca. 90% of the hydrogen was produced by coal gasification. In 1985 only 10% of the hydrogen was produced in this way, mainly in South Africa and India.

Currently by far the most important, and in almost all new ammonia synthesis plants utilized, raw materials are natural gas and oil (naphtha). Worldwide in 1985 73% of ammonia production utilized natural gas and 16% naphtha. Hydrogen from water electrolysis on the other hand is currently unimportant at < 1%. In the long term it may increase in importance due to the development of new electrolysis or photovoltaic processes.

1985 worldwide raw material spectrum for NH_3 production:

- Natural gas 73%
- Naphtha 16%
- Coal/coke 10%
- Other < 1%

Manufacture of Mixtures of Hydrogen, Nitrogen and Carbon Monoxide

Steam Reforming (from Natural Gas and Naphtha)

In the steam-reforming process, natural gas or naphtha are converted into hydrogen, carbon monoxide (and carbon dioxide) with steam in the presence of catalysts at temperatures between 700 and 830 °C and pressures of up to 40 bar:

$$C_nH_{2n+2} + n\,H_2O \longrightarrow n\,CO + (2n+1)H_2$$

This reaction is carried out in so-called "primary reformers", which consist of a number of perpendicularly arranged catalyst-filled tubes, which are externally heated e.g. by the combustion of natural gas (allothermic process).

Steam reforming:

- Desulfurization of the raw materials by hydrogenation on CoO- or NiO- and MoO_3-containing catalysts at 350 to 450 °C
- Adsorption of the hydrogen sulfide formed on zinc oxide:
  $H_2S + ZnO \rightarrow ZnS + H_2O$
- Dissociation of methane or naphtha with steam in primary reformers at 700 to 830 °C under pressure on NiO-Al_2O_3(-MgO) catalysts
- Conversion of methane at 1000 to 1200 °C to hydrogen and carbon monoxide on Ni-containing catalysts in secondary reformers
- Adjustment of the stoichiometric hydrogen-nitrogen ratio by feeding air into the secondary reformer

The principal component of the plants is the nickel catalyst coated onto carrier materials such as α-aluminum oxide or magnesium-aluminum spinels and reduced by hydrogen to nickel, the actual catalyst, under the reaction conditions obtaining in the steam reforming process. Nickel catalysts are easily poisoned e.g. by sulfur compounds. Therefore the sulfur-containing raw materials, natural gas and naphtha, have to undergo prior desulfurization. This is accomplished by contacting the raw materials with cobalt- or nickel-molybdenum oxide-containing catalysts in the presence of hydrogen at 350 to 450 °C. This reduces the sulfur-carbon compounds to hydrogen sulfide which is adsorbed on zinc oxide.

After desulfurization, steam is added and the mixture heated to 480 to 550 °C, before it is fed into the primary reformer.

The gas leaving the primary reformer contains between 7 and 10% methane, depending upon the operating conditions. This is removed in so-called "secondary reformers" in which the gas leaving the primary reformer is reacted with steam and added preheated air in nickel catalyst-filled shaft furnaces at temperatures of 1000 to 1200 °C. The quantity of air is adjusted to give the nitrogen to hydrogen ratio required for the stoichiometry of the ammonia synthesis. The residual methane content is ca. 0.2 to 0.3 % by volume.

Partial Oxidation of Heavy Fuel Oil

In partial oxidation, the raw materials, e.g. gaseous and liquid hydrocarbons with heavy fuel oil being preferred, are oxidized to hydrogen and carbon monoxide with insufficient oxygen for total combustion:

$$2\,C_nH_{2n+2} + nO_2 \rightarrow 2\,nCO + 2(n+1)H_2$$

If oxygen-enriched air is used, its quantity is adjusted to give the nitrogen to hydrogen ratio required for the stoichiometry of the ammonia synthesis.

The partial oxidation is autothermal and in contrast to steam reforming does not require a catalyst. It takes place at temperatures between 1200 and 1500 °C and pressures of 30 to 40 bar (exceptionally 80 bar). The soot formed as a byproduct has to be scrubbed out with water and recycled in a fairly complex process.

Partial oxidation:

- Autothermal combustion of different crude oil fractions at 1200 to 1500 °C and 30 to 40 bar in a deficiency of oxygen
- Removal of the soot byproduct

The advantage of the partial oxidation process is that sulfur does not interfere, making desulfurization of the raw materials unnecessary. A disadvantage compared with the steam reforming process, is the necessity of an air liquification plant for oxygen extraction.

Advantage: desulfurization of the raw materials unnecessary

The Shell and Texaco processes operating at pressures of 30 to 90 bar are successful in large scale production.

Coal Gasification

A mixture of hydrogen, carbon monoxide, carbon dioxide, methane and sometimes nitrogen is formed upon the partial oxidation of coal (hard or soft coal) with oxygen or air and steam at high temperatures. The main reaction taking place during this gasification is the reduction of water with carbon to hydrogen and carbon monoxide

Coal gasification:

- Coal autothermally reacted with O_2 and steam at high temperatures

$$C + H_2O \rightleftharpoons H_2 + CO$$

and the exothermic partial combustion of carbon to carbon monoxide:

$$2C + O_2 \rightleftharpoons 2CO$$

Under these reaction conditions water and carbon monoxide react forming hydrogen and carbon dioxide, and methane is formed by the reduction of carbon monoxide or carbon with hydrogen.

All industrial processes are exothermic, ca. 1/3 of the coal used being combusted to attain the required high reaction temperatures. This is also the case for the Lurgi pressure gasification process carried out in a mechanically agitated solid bed at ca. 1200 °C (as used e.g. in Sasolburg in South Africa) and for the classical Winkler process operating with a pressureless fluidized bed at 800 to 1100 °C. The currently most widely used process is the Koppers-Totzek process in which coal

- Processes used industrially:
- Lurgi
- Koppers-Totzek
- Winkler

is used in the form of flyash (atmospheric pressure, 1400 to 1600 °C).

These gasification processes are, on the basis of invested capital and energy consumption, still inferior to the processes for producing hydrogen-carbon monoxide mixtures from hydrocarbons. Intensive research is currently being carried out in West Germany, Great Britain and the USA to improve these processes further. The prospects are particularly favorable for processes producing synthesis gas at medium pressure, which does not require further compression. A further improvement in the economics of the process is hoped for by utilizing nuclear process heat in (allothermal) coal gasification plants e.g. with 950 °C helium from high temperature nuclear reactors (Sec. 6.4.3.1).

Conversion of Carbon Monoxide

CO oxidation to H_2 and CO_2 with steam:

high temperation oxidation at 350 to 380 °C on

- Iron-chromium oxide catalysts or
- Sulfur-insensitive Co/Mo-containing catalysts

The next step in the manufacture of synthesis gas is the removal of carbon monoxide by oxidizing it with steam to carbon dioxide, the steam being reduced to hydrogen:

$$CO + H_2O \rightarrow CO_2 + H_2, \Delta H = -41 \text{ kJ/mol}$$

As this water gas conversion reaction is exothermic, low temperatures favor the formation of carbon dioxide and hydrogen. The carbon monoxide content in the reaction gas is reduced to ca. 3 to 4% by vol. by reaction on reduced iron-chromium oxide catalysts at 350 to 380 °C. Since these catalysts are sensitive to impurities, cobalt- and molybdenum-(sulfide)-containing catalysts are used for sulfur contents > 1 g/m^3.

Low temperature oxidation at 200 to 250 °C on very sulfur-sensitive CuO/ZnO catalysts

If the reaction gas contains less than 0.1 ppm sulfur, the oxidation is carried out in the presence of copper oxide-zinc oxide catalysts at temperatures of 200 to 250 °C (low temperature oxidation). In this case the residual carbon monoxide content is less than 0.3% by vol.

Gas mixtures from the steam reforming process have already been desulfurized at the raw materials (natural gas and naphtha) stage due to the sensitivity of the steam reforming catalysts to poisoning by sulfur compounds. They are therefore either directly subjected to low temperature conversion or preferably worked up by a combination of high temperature and low temperature conversion. On the other hand, the reaction gases resulting from partial oxidation contain sulfur compounds as no preliminary purification has been carried out. Such gas mixtures can thus only be subjected to high temperature conversion with sulfur-insensitive catalysts, the hydrogen sulfide formed being subsequently removed together with the carbon dioxide.

Utilization of the different oxidation processes dependent upon sulfur content of the gas mixture:

- From steam reforming:
  low temperature or a combination of high temperature and low temperature oxidation is used
- From partial oxidation:
  Only high temperature oxidation can be utilized

Removal of Carbon Dioxide and Hydrogen Sulfide

In the next step the carbon dioxide, mostly from the conversion reaction, and hydrogen sulfide, if present, are removed from the gas mixture. This is accomplished by either physical or chemical absorption in appropriate solvents.

Removal of the acidic gases CO_2 and H_2S by physical or chemical absorption

In physical absorption processes such as the Rectisol process (carried out at low temperatures with methanol as the solvent) the gas mixture under pressure is brought into contact with solvent in absorption columns, the solvent being regenerated by pressure release or high temperature stripping. In this process carbon dioxide and hydrogen sulfide can be jointly scrubbed or hydrogen sulfide can be selectively absorbed. The Fluor-Solvent (propylene carbonate), Purisol (N-methylpyrrolidone) and Selexol [poly(ethyleneglycol dimethylether)] processes operate with lower vapor pressure solvents than that used in the Rectisol process.

Physical absorption:

with methanol, propylene carbonate, N-methylpyrrolidone, poly(ethyleneglycol dimethylether)

Chemical absorption processes use different absorption agents e.g. organic amines such as mono-, di- and triethanolamine, N-methyldiethanolamine or diisopropanolamine (Shell). BASF's alkazide process uses potassium monomethylaminopropionate. Aqueous solutions of potassium carbonate with added corrosion inhibiting and reaction activating agents (e.g. Benfield process) are widely used as absorption agents.

Chemical absorption:

with mono-, di- and triethanolamine, N-methyldiethanolamine, diisopropanolamine, potassium monomethylaminopropionate, potassium carbonate solutions

Combinations of physical and chemical absorption are also used as in Shell's Sulfinol process in which a mixture of diisopropanolamine and sulfolane in water is utilized.

For H_2S-free gases:
chemical scrubbing

For H_2S-containing gases:
physical absorption alone or in combination with chemical absorption

Chemical scrubbing is preferred for the hydrogen sulfide-free gases from the steam reforming process, with physical absorption alone or in combination with chemical absorption being preferred for the hydrogen sulfide-containing gases from the partial oxidation process.

Final Purification of Synthesis Gases

Final purification of synthesis gas after high temperature conversion to remove oxygen-containing impurities:

- Prior removal of H_2O, CO_2 and hydrocarbons
- Scrubbing with liquid N_2 under pressure and low temperatures

To protect the ammonia synthesis catalysts from premature poisoning, the synthesis gas has to be subjected to further intensive purification to remove completely the oxygen-containing compounds present, such as carbon monoxide, carbon dioxide, oxygen and water, and the residual hydrogen sulfide.

Large quantities of carbon monoxide, such as are present after high temperature conversion (3 to 5% by vol.), are removed in liquid nitrogen-filled columns at $-196\,^\circ C$ and increased pressure (up to 80 bar). To prevent the formation of ice the residual quantities of water and carbon dioxide are adsorbed on zeolites before scrubbing with liquid nitrogen. Methane and other hydrocarbons are condensed out. Liquid nitrogen scrubbing is also used to supply the nitrogen shortfall in the gas mixture, particularly when the gasification reaction for the production of synthesis gas is carried out with pure oxygen.

- N_2 introduced into the gas mixture during liquid nitrogen scrubbing

Residual CO in synthesis gas after low temperature conversion removed by methanation on Ni-catalysts under pressure at high temperatures

Nitrogen-hydrogen mixtures resulting from low temperature conversion contain only 0.1 to 0.3% by vol. of carbon monoxide. In this case the ca. 0.01 to 0.1% by vol. of carbon monoxide and carbon dioxide present after carbon dioxide scrubbing is hydrogenated to methane (methanation) in the presence of nickel catalysts in an exothermic reaction at 30 bar and 250 to 350 °C:

$$CO + 3H_2 \longrightarrow CH_4 + H_2O$$

$$CO_2 + 4H_2 \longrightarrow CH_4 + 2H_2O$$

Composition of synthesis gas after methanation:

- H_2 74.0%
- N_2 24.7%
- CH_4 1.0%
- Ar 0.3%
- $CO + CO_2$ < 10 ppm

The water formed in this reaction is removed from the synthesis gas by condensation or adsorption on zeolites. After this treatment the carbon monoxide and carbon dioxide contents are less than 10 ppm.

In some ammonia plants the residual methane (ca. 1% by vol.) and some of the residual argon (ca. 0.3% by vol.) are removed by low-temperature condensation (Braun purifier concept). Excess nitrogen from the secondary reforming process is also removed in this way.

Removal of CH_4, Ar (partially) and excess N_2 by condensation

In newer plants the classical (energy intensive) gas purification processes have been replaced by pressure swing adsorption on synthetic zeolites. Synthesis gases purified using this process contain less than 1 ppm of impurities and result in particularly high yields of ammonia.

1.4.1.2.3 Ammonia Synthesis Catalysts

The catalyst is the heart of every ammonia plant. The choice of ammonia catalyst is largely dictated by the synthesis conditions (pressure, temperature, gas throughput). The design of the ammonia reactor, the synthesis gas purity requirements and last but not least the ammonia yield are crucially influenced by the choice of catalyst.

The catalysts currently used are largely analogous to the first BASF-developed catalysts. They essentially consist of α-iron with small quantities of nonreducible metal oxides (promoters).

NH_3 synthesis catalysts consist of promoter-containing α-iron

There are structural and electronic promoters. The former (e.g. aluminum oxide, silicon dioxide) protect the iron particles of the catalyst against sintering by the formation of aluminosilicates or improve the resistance of the catalyst to sulfur-, and chlorine-compounds (calcium oxide). On the other hand, electronic promoters such as K_2O increase its specific activity in the presence of acidic and amphoteric oxides such as aluminum oxide and silicon dioxide, while reducing its temperature stability. Some catalysts also contain lithium-, beryllium-, vanadium- and uranium-oxides as promoters.

Promoters increase the activity (e.g. K_2O), lifetime, temperature stability of the catalyst (e.g. Al_2O_3, SiO_2) and reduce their sensitivity to poisoning (e.g. CaO)

Typical NH_3-catalyst composition (starting product):

- Fe_3O_4 94.3% by wt.
- K_2O 0.8% by wt.
- Al_2O_3 2.3% by wt.
- CaO 1.7% by wt.
- MgO 0.5% by wt.
- SiO_2 0.4% by wt.

The catalysts are manufactured by fusing a mixture of magnetite (Fe_3O_4) and the promoters at temperatures of 1600 to 2000 °C in an electric furnace or an electric arc furnace, cooling rapidly, pulverizing and sieving. A particle size of 6 to 10 mm is normally required, but there are also ammonia plants for which 1 to 2 mm

Manufacture of catalysts by

- Fusing of magnetite and promoter oxides
- Cooling
- Pulverizing
- Sieving

particles are preferred. Plants with smaller catalyst particles require increased gas speeds to compensate for the higher pressure losses.

Reduction of magnetite:

- Usually by synthesis gas in the pressure reactor
- Recently by prereduction in separate units

The subsequent reduction of magnetite is crucially important for the quality of the catalyst. It is normally carried out with synthesis gas in the pressure reactor of the ammonia production plant at not too high pressures (70 to 300 bar, depending upon the plant type) and at temperatures between 350 to 400 °C. Highly porous α-iron is formed according to the equation:

$$Fe_3O_4 + 4\,H_2 \longrightarrow 3\,Fe + 4\,H_2O$$

The oxidic promoter additives are not themselves reduced, but their presence decreases the magnetite reduction speed. The concentration of the water produced must be kept low to prevent its coming in contact with the reduced catalyst and thereby causing premature aging.

As a result high gas velocities (ca. 5000 to 15 000 h^{-1}) are used. Modern ammonia furnaces contain up to 100 t of catalyst and the reduction lasts several days.

Prereduced catalysts can be activated in less than 2 days at much lower temperatures

In the last few years prereduced catalysts have been preferred which are reduced in separate plants. These catalysts are pyrophoric and are stabilized by exposure to a nitrogen atmosphere containing ca. 0.2% oxygen at 50 to 95 °C. Only a short reaction time in the ammonia kiln is necessary to reduce this partially oxidized catalyst.

Catalyst poisons are:

O-, S-, P- and As-compounds, with hydrocarbons and other inert gases also interfering

These catalysts are extremely sensitive to gaseous oxygen-, sulfur-, phosphorus- and chlorine compounds. These reduce the active surface of the catalysts by chemisorption and hence their activity. Catalysts poisoned with oxygen compounds can be regenerated by reduction with pure synthesis gas. However regeneration of sulfur, phosphorus and arsenic compound-poisoned catalysts under ammonia synthesis conditions is not possible due to the stronger bonds formed.

The lifetime of current catalysts can be up to ten years or more, when highly purified gases and low synthesis pressures are used.

Hydrocarbons such as methane and other inert gases such as argon also interfere, since they interact with the catalyst surface and hinder the diffusion of nitrogen and hydrogen into the catalyst pores.

The efficiency of new catalysts on the basis of intermetallic alloys (e.g. $Fe_{91}Zr_9$), promoted ruthenium as well as transition metals linked to phthalocyanine have still to be demonstrated.

1.4.1.2.4 Reaction of Synthesis Gas to Ammonia

General Information

The industrial reaction of synthesis gas to ammonia in pressure reactors takes place in a cyclic process in which the ammonia formed is removed from the reaction gas and the unreacted synthesis gas returned to the reactor. In addition to the ammonia formed, inert gases and the liberated reaction heat have to be continuously removed from the cyclic process. The excess heat of the product gas is used to heat the feed synthesis gas to the reaction temperature in a heat exchanger integrated into the reactor. Additional waste heat can be utilized for steam generation. The pressure loss in the synthesis gas due to its passage through the synthesis loop is compensated for and the fraction of synthesis gas converted replaced by fresh ("fresh gas") compressed synthesis gas.

NH_3 synthesis is a cyclic process:

- removal of:
  - NH_3
  - Inert gas
  - Reaction heat (utilized for heating up cold synthesis gas)
- recycling of unused synthesis gas into the reactor
- supply of fresh gas
- compensation of pressure losses

The most important part of an ammonia synthesis plant is the pressure reactor, which is filled with catalyst and in which ammonia formation takes place at temperatures between 400 and 500 °C. A maximum temperature of 530 °C must not be exceeded, otherwise catalyst damage will ensue.

Compression is carried out either by piston pumps (in the case of plants with capacities < 600 t/d) or centrifugal compressors. Plants with capacities of up to 1500 t/d can be economically operated in the pressure range of 150 to 250 bar, for which only a centrifugal compresser is able to deliver the required compression of fresh- and cycling-gas.

Compression with the help of:

- Centrifugal compressors
- Piston compressors (for smaller plants)

Reactor Types

Reactor classes:

- Tubular reactors
  - cocurrent
  - countercurrent
- Multiple bed reactors
  - Cooling by cold gas
  - Heat exchange cooling
  - Catalyst arranged in separate layers

In principle there are two classes of reactors: tubular reactors and multiple bed reactors. The former group is split into cocurrent and countercurrent reactors, the latter into those with cold gas cooling and heat-exchange cooling. In the frequently used Casale reactor, a countercurrent reactor, the heat is transferred from the hot reaction gas to the cold synthesis gas by a ring heat-exchanger round the whole catalyst bed. This reactor type is started up by heating the synthesis gases to 400 °C with a centrally positioned electrically heated rod just before their entry into the catalyst mass. In other reactors this heating up takes place indirectly outside the reactor using heat from oil or gas combustion.

Most popular reactor type:

multiple bed reactors directly cooled with cold gas (quench reactor)

In cross flow (layer) reactors catalyst beds, gas flow is radial

In multiple bed reactors the catalyst is arranged in separate beds. Between the beds the gas is either cooled with a heat exchanger or by adding cold synthesis gas. The Uhde-Chemie Linz reactor belongs to the first category. Reactors directly cooled by the addition of cold gas (so-called "quench reactors") are the most widely used type. A representative of this category is the Kellogg quench reactor. Multiple bed reactors in which the gas stream passes radially through the catalyst beds (cross flow reactors) are frequently used (Haldor-Tapsoe). They are cooled by heat exchanger or with cold gas. There are also horizontal versions of such cross flow reactors (Uhde-Chemie Linz and Kellogg) which are notable for their lower pressure losses.

NH_3 synthesis proceeds in reactors with the catalyst arranged in separate containers. Cooling is with cold synthesis gas

In all reactors the catalyst is arranged in one or more containers in the interior of the up to 30 m long reactor tube. The temperature control is so designed that the temperature of the synthesis gas entering the reactor is higher than that of the gases exiting the reactor. This temperature management is assisted by a separate stream of cold synthesis gas.

Ammonia Separation

NH_3 separation from the cycling gas by condensation at low temperatures

The 0 to 15% by volume of ammonia present in the cycling gas (in certain cases up to 35%) is mainly separated by condensation at low temperatures (−25 °C) in which water or air cooling is assisted by the evaporation of liquid ammonia. The evaporated ammonia is either utilized in chemical products derived from ammonia or is liquified by compression and subsequent cooling. At low synthesis pressures and when there is a possibility of utilizing aqueous ammonia, the ammonia is separated by absorption in water. The water vapor thus entrained in the cycling gas is removed with liquid ammonia in an adjoining scrubber to avoid poisoning the catalyst with water vapor.

Before renewed compression and reentry into the ammonia reactor the inert gases (mainly argon) present in the so-called purge gas (unreacted hydrogen and nitrogen) have to be removed e.g. by the membrane separation process, pressure swing adsorption or cryogenic fractionation.

1.4.1.2.5 Integrated Ammonia Synthesis Plants

In modern ammonia plants the individual process steps of gas production, conversion, synthesis gas purification and ammonia synthesis are carried out in single-train plants. Plants for synthesis gas production e.g. steam reforming (Fig. 1.4.1-1) or partial oxidation (Fig. 1.4.1-2) operate economically, since personnel requirements are low and the energy exchange between heat-consuming and heat-producing process steps is optimal, so that a higher level of waste heat utilization is achieved. In addition, steam is produced from the flue gas of the primary reformers e.g. in steam reforming units or in waste heat boilers connected to the secondary reformer, high temperature conversion and methanation units. This steam is used both as a gasification agent and as a propellant for the compressors and pumps.

Integrated NH_3 plants include:

- Gas production
- CO and CO_2 conversion
- Synthesis gas purification

Single-train plants economical, due to the optimal utilization of waste heat and low personnel requirements

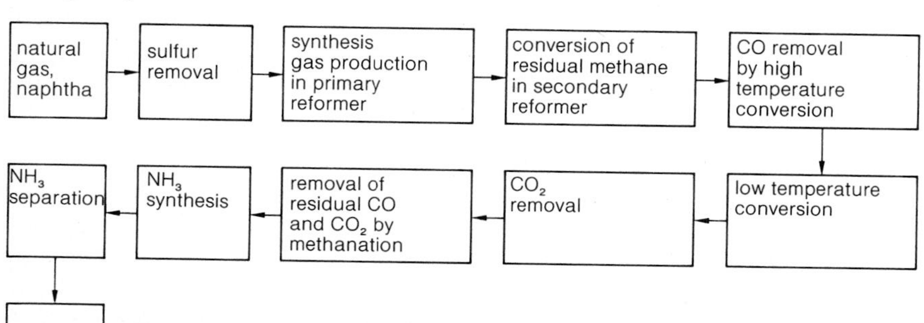

Fig. 1.4.1-1. Block schema of a single-train plant for the synthesis of ammonia using the steam reforming process.

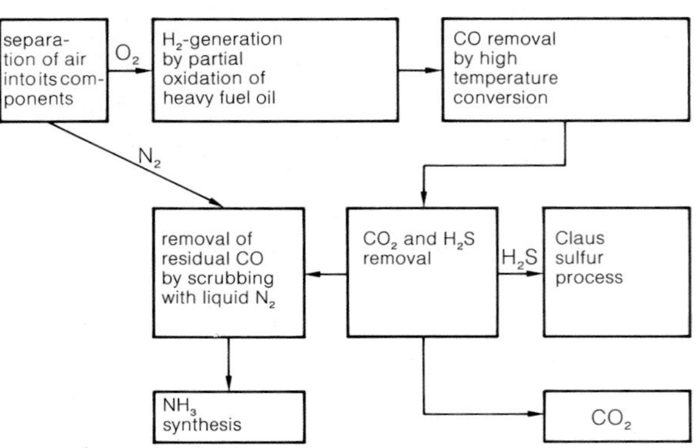

Fig. 1.4.1-2. Block schema of a single-train plant for the synthesis of ammonia using the partial oxidation process.

Economics of NH_3 production plants very dependent upon their size

The energy requirements for ammonia synthesis can be reduced from 80 to 90 GJ/t ammonia (for the classical Haber-Bosch process) to less than 30 GJ/t ammonia by the consistent utilization of waste heat, reducing the heat requirements (e.g. in the removal of carbon dioxide), optimization of the ammonia reactor design and by using purer synthesis gas. This is particularly the case for the currently favored plants with capacities of 1460 t/d. Plants with capacities < 600 t/d on the other hand are hardly competitive due to their poorer utilization of energy.

NH_3 stored and transported at various pressures and temperatures

In single-train plants it is necessary to store a certain amount of ammonia to cover production breakdowns. The ammonia is either stored at atmospheric pressure and low temperatures ($-33\,^\circ$C), under moderate pressure at $+5$ to $-5\,^\circ$C or at the site temperature and the corresponding equilibrium pressure. Ammonia is also transported under these different conditions.

1.4.1.3 Ammonia Applications

NH_3 used in the manufacture of:

- HNO_3
- Urea
- Ammonium phosphates
- Ammonium sulfate
- Organic basic chemicals (amides, nitriles, amines)

The most important chemicals manufactured from ammonia are nitric acid, urea, ammonium sulfate and ammonium phosphates. In organic chemistry ammonia is required for the synthesis of amides, nitriles and amines (e.g. adipic acid, melamine, acrylonitrile).

By far the most important consumer of ammonia is the fertilizer industry (80% of the world production is utilized in this sector). The plastics industry is next, consuming ca. 10% for the production of poly-urethanes, urea-formaldehyde resins, nylon, poly-(acrylonitrile) etc. Ca. 5% of the ammonia production is worked up into explosives and the rest is used in the manufacture of plant protection agents and other, mainly organic, chemicals.

Important end products are:

- Fertilizers (ca. 80% of NH_3 consumption)
- Plastics and fibers
- Explosives

References for Chapter 1.4.1: Ammonia

General Information:

Ullmann's Encyclopedia of Industrial Chemistry, 5th Ed., Vol. A2, pp 143–231, Verlag Chemie, Weinheim-Deerfield Beach-Basel (1985).

Kirk-Othmer: *Encyclopedia of Chemical Technology*, 3rd Ed., Vol. 2, pp 470–516, John Wiley & Sons Inc., New York-Chichester-Brisbane-Toronto (1978).

Specific Information:

Chemical Economics Handbook, Stanford Research Institute, Menlo Park, California, USA: "Nitrogen chemicals – report" (Feb. 1986); "Nitrogen chemicals – marketing research report" (July 1983).

Informations Chimie No. 269, "Ammoniaque" (ammonia), pp 99–122 (1986).

Chemical Week, pp 26–32 (July 1986).

"Ammonia producers find new ways to cut energy", *Chemical Engineering*, pp 14–17 (April 28, 1986).

"New catalysts herald cuts in costs of ammonia", *ECN Process Supplement*, pp 4–8 (Nov. 1985).

J. Whitehead, "Soviet ammonia presence grows", *Chemical Age*, pp 12–13 (May 30, 1980).

G. W. Bridger, R. E. Gadsby and D. E. Ridler, "High-pressure ammonia synthesis and plant operation", in *A Treatise on Dinitrogen Fixation*. pp 291–332, John Wiley & Sons Inc., New York-Chichester-Brisbane-Toronto (1979).

K. Stork, C. F. Wiles and G. K. Mathur, "Ammonia production using the Texaco coal gasification process", *Nitrogen* **160,** pp 39–50 (1986).

D. Keens, "Integrating ammonia and urea production", *Inst. Chem. Eng. Symp. Ser.* **74,** pp 55–65 (1982).

J. S. Merriam and K. Atwood, "Ammonia synthesis catalysts in industrial practice", *Appl. Ind. Cata.* **3,** pp 113–129 (1984).

D. A. Waitzman, "Ammonia from coal: a technical/economic review", *Chem. Eng.*, pp 69–71 (Jan. 30, 1978).

1.4.2 Hydrazine

1.4.2.1 Economic Importance

Hydrazine is commercially available as aqueous solutions and, in smaller quantities, in the form of salts. Its concentrated solution contains 64% by weight of hydrazine, in which the molar ratio of hydrazine to water is 1:1, hence the name "hydrazine hydrate". The capacity statistics in Table 1.4.2-1 relate to "hydrazine hydrate". Anhydrous hydrazine is not commercially available.

1.4.2.2 Manufacture of Hydrazine

Hydrazine is formed in a multitude of chemical reactions. Only a few processes have achieved commercial importance. These all oxidize ammonia or urea, an ammonia derivative, to hydrazine. Sodium hypo-

Manufacture of hydrazine:

oxidation of ammonia with sodium hypochlorite or hydrogen peroxide

chlorite or hydrogen peroxide is used as the oxidizing agent. Certain processes (Bayer-, H_2O_2-processes) operate in the presence of ketones.

1986 Worldwide hydrazine capacity:

68.4×10^6 t/a

Table 1.4.2-1. Hydrazine Capacities in 1986 (in 10^3 t/a). (Capacities given for the Comecon Countries are estimated.)

	Firms	Capacities	Process
USA	Olin	5	Raschig
		10	Bayer
	Mobay	10	Bayer
West Germany	Bayer	10	Bayer
Great Britain	Fisons	3	Raschig
France	Atochem	10	H_2O_2
Japan	Mitsubishi Gas	4	Raschig
	Toyo Chemical	3	Raschig
	Otsuka Kagaku	3	Raschig?
	Nippon Hydrazine	2.4	Raschig
India	IDPL/Pfizer	0.5	Raschig
USSR		3	Raschig
East Germany		0.5	Raschig
P.R.China		4	Raschig
Capacity worldwide:		68.4	

1.4.2.2.1 Raschig Process

Raschig process:

oxidation of ammonia with sodium hypochlorite to hydrazine and sodium chloride. Distillative separation of hydrazine from solid sodium chloride. Yield ca. 70%

In the Raschig process, ammonia is oxidized with sodium hypochlorite:

$$2NaOH + Cl_2 \longrightarrow NaOCl + NaCl + H_2O$$

$$NaOCl + NH_3 \longrightarrow NH_2Cl + NaOH$$

$$NH_2Cl + NaOH + NH_3 \longrightarrow N_2H_4 + NaCl + H_2O$$

$$2NaOH + Cl_2 + 2NH_3 \longrightarrow N_2H_4 + 2NaCl + 2H_2O$$

The sodium hypochlorite is obtained as a ca. 4.7 mol/L solution by mixing chlorine and sodium hydroxide with cooling in a molar ratio of 1:2 (see Fig. 1.4.2-1). This is diluted to ca. 1 mol/L and reacted with an aqueous ammonia solution (ca. 15%) at temperatures around 0°C (with cooling) forming chloramine and sodium hydroxide. The yield is almost quantitative.

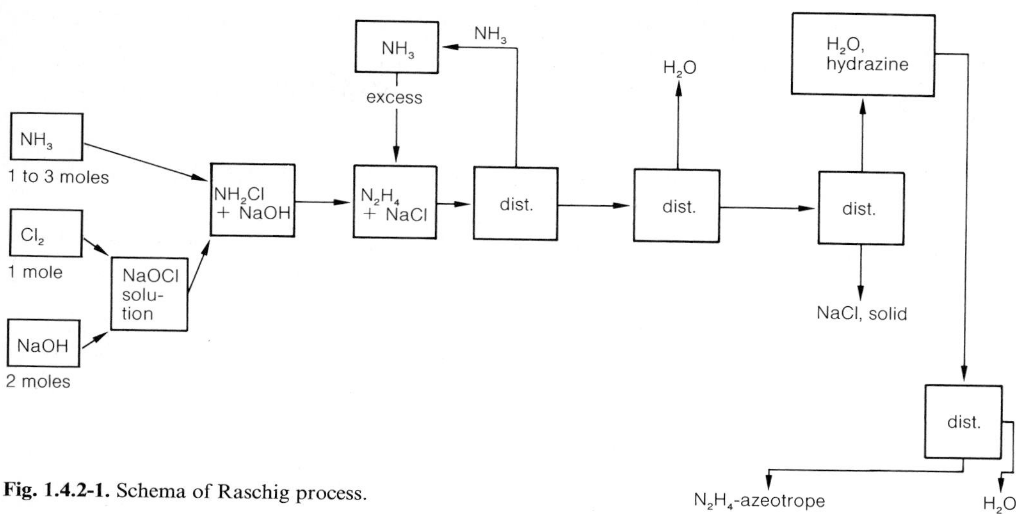

Fig. 1.4.2-1. Schema of Raschig process.

The chloramine solution is then reacted, at ca. 130 °C and under pressure, with a 20 to 30-fold molar excess of anhydrous ammonia. Subsequently, the excess ammonia is removed from the reaction mixture and recycled. Water and the hydrazine-water-azeotrope (b.p. 120.5 °C) are evaporated off leaving solid sodium chloride. The aqueous hydrazine solution obtained is concentrated by distillation. Ca. 70% of the theoretical yield is obtained. Important side reactions during the synthesis are:

- The reaction of chloramine with the hydrazine formed:

$$2NH_2Cl + N_2H_4 \longrightarrow N_2 + 2NH_4Cl$$

  This reaction is particularly catalyzed by copper. As preventive measures complexing agents such as ethylenediamine-tetra-acetic acid (EDTA) are added and a large excess of ammonia is used.

- The decomposition of hydrazine during its evaporation from the solid sodium chloride.

If hydrazine hydrate is not required, hydrazine can be isolated from the Raschig synthesis reaction mixture as the sparingly soluble hydrazine sulfate ($N_2H_6^{2+}SO_4^{2-}$, solubility in water: 2.96 g/L).

Urea process:

oxidation of urea with sodium hypochlorite to hydrazine, sodium chloride and sodium carbonate. Working up as in the Raschig process. Yield: 60 to 70%

1.4.2.2.2 Urea Process

In this process a mixture of urea, sodium hypochlorite and sodium hydroxide is converted into hydrazine, sodium chloride and sodium carbonate.

$$H_2NCONH_2 + NaOCl + 2NaOH \longrightarrow N_2H_4 + NaCl + Na_2CO_3 + H_2O$$

The reaction components are mixed in the cold in the above mole ratios and the mixture is rapidly heated to 100°C. The working up follows the Raschig process procedure, giving 60 to 70% of the theoretical yield. The advantages of this process are the avoidance of a large excess of ammonia and working at atmospheric pressure, which must be set against the serious disadvantage of utilizing two moles of sodium hydroxide per mole of hydrazine. Furthermore, the production of a sodium chloride-sodium carbonate-mixture poses an additional ecological problem. It is not known whether this process is still commercially operated.

Bayer process:

oxidation of ammonia with sodium hypochlorite in the presence of acetone to acetone azine and sodium chloride, distillative separation of the acetone azine from the sodium chloride solution, hydrolysis of the acetone azine to hydrazine and acetone. Hydrazine yield: 80 to 90%

1.4.2.2.3 Bayer Process

From a number of hydrazine syntheses based on the oxidation of ammonia with sodium hypochlorite in the presence of ketones (acetone, methyl ethyl ketone) (see Fig. 1.4.2-2), only that from Bayer A.G. appears to have graduated to a commercial process.

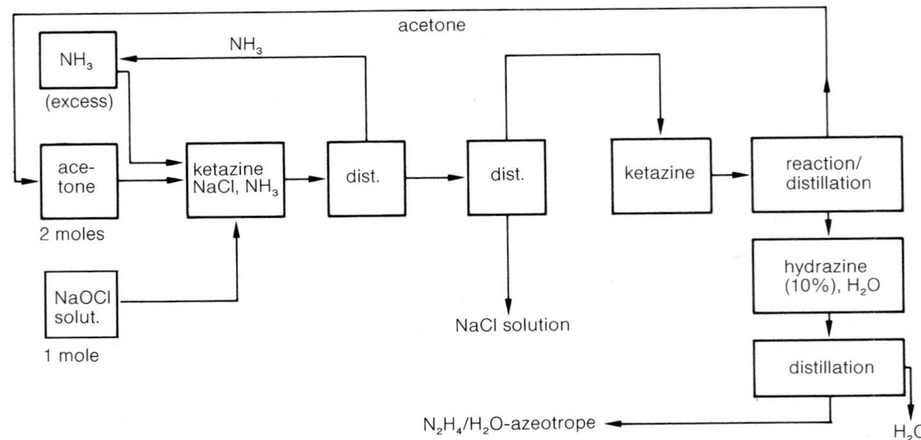

Fig. 1.4.2-2. Schema of hydrazine manufacture by the Bayer process.

The two main reactions are the formation and hydrolysis of acetone azine:

$$NaOCl + 2\,NH_3 + 2\,CH_3COCH_3 \longrightarrow (CH_3)_2C{=}N{-}N{=}C(CH_3)_2 + NaCl + 3\,H_2O$$

$$(CH_3)_2C{=}N{-}N{=}C(CH_3)_2 + 2\,H_2O \longrightarrow 2\,CH_3COCH_3 + N_2H_4$$

The formation of azine is not a means of trapping with acetone the hydrazine formed in the Raschig process. In fact the reaction proceeds by way of two distinct intermediates: dimethyl oxazirane and acetone hydrazone, as follows:

$$NaOCl + NH_3 + CH_3COCH_3 \longrightarrow (H_3C)_2C(NH)(O) + NaCl + H_2O$$

$$(H_3C)_2C(NH)(O) + NH_3 \longrightarrow (CH_3)_2C{=}N{-}NH_2 + H_2O$$

$$(CH_3)_2C{=}N{-}NH_2 + CH_3COCH_3 \longrightarrow (CH_3)_2C{=}N{-}N{=}C(CH_3)_2 + H_2O$$

Sodium hypochlorite solution (ca. 1.5 mol/L), ammonia and acetone in a molar ratio of 1:15 to 20:2 are reacted together at 35 °C. A solution results consisting of 5 to 7% by weight of acetone azine, sodium chloride and excess ammonia. This ammonia is distilled off and returned to the reaction. Next, the acetone azine-water-azeotrope (b.p. 95 °C) is distilled off from the residual sodium chloride solution. Herein lies the essential difference from the Raschig process in which the hydrazine water mixture has to be separated from *solid* sodium chloride.

Finally, the acetone azine is hydrolyzed with water in a reaction distillation tower into acetone (head product) and a 10% aqueous hydrazine solution (sump product) at temperatures up to 180 °C and pressures of 8 to 12 bar. This solution is concentrated to 64% by weight of hydrazine. The hydrazine yield is 80 to 90%, based on the hypochlorite utilized.

H_2O_2 process:

oxidation of ammonia with hydrogen peroxide in the presence of methyl ethyl ketone and acetamide as catalyst to ketazine, separation of ketazine and hydrolysis to hydrazine and ketone

1.4.2.2.4 H_2O_2 Process

This process is similar to the Bayer process except that hydrogen peroxide is used as the oxidizing agent and the ketone used is methyl ethyl ketone:

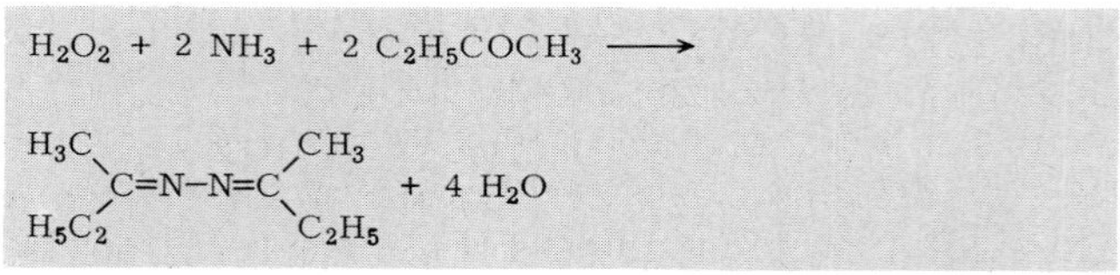

$$H_2O_2 + 2\,NH_3 + 2\,C_2H_5COCH_3 \longrightarrow (H_3C)(H_5C_2)C{=}N{-}N{=}C(CH_3)(C_2H_5) + 4\,H_2O$$

Since the reaction with hydrogen peroxide is too slow, a mixture of acetamide and sodium hydrogen phosphate is added as a catalyst.

The methyl ethyl ketone azine formed which is poorly soluble in water, is separated and hydrolyzed to hydrazine and ketone as in the Bayer process.

This process is operated commercially. However, little has been published about the reaction conditions: reactant and activator (acetamide, phosphate) concentrations, solubilizer used, residence time, temperature and auxilliary circuits, and the yield obtained. The intermediate is probably not the ketimine derivative of the ketone as described in the literature, but the dioxirane derivative of the ketone, which subsequently reacts further to the oxazirane:

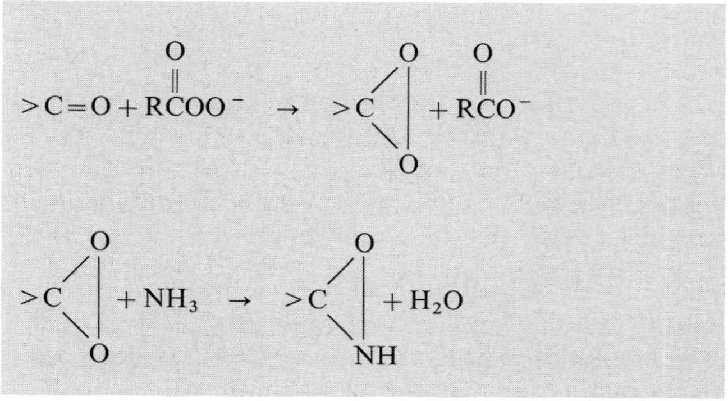

$$>C{=}O + R\overset{\overset{\large O}{\|}}{C}OO^- \rightarrow >C\langle{}^{O}_{O}| + R\overset{\overset{\large O}{\|}}{C}O^-$$

$$>C\langle{}^{O}_{O}| + NH_3 \rightarrow >C\langle{}^{O}_{NH}| + H_2O$$

The advantages of this process over the Bayer and Raschig processes are the low energy consumption and the absence of a sodium chloride byproduct.

1.4.2.3 Applications for Hydrazine

Applications for hydrazine:

- As a corrosion inhibitor
- In the manufacture of:
  - blowing agents
  - polymerization initiators
  - herbicides
  - pharmaceuticals

The presence of hydrazine in ppm quantities in boiler feed water prevents corrosion, due to its promotion of a protective magnetite layer. Derivatives of hydrazine are particularly important as blowing agents in the manufacture of foam plastics and rubbers, as free radical polymerization initiators, as herbicides and as pharmaceutical products. The hydrazine products used as blowing agents or polmerization initiators decompose upon heating into nitrogen and free radicals: the nitrogen works as the blowing agent, the free radicals as polymerization initiators.

Among the commonly used blowing agents are:

- Azodicarbonamide
- Benzene sulfonic acid hydrazide

A typical polymerization initiator is α,α′-azobisisobutyronitrile.

Azodicarbonamide can, for example, be produced as follows:

$$2\,NH_2CONH_2 + N_2H_4 + H_2SO_4 \longrightarrow NH_2CONHNHCONH_2 + (NH_4)_2SO_4$$

$$NH_2CONHNHCONH_2 + Cl_2 \longrightarrow NH_2CON{=}NCONH_2 + 2\,HCl$$

Important hydrazine-based herbicides are Weedazol®, Sencor®, and Goltix®:

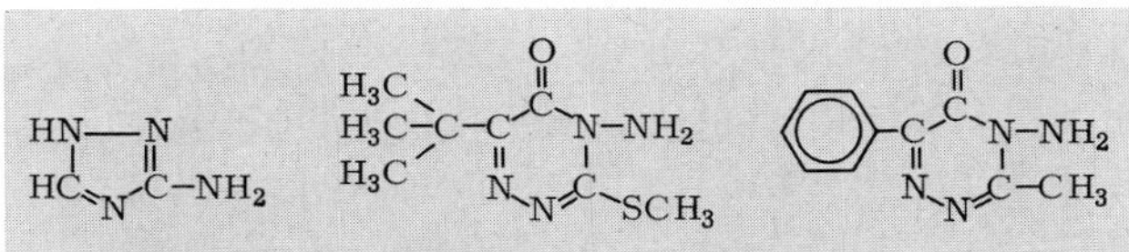

Weedazol® Sencor® Goltix®

A typical hydrazine-based pharmaceutical product is the anti-tubercular agent Neoteben®

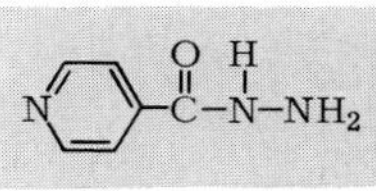

References for Chapter 1.4.2: Hydrazine

Reviews:

Kirk-Othmer: *Encyclopedia of Chemical Technology.* 3rd Ed., Vol. 12, pp 734–771, John Wiley & Sons, Inc., New York, Chichester, Brisbane, Toronto (1980).
Ullmanns Encyklopädie der technischen Chemie. 4th Ed., Vol. 13, pp 95–107, Verlag Chemie, Weinheim-Deerfield Beach-Basel (1977).

Commercial Information:

Chem. Eng. News, pp 23–24 (Aug.23, 1982).
Chem. Week, p. 13 (May 26, 1982).

H_2O_2 Process:

Chem. Ind. **XXXIII,** pp 673–674 (Nov. 1981).
Chem. Eng. News, p. 32 (Nov. 2, 1981).
Chem. Week, p. 28 (Jan. 6, 1982).
R. Jeyaraman and R. W. Murray, "Production of arene oxides by the caroate-acetone system (dimethyldioxirane)", *J. Am. Chem. Soc.* **106,** pp 2462–2463 (1984).
Eur. Chem. News, p. 4 (June 2, 1986).

1.4.3 Hydroxylamine

1.4.3.1 Economic Importance and Applications

Hydroxylamine: more than 97% used in the production of caprolactam. World-wide capacity in 1983 at least 650×10^3 t/a

The commercial production of hydroxylamine is coupled to its use (up to 97%) in the manufacture of caprolactam, a precursor in the manufacture of polyamide 6 (however, not all caprolactam is manufactured using it):

(H)=O + NH_2OH ⟶ (H)=NOH + H_2O

↓

(H)<C=O, NH> (caprolactam) ⟶ Polyamide 6

In 1983 the caprolactam capacity worldwide was $> 2850 \times 10^3$ t/a. Most plants operate with hydroxylamine as an intermediate. The hydroxylamine capacities are not individually documented, but in total must be at least 650×10^3 t/a (calculated on the basis of NH_2OH) in order, assuming 100% synthesis yield, to provide for the capacities of caprolactam plants operating with hydroxlamine.

Hydroxylamine is used to a lesser extent as a reducing agent and in the manufacture of oximes for use in paints (skin prevention agents), pharmaceuticals and pesticides.

1.4.3.2 Manufacture

There are three commercially operated processes for the manufacture of hydroxylamine:

- (modified) Raschig process
- NO reduction process (BASF, Inventa)
- Nitrate reduction process (DSM/HPO-Stamicarbon)

Production of hydroxylamine:

- Raschig process
- NO reduction process
- Nitrate reduction process

1.4.3.2.1 Raschig Process

Nitric oxide from ammonia combustion is reacted with ammonium hydrogen carbonate solution and air:

$$2NO + 0.5O_2 + 2NH_4HCO_3 \longrightarrow 2NH_4NO_2 + 2CO_2 + H_2O$$

The resulting ammonium nitrite solution is reacted with sulfur dioxide at 0 to 5 °C and at pH 2 to 4.5 forming ammonium hydroxylamine disulfonate:

$$NH_4NO_2 + 2SO_2 + NH_3 + H_2O \longrightarrow HON(SO_3NH_4)_2$$

The yield based on nitrite is ca. 90%. Finally, the disulfonate is hydrolyzed at 100 °C in a relatively slow reaction:

$$2HON(SO_3NH_4)_2 + 4H_2O \longrightarrow (NH_3OH)_2SO_4 + (NH_4)_2SO_4 + 2NH_4HSO_4$$

The hydroxylamine content in this solution amounts to ca. 70 g/L, excluding ammonium sulfate. In an integrated caprolactam manufacturing plant using the Raschig process for hydroxylamine production, ca. 1.8 kg ammonium sulfate is produced per kg caprolactam.

Raschig process:

1st step: Preparation of ammonium nitrite from ammonium hydrogen carbonate, nitric oxide and oxygen (in solution)

2nd step: Formation of hydroxylamine disulfonate by reaction with sulfur dioxide

3rd step: Hydrolysis to hydroxylamine sulfate and ammonium sulfate

Ammonium sulfate byproduct: 1.8 kg/kg caprolactam

1.4.3.2.2 Nitrogen(II) Oxide Reduction Process

Nitrogen(II) oxide in 80 to 90% yield is produced by the catalytic combustion of ammonia in oxygen in the presence of water vapor on a platinum-rhodium catalyst:

$$2NH_3 + 2.5O_2 \longrightarrow 2NO + 3H_2O$$

NO reduction process:

1st step: Production of NO by the combustion of ammonia with pure oxygen

2nd step: Reduction of NO to hydroxylamine sulfate with hydrogen on a precious metal catalyst suspended in sulfuric acid

In the BASF process this is reduced with hydrogen to hydroxylamine on a platinum catalyst on carbon substrates suspended in dilute sufuric acid. In the Inventa process palladium is used.

$$2NO + 3H_2 + H_2SO_4 \longrightarrow (NH_3OH)_2SO_4$$

Ammonium sulfate byproduct: ca. 0.8 kg/kg caprolactam

The expected formation of ammonia in a side reaction is largely suppressed by process control measures, e.g. high hydrogen pressure, low hydrogen ion concentration and low temperature, and through a selective poisoning of the catalyst. The yield based on nitrogen(II) oxide is ca. 90%.
The quantity of ammonium sulfate byproduct in this process is ca. 0.8 kg per kg of caprolactam produced. The hydroxylamine content in the solution amounts to ca. 115 g/L.

1.4.3.2.3 Nitrate Reduction Process (DSM/HPO-Stamicarbon)

Nitrate reduction process:

1st step: Reduction of an ammonium nitrate- phosphoric acid solution with hydrogen on a suspended precious metal catalyst to hydroxylamine phosphate solution

2nd step: Reaction of this solution with cyclohexanone in toluene (two phases) to cyclohexanone oxime and separation of the aqueous phase

3rd step: Regeneration of the aqueous "recycled solution" to ammonium nitrate-phosphoric acid solution by introduction of nitrogen oxides and oxygen

Ammonium nitrate solution is reduced to hydroxylamine phosphate with hydrogen under pressure on a suspended precious metal catalyst (substrate: activated carbon) in the presence of phosphoric acid (HPO: Hydroxylamine Phosphate Oxime):

$$2H_3PO_4 + NH_4NO_3 + 3H_2 \longrightarrow (NH_3OH)H_2PO_4 + NH_4H_2PO_4 + 2H_2O$$

After separating off the catalyst, the solution is directly reacted with a solution of cyclohexanone in toluene forming cyclohexanone oxime:

$$(NH_3OH)H_2PO_4 + NH_4H_2PO_4 + \langle H \rangle{=}O \longrightarrow \langle H \rangle{=}NOH + H_3PO_4 + NH_4H_2PO_4 + H_2O$$

After the separation of residual organic substances, the aqueous phase is reacted with nitrogen oxides and atmospheric oxygen and reused as "recycled solution":

$$2H_3PO_4 + 2NH_4H_2PO_4 + 2NO + 1.5O_2 + H_2O \longrightarrow 4H_3PO_4 + 2NH_4NO_3$$

Ammonium ions formed in the side reactions during the hydrogenation are oxidized with nitrogen oxides to nitrogen in this step. The yield based on the ammonia utilized is ca. 58%. There are no ammonium salt byproducts.

No ammonium salt byproducts

References for Chapter 1.4.3: Hydroxylamine

Reviews:

Ullmanns Encyklopädie der technischen Chemie. 4th Ed., Vol. 13, pp 169–172, Verlag Chemie, Weinheim-Deerfield Beach-Basel (1977).

Inf. Chim., No. **187**, pp 139–145 (March 1979).
Inf. Chim., No. **212**, pp 201–208 (April 1981).

Commercial Information:

Caprolactam, *Chemical Economics Handbook.* Stanford Research Institute. Menlo Park, California, USA (Aug. 1980 and Aug. 1983).

Technical Details:

Nitrogen, No. **108**, pp 33–37 (July/Aug. 1977).

1.4.4 Nitric Acid

1.4.4.1 Economic Importance

Nitric acid is one of the most important inorganic chemicals. Quantitywise it is one of the top ten industrial chemicals.

The worldwide production of nitric acid currently amounts to well over 30×10^6 t/a (as 100% acid). This is distributed regionally as follows (Table 1.4.4-1):

Worldwide nitric acid production: $> 30 \times 10^6$ t/a

Table 1.4.4-1. 1983 Nitric Acid Production in the World Regions in 10^6 t HNO_3.

World*	North America	South America	West Europe	East* Europe	Asia*/ Oceania	Africa
28.8	7.9	0.4	14.1	4.9	1.4	0.1

* without the USSR and P.R. China

The USA production was 7.1×10^6 t in 1985, with 2.9×10^6 t in West Germany. The capacities are far higher, being in 1986 9.7×10^6 t/a HNO_3 in the USA and 21.1×10^6 t/a in Western Europe with more than 3×10^6 t/a in each of France, Great Britain and West Germany. There are plants capable of producing up to 1380 t/d in the USA.

Nitric acid mostly used as 50 to 70% HNO_3, only 20% as highly concentrated 98–100% HNO_3

Nitric acid is mostly produced as 50 to 70% HNO_3. Ca. 20% is processed to 98–100% highly concentrated nitric acid, which is e.g. used for nitration in organic chemistry.

1.4.4.2 Manufacture

1.4.4.2.1 Fundamentals of Nitric Acid Manufacture

Nitric acid is currently almost exclusively produced by the catalytic oxidation of ammonia using the Ostwald process (1902). The reaction of sodium nitrate (the only nitrate occurring naturally in large quantities) with sulfuric acid has not been economic since the emergence of the Haber-Bosch ammonia synthesis process shortly before the 1st World War. The previously developed direct synthesis from nitrogen(II) oxide produced from nitrogen and oxygen at temperatures $> 2000\,^\circ C$ (arc process, thermal NO-synthesis) is also no longer operated industrially.

HNO_3 manufactured in a 3-step process from NH_3:

(1) $NH_3 \rightarrow NO$
(2) $NO \rightarrow NO_2$ (or N_2O_4)
(3) $NO_2 \xrightarrow{+H_2O} HNO_3$

The process for nitric acid manufacture from ammonia consists of three exothermic reaction steps as shown schematically in Fig. 1.4.4-1:

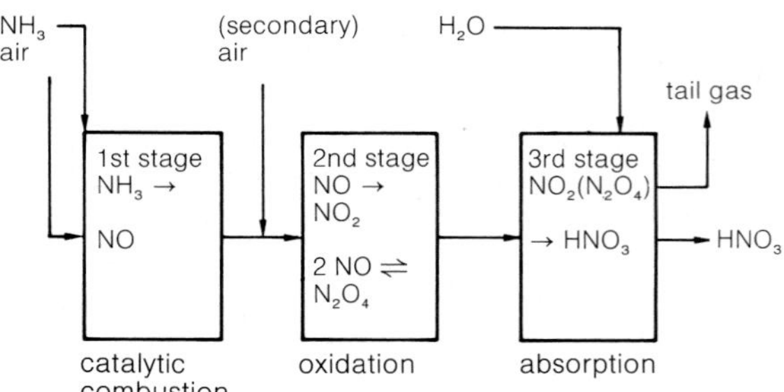

Fig. 1.4.4-1. Schema of the Ostwald process for the manufacture of nitric acid from ammonia.

The overall reaction can be described by:

$$NH_3 + 2\,O_2 \rightarrow HNO_3 + H_2O, \Delta H = -369 \text{ kJ/mol}$$
(reaction heat is for 60% acid)

The three reaction steps are based on very different physical chemical relationships which appreciably influence the operation of the process. The third step is normally followed by tail gas purification.

Catalytic Combustion of Ammonia:

The combustion of ammonia with atmospheric oxygen to nitrogen(II) oxide (NO) is carried out in the presence of a catalyst at 820 to 950 °C and 1 to 12 bar pressure:

$$4\,NH_3 + 5\,O_2 \rightarrow 4\,NO + 6\,H_2O,\ \Delta H = -904\ kJ/mol$$

This reaction is one of the most efficient catalytic processes in industrial chemistry, having an extremely short reaction time (10^{-11} s) and a high selectivity. The oxidation of ammonia benefits slightly from pressure reduction, since less nitrogen and nitrogen(I) oxide (N_2O) is then produced:

$$4\,NH_3 + 3\,O_2 \rightarrow 2\,N_2 + 6\,H_2O,\ \Delta H = -1268\ kJ/mol$$

$$4\,NH_3 + 4\,O_2 \rightarrow 2\,N_2O + 6\,H_2O,\ \Delta H = -1105\ kJ/mol$$

Operation at higher pressures reduces the plant size and hence its cost, but at the expense of increased catalyst losses due to the need for higher reaction temperatures to compensate for reduced yield. The yield is generally 94 to 98% (e.g. 97 to 98% at 1 bar, 95 to 96% at 5 bar, 94% at 8 to 10 bar). The ammonia content in the combustion mixture is kept below the lower explosion limit for ammonia-air mixtures (15.5 % by volume at 1 bar) i.e. ≤ 13 % by volume at atmospheric pressure and appreciably less with medium and high pressure plants (the lower explosion limit decreasing with increasing temperature).

NO-yield in ammonia combustion is between 94 and 98% depending upon temperature, pressure and flow rate

Precise NH_3 concentration in NH_3/air mixtures is 9.5 to 13 % by volume depending upon the pressure

The ammonia oxidation catalyst is usually a platinum alloy gauze containing 5 to 10% rhodium, or optionally 5% palladium, with a diameter of up to 4 m (with 1024 meshes/cm^2 and a wire thickness of 0.06 to 0.076 mm, the latter for higher pressures). The higher the pressures and flow rates the larger the number of gauzes incorporated into the reactor (up to 50). Under operating conditions, particularly at higher pressures

Pt catalysts for NH_3 oxidation:

- Pt/Rh 90:10
  or 95: 5
- Pt/Rh/Pd 90:5:5

Contact time ca. 10^{-3} s

and operating temperatures, mechanical abrasion or evaporation of the platinum dioxide (PtO_2) formed as an intermediate takes place resulting in precious metal losses of 0.05 to 0.45 g/t HNO_3 (100%). Up to 80% of this precious metal can be recovered by adsorption on marble chips or on palladium-gold gauzes. The annual worldwide platinum gauze replacement requirement currently amounts to ca. 52 t.

During NH_3 oxidation Pt losses occur which increase with increasing pressure. Up to 80% recoverable

Oxidation of Nitrogen(II) Oxide:

The hot nitrogen(II) oxide-containing gas from the combustion step (e.g. with ca. 10 to 12% NO) is cooled, the heat content being utilized for steam production or waste gas-heating. It is then reacted with additional atmospheric oxygen (secondary air) to nitrogen(IV) oxide (NO_2):

$$2\,NO + O_2 \rightarrow 2\,NO_2,\ \Delta H = -114\ kJ/mol$$

NO oxidation and NO_2 dimerization favored by low temperatures and high pressures

This reaction is favored by low temperatures, the temperature coefficient of the rate constant being negative, and still more strongly by increased pressure due to the volume reduction during the reaction. Dimerization to dinitrogen tetroxide is also promoted by low temperatures and high pressures.

$$2\,NO_2 \rightleftharpoons N_2O_2,\ \Delta H = -57\ kJ/mol$$

H_2O content of the combustion gases removed by cooling. Acid condensate contains 2 to 50% HNO_3

Nitrogen(II) oxide oxidation partly takes place in the waste heat boiler, due to reaction with the excess oxygen present in the combustion gases from the ammonia oxidation, and partly with added air in absorption columns mostly operated at high pressures or in a preceding oxidation tower. The higher the pressure in the combustion step the greater the amount of nitrogen(IV) oxide formed during the cooling of the combustion gases. This reacts with the reaction water forming nitric acid, the HNO_3 concentration in these acid condensates being 2 to 50%.

Conversion of Nitrogen(IV) Oxide into Nitric Acid:

The nitrogen(IV) oxide-dinitrogen tetroxide containing mixture (so-called nitrous gases) obtained from the oxidation of nitrogen(II) oxide is reacted with water to nitric acid, nitrogen(II) oxide and nitrous acid in the third step as follows:

$$3\ NO_2 + H_2O \rightarrow 2\ HNO_3 + NO, \Delta H = -73\ kJ/mol$$

$$N_2O_4 + H_2O \rightarrow HNO_3 + HNO_2, \Delta H = -65\ kJ/mol$$

NO_2/N_2O_4-absorption in water is a function of:

- Temperature (as low as possible)
- Pressure (as high as possible)
- Contact time (reaction volume as high as possible)
- Mass transfer
- Quantity of H_2O present

The latter is reoxidized with the atmospheric oxygen present either in the liquid or vapor phase.

The absorption of nitrous gases in the process water is favored by low temperatures, high pressures and long contact times. The amount of process water-containing acid condensate is among other things dependent upon the desired nitric acid concentration. Higher pressures enable the manufacture of up to 70% nitric acid, since with comparatively little process water an almost complete (99%) absorption can be achieved with a very low nitrous gas concentration in the tail gases. At atmospheric pressure only 45 to 50% nitric acid can be produced.

The most important absorption parameter is pressure: at high pressures (up to 15 bar) concentrated HNO_3 can be produced and tail gas purification can be dispensed with

1.4.4.2.2 Plant Types

There are basically two types of processes for the manufacture of nitric acid i.e. monopressure processes in which ammonia combustion and absorption of nitrogen oxides take place at the same pressure and dual pressure processes in which the pressure in the first stage is lower than that in the second stage.

Modern HNO_3 plants:

- Monopressure plants
  - Medium pressure 3 to 6 bar
  - High pressure 8 to 15 bar
- Dual pressure plants combustion at 4 to 6 bar oxidation/absorption at 8 to 10 bar

In addition nitric acid plants can be categorized on the basis of operating pressure: low pressure-(L)-, medium pressure-(M)- and high pressure-(H)-plants which operate in the following pressure ranges: 1 to 2 bar, 3 to 7 bar and 8 to 12 bar, occasionally up to 15 bar. The plant type can be characterized by the pressure ranges for ammonia combustion and oxidation/absorption.

The tendency in the nitric acid industry is to ever larger plants (capacities up to 1500 t 100% HNO_3/d are currently possible) and to ever higher pressures both in the combustion and in the absorption stages to solve the emission problems. The developments in the USA and in Western Europe are somewhat different: plants in the USA being ca. 90% monopressure/high pressure

Plant type selection on the basis of

- Local factors
- Raw material and energy costs
- Investment costs
- Emission regulations

plants (H/H-type), whereas in Western Europe medium pressure and medium/high pressure plants are common. The advantages of the former are: relatively low plant costs, lower space requirements and satisfactory emission values. The disadvantages of higher catalyst consumption, higher energy consumption (due to gas compression) and lower nitrogen(II) oxide yield (75 to 80% of the nitric acid manufacturing costs being ammonia costs) are less important in the USA than in Western Europe due to the relative cheapness of energy and ammonia (from natural gas). In the many medium pressure and medium/high pressure plants in Western Europe combustion is carried out at comparatively low pressures, enabling high nitrogen(II) oxide yields, low catalyst losses and low compression costs to be achieved and the absorption/emission problems are satisfactorily solved by combination with high pressure absorption units. Older units, particularly the L/L-type, are fitted with catalytic tail gas purification units for environmental reasons.

Important elements in HNO_3 manufacturing costs:

- Ammonia price
- Maintenance costs
- Amortization
- Personnel costs
- Energy costs

High pressure plants:

compressor costs make up almost half of the total investment costs in the USA

1.4.4.2.3 Process Description

Air for NH_3 combustion has to be carefully prepurified to avoid deactivation of the Pt-catalyst

The first stage of a nitric acid plant consists of an ammonia evaporator and air compressor whose products are mixed, prewarmed, carefully prepurified with gas filters to remove catalyst poisons and then fed into the combustion reactor in which the gas flows downwards through the platinum catalyst gauzes suspended in it. The operating lifetime of the catalysts varies from 3 to 18 months depending upon the combustion system. The hot reaction gases with a temperature of ca. 900 °C are passed into a heat exchanger (waste heat boiler) in which they are cooled to ca. 150 °C [composition e.g. in the L/M-plant type: 10 to 12% NO, 17 to 20% H_2O, 2 to 5% O_2 (by volume) with the remainder being made up of N_2, inert gases and a small amount of NO_2]. The steam generator integrated into the plant produces steam at e.g. 400 °C/40 bar which is mostly fed into a local network. This substantially reduces the manufacturing costs of nitric acid. Nitric acid plants with appropriate equipment can supply energy despite the very high energies required for compression particularly in high pressure plants. Much of the heat liberated in the three exother-

Utilization of the heat content of the 900 °C reaction gases for steam generation reduces the manufacturing costs of nitric acid

Compression energy can be partly recovered by gas expansion driven turbines. Tail gas heating improves energy recovery

mic reactions is however at such low temperatures (oxidation of nitrogen(II) oxide, absorption of nitrous gases in water) that it can not be utilized economically.

The nitrogen(II) oxide-containing combustion gas is further cooled in a gas cooler to ca. 20 to 30 °C, whereupon most of the water vapor is condensed as a nitric acid-containing acid condensate. This is fed into the absorption towers as part of the process water. The virtually dry gas is oxidized by mixing with secondary air (exhaust air) and also fed into the absorption plant. In dual pressure plants the reaction gas is compressed to the absorption pressure in an acid resistant nitrous gas compresser and the heat of compression removed by heat exchangers and coolers (heating up the tail gases). Part of the compression energy is recovered by turbines driven by heated tail gases.

Acid condensate obtained upon cooling the NO- and H_2O-containing combustion gases is fed into the absorption towers

The reaction gases are fed into the absorption towers from below and the nitrous gases scrubbed from the reaction gas with cooled process water in countercurrent under pressure. In this process the nitrogen(II) oxide formed by the disproportionation of nitrogen(IV) oxide is also reoxidized. The size of the absorption volume depends on this reaction. Most of the large volume absorption towers are built with chromium-nickel steel and filled with Raschig rings or with bubble or sieve plates which are sprayed from above with water [which is precooled by supplying the heat for ammonia evaporation (combustion stage)] forming concentrated nitric acid at the bottom of the tower. The acid concentration depends upon the quantity of water fed in at the absorption stage: e.g. ca. 0.2 m^3/t is required for 65% acid. 77.8% nitric acid can theoretically be formed when acid condensate is solely used as the absorption agent (1.5 mol H_2O produced per mol NH_3).

Mass transfer systems

- Raschig rings
- Bubble plates
- Sieve plates

Without additional water in the absorption tower 78% acid theoretically possible

The emission of nitrogen oxides increases when high acid concentrations and low pressures are used. The previously utilized very economic low pressure absorption plants may, therefore, no longer be built.

At low pressures only a weak acid can be produced, due to excessive emission of NO_x at low volumes of process water

A nitrous gas compressor is not necessary in monopressure plants. In plants operating at higher pressures e.g. H/H- and M/H-types the absorption volume is drastically reduced (at 10 bar only 1/10th of that needed at 4.6 bar) so that only one absorption tower is required.

The residual gas is heated up with the aid of compression heat to e.g. 250 to 300 °C to improve energy recovery and released into the atmosphere by way of expanded gas driven turbines (with or without gas purification).

1.4.4.2.4 Manufacture of Highly Concentrated Nitric Acid

The 50 to 70% nitric acid produced in conventional nitric acid plants is suitable for industrial purposes e.g. for the manufacture of fertilizers, the synthesis of ammonium nitrate for example, requiring ca. 60% nitric acid. However, for nitration reactions in organic chemistry a highly concentrated (ca. 98 to 99%) nitric acid is required. Since nitric acid forms an azeotrope with water at 68.8% nitric acid, the water cannot be separated from the acid by simple distillation.

Highly concentrated (98 to 99%) nitric acid mainly utilized for nitration

Highly *c*oncentrated *n*itric *a*cid (CNA) can be manufactured directly by reaction with additional nitrogen oxides using *d*irect *s*trong *n*itric (DSN) processes or indirectly by extractive distillation. The former processes predominate in Europe and the latter in the USA.

Manufacture of CNA's:

- Direct (strong nitric) processes (variations of the normal HNO_3 synthesis)
- Indirect (extractive distillation) processes (H_2O removal from dilute HNO_3)

Direct Strong Nitric Processes

In the direct strong nitric processes, which are operated in many variants, the nitrous gases resulting from the catalytic combustion of ammonia and oxidation of the resulting nitrogen(II) oxide are either separated and the dinitrogen tetroxide reacted with oxygen and water forming nitric acid, or dissolved in concentrated nitric acid and the superazeotropic acid distilled. The latter is the more economic even compared with the indirect processes, because no pure oxygen production is necessary and the investment costs are lower.

Direct strong nitric processes:

- Oxidation of N_2O_4 with pure oxygen in the presence of H_2O at high pressures
- Absorption of NO_2/NO in concentrated HNO_3 producing superazeotropic acid followed by distillation (normally more economic)

In the Uhde (CNA) process the ammonia combustion is carried out at atmospheric pressure. The reaction water largely separates off as acid condensate when the combustion gases are cooled and the nitrogen(II)

Direct CNA processes often also supply:

- Weak acid (in any desired concentration)
- Pure N_2O_4

oxide is compressed with secondary air to e.g. 1.6 bar. Oxidation takes place in oxidation towers and the residual nitrogen(II) oxide is converted to nitrogen(IV) oxide in a postoxidative reaction with highly concentrated nitric acid:

$$2\,HNO_3 + NO \longrightarrow 3\,NO_2 + H_2O$$

After cooling and removing residual water, the nitrogen(IV) oxide is physically absorbed in highly concentrated deep frozen nitric acid and is thereby separated from the other components of the gas (nitrogen, residual oxygen). The acid is then distilled producing 98 to 99% nitric acid (sump product) and nitrogen(IV) oxide (head product). The latter is liquified by deep freezing, which almost completely dimerizes it to dinitrogen tetroxide, which can, if desired, be partly separated off as a byproduct. The liquid dinitrogen tetroxide is stirred with dilute nitric acid and fed into a reactor with a high pressure centrifugal pump in which the tetroxide/nitric acid/water mixture is oxidized with pure oxygen at a pressure of 50 bar to highly concentrated nitric acid:

$$N_2O_4 + H_2O + 1/2\,O_2 \longrightarrow 2\,HNO_3$$

The gas emerging from the absorption column is scrubbed with acid condensate (ca. 2% nitric acid from the ammonia combustion see Sec. 1.4.4.2.1), which reduces the nitrous gas concentration to < 200 ppm, and is then vented. The consumption figures for this process are summarized in the notes in the adjoining margin.

Consumption figures per t 100% HNO_3 for the direct CNA processes with added O_2:

- 0.282 t NH_3
- 125 m^3 O_2
- 285 kWh electricity
- 200 m^3 cooling water ($\Delta T = 7\,°C$)
- 0.6 t excess steam

In Davy McKee's Sabar process (**S**trong **A**cid **B**y **A**zeotropic **R**ectification) the nitrous gases from the oxidation of nitrogen(II) oxide are absorbed in azeotropic (ca. 68 to 69%) nitric acid in the presence of atmospheric oxygen (at 6 to 13 bar) and superazeotropic acid is formed:

$$2\,NO_2 + 1/2\,O_2 + H_2O \longrightarrow 2\,HNO_3$$

The acid is degassed by blowing out with secondary air and distilled (the head product consisting of highly concentrated acid and the sump product of the azeotropic acid). The azeotropic acid is recycled. The concentration of nitrous gas in the tail gas is < 500 ppm. In the Sabar process the manufacture of low or medium concentration acid is possible in addition to highly concentrated acid. The former are obtained by adding acid condensate from the combustion of ammonia to the sump of the nitric acid tower.

The Conia process operates with small units at ca. 5 bar and produces highly concentrated nitric acid, weak acid in any desired concentration and dinitrogen tetroxide. In this process most of the combustion water from ammonia combustion is removed by condensation as 30 to 40% acid condensate. After the oxidation of nitrogen(II) oxide, part of the nitrogen(IV) oxide formed is reacted in countercurrent with acid condensate, additional process water and atmospheric oxygen to medium concentrated nitric acid (50 to 70%) in the usual way. The remaining nitrogen(IV) dioxide is processed to highly concentrated acid using DSN-technology.

Indirect Extractive Distillation Processes

Highly concentrated nitric acid manufacture by indirect processes:

dehydration of water-containing HNO_3 with concentrated H_2SO_4 or $Mg(NO_3)_2$ solutions

Of the various indirect processes for the manufacture of highly concentrated acid only two are industrially important: the sulfuric acid process and the magnesium nitrate process.

In the sulfuric acid process, which poses acute corrosion problems, medium concentrated nitric acid is first produced using conventional methods (e.g. in a M/M-type unit) as in the magnesium nitrate process. In a concentrating tower concentrated sulfuric acid is fed in at the head. During the extractive distillation diluted sulfuric acid accumulates in the sump and 99% nitric acid is driven off. The diluted sulfuric acid is then concentrated by vacuum distillation and recycled.

In the USA:
mainly indirect extractive distillation processes;

in Europe:
mainly direct strong nitric processes

In the magnesium nitrate process weak acid is distilled with 72% magnesium nitrate solution, whereupon highly concentrated sulfate-free nitric acid is driven off at the head of the dehydration tower. The sump product is then concentrated by vacuum distillation.

1.4.4.2.5 Tail Gases from Nitric Acid Manufacture

In the manufacture of nitric acid up to 3300 m^3 of water vapor-saturated tail gases (residual and end gases) are produced per t 100% nitric acid. They contain up to 97% nitrogen, 1% inert gases, 2 to 4% oxygen (by volume) and, depending upon plant type, 150 to > 2000 ppm of nitrogen oxides NO_x (NO, NO_2 and traces of N_2O) calculated on the basis of nitrogen(II) oxide. The higher nitrogen oxide emission values apply to low pressure and medium pressure plants. The tail gas quantities can be drastically reduced by carrying out the ammonia combustion and nitrogen(II) oxide oxidation with pure oxygen, but this is uneconomic.

Tail gases from HNO_3 plants per t 100% HNO_3:

- Up to 3300 m^3
- NO_x-content: 150 to > 2000 ppm

Older nitric acid plants are identifiable by the yellow-brown color of their tail gas emissions. This color is due to their nitrogen(IV) oxide content (ca. 40 to 50% of the NO_x-content). The tail gases from modern plants with high pressure absorption steps, optimum mass transfer plates (sieve plate technology), sufficiently large absorption volumes (and hence long contact times) and low cooling water temperatures are virtually colorless and do not pose any problems, because the nitrogen oxide content is < 200 ppm. The mandatory emission limits for nitrogen oxides (usually 200 ppm for new plants) can thus be fulfilled with new plants without an additional scrubbing step. The achievement of low NO_x values of e.g. 100 to 200 ppm is a considerable advance and can only be achieved by optimizing the absorption step, since subsequent nitrogen oxide removal from the tail gas is very expensive (necessitating the use of sensitive precious metal catalysts, additional fuel and energy requirements and may cause other environmental problems). In modern medium pressure plants (monopressure plants with pressures of 5 to 6 bar), however, the nitrogen oxide content in the tail gas is up to 500 ppm and therefore a postpurification step is necessary to attain emission values below 200 ppm. The installation of low pressure absorption units has virtually ceased due to their high emission values (≥ 2000 ppm). The nitrogen oxide emissions of nitric acid plants currently account for ca. 1% of all NO_x-emissions (NO_x emissions come predominantly from power stations

NO_2 content of tail gases (40 to 50% of the NO_x) causes brown coloration (NO is colorless)

No tail gas problems with CNA plants and modern high pressure absorption plants with

- Optimum mass transfer plates
- Effective cooling
- Sufficiently large reaction volumes (NO_x: 100 to 200 ppm)

and motor vehicles), although local emissions from old nitric acid plants can be a problem.

Subsequent removal of NO_x often more complicated than absorption technology improvement

Tail gases from old nitric acid plants and newer medium pressure plants can be purified using several processes:

Purification of tail gases from (old) low pressure, medium pressure and nonoptimized high pressure plants:

- Wet processes
- Dry processes (thermal, catalytic, adsorptive)

Alkaline scrubbing:

nitrate and nitrite formation pose disposal problems

- Alkali scrubbing e.g. with sodium hydroxide, ammonia solutions or urea solutions. Nitrogen oxide values of 100 ppm can be achieved but the disposal of the nitrites and nitrates formed can pose problems.
- Reductive postcombustion of the tail gases with natural gas, hydrogen, fuel oil, carbon monoxide and other reducing agents in which the nitrogen oxides are reduced either thermally at > 1000 °C or catalytically at 170 to 600 °C to colorless nitrogen(II) oxide or nitrogen. This process is widely used with high pressure plants in the USA. The susceptibility of the catalysts to poisoning, the additional fuel requirements (particularly for tail gases with high oxygen contents) and the secondary emission of e.g. carbon monoxide and hydrogen cyanide are drawbacks with this process.

NO_x-reduction with hydrocarbons:

- Expensive
- Hydrocarbon emissions

- Selective catalytic reduction with ammonia is more favored:

NO_x-reduction with NH_3:

- Inexpensive
- Not suitable for low pressure absorption units

$$4\,NH_3 + 6\,NO \longrightarrow 5\,N_2 + 6\,H_2O$$

$$8\,NH_3 + 6\,NO_2 \longrightarrow 7\,N_2 + 12\,H_2O$$

In the process developed by BASF, Gulf Oil, Mitsubishi etc. this takes place after the absorption stage at 220 to 350 °C on vanadium(V) oxide or precious metal catalysts at the pressure of the residual gas. Preferably equimolar mixtures of nitrogen(II) and nitrogen(IV) oxides are thereby reduced to nitrogen and water:

$$NO + NO_2 + 2\,NH_3 \longrightarrow 2\,N_2 + 3\,H_2O$$

The residual nitrous gas content is < 200 ppm and that of ammonia 10 to 20 ppm. This process is mainly utilized with monopressure/medium pressure plants (M/M-type) and has made this process competitive with dual pressure processes of the M/H-type. It is unsuitable for low pressure absorption units (< 3 bar) for energy reasons. In the BASF process the total yield of nitrogen is slightly reduced e.g. in the M/M-process from 96 to 95.3%.

- Adsorption of nitrogen oxides at low temperatures on molecular sieves (e.g. Pura-Siv-N-process from Union Carbide). Desorption occurs at higher temperatures (250 to 300 °C), the liberated nitrogen oxides being returned to the nitric acid process.

1.4.4.3 Nitric Acid Applications

Most of the nitric acid is utilized in the form of 50 to 70% acid for the manufacture of nitrogen-containing fertilizers, particularly ammonium nitrate. Ammonium nitrate is also utilized as an explosive e.g. in mining. It is notable for its low explosion temperature (safety explosive). Other industrially important nitrates are sodium nitrate, a special fertilizer and oxidizing agent in the glass and enamel industries, and calcium and potassium nitrates, also fertilizers. Nitric acid is also utilized in the digestion of raw phosphate (Odda process, not operated in the USA, see Sec. 2.1.2.4):

$$Ca_5[(PO_4)_3(F)] + 10\ HNO_3 \longrightarrow 5\ Ca(NO_3)_2 + 3\ H_3PO_4 + HF$$

Worldwide 75 to 85% of the nitric acid produced is utilized in the fertilizer sector, in the USA ca. 60 to 70%. In the USA in 1985, ca. 19% of the nitric acid was consumed in the explosives sector, 9% for the manufacture of adipic acid (fiber and plastics precursor) and 7% for the production of nitrobenzene and dinitrotoluene. Nitric acid is also used for pickling stainless steel, for other metallurgical purposes and in the nuclear fuel cycle.

Estimated USA applications spectrum for nitric acid in 1985:

- 76% for ammonium nitrate of which
  - 80% was for fertilizers
  - 19% was for explosives
  - ca. 1% was for N_2O
- 9% for adipic acid (fiber and plastics precursor)
- 3.5% for dinitrotoluene (for toluene-diisocyanate)
- 3.5% for nitrobenzene (for aniline)
- 2% for sodium, potassium and calcium nitrates
- 1% for nitro-compounds for explosives
- ca. 5% for miscellaneous purposes

References for Chapter 1.4.4: Nitric Acid

General Information:

Kirk-Othmer: *Encyclopedia of Chemical Technology,* 3rd Ed., Vol. 15, pp 857–870, John Wiley & Sons Inc., New York-Chichester-Brisbane-Toronto (1981).
Ullmanns Encyklopädie der technischen Chemie, 4th Ed., Vol. 20, pp 306–358, Verlag Chemie, Weinheim-Deerfield Beach-Basel (1981).
S. P. S. Andrew, *Spec. Publ. Chem. Soc. Series* **31,** pp 221–228 (1977).

Specific Information:

Commercial:

Chemical Economics Handbook, Stanford Research Institute, Menlo Park, California, USA: "Nitrogen chemicals – Marketing research report" (July 1983).
Chemical Economics Handbook, Stanford Research Institute, Menlo Park, California, USA: "Manual of current indicators – supplemental data" (June 1986).
Industrial Statistics Yearbook. 1983 Ed., Vol. II, United Nations, New York, p. 405 (1985).
Directory of Chemical Producers. 8th Ed., Vol. 2, pp 1485–1487, SRI International, Menlo Park, California, USA (1985).

Manufacture:

Fert. Sci. Technol. Ser., **4** (Nitric Acid Fert., Nitrates), pp 31–129 (1985).
B. C. Mandelik and W. Turner, "How to operate nitric acid plants", *Hydrocarbon Processing* **56,** pp 175–177 (July 1977).
L. Hellmer, "Strong-nitric-acid process features low utility costs", *Chemical Engineering,* 8 Dec., pp 98–99 (1975).
Anon., *Nitrogen,* **No. 106,** pp 35–39 (March/April 1977).
C. A. Lawson, *Fert. Ind. Environ. Concerns,* Technol. Transfer Seminar, Toronto 1978, paper No. 2.

1.5 Phosphorus and Its Compounds

1.5.1 Phosphorus and Inorganic Phosphorus Compounds

1.5.1.1 Raw Materials

Most important raw material for phosphor chemistry: apatite $Ca_5(PO_4)_3(\mathbf{F}, OH, Cl)$

The most important group of minerals for the manufacture of phosphorus and phosphorus compounds are the apatites, which have the general formula

$$Ca_5(PO_4)_3(\mathbf{F}, OH, Cl)$$

Of these, the commercially most important are the fluorapatites. The aluminum phosphate deposits in Africa and South America and guano are much less important.

Estimated apatite reserves:

$> 67 \times 10^9$ t, a bottleneck in supply not in sight in the foreseeable future

Phosphate ores, and especially sedimentary apatites, are widely distributed. There are massive beds particularly in the USA and Africa. An estimate of the reserves is given in Table 1.5.1-1. The term "reserves" not only takes into account the available deposits, but also the economics of exploitation, the available infrastructure etc. The figures given in this table are valid only under the proviso that they are subject to change.

Table 1.5.1-1. Reserves of Phosphate Ores in 10^9 t (estimated, 1982).

World	North America	South America	Western Europe	Eastern Europe	Africa	Western Asia	Eastern Asia	Oceania
67.1	12.9	2.8	1.0	4.0	33.7	2.9	7.1	2.7

From the present viewpoint, a bottleneck in the supply of phosphate ores is not to be expected in the foreseeable future. Ca. 90% of the apatites mined is used for the production of fertilizers.

The sedimentary apatite deposits are in quantity much more important than the magmatic ones.

Apatite deposits

- Sedimentary
- Magmatic

The magmatic deposits (Kola: USSR, Phalaborwa: South Africa) are extracted from open cast mines or underground mines. They have a high phosphate content and are free of organic impurities and are separated from the gangue by flotation.

The sedimentary deposits in the USA are exploited by open cast mining and those in North Africa by open cast mining and underground mining. Beds are exploited with phosphorus pentoxide contents down to 5%. A marketable material is obtained by subsequent beneficiation. The individual beneficiation steps are: grinding, washing, sieving, magnetic separation of iron oxides, flotation to remove quartz and calcination to remove organic impurities. A uniform marketable quality is finally obtained by mixing.

Data regarding the phosphorus content in apatites are given either in % P_2O_5 or in % $Ca_3(PO_4)_2$ (bone phosphate of lime, BPL). 1% Phosphorus pentoxide is equivalent to 2.19% BPL. Table 1.5.1-2 gives a survey of the phosphorus pentoxide contents of raw phosphates from different deposits.

Content data for apatites:

in % P_2O_5 or in % $Ca_3(PO_4)_2$ (BPL)
1% P_2O_5 = 2.19% BPL

Table 1.5.1-2. P_2O_5 Content of Raw Phosphates in % by Weight.

Kola	Phalaborwa	Florida	Idaho	Morocco
39.0	38.4	30.4	31.2	36.8

Marketable raw phosphate contains > 30% P_2O_5

Currently marketed raw phosphate generally contains more than 30% phosphorus pentoxide.

Fluorine content of apatites:

between 2.3 and 4.8% by weight
Some apatites contain uranium

The fluorine contents of the raw phosphates mentioned in Table 1.5.1-2 are between 2.3 and 4.8% by weight. The raw phosphate contains numerous other elements particularly aluminum, iron and magnesium, but also rare earths and occasionally uranium all in the form of compounds.

The worldwide phosphate ore production in 10^6 t is given in Table 1.5.1-3.

World production of phosphate ore in 10^6 t:

1982: 125
1983: 135
1984: 146

Table 1.5.1-3. World Phosphate Ore Production 1982 to 1984 in 10^6 t (selected areas) (e = estimated).

	1982	1983[e]	1984[e]
World	124.8	135.2	146.1
USA	37.4	42.6	48.4
Mexico	0.4	0.4	0.4
Central and South America	2.8	3.2	3.5
Africa	30.1	33.7	35.5
– Morocco	17.9	20.1	21.2
– Tunisia	4.2	5.9	5.3
Western Asia	9.0	10.2	12.4
– Jordan	4.4	4.7	6.3
Eastern Asia	14.1	14.1	15.0
Western Europe	0.4	0.5	0.5
Australia and Oceania	2.9	2.8	2.8
Eastern Europe	26.6	27.7	28.0

1.5.1.2 Products

1.5.1.2.1 Phosphoric Acid

Economic Importance and Applications

More than 75% of apatite mined converted into phosphoric acid

Phosphoric acid is produced both by the digestion of apatite with sulfuric acid (wet-process phosphoric acid) and by the combustion of phosphorus ("furnace phosphoric acid"). Most (> 75%) of the mined apatite is used in the manufacture of phosphoric acid. The proportion of furnace acid is progressively decreasing in favor of the wet-process acid, due to the high energy requirements of the phosphoric acid manufacture by phosphorus combustion. The use of furnace acid for fertilizer production, previously customary in the USA, has almost ceased. In those application areas in which high purity phosphoric acid is required (e.g. phosphates for detergents) postpurified wet-process acid is being increasingly used. The phosphoric acid capacities (1985) are given in Table 1.5.1-4.

Table 1.5.1-4. Phosphoric Acid Capacities in 10^3 t/a P_2O_5 (1985).

Area	Capacity available	Capacity available and planned
World	37 612	43 098
North America	13 091	14 136
– USA	11 109	11 802
– Canada	1 262	1 310
– Mexico	720	1 120
West Europe	4 734	4 994
– West Germany	330	330
Central and South America	961	1 301
– Brazil	832	952
Africa	4 374	6 049
– Morocco	1 520	2 840
– Tunisia	1 100	1 430
– South Africa	882	907
Middle East	1 326	1 506
Asia and Australia	2 708	3 906
– Japan	500	500
Comecon Countries	9 918	10 610
– USSR	6 746	6 876

1985 World wide phosphoric acid capacity:

$> 37.6 \times 10^6$ t/a,
planned: 5.5×10^6 t/a

The phosphoric acid production in the USA was 7.7×10^6 t in 1982, 9.0×10^6 t in 1983 and 9.8×10^6 t in 1984 (P_2O_5 content), of which 0.635×10^6 t was furnace acid in 1984. In 1984,the world capacity for furnace acid was 2.7×10^6 t/a P_2O_5 i.e. ca. 8% of the phosphoric acid capacity. However, only 1.2×10^6 t were produced in 1983.

By far the most important class of products manufactured from phosphoric acid is fertilizer. In the USA, 90% of the wet-process acid is employed directly or indirectly in this sector, compared with ca. 50% in West Germany. Other applications are e.g. in the production of:

- Phosphates for detergents
- Industrial cleaners
- Water treatment agents
- Nutrition supplement for animal feedstuffs
- Flame retarding agents

Applications for phosphoric acid:

- Main application fertilizers (> 90% of wet-process acid in the USA, ca. 50% in W. Germany)
- Manufacture of
  - phosphates for detergents
  - industrial cleaners
  - water treatment agents
  - nutrition supplement for animal feedstuffs
  - flame retarding agents

Phosphoric acid directly used for the treatment of metals

Phosphoric acid itself is, amongst other things, used for the treatment of metals. Furnace acid is generally used for applications in the food sector. Phosphates for washing powder are manufactured from both furnace and purified phosphoric acid.

Manufacture of Phosphoric Acid

Production of Wet-Process Phosphoric Acid

"Wet-process phosphoric acid":

by reaction of apatite with sulfuric acid and separation of the calcium sulfate formed as the dihydrate or the hemihydrate

Wet process. The reactions in the digestion of apatite with sulfuric acid are complex. The calcium phosphate portion of apatite reacts with sulfuric acid as follows:

$$Ca_3(PO_4)_2 + 3H_2SO_4 \longrightarrow 3CaSO_4 + 2H_3PO_4 ,$$

Fluoride is mostly removed as gaseous silicon tetrafluoride and recovered in the form of hexafluorosilicic acid

in which, depending on reaction conditions, the calcium sulfate either precipitates as the dihydrate or the hemihydrate (see below).

The "fluoride part" of the apatite is either removed as gaseous silicon tetrafluoride in the presence of silica and absorbed in scrubbers as aqueous hexafluorosilicic acid solution, or largely removed with the calcium sulfate as the sparingly soluble sodium hexafluorosilicate by adding a sodium compound. Part of the fluoride remains in the acid.

The other components of apatite (iron, aluminum, uranium) partially pass into solution as salts and are partially precipitated with the calcium sulfate. Any carbonate present produces carbon dioxide during digestion. If sedimentary noncalcined apatite is employed, the phosphoric acid obtained is colored black by the organic impurities.

In principle, there are two different digestion processes:

Wet process:

- Process in which calcium sulfate dihydrate is formed and separated
- Process in which calcium sulfate hemihydrate is produced as an intermediate or end product

- That in which calcium sulfate dihydrate is produced and separated
- That in which calcium sulfate hemihydrate is formed as an intermediate or an end product

Three variants in the hemihydrate process:

Three variants of the hemihydrate process are operated. These vary in importance:

Variant 1: Calcium sulfate precipitates out immediately as the hemihydrate and is filtered off.
Variant 2: Calcium sulfate precipitates out initially as the dihydrate and is then recrystallized to the hemihydrate.
Variant 3: Calcium sulfate precipitates initially as the hemihydrate and is then recrystallized to the dihydrate.

- Variant 1: Hemihydrate formed and filtered off
- Variant 2: Dihydrate formed and recrystallized to hemihydrate
- Variant 3: Hemihydrate formed and recrystallized to dihydrate. Recrystallization in situ or after filtering off

The recrystallization in variant 3 can be carried out either after filtering off the calcium sulfate hydrate or in situ.

Those processes yielding anhydrite (anhydrous $CaSO_4$) have little commercial importance.

The crucial step in the decomposition of apatite is the formation of calcium sulfate. Its properties, in particular its ability to be filtered, are very important for throughput optimization. The incorporation of phosphate into its crystal lattice reduces the yield and can make its use in the building industry impossible.

The choice of process depends upon a number of factors such as

- Raw material price
- Source of the apatite
- Energy costs and availability at the plant site
- Possibilities regarding the further utilization of the calcium sulfate byproduct

Factors in the choice of digestion process:

- Raw material price
- Apatite source
- Energy costs and availability on site
- Possibilities regarding utilization of the calcium sulfate byproduct

The dihydrate process is the most widely used. The process parameters of the different processes are compared in Table 1.5.1-5.

Table 1.5.1-5. Apatite Digestion, Process Comparison.

Process	Digestion temperature °C	Recrystallization temperature °C	Acid concentration (% P_2O_5)	P_2O_5 yield (%)	$CaSO_4$ usable in the building industry
Dihydrate	71 to 85	–	28 to 32	95	no
Hemihydrate Variant 1	91 to 99	–	45 to 50	91 to 94	no
Hemihydrate Variant 2	62 to 68	93 to 99	33 to 38	97	yes
Hemihydrate Variant 3	91 to 99	60 to 19	40 to 50	96 to 98	yes

Process developers:

Dihydrate: Prayon, Dorr, St. Gobain/Rhone Progil, Fisons, Kellog-Lopker amongst others

Hemihydrate, Variant 1: Fisons, Occidental

Hemihydrate, Variant 2: Prayon, Nissan, Mitsubishi, Fisons amongst others

Hemihydrate, Variant 3: Breyer/Heurty, Dorr, Fisons

Essential parameters for modern plants using the *dihydrate process* are:

- Reaction temperature of ca. 80 °C
- Separate feed for ore and sulfuric acid
- Vacuum evaporation cooling to ensure a constant reaction temperature (exothermic reaction!)
- Monitoring of sulfuric acid (sulfate) concentration and if necessary recycling part of the dihydrate

These measures are particularly aimed at a uniform growth of the dihydrate crystals which is indispensable for efficient filtration.

Dihydrate process:

- Most widely operated process
- Supplies phosphoric acid with 28 to 32% P_2O_5
- Gypsum must generally be dumped

The phosphoric acid obtained in the dihydrate process has a concentration of 28 to 32% (as P_2O_5). The yield, based on the phosphorus content of the apatite, is ca. 95%. Modern plants have reactors with capacities of ≥ 1000 t phosphorus pentoxide per day. Fig. 1.5.1-1 shows the (simplified) flow sheet of a dihydrate plant.

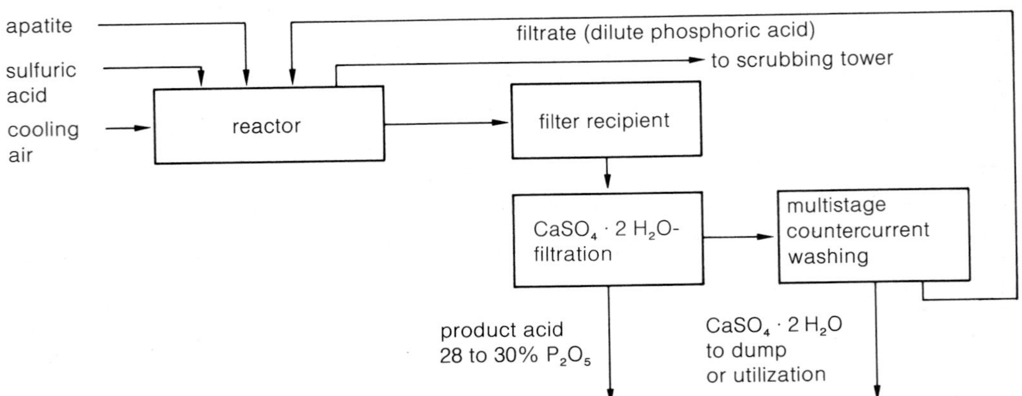

Fig. 1.5.1-1. Flow sheet for phosphoric acid manufacture using the dihydrate process.

The three variants of the hemihydrate process produce different results:

With *variant 1* (in which temperatures of 90 to 100 °C are used and in which the hemihydrate is directly filtered off) a relatively concentrated acid is obtained, but the yield of phosphorus pentoxide is low.

Hemihydrate processes, variant 1:

- Produces phosphoric acid with 45 to 50% P_2O_5
- P_2O_5 yield relatively low
- Calcium sulfate not marketable

With *variant 2* (in which dihydrate is initially obtained at 60 to 70 °C and the filter cake resulting from filtration is recrystallized to the hemihydrate by slurrying in sulfuric acid at temperatures > 90 °C) a very high yield of phosphorus pentoxide is obtained and a phosphoric acid with a content of 33 to 38% (as P_2O_5).

Variant 2:

- Produces phosphoric acid with 33 to 38% P_2O_5
- P_2O_5 yield very good
- Calcium sulfate marketable

With *variant 3* (in which the hemihydrate is formed at 90 to 100 °C and changes into the dihydrate upon cooling to 50 to 60 °C) the phosphorus pentoxide content in the dihydrate is less than 0.3% due to the recrystallization and the yield is higher than with the dihydrate process. The Nissan Process is typical of this variant.

Variant 3:

- Produces phosphoric acid with 40 to 50% P_2O_5
- P_2O_5 yield very good
- Calcium sulfate marketable

The hemihydrate-dihydrate conversion can also be carried out after filtering off the hemihydrate. The acid concentration is then as high as in variant 1 (> 42% P_2O_5).

The hemihydrate is subsequently slurried in dilute acid and recrystallized to the dihydrate. The process calls for very uniformly-shaped hemihydrate crystals and the recycling of the acid is very complicated. This process uses a particularly small amount of energy because it is unnecessary to concentrate the acid produced. This is an advantage. In addition, this process uses phosphate ore which does not have to be so finely ground as for the dihydrate process. A comparison of the classical dihydrate process and the Fisons hemihydrate process, with filtration and conversion of the hemihydrate, is given in Table 1.5.1-6.

Table 1.5.1-6. Comparison of the Consumption Figures for a Dihydrate and a Hemihydrate Process (Variant 3 with filtration of the hemihydrate and conversion to the dihydrate) per t of P_2O_5 Produced.

Charge	Dihydrate process	Hemihydrate process
t phosphate ore	3.45	3.38
t sulfuric acid	2.8	2.74
m^3 cooling water	37.5	23.5
m^3 process water	7.0	7.0
t steam	2.0	0.2
kWh electricity	120	130

Fig. 1.5.1-2 shows the flow sheet of the Fisons hemihydrate process.

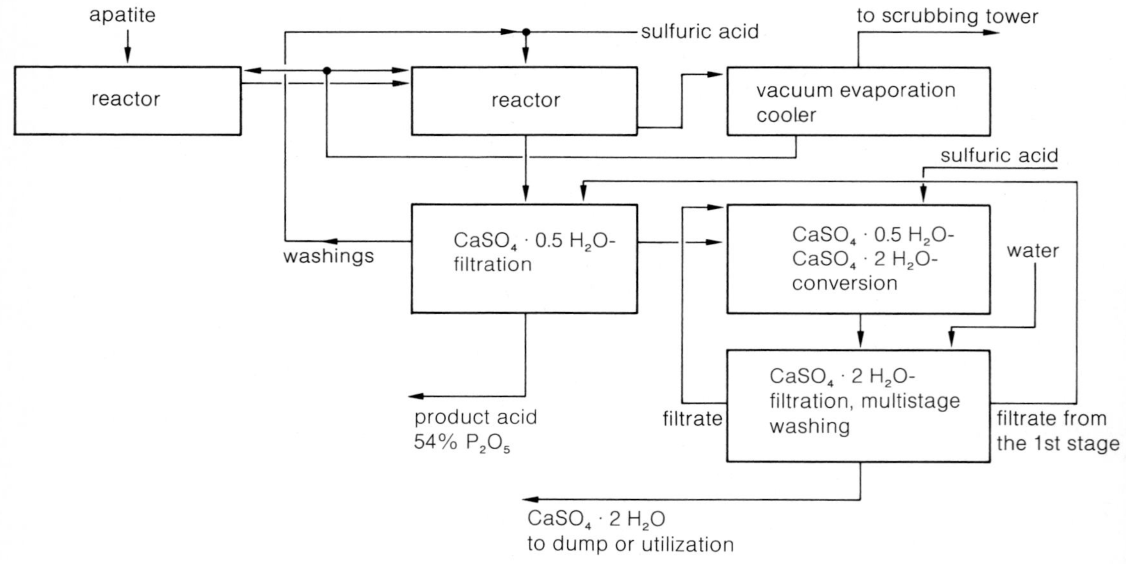

Fig. 1.5.1-2. Flow sheet of phosphoric acid manufacture using the hemihydrate process (Variant 3).

Acids may have to be concentrated:

- To 40 to 54% P_2O_5 for fertilizers
- To 52 to 72% for transportation

Concentration of acid. The phosphoric acid produced in several of the processes mentioned above may, depending upon the application (currently mainly fertilizer manufacture), have to be concentrated. Fertilizer production requires acids containing between 40 and 54% phosphorus pentoxide. For transportation purposes the acid concentration is increased to 52 to 72%.

Several of the evaporation processes (submerged burner, vacuum evaporation etc.) are technically fairly demanding because of the high corrosiveness of the acid, the formation of precipitates and the release of acid-containing gases (fluorine compounds and phosphoric acid mist). Furthermore, they are very energy intensive.

Purification of wet-process acid. Wet-process acid is very impure. During concentration and subsequent standing, wet-process acid deposits a considerable fraction of its insoluble impurities as sediment. These "deslimed" acids are transportable.

Purification of wet-process acid by:

- Concentration and desliming
- Precipitation of such interfering ions as copper, arsenic, sulfate
- Liquid-liquid extraction yielding a pure phosphoric acid and in some processes a residue of more strongly contaminated acid. Yield of pure phosphoric acid up to 98%

Table 1.5.1-7 gives the average composition of such acids.

Table 1.5.1-7. Average Composition of US Wet-process Acids after Concentration and Desliming in % by weight.

P_2O_5	Fe_2O_3	Al_2O_3	MgO	SO_4	F	Organic compounds	Water insoluble
53.1	1.3	1.6	0.6	3.7	0.9	0.7	2.0

Further concentration to acids with phosphorus pentoxide contents of ca. 67 to 72% reduces the fraction of some impurities still further e.g. fluoride content is reduced to ca. 0.4%.

A much more extensive purification is possible through:

- Precipitation of interfering ions
- Multistage liquid-liquid countercurrent extraction of the phosphoric acid

In precipitation purification, heavy metal ions such as copper or arsenic are precipitated as their sulfides and sulfate as barium sulfate.

In liquid-liquid extraction three types of extraction agents are used:

- Fully water miscible, such as methanol or isopropanol
- Partially water miscible, such as butanol, isoamyl alcohol, n-heptanol
- Water immiscible, such as tri-n-butyl phosphate or di-isopropyl ether.

Upon mixing the impure phosphoric acid with the organic extraction agent two phases are always formed: one consists of a solution of purer phosphoric acid in the organic extraction agent and the other of an aqueous phosphoric acid solution together with the impurities. Pure acid can be recovered from the organic phase by stripping with water, distilling off the solvent or temperature variation. Depending upon the process used, phosphoric acid yields between 85 and 98% are obtained.

The quality of the acids produced is between technical and food grades, depending on the process used. The strongly contaminated residual acid remaining in some processes can still be used for fertilizer manufacture.

World consumption of phosphoric acid purified by precipitation or extraction in 1983: 1.2×10^6 t P_2O_5

World consumption of purified phosphoric acid (using precipitation and liquid-liquid extraction) was 1.2×10^6 t in 1983. Hoechst (FRG) has a plant capacity of 75×10^3 t/a and Albright and Wilson (GB) one of 90×10^3 t/a for purifying phosphoric acid by extraction. The precipitation purification capacity at Hoechst (FRG) is 111×10^3 t/a.

Uranium and cadmium can be separated from wet-process phosphoric acid by extraction

Processes for purifying phosphoric acid by impurity extraction are not being commercially operated at the moment. However, it is possible to separate uranium and cadmium by extraction. A combined precipitation-flotation process has been proposed for the removal of cadmium.

Manufacture of Furnace Phosphoric Acid

"Furnace phosphoric acid":

- Combustion of white phosphorus
- Hydration of the phosphorus pentoxide formed

Furnace phosphoric acid is obtained by the combustion of white phosphorus in air and the absorption of the resulting phosphorus pentoxide in water (in practice dilute phosphoric acid).

$$4P + 5O_2 \longrightarrow P_4O_{10}$$

$$P_4O_{10} + 6H_2O \longrightarrow 4H_3PO_4$$

Two processes are commercially operated: the "IG" and the "TVA" processes. In the TVA process combustion and absorption take place in separate towers and in the IG process in a single tower. In these processes the walls of the towers are protected from the hot phosphorus flame by recirculating phosphoric acid. The water for the formation of phosphoric acid is added to this recirculating acid, which also removes the heat of reaction and transports it to a cooler. It is from this solution that the product acid is continuously extracted. The mist is removed from the exhaust gases (oxygen content 6 to 8%) with, for example, venturi scrubbers. The material for these plants has to fulfill demanding specifications.

In this way (poly-)phosphoric acids are produced with 54.5 and 61.5% phosphorus pentoxide contents. (In the USA, polyphosphoric acids with contents of 76 or 84% are produced to save transport costs. These are diluted at the site of their utilization before use). Furnace phosphoric acids are very pure. If required, the arsenic content can be reduced by precipitation below 0.1 ppm by hydrogen sulfide treatment.

1.5.1.2.2 Phosphoric Acid Salts

The phosphoric acid salts used in fertilizers (calcium hydrogen phosphate, ammonium phosphate and mixtures of different phosphoric acid salts) are dealt with in Section 2.1: Fertilizers. Of those phosphates commercially produced for nonfertilizer applications the following are discussed below:

- Sodium ortho-, di- and polyphosphates
- Ammonium phosphates
- Tetrapotassium diphosphate
- Calcium phosphates

Phosphoric acid salts used in nonfertilizer applications:

sodium ortho-, di- and polyphosphates; ammonium phosphates; tetrapotassium diphosphate; calcium phosphates.

Economic Importance and Applications

Sodium Phosphate

In terms of quantity, pentasodium triphosphate [sodium tripolyphosphate (STPP)], which is primarily used in detergents and cleaning agents, is the most important sodium phosphate.

Table 1.5.1-8 gives the pentasodium triphosphate capacities for 1985.

Most important sodium phosphate, pentasodium triphosphate:

- USA capacity (1984): 1030×10^3 t/a
- W. Europe capacity (1985): 1635×10^3 t/a

Main application: in detergents and cleaning agents

Table 1.5.1-8. Capacities of Pentasodium Triphosphate in 1985 in 10^3 t/a.

W. Europe	W. Germany	USA*	Japan
1635	305	1030	185

* 1984 capacity

Reduction in USA production of pentasodium triphosphate from

1970: 1096×10^3 t/a to
1985: 550×10^3 t/a for ecological reasons

The reduction in sodium tripolyphosphate production in the 1970's, particularly in the USA, was due to the passing on ecological grounds of laws reducing the pentasodium triphosphate content in household detergents (Table 1.5.1-9).

Table 1.5.1-9. Sodium Tripolyphosphate Production in the USA, 1970 to 1985, in 10^3 t/a.

1970	1975	1980	1982	1983	1984	1985[e]
1096	698	660	590	607	612	550

[e]: estimated

The USA production of mono-, di- and trisodium phosphate in 1984 was 104×10^3 t.

Applications for mono-, di-, and trisodium phosphates:

- Metal cleaning
- Phosphatizing
- Boiler water treatment
- For buffer systems
- In the food sector
- Nutritional supplement in animal feedstuffs

for disodium dihydrogen diphosphate:

- As baking powder

for tetrasodium diphosphate:

- In industrial cleaning products

for higher molecular weight sodium polyphosphates:

- In food manufacture, for pigment suspension stabilization, in leather tanning

Applications: Mono-, di- and trisodium phosphates are used in industrial and metal cleaning, for phosphatizing, in boiler water treatment and for buffer systems e.g. in the textile industry. Disodium hydrogen phosphate is also used as a nutritional supplement in foods and in animal feedstuffs.

Disodium dihydrogen diphosphate is used as baking powder and tetrasodium diphosphate in industrial cleaning products.

Higher molecular weight sodium polyphosphate is used in the production of reconstituted cheese, condensed milk, sausages, for the stabilization of pigment suspensions and in leather tanning.

Ammonium Phosphates

Ammonium phosphate production for nonfertilizer applications in the USA was 36×10^3 t/a in 1976. More recent statistics are not available.

Applications: Their main applications are in fertilizers (see Section 2.1). In addition, they are used in fire protection e.g. as an ingredient in fire extinguisher agents for fighting forest fires, as flame retardants in paper, textiles and poly(urethanes), in the prevention of afterglow in matches and as a component of intumescent paints (flame retarding coatings). They are used in small quantities in animal, mainly cattle, feedstuffs as a nutritional supplement.

Ammonium phosphates (main application in fertilizers) applications in:

- Fire protection
- Intumescent paints
- Animal feedstuffs

Tetrapotassium Diphosphate

The USA production was 30×10^3 t and the capacity 80×10^3 t/a in 1984. The capacity in Western Europe was 49×10^3 t/a in 1985.

Applications: It is used in liquid cleaners due to its high water solubility. The same ecological reservations exist as with sodium tripolyphosphate.

Tetrapotassium diphosphate:

USA production in 1984: 30×10^3t.

Applications: in liquid cleaners

Calcium Phosphates

The main use of calcium phosphates outside the fertilizer sector (see Section 2.1) is in animal feedstuffs as a nutritional supplement. The products concerned are

Monocalcium phosphate – $CaH_4(PO_4)_2$ (calcium phosphate)
Dicalcium phosphate – $CaHPO_4$ (calcium hydrogen phosphate)
} or hydrates
Defluorinated apatite – $Ca_5(OH)(PO_4)_3$

Table 1.5.1-10 gives the calcium phosphate capacities for animal nutrition purposes in 1980.

Calcium phosphate for the animal nutrition sector: World capacity 1980: 3971×10^3 t/a

Other applications:

- As baking powder
- As cleaning agent in toothpastes

Table 1.5.1-10. Calcium Phosphate Capacities for Animal Nutrition Purposes in 10^3 t/a for 1980.

World	North America	USA	W. Europe & Israel	West Germany	Far East	Africa
3971	1676	1540	1912	355	170	213

60% of the USA capacity is for the production of dicalcium phosphate. In animal nutrition applications calcium phosphate is also mixed with the phosphates of other metals, particularly sodium and magnesium.

Other applications for calcium phosphates are as baking powder (monocalcium phosphate) and in toothpastes (dicalcium phosphate dihydrate). The fluoride-inert dicalcium phosphate is used in fluoride-containing toothpastes. The consumption of calcium phosphates in the food and toothpaste sectors in the USA was ca. 20×10^3 t in 1984.

Manufacture of Phosphates

Sodium Phosphates

Sodium monophosphates: from phosphoric acid and sodium carbonate or sodium hydroxide

Sodium monophosphates. They are produced by the reaction of furnace or (postpurified) wet-process acid with sodium carbonate or sodium hydroxide.

$$H_3PO_4 + NaOH \longrightarrow NaH_2PO_4 + H_2O$$

$$H_3PO_4 + 2NaOH \longrightarrow Na_2HPO_4 + 2H_2O$$

$$H_3PO_4 + 3NaOH \longrightarrow Na_3PO_4 + 3H_2O$$

Sodium hydroxide is required for the production of trisodium phosphate, as sodium carbonate is insufficiently alkaline. The anhydrous and hydrated salts are obtained using conventional techniques. Particular temperatures must be precisely maintained during the dehydration of the two hydrogen phosphates to avoid condensation to the di- or polyphosphates.

Sodium diphosphates: from sodium hydrogen phosphates by thermal dehydration

Sodium di- and polyphosphates. Diphosphates: Both disodium dihydrogen diphosphate and tetrasodium diphosphate are commercially manufactured. They are obtained by heating sodium hydrogen phosphates:

$$2\,NaH_2PO_4 \xrightarrow{245\,^\circ C} Na_2H_2P_2O_7 + H_2O$$

$$2\,Na_2HPO_4 \xrightarrow{300 \text{ to } 900\,^\circ C} Na_4P_2O_7 + H_2O$$

Both the reaction time and temperature must be limited during production to prevent further condensation to polyphosphates.

The dehydration can be carried out in rotary kilns or in spray towers.

Polyphosphates: Pentasodium triphosphate is obtained by the dehydration of sodium dihydrogen phosphate and disodium hydrogen phosphate in a 1:2 ratio:

$$2\,Na_2HPO_4 + NaH_2PO_4 \xrightarrow{300\text{ to }550\,^\circ C} Na_5P_3O_{10} + 2\,H_2O$$

Pentasodium triphosphate:

$2\,Na_2HPO_4 + NaH_2PO_4 \xrightarrow{300\text{ to }550\,^\circ C} Na_5P_3O_{10} + 2\,H_2O$

Industrially, a phosphoric acid solution is used which is neutralized with the appropriate quantity of sodium hydroxide or sodium carbonate. The dehydration of the orthophosphate solution takes place in one or more stages in spray towers or rotary kilns.

The anhydrous sodium tripolyphosphate produced has a high and a low temperature modification, which react with water at different speeds to the hexahydrate. The ratio of the two modifications depends upon the dehydration temperature, which is varied according to the envisaged application.

Higher molecular weight polyphosphates: Oligomeric polyphosphates with n = 2 (tetrametaphosphate) to n = ca.25 (Graham's salt) can be obtained in the dehydration process by increasing the fraction of the monosodium compound in the disodium hydrogen phosphate-sodium dihydrogen phosphate mixture.

The manufacture of these long chain polyphosphates takes place according to the equation below:

$$2\,Na_2HPO_4 + n\,NaH_2PO_4 \xrightarrow{600\text{ to }800\,^\circ C} (n+1)\,H_2O + (NaO)_2P(O) - [OP(O)(ONa)]_n - OP(O)(ONa)_2\,,$$

n: 2 to ca. 25 .

Higher molecular weight (long chain) polyphosphates:

$(NaO)_2P(O) - [OP(O)(ONa)]_n - OP(O)(ONa)_2$

n: 2 to ca. 25

Upon heating at temperatures above 250 °C sodium dihydrogen phosphate is converted by way of disodium dihydrogen diphosphate (see above) to the high molecular weight Maddrell's salt (n > 1000).

The manufacture of sodium polyphosphates is carried out in vessels clad with zirconium silicate bricks, due to the high corrosivity of the melts. After reaction completion the melts are quenched on cooling rollers.

In a process developed by Hoechst AG (FRG), the polymeric phosphates are obtained by phosphorus combustion in air in the presence of sodium hydroxide in a graphite-clad tower. All the energy for the process is provided by the phosphorus combustion.

Ammonium Phosphate

Ammonium dihydrogen phosphate and diammonium hydrogen phosphate:

from ammonia and phosphoric acid

Ammonium polyphosphate $[NH_4PO_3]_n$:

from phosphoric acid and urea or diammonium phosphate and P_4O_{10}

Industrially, three compounds are used: ammonium dihydrogen phosphate, diammonium hydrogen phosphate and ammonium polyphosphate. The two hydrogen phosphates are obtained from ammonia and pure phosphoric acid:

$$H_3PO_4 + NH_3 \longrightarrow NH_4H_2PO_4$$

$$H_3PO_4 + 2NH_3 \longrightarrow (NH_4)_2HPO_4$$

If the ammonia is added stepwise, wet-process acid can also be used. Diammonium hydrogen phosphate has a noticeable partial pressure of ammonia at room temperature. The tendency of ammonium hydrogen phosphates to split off ammonia at higher temperatures means that higher molecular weight ammonium polyphosphates cannot be produced using the hydrogen monophosphates, in analogy with the sodium polyphosphates. Industrially, they are produced from urea and phosphoric acid, or diammonium phosphate and P_4O_{10}, in an ammonia atmosphere:

$$n\,H_3PO_4 + n\,H_2NCONH_2 \longrightarrow [NH_4PO_3]_n + n\,CO_2 + n\,NH_3$$

Potassium Phosphates

Tetrapotassium diphosphate:

from dipotassium hydrogen phosphate by thermal dehydration

Only tetrapotassium diphosphate (due to its high water solubility) has major industrial importance. It is obtained from dipotassium hydrogen phosphate in an analogous way to the sodium compound:

$$2\,K_2HPO_4 \xrightarrow{350\text{ to }400\,^\circ C} K_4P_2O_7 + H_2O$$

Calcium Phosphates

Industrially, the most widely used calcium hydrogen phosphates:

$CaH_4(PO_4)_2$ monocalcium phosphate (calcium phosphate)

and

$CaHPO_4$ dicalcium phosphate (calcium hydrogen phosphate)

are manufactured by reacting calcium oxide, calcium hydroxide or other calcium salts with pure or purified phosphoric acid:

$$CaO + 2\,H_3PO_4 \rightarrow CaH_4(PO_4)_2 + H_2O$$

$$CaO + H_3PO_4 \rightarrow CaHPO_4 + H_2O$$

(idealized equations)

These products are produced in both anhydrous and hydrated forms.

Defluorinated apatite is produced by heating apatite in a fluidized bed or in a rotary kiln at temperatures of ca. 1500 °C in the presence of phosphoric acid or silica and steam.

Calcium hydrogen phosphates:

from calcium oxide or hydroxide and phosphoric acid

1.5.1.2.3 Phosphorus

Economic Importance and Applications

The USA production of phosphorus was ca. 570×10^3 t in 1969, 332×10^3 t in 1983 and 348×10^3 t in 1984. This decline has been mainly caused by increasing energy costs, which make the production of phosphoric acid for fertilizer from elemental phosphorus increasingly uneconomic. In addition, the incorporation of phosphates in household detergents has decreased due to ecological reasons.

Table 1.5.1-11 gives a survey of the phosphorus production capacities in different countries and regions.

Elemental phosphorus:

USA production in 1984:

348×10^3 t

Phosphorus capacities, 1984 in 10^3 t/a:

World: 1443
USA: 412
W. Europe: 200
USSR: 615

Table 1.5.1-11. Phosphorus Capacities in 1984 in 10^3 t/a.

World	USA	Canada	Mexico	Western Europe	West Germany	The Netherlands	Comecon	USSR	South Africa	India
1443	412	90	10	200	80	90	715	615	6	10

Applications:

White phosphorus

White phosphorus is used for the manufacture of:

- Pure phosphoric acid
- Phosphorus sulfides
- Phosphorus oxides
- Phosphorus halides
- Red phosphorus

The use spectrum of white phosphorus in the USA in 1984 (consumption: 336×10^3t) is given in Table 1.5.1-12:

Table 1.5.1-12. Application Spectrum of White Phosphorus in Terms of End Products in the USA in 1984 in % by Weight.

	USA
furnace acid and its salts	86
PCl_3, P_2S_5, P_2O_5	12.1
sodium hypophosphite	0.5
rest	1.4

Red phosphorus

Red phosphorus used for

- Striking surface of matchboxes
- Pyrotechnics
- Flame retardation e.g. of polyamides

Red phosphorus is used in the striking surface of matchboxes, for pyrotechnical purposes and as a flame retarding agent e.g. in poly(amides).

Manufacture of White Phosphorus

White phosphorus manufactured from apatite by its electrothermal reduction with coke in the presence of SiO_2

"White phosphorus", the industrially far more important form of elemental phosphorus, is produced by the electrothermal reduction of apatite with coke in the presence of gravel (silica). The main reaction can probably be represented by the equations:

Reactions:

- $Ca_3(PO_4)_2 + 5CO \rightarrow 3CaO + 5CO_2 + 2P$
- $CO_2 + C \rightarrow 2CO$
- $CaO + SiO_2 \rightarrow CaSiO_3$
- Formation of ferrophosphorus (Fe_2P)

$$Ca_3(PO_4)_2 + 5CO \xrightarrow{>1400\,°C} 3CaO + 5CO_2 + 2P$$
$$5CO_2 + 5C \longrightarrow 10CO$$
$$\overline{Ca_3(PO_4)_2 + 5C \longrightarrow 3CaO + 5CO + 2P}$$

Carbon monoxide is regenerated from carbon dioxide and coke. The silica acts as a flux, converting the calcium oxide formed into a low melting slag as follows:

$$CaO + SiO_2 \longrightarrow CaSiO_3$$

This slag also takes up most of the minor constituents introduced into the process (with the exception of iron) and part of the fluoride. The iron is converted into ferrophosphorus with the approximate composition Fe_2P.

The raw materials used have to have a minimum size (1 to 2 cm or more) to ensure an adequate permeability to gases in the phosphorus reduction furnace. Gravel and coke (smelting coke with preferably < 10% ash) can be obtained in such sizes, but the apatite for this application is finer and has to be agglomerated. This can be achieved by, for example, pelletization with a pan granulator using the dust from the electrostatic filters (see below) as a binder. These pellets still need to be dried and sintered, for which the carbon monoxide containing flue gases can be used as an energy source.

Apatite has to be pregranulated and sintered to ensure an adequate permeability to gases in the furnace

Modern reduction plants consist of three units:

- An electrothermal reactor
- A gas purifier
- A phosphorus condenser

Modern reduction plants consist of:

- An electrothermal reactor
- A gas purifier
- A phosphorus condenser

Fig. 1.5.1-3 shows a process flow chart. The furnace, which is circular or triangular with rounded corners,

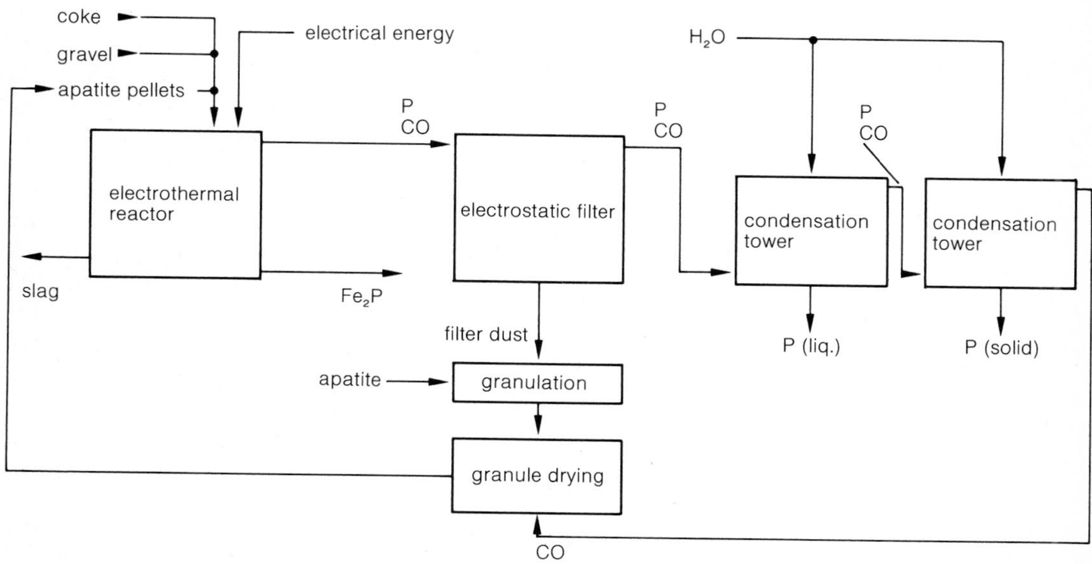

Fig. 1.5.1-3. Flow chart for the electrothermal manufacture of phosphorus.

is clad with carbon blocks at the bottom and with firebricks above. The symmetrically placed carbon electrodes (Soederberg self-baking electrodes or, particularly in the USA, preformed carbon electrodes) have a diameter of 1.3 to 1.5 m in a modern furnace. The electricity consumption is 50 to 70 MW.

The feed shutes for the raw materials are placed symmetrically round the electrodes. The tap hole for the ferrophosphorus is positioned right at the bottom of the furnace, with the tap hole for the less dense calcium silicate slag slightly above it. Carbon monoxide and phosphorus vapor leave by the gas outlet at the top of the furnace.

The gas purification (hot electrostatic precipitators) has to be operated above 280 °C to prevent phosphorus condensation. Ca. 0.04 kg of dust is separated per m^3 of gas.

In modern plants the condensation of phosphorus is carried out in two steps. In the first condenser tower 50 to 60 °C hot water is sprayed from above and meets the phosphorus vapor being transported countercurrently, whereupon the phosphorus condenses to a liquid. In the second tower 10 to 25 °C water is used and solid phosphorus is formed. This is melted from time to time and, as with the phosphorus from the first tower, run off into a holding tank. The liquid phosphorus is stored under water, drawn off and transported.

Byproducts:

- Carbon monoxide – valuable energy carrier
- Calcium silicate slag – limited use in road building
- Ferrophosphorus – usable in metallurgy

The byproducts can, in part, be further utilized: carbon monoxide is a valuable energy source and can, for example, be used in the sintering of the apatite pellets. The calcium silicate slag has, however, only very limited marketability e.g. in road building. The ferrophosphorus contains 15 to 28% phosphorus and is, when the silicon content is below 3%, usable in metallurgy. Ferrophosphorus from plants in the western USA contains 7 to 14% vanadium (as V_2O_5). This ferrophosphorus is processed to vanadium pentoxide or vanadium chloride and represents the second most important source of vanadium in the USA. The phosphorus contained in the ferrophosphorus is simultaneously converted into nonutilizable byproducts.

Phosphorus-containing sludge and process water are also produced which are difficult to dispose of.

The production of one ton of phosphorus entails the consumption of (figures in [ ] apply to USA):

- 8.0 t Florida apatite (31% P_2O_5) [10.625 t apatite (24 to 31% P_2O_5)]
- 2.8 t gravel (97% SiO_2) [2.75 t]
- 1.25 t coke (90% carbon) [1.625 t]
- 0.05 t Soederberg electrodes [0.02 t electrodes]
- ca. 13 MWh of electrical energy [14.6 MWh]

Consumed per t phosphorus produced:

- 12.1 t raw materials
- ca. 13 MWh

Ca. 8 t solid byproducts formed

and the production of the following quantities of by-products:

- 7.7 t calcium silicate slag (90% calcium silicate)
- 0.15 t ferrophosphorus (22% phosphorus) [0.26 t (22.5% P)]
- 0.1 t filter dust (20% P_2O_5)
- 2500 m^3 flue gas (85% carbon monoxide)

Furnaces with an effective power of 70 MW can therefore produce a maximum of 5.4 t per hour.

Modern furnaces produce a maximum of 5.4 t per hour

Manufacture of Red Phosphorus

In terms of quantity produced, red phosphorus is less important than white phosphorus. The West European capacity for red phosphorus was ca. 3.8×10^3 t/a in 1985 [of which Hoechst A.G. (FRG), the largest producer in the world, has 3×10^3 t/a]. That in Japan was 1.2×10^3 t/a. The USA, which does not produce red phosphorus, imported ca. 4×10^3 t/a in 1984 mainly for military purposes.

Red phosphorus:

- West European capacity in 1985: 3.8×10^3 t/a
- Japanese capacity in 1985: 1.2×10^3 t/a
- USA imports in 1984: ca. 4×10^3 t/a
- Manufacture by exothermic conversion of white phosphorus semicontinuously in a ball mill at 350 °C, grinding under water, filtering, drying, stabilizing e.g. with magnesium oxide

The transformation of white into red phosphorus is an exothermic reaction producing red phosphorus as a solid product, making heat removal from the reaction difficult. In a modern process developed by Hoechst, the conversion is carried out semicontinuously in a ball mill at 350 °C. The mills used have a volume of e.g. 5 m^3 filled with 3.7 t of iron shot of different diameters. The white phosphorus is fed into the ball mill at such a speed as to maintain the reaction temperature through the heat of reaction. The ball mill is at most only half filled. After tempering and cooling, water is added and the red phosphorus ground to a suspension. Heating with sodium hydroxide removes the residual white phosphorus. The red phosphorus is then filtered and dried under inert gas. The red phosphorus can be stabilized against autooxidation by mixing in small quantities of magnesium oxide. Red phosphorus is twice as expensive as white phosphorus.

1.5.1.2.4 Products Manufactured from Phosphorus

Phosphorus Pentoxide

P_4O_{10}
Manufactured from white phosphorus:

$4P + 5O_2 \rightarrow P_4O_{10}$

About 85% of the white phosphorus produced is burnt to phosphorus pentoxide:

$$2\,P + 2.5\,O_2 \rightarrow P_2O_5 \text{ or } 4\,P + 5\,O_2 \rightarrow P_4P_{10}$$

This is mainly directly hydrolyzed to phosphoric acid. Only several thousand tons of phosphorus pentoxide are produced worldwide annually.

The USA capacity for phosphorus pentoxide in 1984 was 8.6×10^3 t/a, the production 6.4×10^3 t and the consumption 8.5×10^3 t. The consumption in Western Europe is estimated to be 5×10^3 t.

Phosphorus pentoxide is produced using burners similar to those used in the production of phosphoric acid. Here dry gas has to be used, the walls of the combustion chamber being cooled from the outside with water. After leaving the burning chamber, the gases are cooled in a so-called "barn", a large settling chamber in which ca. 95% of the phosphorus pentoxide formed is condensed. At temperatures between 170 and 200 °C a denser and more crystalline product is obtained than when the chamber is water cooled. The tail gases which contain phosphorus pentoxide can, for example, be utilized in a phosphoric acid plant.

Applications:

- Mainly for manufacture of pure phosphoric acid
- As a drying agent
- In dehydration reactions in organic chemistry
- For improving the properties of asphalt
- In the production of additives for plastics and lubricating oils

Applications: Phosphorus pentoxide is used as a drying agent, in dehydration reactions in organic chemistry, for the improvement of the properties of asphalt, and in the manufacture of surface active substances for plastics' and lubricating oil additives.

Phosphorus Pentasulfide

Of the large number of binary phosphorus sulfides, only phosphorus pentasulfide is commercially important. The USA capacity in 1985 was 120×10^3 t/a and that in Western Europe 66×10^3 t/a (Hoechst AG accounting for 30×10^3 t/a). The USA and Western Europe consumed 68×10^3 t and 31×10^3 t respectively in 1985.

Phosphorus pentasulfide is manufactured by the exothermic reaction of liquid phosphorus with liquid sulfur at temperatures above 300 °C:

$$2\,P + 5\,S \rightarrow P_2S_5 \text{ or } 4\,P + 10\,S \rightarrow P_4S_{10}$$

The reaction product can either be directly poured onto cooling rollers or first purified by distillation (b.p. 513 to 515 °C). If phosphorus pentasulfide is required with a satisfactory color, organic impurity-free phosphorus and very pure sulfur have to be used. The former can be obtained by sulfuric acid extraction and the latter from natural gas purification.

P_4S_{10} manufacture:

$$4P + 10S \xrightarrow{> 300\,^\circ C} P_4S_{10}$$

Purification by e.g. distillation

Applications: Phosphorus pentasulfide is used in the manufacture of insecticides (ca. 40%), of lubricating oil additives (ca. 50%) and of flotation agents.

Applications:

in the manufacture of
- Insecticides
- Lubricating oil additives
- Flotation agents

Phosphorus Halides

Phosphorus trichloride

The USA capacity for phosphorus trichloride was 156×10^3 t/a in 1985 and that in Western Europe 115×10^3 t/a. The USA consumption of phosphorus trichloride in 1985 was 84×10^3 t.

Phosphorus trichloride capacities in 1985:
- USA: 156×10^3 t/a
- Western Europe: 115×10^3 t/a

Phosphorus trichloride is manufactured by the exothermic reaction between phosphorus and chlorine:

$$2\,P + 3\,Cl_2 \longrightarrow 2\,PCl_3$$

PCl_3

Manufacture:

$$2P + 3Cl_2 \rightarrow 2PCl_3$$

purification by distillation

This reaction can be carried out in a number of ways. Chlorine can be passed into a suspension of phosphorus in phosphorus trichloride. The heat liberated during the reaction is then sufficient to evaporate off the phosphorus trichloride, which is condensed in reflux condensers and in part returned to the reaction mixture. Part of the distillate is run off and purified by fractional distillation. Direct reaction of stoichiometric quantities of phosphorus and chlorine in a burner is also possible with subsequent working up. Commercial phosphorus trichloride is more than 99.7% pure and does not contain elemental phosphorus.

Applications:

in the manufacture of

- Phosphorus oxychloride
- Intermediates for crop protection products
- Phosphorous acid
- Flame retardants
- Stabilizers for plastics

Applications: Phosphorus trichloride is a raw material in the manufacture of phosphorus oxychloride (33%), intermediates for crop protection products (30%), phosphorous acid (10%), flame retardants (8%) and stabilizers for plastics (6%) (USA consumption in %).

Phosphorus pentachloride

PCl_5

Manufacture:

$PCl_3 + Cl_2 \rightarrow PCl_5$

Phosphorus pentachloride is manufactured continuously in lead-clad towers in which phosphorus trichloride is fed in from above and chlorine from below:

$$PCl_3 + Cl_2 \longrightarrow PCl_5$$

The phosphorus pentachloride formed sinks to the bottom and is removed by means of a screw conveyor.

Utilized:

as a chlorination agent in organic chemistry

Applications: Phosphorus pentachloride is mainly used as a chlorination agent in organic chemistry.

Phosphorus oxychloride

$POCl_3$

USA capacity 1985:

60×10^3 t/a

The USA capacity for phosphorus oxychloride was 60×10^3 t/a in 1985 and its consumption 30×10^3 t.

Manufacture:

$2PCl_3 + O_2 \xrightarrow{50 \text{ to } 60\,^\circ C} 2POCl_3$

It is manufactured by reacting pure phosphorus trichloride with oxygen at ca. 50 to 60 °C with cooling:

$$2\,PCl_3 + O_2 \xrightarrow{50 \text{ to } 60\,^\circ C} 2\,POCl_3$$

This is a free radical reaction, which can be inhibited by small quantities of sulfur, sulfur compounds, iron, copper etc. The reaction product can be further purified by fractional distillation.

The extent to which phosphorus oxychloride is still manufactured from phosphorus pentoxide and phosphorus pentachloride (from phosphorus trichloride and chlorine) according to:

$$P_2O_5 + 3\,PCl_3 + 3\,Cl_2 \longrightarrow 5\,POCl_3$$

is unknown.

Applications: Phosphorus oxychloride is mainly used in the production of aliphatic and aromatic phosphoric acid esters.

Applications:

mainly in the production of aliphatic and aromatic phosphoric acid esters

Phosphorus sulfochloride (thiophosphoryl chloride)

Phosphorus sulfochloride has not been produced in the USA since 1980. However, in Western Europe there are a number of manufacturers. It can be either produced by the reaction of phosphorus trichloride with sulfur in an autoclave at 180°C or by passing phosphorus trichloride vapor through molten sulfur:

$$PCl_3 + S \longrightarrow PSCl_3$$

$PSCl_3$

Manufacture:

$PCl_3 + S \rightarrow PSCl_3$

Catalysts such as e.g. aluminum chloride reduce the reaction temperature so much that the reaction can be carried out in phosphorus sulfochloride solution. Purification is by distillation.

Applications: Phosphorus sulfochloride is mainly used in the manufacture of thiophosphoric acid ester chlorides (starting materials for crop protection products).

Use:

Manufacture of thiophosphoric acid ester chlorides (starting materials for crop protection products)

Acids and Salts of Phosphorus with $P^{<5+}$

Hypophosphorous acid and hypophosphites

The USA sodium hypophosphite capacity was 8.6×10^3 t/a in 1985, that in Western Europe $> 9 \times 10^3$ t/a and that in Japan 2.8×10^3 t/a. Hypophosphites, the salts of hypophosphorous acid, are produced by reacting white phosphorus with boiling calcium hydroxide, according to the equation:

$$2P_4 + 3Ca(OH)_2 + 6H_2O \longrightarrow 2PH_3 + 3Ca(H_2PO_2)_2$$

$Ca(H_2PO_2)_2$

Manufacture:

$2P_4 + 3Ca(OH)_2 + 6H_2O \rightarrow 2PH_3 + 3Ca(H_2PO_2)_2$

Hydrogen and calcium phosphite are formed as byproducts. The analogous reaction with sodium hydroxide

H_3PO_2

Manufacture:

from the calcium salt and sulfuric acid

Hypophosphite mainly used for electroless nickel plating

yields sodium hypophosphite. Calcium hypophosphite reacts with sulfuric acid to produce hypophosphorous acid.

Applications: More than 95% of the sodium hypophosphite is used for electroless nickel plating (Kanigen® Process).

Phosphorous acid

H_3PO_3

Manufacture:

$PCl_3 + 3H_2O \rightarrow H_3PO_3 + 3HCl$

In the USA in 1984, 7.7×10^3 t of phosphorus trichloride was consumed in the manufacture of acid chlorides which theoretically should yield 4.6×10^3 t of phosphorous acid as a byproduct. There are a number of companies in the USA which process this product to pure phosphorous acid. Much more than 6×10^3 t of phosphorous acid is produced by the hydrolysis of phosphorus trichloride. In a process developed by Hoechst AG (FRG) phosphorous acid is efficiently produced by spraying phosphorus trichloride with excess steam, in which the hydrogen chloride formed and the excess water are driven off by the heat of reaction:

$$PCl_3 + 3H_2O \longrightarrow H_3PO_3 + 3HCl$$

Impure phosphorous acid is formed when phosphorus trichloride is used in the manufacture of organic acid chlorides.

Applications:

in the manufacture of

- Basic lead phosphite
- Phosphonic acids

and as a reducing agent

Applications: Phosphorous acid is used in the manufacture of basic lead phosphite (PVC stabilizer), aminomethylene phosphonic acids and hydroxyethane diphosphonic acid (Sec. 1.5.2.2.5) and as a reducing agent.

References for Chapter 1.5.1: Phosphorus and Inorganic Phosphorus Compounds

Reviews:

J. R. van Wazer: *Phosphorus and its Compounds, Vol. 1 – Chemistry, Vol. 2 – Technology, Biological Functions and Applications.* Interscience Publishers, Inc., New York-London (1958/1961).

Phosphoric Acid. Part I and II, A. V. Slack (ed.), Marcel Dekker Inc., New York (1968).

R. Noyes: *Phosphoric Acid by the Wet Process,* Noyes Development Corp., Park Ridge, N. J., USA-London (1967).

Kirk-Othmer: *Encyclopedia of Chemical Technology.* 3rd Ed., Vol. 17, pp 426–539, John Wiley & Sons, Inc., New York-Chichester-Brisbane-Toronto-Singapore (1982).

Ullmanns Encyklopädie der technischen Chemie. 4th Ed., Vol. 18, Verlag Chemie, Weinheim-Deerfield Beach-Basel (1979), pp 283–300: "Phosphor" (Phosphorus), pp 301–364: "Phosphoroxide, Phosphorsäuren, Phosphate" (Phosphorus oxides, phosphorus acids, phosphates), pp 365–375: "Phosphor-Verbindungen, anorganische, sonstige" (Miscellaneous inorganic phosphorus compounds).

Apatite as a Raw Material:

Chemical Economics Handbook, Stanford Research Institute, Menlo Park, California, USA: Phosphate Rock (Jan. 1986).

Phosphoric Acid:

Commercial Information:

Inf. Chim., No. **271**, pp 249–263 (April 1986).

Chemical Economics Handbook, Stanford Research Institute, Menlo Park, California, USA: Industrial Phosphates (July 1986).

Chemical Profile, Phosphoric Acid, Schnell Publishing Company Inc. (Jan 1, 1984).

Technical Information:

"Phosphoric acid process proven for large-capacity plants", *Chemical Engineering,* pp 66–67 (Sept. 6, 1982).

"Thermische und Naß-Phosphorsäure konkurrieren am Markt für technische Phosphorsäure" (Furnace and Wet-process Phosphoric Acid Compete for Technical Phosphoric Acid Market), *Chem. Industrie* **XXXVI,** pp 265–266 (May 1984).

Chem. Week, p. 25 (Feb. 20, 1985).

Chem. Eng. News, p. 20 (April 22, 1985).

Phosphoric Acid Salts:

Commercial Information:

Chemical Economics Handbook, Stanford Research Institute, Menlo Park, California, USA: Industrial Phosphates (July 1986).

Chemical Profile, Sodium Tripoly, Schnell Publishing Company, Inc. (April 1, 1984).

Technical Information:

"STPP: The Controversy Continues", *Phosphorus & Sulfur* **129,** pp 23–27 (Jan.–Feb. 1984).

"Agrico's new wrinkle to make MAP", *Chem. Week,* pp 20–21 (Nov. 11, 1982).

Elemental Phosphorus:

Commercial Information:

Chemical Economics Handbook, Stanford Research Institute, Menlo Park, California, USA: Phosphorus and Phosphorus Chemicals (Feb. 1986).

Chemical Profile, Phosphorus, Schnell Publishing Company, Inc. (July 1, 1985).

Technical Information:

G. Breil, "Elektrothermische Erzeugung von Phosphor" (Electrothermal Production of Phosphorus), *Chem.-Ing.-Tech.* **42,** pp 213–218 (1970).

DOS 2 907 059 (Hoechst AG) (red phosphorus).

Phosphorus Chlorides, Oxides and Sulfides:

Commercial Information:

Chemical Economics Handbook, Stanford Research Institute, Menlo Park, California, USA: Phosphorus and Phosphorus Chemicals (Feb. 1986).

Chemical Profile, Phosphorus Oxychloride, Phosphorus Pentasulfide, Phosphorus Trichloride, Schnell Publishing Company, Inc. (July 1, 1985).

Sodium Hypophosphite, Phosphorous Acid:

Commercial Information:

Chemical Economics Handbook, Stanford Research Institute, Menlo Park, California, USA: Phosphorus and Phosphorus Chemicals (Feb. 1986).

Technical Information:

Chem. Week (Aug. 24, 1983).

1.5.2 Organic Phosphorus

The following classes of substances will be considered:

- Esters of phosphoric and thiophosphoric acid and their derivatives
- Phosphorous acid esters
- Phosphonic acids

1.5.2.1 Economic Importance

Estimated 1986 capacities for triaryl phosphoric acid esters:

- Europe, 50×10^3 t/a
- USA, 80×10^3 t/a
- Japan, 15 to 20×10^3 t/a

Relatively little commercial information is available about organic phosphorus compounds. The 1986 capacities for the triaryl esters of phosphoric acid, the class of products with the largest turnover in quantity, were estimated to be 50×10^3 t/a in Europe, 80×10^3 t/a in the USA and 15 to 20×10^3 t/a in Japan.

USA production of phosphorus-containing insecticides:

- 1982: 55×10^3 t
- 1987: 60 to 70×10^3 t (estimated)

Production of phosphorus-containing herbicides (excl. Glyphosate)in the USA:

- 1983: 2.7×10^3 t
- 1987: 2.3 to 4.5×10^3 t (estimated)

55×10^3 t of phosphorus-containing insecticides (calculated on the basis of 100% active ingredients) were produced in the USA in 1982, with 60 to 70×10^3 t being predicted for 1987. The corresponding figures for phosphorus-containing herbicides are 2.7×10^3 t in 1983, with 2.3 to 4.5×10^3 t being predicted for 1987. These figures do not include the most important herbicide, Glyphosate (phosphonomethylglycine).

1983 production of trialkyl phosphates in the USA

- 6×10^3 t

USA production of triaryl phosphites:

- 1983: 5.4×10^3 t
- 1987: 6.9×10^3 t (estimated)

6×10^3 t of trialkyl phosphates and 5.4×10^3 t of triaryl phosphites were produced in the USA in 1983.

1.5.2.2 Manufacture of Organic Phosphorus Compounds

1.5.2.2.1 Phosphoric Acid Triesters

Triaryl phosphates

Triaryl phosphates:

- Triphenyl phosphate
- Diphenyl-cresyl phosphate
- Tricresyl phosphate
- Mixed (isopropyl-phenyl)-phenyl phosphates

Triphenyl phosphate, diphenyl-cresyl phosphate, tricresyl phosphate and mixed (2-propyl-phenyl)-phenyl phosphates belong to this class of substances. They are manufactured by reacting phosphorus oxychloride with a slight excess of the corresponding phenols or phenol mixtures above 140 °C:

Manufactured by reaction of phosphorus oxychloride with phenols in the presence of catalysts above 140 °C

$$3\ R{-}C_6H_4{-}OH + POCl_3 \longrightarrow P(O)(O{-}C_6H_4{-}R)_3 + 3\ HCl$$

R: H, CH_3, $(CH_3)_2CH$

Ortho-cresol may not be used for toxicological reasons.

Mixed (2-propyl-phenyl)-phenyl phosphates are produced by first alkylating phenol with propene, then reacting the resulting phenol mixtures with phosphorus oxychloride in the presence of catalysts (potassium or magnesium salts). As the conversion increases the temperature increases driving off the hydrogen chloride formed. The excess phenol is distilled off at the end of the reaction. The phosphoric acid esters can be purified by distillation. Final traces of phenol and ester acids can be removed by washing with very dilute sodium hydroxide.

Diaryl-alkyl phosphates

Diphenyl-butyl-, diphenyl-(2-ethylhexyl)- and diphenyl-isodecyl-phosphates are produced commercially. Manufacture is in two steps: first the alcohol is converted to the alkyl ester dichloride by reaction with excess phosphorus oxychloride:

Manufacture of diaryl-alkyl phosphates:

1st step: Reaction of the alcohol with phosphorus oxychloride to the ester dichloride $ROP(O)Cl_2$

2nd step: Reaction of the ester dichloride with aqueous sodium phenolate solution to the triester

$$ROH + POCl_3 \rightarrow ROP(O)Cl_2 + HCl$$

R: butyl, 2-ethylhexyl, isodecyl

After distilling off the excess phosphorus oxychloride, the alkyl ester dichloride is then converted to the triester by reaction with an aqueous solution of sodium phenolate:

$$ROP(O)Cl_2 + 2NaOC_6H_5 \longrightarrow ROP(O)(OC_6H_5)_2 + 2NaCl$$

Trialkyl phosphates

These are manufactured by reacting phosphorus oxychloride with excess alcohol, in particular ethanol, butan-1-ol, butan-2-ol and 2-ethyl-hexan-1-ol:

$$(3 + n)ROH + POCl_3 \rightarrow P(O)(OR)_3 + (nROH \cdot 3\,HCl)$$

R: ethyl, butyl, isobutyl, 2-ethylhexyl

Trialkyl phosphates manufactured by reacting phosphorus oxychloride with excess alcohol and then neutralizing with aqueous alkali

An excess of alcohol is necessary to suppress the hydrochloric acid-induced cleavage of the triester formed. The product is worked up by neutralizing with aqueous alkali and phase separation. (In the case of triethyl phosphate an extraction agent has to be used, due to its solubility in water.) The excess alcohol is distilled off from the organic phase and triethyl- and tributylphosphates are purified by distillation. Dibutyl- and di-(2-ethylhexyl)phosphates, which are formed in small quantities as byproducts, separate from the aqueous phase upon acidification.

Tri-(chloroalkyl) phosphates manufactured from phosphorus oxychloride and oxiranes in the presence of catalysts

Phosphoric acid chloroalkyl esters are obtained by reacting oxiranes with phosphorus oxychloride in the presence of catalysts (aluminum trichloride, titanium tetrachloride) in an exothermic reaction:

$$POCl_3 + 3\,RCH{-}CH_2\ (\text{epoxide, ring O}) \longrightarrow P(O)(OC(H)(R){-}CH_2Cl)_3$$

$R = H, CH_3, CH_2Cl$

The resulting esters can only be worked up by washing, first with acid to remove the catalyst and then with alkali to remove the ester acids, since they decompose easily at high temperatures.

1.5.2.2.2 Phosphorus(V) Ester Acids

Phosphorus(V) ester acids manufactured:

- As byproducts in the manufacture of aliphatic triesters
- From phosphorus pentoxide and alcohols
- From poly(phosphoric acid) and alcohols

The types of ester acids available are:

- Diesters, obtained, for example, as byproducts in the manufacture of trialkyl esters (see above)
- Mixtures of mono- and diesters by the reaction of phosphorus pentoxide with alcohols:

$$P_2O_5 + 3\,ROH \longrightarrow (RO)_2P(O)OH + ROP(O)(OH)_2$$

- Mixtures of monoesters with free phosphoric acid by reacting alcohols with poly(phosphoric acid).

A large number of different compounds are manufactured using alcohols from methanol to octadecanol. Ester acids are also marketed as their salts e.g. with aminoalcohols.

1.5.2.2.3 Thiophosphoric Acid Derivatives

Dithiophosphoric acid diesters are obtained by reacting phosphorus pentasulfide with alcohols or phenols:

$$P_2S_5 + 4\,ROH \rightarrow 2(RO)_2P(S)SH + H_2S$$

R: CH_3, C_2H_5, C_nH_{2n+1} ($n \geq 5$), cresyl

Manufacture of thiophosphoric acid derivatives:

- Dialkyl dithiophosphoric acids, $(RO)_2$-P(S)SH, from phosphorus pentasulfide and alcohols or phenols
- Dialkyl thiophosphoric acid chloride $(RO)_2P(S)Cl$ from dialkyl dithiophosphoric acid by chlorination or from phosphorus sulfochloride and sodium alcoholate

This exothermic reaction has to be carried out with cooling.

Thiophosphoric acid ester chlorides are important starting materials in the production of crop protection agents and are obtained by the chlorination of dialkyl dithiophosphoric acids (R: methyl, ethyl):

$$(RO)_2P(S)SH + 1.5\,Cl_2 \longrightarrow (RO)_2P(S)Cl + 0.5\,S_2Cl_2 + HCl$$

R: CH_3, C_2H_5

They can also be obtained from phosphorus sulfochloride:

$$PSCl_3 + 2\,NaOR \longrightarrow (RO)_2P(S)Cl + 2\,NaCl$$

1.5.2.2.4 Phosphorous Acid Esters

Triaryl phosphites

These are obtained, analogously with the phosphoric acid esters, by reacting phosphorus trichloride with phenols:

Manufacture of triaryl phosphites:

from phosphorus trichloride and phenols

$$PCl_3 + 3\,R\text{-}C_6H_4\text{-}OH \longrightarrow P(O\text{-}C_6H_4\text{-}R)_3 + 3\,HCl$$

R: mainly H, $i\text{-}C_9H_{19}$

Triphenyl phosphite is purified by distillation.

Trialkyl phosphites

Manufacture of trialkyl phosphites:

from phosphorus trichloride and alcohols in the presence of bases

Due to its extreme sensitivity to hydrochloric acid, the synthesis of trialkyl phosphites can only be carried out in the presence of molar quantities of inorganic or organic bases:

$$PCl_3 + 3\,ROH + 3\,B \rightarrow P(OR)C_3 + 3\,B \cdot HCl$$

R: mainly CH_3, C_2H_5;
B: ammonia, aniline, dimethylamine etc.

The reaction is strongly exothermic and cooling is necessary. The processes operated differ in their reaction conditions (solvent, temperature), the base used and in the working up of the reaction mixture. The esters can be purified by distillation.

Manufacture of tri(chloroalkyl) phosphites:

from phosphorus trichloride and oxiranes

Chloroalkyl esters are obtained by reacting phosphorus trichloride with oxiranes, if necessary in the presence of catalysts:

$$PCl_3 + 3\,RCH\underset{O}{-}CH_2 \longrightarrow P(O\overset{H}{\underset{R}{C}}-CH_2Cl)_3$$

R: mainly H_3, CH_3

Dialkyl phosphites

Manufacture of dialkyl phosphites:

from phosphorus trichloride and alcohols in the absence of bases

Dialkyl phosphites are obtained by reacting alcohols with phosphorus trichloride in the absence of bases:

$$PCl_3 + 3\,ROH \rightarrow (RO)_6P(O)H + RCl + 2\,HCl$$

R: mainly CH_3, C_2H_5

In this strongly exothermic reaction, the hydrogen chloride must be removed as rapidly as possible to prevent dealkylation of the esters formed. The dialkyl phosphites can be purified by distillation.

1.5.2.2.5 Phosphonic Acids

Four types of phosphonic acids are produced commercially:

- (Aminoalkane)poly(phosphonic acids)
- (Hydroxyalkane)diphosphonic acids, in particular 1-hydroxyethane-1,1-diphosphonic acid
- Phosphonocarboxylic acids, in particular 2-phosphonobutane-1,2,4-tricarboxylic acid
- Phosphonomethylglycine

(Aminoalkane)poly(phosphonic acids) are obtained by reacting ammonia or amines with formaldehyde and phosphorous acid in concentrated hydrochloric acid:

(Aminoalkane)poly(phosphonic acids) manufactured:

from ammonia or amines with formaldehyde and phosphorous acid in hydrochloric acid solution

$$NH_3 + 3CH_2O + 3H_3PO_3 \xrightarrow{HCl} N[CH_2P(O)(OH)_2]_3 + 3H_2O$$

$$H_2NCH_2CH_2NH_2 + 4CH_2O + 4H_3PO_3 \xrightarrow{HCl} [(HO)_2P(O)CH_2]_2NCH_2CH_2N[CH_2P(O)(OH)_2]_2 + 4H_2O$$

They can be purified by crystallization.

(Hydroxyethane)diphosphonic acid is produced by reacting acetic anhydride with phosphorous acid:

1-hydroxyethane-1,1-diphosphonic acid manufactured:

from phosphorous acid and acetic anhydride

$$2H_3PO_3 + (CH_3CO)_2O \longrightarrow CH_3C(OH)[P(O)(OH)_2]_2 + CH_3COOH$$

The reaction proceeds by way of intermediates.

2-phosphonobutane-1,2,4-tricarboxylic acid is manufactured in a three step process: first the phosphonosuccinic acid ester is prepared from dimethyl maleate and dimethyl phosphite in the presence of a sodium methylate catalyst:

Manufacture of 2-phosphonobutane-1,2,4-tricarboxylic acid:

1st step: Production of phosphonosuccinic acid ester from dimethyl phosphite and dimethyl maleate

2nd step: Reaction with methyl acrylate to pentamethyl ester of phosphonobutane-tricarboxylic acid

3rd step: Saponification to the free acid

$$\begin{matrix} HC{-}COOCH_3 \\ \| \\ HC{-}COOCH_3 \end{matrix} + (CH_3O)_2P(O)H \xrightarrow{[NaOCH_3]} (CH_3O)_2P(O){-}\overset{H}{\underset{|}{C}}{-}COOCH_3 \quad H_2C{-}COOCH_3$$

Reaction of this ester with methyl acrylate in the presence of catalytic quantities of sodium methylate then yields the pentamethyl ester of phosphonobutanetricarboxylic acid:

$$(CH_3O)_2P(O)\overset{H}{\underset{H_2C-COOCH_3}{C}}-COOCH_3 + CH_2{=}CHCOOCH_3 \xrightarrow{[NaOCH_3]}$$

$$(CH_3O)_2P(O)\overset{CH_2CH_2COOCH_3}{\underset{H_2C-COOCH_3}{C}}-COOCH_3$$

Saponification finally produces the free acid.

Various processes have been described for the manufacture of phosphonomethylglycine.

1.5.2.3 Applications of Organic Phosphorus Compounds

1.5.2.3.1 Phosphoric Acid Triesters

Uses for triaryl phosphates:

- Mainly as flame-retarding plasticizers
- Hydraulic fluids

Triphenyl phosphate is added to cellulose acetate and poly(phenylene oxide) to endow them with flame-retarding properties. Tricresyl phosphate, triphenyl phosphate and (2-propyl-phenyl)-phenyl phosphate mixtures are used as plasticizers in poly(vinylchloride) for the same reason. These relatively expensive compounds compete with mixtures of phthalic acid esters (as plasticizers) and antimony trioxide (as flame-retarding agent). Diphenyl-(2-ethylhexyl) phosphate is used as a plasticizer in poly(vinylchloride) utilized for wrapping food. Diphenyl-butyl phosphate is used as a hydraulic fluid.

Uses for trialkyl phosphates:

- In many sectors

Triethyl phosphate is used as a catalyst in the manufacture of ketene from acetic acid and for reducing the decomposition tendency of hydroperoxides.

Tributyl phosphate [in common with tri(2-butyl) phosphate] is used as an antifoaming agent e.g. in the paper industry. It is also used as a hydraulic fluid, as an extraction agent in the Purex process for uranium and plutonium and for the extraction of pure phosphoric acid from wet-process acid. Tris(2-ethylhexyl) phosphate is a plasticizer for poly(vinylchloride) and serves as a cosolvent in the anthraquinone process for the manufacture of hydrogen peroxide. Chlorine-containing esters, particularly tris-(2-chloroethyl)- and tris-(2-chloroisopropyl)-phosphate, are incorporated into polyurethane foams to reduce their flammability.

1.5.2.3.2 Phosphorus(V) Ester Acids

Phosphorus(V) ester acids have many uses as the following examples illustrate:

- Di-(2-ethylhexyl)-phosphate is used as an extraction agent in the recovery of uranium from wet-process acid and of zinc from aqueous solutions and in the form of its alkanolamino salt as an antistatic agent in textiles.
- Mixtures of mono- and dibutylphosphate are present in industrial cleaning agents.
- Monooctadecyl phosphate is used, as its triethanolamine salt, in, for example, dishwashing liquids.

Applications for phosphorus(V) ester acids:

- Extraction agents
- Cleaning agent additives
- Antistatic additives

1.5.2.3.3 Thiophosphoric Acid Derivatives

The lower dialkyl dithiophosphoric acids and the thiophosphoric acid diester chlorides are important starting materials for crop protection agent manufacture.

Typical examples are the manufacture of Malathion® by the reaction of O,O-dimethyl dithiophosphoric acid with dimethyl maleate:

$$(CH_3O)_2P(S)SH + \begin{matrix} HC{-}COOC_2H_5 \\ \| \\ HC{-}COOC_2H_5 \end{matrix}$$

$$\xrightarrow{\text{Alkali}} (CH_3O)_2P(=S)S{-}\begin{matrix} CH{-}COOC_2H_5 \\ | \\ CH_2{-}COOC_2H_5 \end{matrix}$$

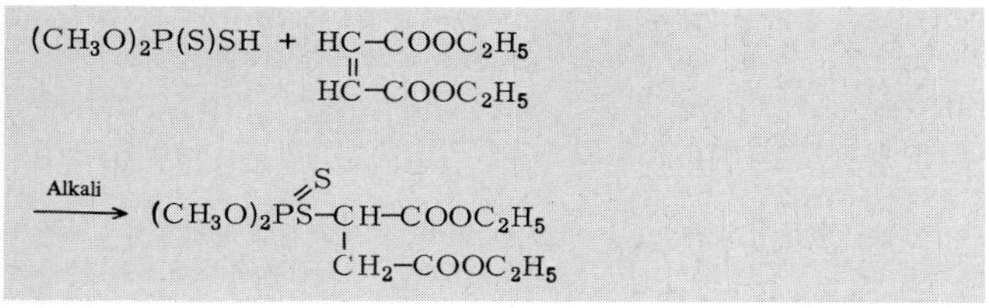

and the manufacture of Parathion(E 605®) by the reaction of the acid chloride of O,O-diethyl dithiophosphoric acid with sodium p-nitrophenolate:

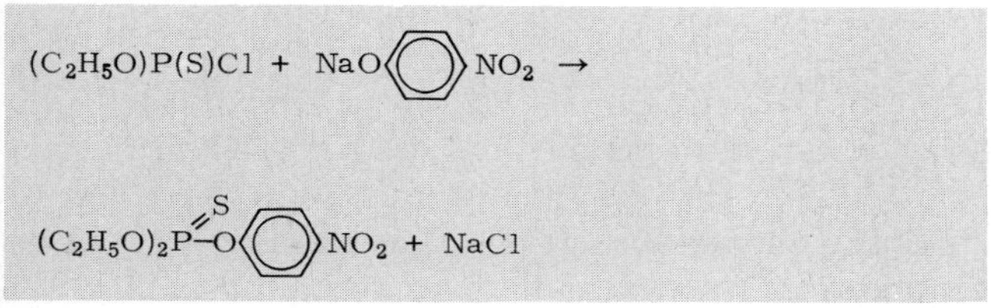

$$(C_2H_5O)P(S)Cl + NaO{-}C_6H_4{-}NO_2 \rightarrow$$

$$(C_2H_5O)_2P(=S){-}O{-}C_6H_4{-}NO_2 + NaCl$$

Applications for thiophosphoric acid derivatives:

- Starting materials for crop protection agent manufacture
- Lubricating oil additives

The zinc salts of higher dialkyl dithiophosphoric acids are utilized in lubricating oils.

1.5.2.3.4 Phosphorous Acid Esters

Aromatic phosphorous acid esters: antioxidants, stabilizers

Triphenyl phosphite and tris-(nonyl-phenyl) phosphite are used as antioxidants and stabilizers in plastics, rubber and lubricating oils.

Aliphatic phosphorous acid esters: starting materials for insecticides and veterinary products

Trimethyl- and triethylphosphites, as well as dimethyl- and diethyl-phosphites, are important starting materials for the manufacture of insecticides and veterinary products.

Phosdrin®, a phosphoric acid enol-ester, is produced by the reaction of trimethyl phosphite with methyl chloroacetoacetate:

$$P(OCH_3)_3 + H_3C{-}CO{-}\underset{\displaystyle Cl}{\underset{|}{C}}H{-}COOCH_3 \longrightarrow$$

$$(CH_3O)_2P(O)O{-}\underset{\displaystyle CH_3}{\underset{|}{C}}{=}CHCOOCH_3 + CH_3Cl$$

DDVP is produced with trichloroacetaldehyde:

$$P(OCH_3)_3 + Cl_3CCHO \longrightarrow (CH_3O)_2P(O)O\text{-}CH{=}CCl_2 + CH_3Cl$$

Dipterex® is produced from dimethyl phosphite and trichloroacetaldehyde:

$$(CH_3O)_2P(O)H + Cl_3CCHO \longrightarrow (CH_3O)_2P(O)CH(OH)CCl_3$$

Dimethyl phosphite is the starting material for the manufacture of phosphonobutanetricarboxylic acid.

Tris-(chloroethyl) phosphite can be transformed into chloroethanephosphonic acid di-(2-chloroethyl) ester from which the reactive flame protection agent, vinylphosphonic acid di-(2-chloroethyl) ester, can be produced by hydrogen chloride elimination.

1.5.2.3.5 Phosphonic Acids

(Aminoalkyl)phosphonic acids, 1-hydroxyethane-1,1-diphosphonic acid and 2-phosphonobutane-1,2,4-tricarboxylic acid are mainly utilized in substoichiometric quantities (threshold effect) to prevent the precipitation of calcium carbonate from water. These acids are also used in industrial cleaning agents, cooling towers, the extraction of crude oil etc., as well as in the manufacture of phosphonomethylglycine (Roundup®), a nonselective herbicide.

Applications of phosphonic acids:

- Prevention of $CaCO_3$ precipitation from water
- In technical cleaning agents
- In crude oil extraction
- Manufacture of phosphonomethylglycine, a nonselective herbicide

References for Chapter 1.5.2: Organic Phosphorus Compounds

Reviews:

Ullmanns Encyklopädie der technischen Chemie. 4th Ed., Vol. 18, pp 377–398, Verlag Chemie, Weinheim-Deerfield Beach-Basel (1979).

Kirk-Othmer: *Encyclopedia of Chemical Technology*, 3rd. Ed. Vol. 17, pp 490–539, John Wiley & Sons, New York (1982).

Commercial Information:

Chemical Economics Handbook, Stanford Research Institute, Menlo Park, California, USA:

- Insecticides (Jan. 1984)
- Herbicides (Dec. 1985)
- Plasticizers (Sept. 1984)
- Plastic Additives (June 1985)

1.6 Sulfur and Sulfur Compounds

1.6.1 Sulfur

1.6.1.1 Occurrence

Sulfur is widely distributed in the Earth's crust. Elemental sulfur is found in sedimentary deposits (domed deposits of rock salt, gypsum and limestone, mainly in the Gulf of Mexico, Sicily and Poland) or in deposits of volcanic origin (e.g. in East Asian basins along the Philippines and Japan).

Deposits as:

- Elemental sulfur in sedimentary or volcanic deposits
- Sulfates (gypsum)
- Sulfides (inorganic or organic in ores, coal, crude oil and natural gas)
- H_2S in natural gas

In nonelemental form it is found in sulfates (gypsum), in sulfidic ores (e.g. pyrites, copper, zinc, lead, nickel and cobalt sulfides) and in fossil fuels in which it is bonded to both hydrogen and carbon and particularly in coal as inorganic or organic sulfides. Sulfur deposits in natural gas and crude oil are becoming increasingly important as a source of raw materials for sulfur compounds. The known world reserves of sulfur are ca. 2.5×10^9 t.

1.6.1.2 Economic Importance

Most important sources of sulfur:

- Natural gas (contains up to 30% by weight H_2S)
- Frasch sulfur
- Pyrites

together with metallurgy, crude oil, coal, gypsum, tar sands, flue gases etc.

Sulfur is one of the most important raw materials for the chemical industry. It is used as its compounds (particularly sulfuric acid) in many chemical and industrial processes (particularly in the fertilizer industry).

In 1985 54×10^6 t of sulfur was produced worldwide in all forms (Table 1.6.1-1). Elemental sulfur from natural gas and crude oil predominates with 38% (largest producer: Canada), followed by Frasch-sulfur and mined sulfur with 27.8%, sulfur from pyrites and roasting gases from nonferrous metal sulfides with ca. 18%, the rest being distributed among sulfur from hydrogen sulfide, tar sands, gypsum and to an increasing extent sulfur dioxide-containing flue gases (should amount to 10^6 t by 1990).

Table 1.6.1-1. 1985 World Sulfur Production (in 10^6 t).

Country Sources	World	USA	USSR	Canada	Poland	PR China	Japan
Sulfur – all forms	54.0	11.4	9.7	6.7	5.1	2.9	2.5
– Frasch	12.3	5.0	0.9	–	4.4	–	–
– Native (mined)	2.7	–	1.7	–	0.5	0.3	–
– Pyrites	10.0	withheld	3.4	–	–	0.2	0.3
– Byproducts*	29.0 (N,M,C)	6.4 (C,N)	3.8 (N)	6.7 (N)	0.2 (M)	0.4	1.2 (M,C)

* Primarily: N = natural gas, M = metallurgy, C = crude oil

The present shortfall in production over demand in the Western World, currently 2.5×10^6 t sulfur, can most probably only be made up in the next 10 years by importation from Communist Countries (particularly the USSR and Poland).

1.6.1.3 Production

1.6.1.3.1 From Elemental Sulfur Deposits

Extraction:

- From elemental sulfur deposits
- By forcing out with hot water (Frasch process)

Elemental sulfur from deep sedimentary deposits or sulfur-containing lime in the caprocks of salt domes e.g. along the Gulf of Mexico, in Canada or in Iran is extracted using the Frasch process. In this process, superheated water at ca. 165 °C is piped down the outside ring of three concentric pipes to the sulfur-laden strata of the deposit. The sulfur is melted by the hot water and in molten form is forced up through the second ring by compressed air (at ca. 40 bar) fed down through the center pipe. This process can be carried out efficiently down to depths of 50 to 80 m, the yield of a single bore hole being up to 300 t sulfur/d.

If the sedimentary deposits lie near the Earth's surface, the sulfur ore can be mined. Sulfur is produced from sulfur ores by a smelting process in which heat is provided by burning sulfur (Sicily) or by distillation at temperatures above 440 °C. A combination of flotation with subsequent smelting is also operated (Poland).

- Mining with subsequent flotation, smelting or distillation

1.6.1.3.2 From Hydrogen Sulfide and Sulfur Dioxide

Another source of sulfur is the hydrogen sulfide contained in e.g. natural gas, refinery gases, synthesis gas and coking oven gases. It is extracted from the gases by physical and chemical scrubbing (e.g. with ethanolamine), concentrated by regenerating the absorption liquid and converted into elemental sulfur by the widely used Claus process.

- From H_2S (natural gas, crude oil, coal): oxidation by the Claus process

In this process hydrogen sulfide is converted into elemental sulfur by reaction with stoichiometric quantities of air/oxygen and heating gas in an exothermic reaction yielding 60 to 70% sulfur:

$$H_2S + 1.5O_2 = SO_2 + H_2O$$
$$2\,H_2S + SO_2 = 3S + 2H_2O$$
$$3\,H_2S + 1.5O_2 = 3S + 3H_2O, \quad \Delta H = -664\ \text{kJ/mol}$$

After cooling the reaction gases from the combustion chamber to ca. 300 °C in waste heat boilers, the conversion is continued in a reactor filled with an alumina-supported cobalt-molybdenum catalyst giving an overall yield of up to 85%. After condensing the sulfur formed at temperatures below 170 °C, the reaction gases are converted to sulfur in a second (and if necessary a third) reactor at 220 °C on highly reactive alumi-

Claus process:

- Combustion chamber
- Waste heat boiler
- Two catalyst-filled reactors

num oxide with a total yield of 96 or 98% sulfur. The sulfur content in the waste gases must be reduced to ca. 10 ppm by scrubbing.

COPE process:

hydrogen sulfide combustion gas contains 80 to 100% oxygen

In contrast with the classic Claus process in which the oxygen content in the combustion air cannot exceed 30%, it can be increased to 80 to 100% in the COPE process (**C**laus **O**xygen-based **P**rocess **E**xpansion) introduced in large scale plants in 1985. Overheating of the combustion chamber is avoided by mixing with recycled gas from the first sulfur condenser. The COPE process reduces costs by increasing the plant capacity and using cheaper sulfur-rich crude oils. In addition the quantity of waste gases can be reduced by ca. 35% compared with the Claus process.

1.6.1.3.3 From Pyrites

- From pyrites

  by heating in the absence of air (Outokumpu process)

Heating iron pyrites to ca. 1200 °C in the absence of air (Outokumpu process) yields sulfur and liquid iron(II) sulfide.

1.6.1.4 Applications

Applications:

- 85 to 90% for sulfuric acid manufacture
- Rest for production of:
  - SO_2, SO_3, CS_2 and P_2S_5;
  - rubber vulcanization agents,
  - gunpowder, sulfur dyes,
  - putties, sulfur concrete,
  - paper.

Most of the elemental sulfur (ca. 85 to 90%) is consumed in the manufacture of sulfuric acid. Elemental sulfur is also utilized for the manufacture of sulfur dioxide, sulfur trioxide, vulcanization agents, sulfur-containing concrete, gunpowder and fireworks, carbon disulfide, phosphorus pentasulfide, sulfur dyes and rubber vulcanization agents as well as in pharmaceuticals.

1.6.2 Sulfuric Acid

1.6.2.1 Economic Importance

Sulfuric acid is, despite a series of recent substitutions in large scale processes (e.g. TiO_2 manufacture, see Sec. 5.9.2.2.3), still the most important raw material in the chemical industry with steadily increasing world production figures (see Table 1.6.2-1).

Table 1.6.2-1. World Production of Sulfuric Acid 1975–1986 (in 10^6 t 100% H_2SO_4).

	World	USA/ Canada	Latin America	West. Europe	East. Europe	Africa	Asia/ Oceania
1975	109.2	32.1	4.1	25.1	28.1	3.8	16.0
1980	142.0	44.4	6.1	27.8	32.6	8.6	22.5
1986	145.3	34.8	7.7	24.8	38.5	13.1	26.5

1986 sulfuric acid production in 10^6 t 100% H_2SO_4:

- Western World 99.0
- USA/Canada 34.8
- Western Europe 24.8
- Eastern Europe 38.5

strong growth in Latin America and Africa

Regionally the evolution in sulfuric acid production is very different. In Western Europe a decline in production is predicted, whereas further high growth rates are expected in Latin America and Africa. The USA (with 31.2×10^6 t in 1986) and the USSR (with 27.9×10^6 t) produce the highest quantities followed by the People's Republic of China, Japan, France, West Germany and Canada.

1.6.2.2 Starting Materials for Sulfuric Acid Manufacture

H_2SO_4 manufactured from

- SO_2 (fresh acid)
- Waste sulfuric acid

Sulfuric acid is manufactured by oxidizing sulfur dioxide to sulfur trioxide and then reacting the sulfur trioxide with water:

$$SO_2 + 1/2\,O_2 \longrightarrow SO_3 \xrightarrow[\text{(in } H_2SO_4)]{+\ H_2O} H_2SO_4$$

In addition to this "fresh acid" sulfuric acid is produced in increasing quantities by working up waste acid (Sec. 1.6.2.2.2).

1.6.2.2.1 Sulfuric Acid from Sulfur Dioxide

Manufacture of SO_2 by:

- Combustion of S or H_2S
- Roasting of metal sulfides
- Decomposition of waste sulfuric acid and metal sulfates

The starting material, sulfur dioxide, is produced by the combustion of elemental sulfur or hydrogen sulfide or the roasting of pyrites, copper, zinc or lead sulfide or (for environmental protection reasons) the decomposition of waste sulfuric acid or metal sulfates.

Worldwide the most important raw material for sulfuric acid production is elemental sulfur (from the Claus process, the Frasch process or mined from sedimentary deposits). In the USA ca. 90% was produced in 1985 in this way. Waste sulfuric acid from e.g. ni-

Proportions of raw materials in sulfuric acid production in 1985 in %:

	W. Europe	USA*
Sulfur	47	82
Pyrites	35	2
CuS		7
ZnS	17	1.5
PbS, MoS_2		1
Waste H_2SO_4	7	5.6

* estimated

Importance of the different raw materials for SO_2/H_2SO_4:

- Most important raw material is elemental sulfur
- Importance of pyrites has decreased in some regions (for economic and environmental reasons)
- Zinc blende, galena and iron pyrites (supply so-called smelter acid) holding their own
- Utilization of waste sulfuric acid increasing (particularly in West Germany)

Major pyrites processors:

USSR, P.R. China, Spain

tration reactions or oil refineries is becoming increasingly important.

The sulfur dioxide from pyrites (expressed as sulfur) has remained constant over the last few years at ca. 10^7 t, of which in 1985 the USSR accounted for 3.4×10^6 t and the People's Republic of China for 2.2×10^6 t. The Western World (with exploitable deposits in Spain, Portugal, Cyprus, Sweden, Japan and the USA) accounts for ca. 12% of the world production of sulfur dioxide from pyrites (corresponding to 4.4×10^6 t of sulfur). The changeover in the 1960's and 1970's to the environmentally cleaner and cheaper sulfur dioxide production from sulfur combustion plants is now almost complete and in recent years there has again been a slight increase in sulfur dioxide production from pyrites roasters.

Sulfuric acid formed during the smelting of zinc blende, galena and copper pyrites (smelter acid) is a secondary product, its production being strongly dependent upon the demand for the corresponding metal.

Manufacture of Sulfur Dioxide

The combustion of sulfur to sulfur dioxide is exothermic ($\Delta H = -297$ kJ/mol) and is carried out in a combustion chamber by spraying finely divided droplets of hot liquid sulfur (at ca. 145 °C) into dry air (as oxidizing agent). There are three plant types: with pressurized atomizers, binary burners and rotary spinning cups respectively. In the latter hot sulfur at 145 °C is passed down the inner wall of a beaker rotating at ca. 4000 rpm forming a film. The beaker is surrounded by a ring slit into which high velocity air is forced. The atomized sulfur film resulting from a combination of air speed and centrifugal forces is then combusted.

SO_2-production by:

- Combustion of liquid sulfur:
  - in a single stage process (with an overstoichiometric oxygen/air mixture)
  - in a two stage process producing gases free of nitrogen oxides

Single or two stage combustion processes are used. In the former an excess of air is used and in the latter only 90% of the air/oxygen mixture necessary for complete combustion is added in the first stage, the rest being added in the second stage. The advantage of two stage combustion is that the gases leaving the combustion chamber are almost free of nitrous gases.

The sulfur dioxide concentration in the emerging gases is adjusted with air to ca. 8 to 12%. The hot gases are cooled with waste heat boilers and steam superheaters to ca. 450 °C producing wet steam and superheated steam. Purification is unnecessary.

Sulfur dioxide is produced from *sulfidic ores* by roasting. Pyrites, for example, is combusted with air to sulfur dioxide and iron oxide at temperatures at or above 800 °C:

- Roasting of pyrites

$$2\,FeS_2 + 5.5\,O_2 \xrightarrow{\geq 800\,^\circ C} Fe_2O_3 + 4\,SO_2 \qquad \Delta H = -1660\ kJ/mol$$

These reactions are carried out in multiple hearth roasters, rotary kilns or more preferably in fluidized bed furnaces.

In multiple hearth roasters with up to 15 hearths the comminuted roasting charge is agitated with a stirring arm to renew the surface continuously and is transported from one hearth to another. The roasting gases leave the furnace with 9 to 12% sulfur dioxide at 700 to 750 °C and are cooled in a waste heat boiler producing steam.

– in multiple hearth roasters

In rotary kilns the pyrites and air are intimately mixed in countercurrent, the waste heat being dissipated by water-cooled pipes built into the kilns or converted into usable heat.

– in rotary kilns

In fluidized bed furnaces consisting of a cylindrical roasting shaft and a resting chamber, the combustion air enters from below through a grate and reacts at temperatures of 650 to 1100 °C with the sulfidic ore forming a fluidized bed largely consisting of calcined iron oxide. The hot combustion gases containing 14% sulfur dioxide and a little sulfur trioxide are cooled in a waste heat boiler yielding 1.5 t high pressure steam (40 bar) per t pyrites (containing 45 to 50% sulfur).

– in fluidized bed furnaces

Fluidized bed technology is also utilized for the roasting of zinc blende. Lead sulfides are roasted in traveling-grate sinter plants.

- Zinc blende roasted in fluidized bed furnaces
- Lead sulfides roasted in traveling-grate sinter plants

The hot sulfur dioxide-containing combustion gases have to be freed from catalyst poisons and flyash before further reaction to sulfur trioxide can take place. The gases emerging from the waste gas boiler are passed through a cyclone and electrostatic filter to remove any dust. Reaction of the sulfur trioxide present with condensed water to sulfuric acid must be avoided to prevent corrosion. Finally the hot (ca. 350 to 400 °C) combustion gases are cooled to 60 to 80 °C, freed from volatile impurities, particularly arsenic(III) oxide, in acid-resistant venturi scrubbers and dried with sulfuric acid. The resulting ca. 40 °C gas contains 12 to 14% by volume of sulfur dioxide.

Purification of combustion gases by:

- dust removal with a cyclone and an electrostatic filter
- scrubbing with jet scrubbers after cooling

Conversion of Sulfur Dioxide into Sulfuric Acid

The oxidation of sulfur dioxide to sulfur trioxide and its subsequent conversion to sulfuric acid is currently almost exclusively carried out using the contact process, in particular the double contact process. The lead chamber process is no longer important.

Fresh sulfuric acid:

almost exclusively produced by the contact process, in particular the double contact process

The *contact process* is based on the equilibrium between sulfur dioxide and its oxidation product sulfur trioxide:

$$SO_2 + 1/2\,O_2 \xrightleftharpoons{\text{catalyst}} SO_3,\ \Delta H = -99.0\ \text{kJ/mol}$$

This is only attained sufficiently rapidly in the presence of a catalyst. Since this equilibrium shifts in favor of the starting materials with increasing temperature, the process has to be carried out at as low a temperature as possible, this being limited by the operating temperature of the catalyst. A higher sulfur dioxide conversion can be obtained by lowering the concentration of the sulfur trioxide formed (double contact process) or by increasing the operating pressure to 5 bar (Ugine-Kuhlmann process).

SO_3 formation favored by

- As low temperature as possible (lowest operating temperature determined by the operating temperature of the catalyst)
- Lowering the SO_3-concentration
- Increased pressure

Currently vanadium(V) oxide catalysts are almost exclusively used. These in the cold state consist mainly of vanadium(IV) oxide sulfate ($VOSO_4$) and potassium disulfate coated onto a porous carrier (kieselgur or diatomaceous earth). Under operating conditions (410 to 440 °C) the active components are present in the form of a salt melt. The oxidation state changes of vanadium (IV/V) play a decisive role in the process.

Catalyst:

vanadium(V) oxide under operating conditions at 410 to 440 °C

The oxidation of sulfur dioxide to sulfur trioxide mostly proceeds in fixed bed reactors (catalyst converters) with four to five sieve trays containing loosely packed catalyst particles. The sulfur dioxide content of the reaction gas is first adjusted to ca. 10% by volume with dry air. It passes through the catalyst trays from the top to the bottom, being at ca. 410 °C when entering the first tray. The gas emerges from the first tray at ca. 620 °C and has to be cooled to 420 °C before entering the second tray.

Oxidation takes place in catalyst beds

Gas has to be cooled after passing through catalyst beds

In plants in which ca. 35 °C sulfur dioxide-containing gas from roasting processes is utilized, the cold gas is also used as a coolant for the built-in gas/gas-heat exchanger (cold gas process). In this way the gas is heated up to ca. 420 °C. In the case of sulfur dioxide-containing ca. 420 °C gas from sulfur combustion, however, it cannot be used as a coolant and the heat of combustion has therefore to be removed and utilized by evaporators or superheaters for the production of wet steam or superheated steam (e.g. 300 °C, 30 bar) respectively (hot gas process).

Cooling by:
- Heat transfer to cold SO_2 gases from the roasting process
- Steam production (if SO_2 from S combustion utilized)

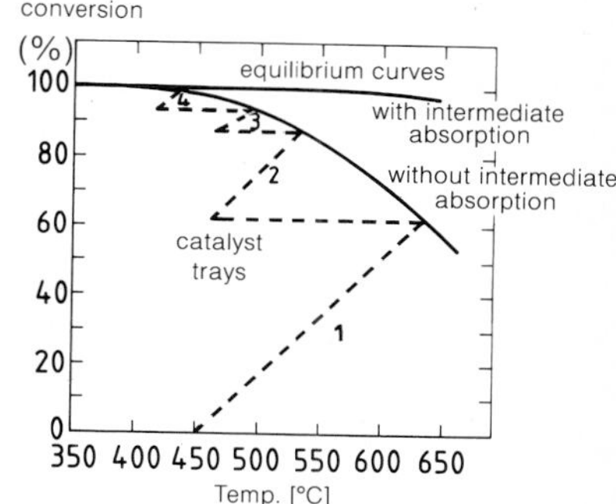

There are two processes for the manufacture of sulfur trioxide from sulfur dioxide on catalyst beds.

In the *single absorption process* (single contact process) the reaction gases are passed through four catalyst beds without intermediate sulfur trioxide absorption, with the gas being cooled to 450 °C after passing through the upper beds and 430 °C after passing through the lower ones. 60 to 63% of the sulfur dioxide has been converted into sulfur trioxide after the first bed, 89 to 90% after the second and a maximum of 97.5% after the fourth.

Higher sulfur dioxide conversions can be achieved using the *double absorption process* developed by Bayer AG. In this process sulfur dioxide is converted into sulfur trioxide as in the single absorption process, but after the third (sometimes after the second) bed the sulfur trioxide in the reaction gas (conversion 90 to 93% with respect to sulfur dioxide) is completely removed by absorption in 98.5 to 99% sulfuric acid. This is achieved by cooling the reaction gas to 180 to 200 °C with a gas cooler and then feeding it into a packed absorption tower in which the sulfur trioxide-containing gas is brought into contact with cooled sulfuric

SO_3 manufacture from SO_2 in catalyst beds:

- Single absorption process:
  - reaction gases passed through trays without intermediate absorption
  - maximum conversion with respect to sulfur dioxide: 97.5%
  - SO_3 after 4th bed absorbed with 98.5 to 99% H_2SO_4 or oleum
- Double absorption process:
  - after the 3rd bed SO_3 removed by absorption in 98.5 to 99% H_2SO_4

- after SO_3 intermediate absorption the residual SO_2 is converted in final contact with a 4th and optionally a 5th bed

acid. The acid is thereby heated up and has to be cooled with air or water coolers before it is returned to the absorption tower. The resulting ca. 1% sulfur dioxide-containing gas is heated up with the ca. 480 °C hot gas from the third tray and passed into the fourth and optionally fifth bed (final contact). This increases the total conversion to > 99.6% with respect to sulfur dioxide.

- yield with double absorption process (with respect to SO_2):

  > 99.6%

- SO_3 after last catalyst bed converted to sulfuric acid in 98.5 to 99% H_2SO_4 or absorbed in oleum

In both absorption processes the sulfur trioxide is, after passing through the final bed and being cooled to 180 to 200 °C, absorbed in 98.5 to 99% sulfuric acid. This occurs either in a countercurrent absorber tower or in a jet scrubber in which acid is sprayed downwards in cocurrent with the reaction gas. This absorption is accompanied by conversion of sulfur trioxide into sulfuric acid.

If the gas emerging from the beds is sprayed with oleum, oleum with a higher sulfur trioxide content is formed. Commercial products usually contain 20, 24, 27 or 35% by weight of sulfur trioxide. They have low freezing points and are therefore suitable for lengthier transportation (for oleum with higher sulfur trioxide contents see Sec. 1.6.4).

Reaction heat liberated in the individual process steps in modern H_2SO_4-plants largely utilized for steam generation

The production of sulfuric acid from elemental sulfur is exothermic in all the reaction steps. In total ca. 535 kJ/mol or 5.4 GJ/t 100% sulfuric acid is produced. Most of this, ca. 60%, is utilized for the production of superheated steam (e.g. 400 °C, 40 bar).

Purification of tail gas unnecessary with double absorption process

The emission of sulfur dioxide from sulfuric acid production plants can be considerably reduced with the double absorption process, making further purification of the tail gases unnecessary. In plants with higher sulfur dioxide contents desulfurization can be carried out by e.g. scrubbing with ammonia or aqueous sodium sulfite/sodium hydrogen sulfite (Wellmann-Lord process), adsorption on moist activated carbon (sulfacid process, Lurgi) and oxidative gas purification (e.g. oxidation with hydrogen peroxide or peroxomonosulfuric acid, peracidox process).

Purification processes for single absorption plants:

- NH_3-scrubbing
- Scrubbing with aqueous Na_2SO_3/$NaHSO_3$
- Adsorption on activated carbon
- Oxidation of SO_2 with H_2O_2 or H_2SO_5

The following less important sulfuric acid production processes are operated in addition to the classical fixed bed reactor process:

- Isothermal oxidation of sulfur dioxide on abrasion-proof catalyst beads in a fluidized bed reactor with built-in cooling grids
- The *moist gas catalysis process* in which wet sulfur dioxide from hydrogen sulfide combustion:

$$2\,H_2S + 3\,O_2 \longrightarrow 2\,SO_2 + 2\,H_2O$$

is converted into sulfuric acid in coking kilns. Due to the high water content only 78% acid can be produced (corrosion problems). This is mostly processed with ammonia from coking kiln cleaning to ammonium sulfate.

Other H_2SO_4 production processes:

- Fluidized bed reactor instead of a fixed bed reactor for the manufacture of SO_3
- Moist gas catalysis process for the oxidation of wet SO_2 (from H_2S), only 75 to 78% sulfuric acid produced

- The *nitrous process* (lead chamber and tower process) has decreased strongly in importance since the development of the catalyst converters. It is only operated in special cases e.g. in the purification of 0.5 to 3% by volume sulfur dioxide-containing waste gases (Ciba-Geigy process). The main disadvantage of this wet catalytic process in which the oxidizing agent is nitrosyl hydrogen sulfate is the low acid concentration attainable (78%).

- Nitrous process currently only used in special processes e.g. for tail gas desulfurization

$$SO_2 + 2\,NOHSO_4 + 2\,H_2O \rightarrow 3\,H_2SO_4 + 2\,NO$$

On the other hand the low operating temperature (max. 80 °C) and the possibility of utilizing gases with very low sulfur dioxide contents are advantageous. A possible renaissance of this process in the future can thus not be ruled out e.g. in connection with flue gas desulfurization.

1.6.2.2.2 Sulfuric Acid from Waste Sulfuric Acid and Metal Sulfates

Ecological considerations have led to increased processing of waste sulfuric acid and to its recycling in particular production processes. The latter is only possible when the concentration of impurities in the refortified acid is not too high. If the amount of impurities is too high, the only option is thermal decomposition.

Process used for waste sulfuric acid recovery dependent upon degree of impurity:

- High level of impurities: decomposition to SO_2
- Low level of impurities: concentration of waste acid and recycling of the H_2SO_4 formed with, if necessary, oxidative removal of organic impurities

When there are few impurities, the acid can be concentrated by evaporation and any organic compounds present removed oxidatively during the evaporation.

Regeneration of waste sulfuric acid is very expensive. Minimum practical concentration for waste acid regeneration 20%

Regeneration of waste sulfuric acid is very expensive, refortified acid always being more expensive than fresh acid manufactured from sulfur or pyrites, the cost increasing as the concentration of the acid to be processed decreases. The concentration should therefore be at least 20%. In the case of more dilute acids neutralization with e.g. lime or sodium hydroxide can be carried out.

Concentration of diluted sulfuric acid energy intensive due to:

- High evaporation enthalpy of H_2O
- Heat of dehydration of H_2SO_4 has to be supplied

The concentration of waste sulfuric acid is extremely energy intensive due to the high heat of evaporation of water and the heat of dehydration of sulfuric acid. Concentration processes in which the heat of evaporation is supplied indirectly are mainly carried out under reduced pressure. (The boiling point of 70% sulfuric acid is 160 °C at 1 bar and 54 °C at 0.01 bar.) Concentration to 96% sulfuric acid at atmospheric pressure requires a temperature of 320 °C. The highest concentration achievable by evaporation is 98.3% (azeotropic composition).

Concentration carried out under:

- Reduced pressure
- Atmospheric pressure

Evaporation in two stages:

- Preconcentration to 65% acid
- Superconcentration to over 90% acid

The evaporation process is divided into two steps: preconcentration from ca. 15 to ca. 65% sulfuric acid and superconcentration from 60 to 70% to over 90% acid. Preconcentration is carried out under atmospheric or reduced pressure. The usual processes are:

Preconcentration with:

- Venturi-concentrators:
  - utilized for the preconcentration of large quantities of diluted H_2SO_4 or for sulfuric acid strongly contaminated with solids
  - use of low level waste heat possible

Venturi reconcentration process: Water can be removed from acid well under its boiling point by subjecting it to a dry gas stream which is continuously in contact with the sulfuric acid being concentrated. The heat required for the evaporation is either supplied directly by hot combustion gases or indirectly by heating the acid e.g. with tail gases from a double contact plant or hot sulfuric acid from sulfur trioxide absorption. The venturi concentrator is preferred for the preconcentration of large quantities of diluted sulfuric acid or sulfuric acid strongly contaminated with solids.

- Submerged-burner process:
  - mainly utilized for salt-containing sulfuric acids from titanium dioxide production or metal pickling processes

Submerged-burner process: Evaporation takes place in an evaporator boiler which is directly heated with combustion gases at 1500 °C. This process is particularly used for the preconcentration of salt-containing diluted sulfuric acids, the salts crystallizing out during the evaporation process and is used e.g. for the treatment of waste sulfuric acid from the production of titanium dioxide and from metal pickling processes.

Forced circulation evaporators: This process is generally operated under reduced pressure and with indirect heating at low temperature levels and units usually consist of multiple stage evaporation systems.

- Forced circulation evaporator:
  - operated at reduced pressures with indirect heating at low temperatures

In superconcentration plants either more temperature resistant materials are utilized or lower pressures are used. Several processes are known:

Superconcentration and purification:

Pauling-Plinke process: In this process sulfuric acid is distilled at atmospheric pressure with indirect heating by natural gas and oil combustion (combustion chamber temperature 800 to 1100 °C). Waste acid is added to boiling 96% sulfuric acid (320 °C) in a cast iron boiler by way of a distillation column. At the column head, water is expelled leaving a 96% acid. By adding nitric acid, which is mostly reduced to nitrogen, organic impurities can be oxidatively removed at high temperatures. This process is predominantly used in Europe.

- Pauling-Plinke process:
  - superconcentration occurs at atmospheric pressure
  - removal of organic impurities by addition of HNO_3

Drum concentrators: Concentrations up to 93% acid are achieved by direct contact with combustion gases. The purification effect is less than in the Plinke process. The sulfuric acid to be concentrated flows through a number of brick-lined chambers which are connected in a cascade through which 600 °C combustion gases flow in countercurrent. These become saturated with water vapor.

- Drum concentrators:
  - smaller purification effect
  - direct heating with hot combustion gases

Bayer-Bertrams process: The central components in this plant are forced circulation evaporators constructed of enameled steel and a falling film evaporator made of quartz (in the second stage), heated either by heat transfer oil or (more directly) by flue gases. The purification effect can be increased by an additional oxidation stage. The thermal efficiency of the installation can be increased by parallel connection of a number of evaporator tubes. The sulfuric acid can be superconcentrated to 98% using this process.

- Bayer-Bertrams process:
  - a falling film evaporator operated at atmospheric pressure, usually with a forced circulation evaporator as a first stage
  - very good purification possible

Some preconcentrated 65% sulfuric acid is also superconcentrated to acid contents of 80 to 90% in venturi-type plants e.g. by passing in 400 °C hot gases.

In another process preheated acid is fed into an enameled steel vessel equipped with a steam-heated tantalum coil. The water is distilled off from the acid under reduced pressure.

• Decomposition of waste sulfuric acid

In the case of strongly contaminated waste acid or acid sludge arising for example in the petrochemical industry and in organic chemical synthesis, a distillative or oxidative removal of impurities is often impossible. In this case it is processed by *thermal decomposition*:

$$2\,H_2SO_4 \longrightarrow 2\,SO_2 + O_2 + 2\,H_2O$$

Concentration of waste sulfuric acid has to be at least 60%

the sulfur dioxide formed being converted into (recycled) sulfuric acid. The waste acid must be concentrated to at least 60% by weight to enable conversion of the sulfur dioxide to sulfur trioxide and to compensate for the dilution of the flue gases when direct heating is used. Depending upon the kind of impurities present, the thermal decomposition can be carried out under reducing or oxidizing conditions, the organic impurities being either reduced to carbon or oxidized to carbon dioxide. The energy required for the sulfuric acid decomposition process is usually provided by direct combustion of gaseous or liquid fuels.

Thermal decomposition of sulfuric acid in oxidizing or reducing conditions, depending on the impurities

Reductive cracking by heating in a reducing atmosphere

Acid waste from petrochemical refining processes is, for example, reductively processed at 200 to 600 °C, the acidic tars or acidic resins added as reducing agents being converted into coke. In this way waste acids with carbon contents of 0.5 to 60% and acid concentrations between 40 and 90% can be refortified.

Oxidative cracking by combustion of auxiliary fuels in an excess of air

Oxidative processes have the advantage that organic impurities can be combusted to carbon dioxide in the reactor itself. The oxidizing atmosphere is obtained by combustion of fuel oil or gaseous fuel in an excess of air. Decomposition takes place in brick-lined furnaces at ca. 1000 °C in which finely dispersed waste acid from a rotary atomizer are mixed with ca. 1800 °C combustion gases. The sulfur dioxide-containing reaction gases emerging at ca. 1000 °C can be cooled in a waste heat boiler e.g. to ca. 350 °C with the production of steam, purified and then used for sulfuric acid production.

The energy required by acid decomposition is high, ca. 6.36 GJ/t 100% sulfuric acid being necessary for the cracking of 70% waste acid at 1000 °C. Almost half of this (46%) is needed for dehydrating the 70% acid, evaporating the water and sulfuric acid and heating the gases from 20 °C to 340 °C. The actual decomposition consumes about 30% of the total energy, the rest being used for heating the reaction products to ca. 1000 °C. The latter is in general recoverable.

Sulfur dioxide from waste sulfuric acid:

environmentally responsible, but energy intensive and thus expensive

Metal sulfates also yield sulfur dioxide upon decomposition. In the Müller-Kühne process, which is currently only operated in Austria, South Africa and several Eastern European countries, gypsum or *calcium sulfate* from the manufacture of phosphoric acid are reacted with coal to sulfur dioxide and calcium oxide in rotary kilns at 700 to 1200 °C:

- Decomposition of $CaSO_4$
  sulfur dioxide manufacture from calcium sulfate and coal with Portland cement as a byproduct decreasing in importance

$$4\,CaSO_4 + 2\,C \rightarrow 4\,CaO + 4\,SO_2 + 2\,CO_2$$

This process is, however, expensive and is only economic if the calcium oxide coproduct is simultaneously utilized for Portland cement manufacture by adding clay and sand to the kiln, the furnace temperature being simultaneously increased to 1400 °C. The sulfur dioxide content of the gases emerging is 7 to 10% by volume.

Another sulfur dioxide source is *iron(II) sulfate* from the manufacture of TiO_2 by the sulfate process (Sec. 5.9.2.2.3) in the form of the water soluble heptahydrate ($FeSO_4 \cdot 7H_2O$), which is difficult to dispose of. Bayer AG has developed a two stage process in which the heptahydrate is converted into sulfur dioxide and iron oxide:

- Decomposition of $FeSO_4 \cdot 7H_2O$
  Sulfur dioxide manufacture from iron sulfate formed during titanium dioxide production contributes to environmental protection

$$2\,FeSO_4 \cdot 7\,H_2O \rightarrow 2\,FeSO_4 \cdot H_2O + 12\,H_2O$$

$$2\,FeSO_4 \cdot H_2O \rightarrow Fe_2O_3 + 2\,SO_2 + 1/2\,O_2 + 2\,H_2O$$

$$\Delta H_{overall} = +601\ \text{kJ/mol}$$

It is first dehydrated to iron(II) sulfate monohydrate in a fluidized bed dryer at ca. 200 °C. The second step

Thermal decomposition of $FeSO_4 \cdot H_2O$ in

- Fluidized bed furnaces
- Rotary kilns
- Multiple hearth roasters

decomposition energy supplied by combustion of S, fuel oil or natural gas

is thermal decomposition at ca. 900 °C in a fluidized bed furnace or rotary kiln, with multiple hearth roasters being preferred. The energy required is supplied by the combustion of fuel oil, natural gas or sulfur. The decomposition gases leave the multiple hearth roasters at ca. 400 °C and are purified. The iron(III) oxide formed can be utilized in the metallurgical or cement industries.

The metal sulfates contained in the dilute H_2SO_4 produced during TiO_2 production can also be separated and decomposed to SO_2

The metal sulfates which precipitate during the production of titanium dioxide pigments in dilute sulfuric acid after concentration of the acid to ca. 65 to 70% with a vacuum evaporator (8 to 15% in 20 to 25% acid), can also be separated and thermally decomposed to sulfur dioxide, metal oxides, water and oxygen in fluidized bed furnaces at 900 to 1000 °C. The same holds for the metal sulfates which are precipitated upon further concentration of the sulfuric acid to 96%. The sulfur dioxide-containing decomposition gases are cooled to ca. 350 to 400 °C in a waste heat boiler with the production of steam and purified.

1.6.2.3 Applications of Sulfuric Acid

Applications spectrum for sulfuric acid in the USA in 1985 (in %):

• Phosphate fertilizers	70
• Petroleum refining	5
• Inorganic chemicals and pigments	4
• Copper leaching	3
• Synthetic rubbers and plastics	2.5
• Industrial organic chemicals	2.5
• Pulp and paper	2
• Others	11

Sulfuric acid is commercially available as 78, 96, 98 and 100% acids as well as products resulting from their dilution. As a key product in the chemical industry it has a multitude of applications. The largest quantities are consumed in the manufacture of phosphorus and nitrogen fertilizers (in 1985 ca. 60% worldwide, 70% in the USA and 44% in Western Europe). In the petrochemical industry it is utilized in the alkylation of isoalkanes with alkenes and in the refining of raw paraffin. In the chemical industry it is used in the production of inorganic chemicals (e.g. hydrofluoric acid, chromic acid, boron(III) oxide, aluminum sulfate and sodium sulfate) and of organic products (such as dyes, explosives, soaps, isocyanates, detergents, plastics, fibers and pharmaceuticals).

1.6.3 100% Sulfur Dioxide

The following processes are used for the manufacture of 100% sulfur dioxide:

- Absorption of sulfur dioxide-containing gases with solvents e.g. water at atmospheric pressure or higher pressures (ca. 5 bar), the absorbed sulfur dioxide being expelled with steam.

100% SO_2 by:

- Absorption in:
  - water
  - organic amines
  - solutions of alkali or alkaline earth hydroxides
  - ammonia
  - alkali salts of weak inorganic and organic acids

- Absorption of sulfur dioxide with strongly alkaline solutions (e.g. with alkali and alkaline earth hydroxides), ammonia and to an increasing extent alkali salts of weak inorganic and organic acids (e.g. sodium carbonate or citrate) from which it is subsequently expelled by adding acid.

  Occasionally aliphatic or aromatic amines are also utilized e.g. N,N-dimethylaniline (DMA-process) or triethanolamine (**U**nion **C**arbide **A**mine **P**rocess). In the Wellman-Lord process an aqueous solution of sodium sulfite and sodium hydrogen sulfite is used as the absorption medium. This process is mainly utilized for the desulfurization of combustion gases from power stations and tail gases from Claus plants.

  Lime and limestone processes (sulfur dioxide neutralization by milk of lime) are employed for flue-gas desulfurization, yielding calcium sulfite (sulfate) sludge or gypsum on complete oxidation:

Desulfurization of power station combustion gases technically feasible by wet and dry purification processes

$$SO_2 + Ca(OH)_2 \longrightarrow CaSO_3 \cdot 1/2\,H_2O + 1/2\,H_2O$$

$$SO_2 + Ca(OH)_2 + 1/2\,O_2 + H_2O \longrightarrow CaSO_4 \cdot 2\,H_2O$$

The quality of the gypsum is, however, inferior to natural gypsum so that it is not possible to substitute it completely for natural gypsum (e.g. in the construction industry). On the other hand there are no disposal problems with dry absorption on activated carbon (BF-process from Bergbau Forschung) producing a gas containing ca. 50 to 60% by volume of sulfur dioxide (the rest being carbon dioxide).

- Condensation of sulfur dioxide from roasting and sulfur combustion gases by increasing the pressure to ca. 5 bar and liquification in liquid ammonia countercurrent coolers.

- Condensation by pressure increase and cooling

- Reaction of sulfur with sulfur trioxide in oleum

- Reaction of SO_3 with S

$$2\,SO_3 + S \longrightarrow 3\,SO_2$$

followed by scrubbing the sulfur dioxide formed with sulfuric acid. In modern processes sulfur trioxide is not used as oleum but as an undiluted liquid, as is sulfur, with 25% oleum or liquid sulfur, for example, as the reaction medium.

The 1985 production of 100% sulfur dioxide was 162×10^3 t in the USA and ca. 65×10^3 t in West Germany.

Applications:

- Manufacture of chemicals
- Cellulose production
- Dressing of ores
- Mineral oil processing
- Disinfection and preservation agent
- Agricultural products

Applications: Ca. 50% of the liquid sulfur dioxide produced is utilized in the manufacture of sulfur-containing chemicals such as sulfites, hydrogen sulfites, thiosulfates, dithionites, salts of hydroxyalkanesulfinic acids and alkane sulfonates. It is also utilized in cellulose manufacture, in the chemical dressing of manganese ores, in the removal of sulfur-containing impurities from mineral oils, for disinfection and preservation in the food industry and for the treatment of water.

1.6.4 100% Sulfur Trioxide

100% SO_3:

- Manufacture from oleum by distillation and liquification

Manufacture: More highly concentrated oleum (with e.g. 65% by weight SO_3) and pure sulfur trioxide are industrially produced by distillation of oleum with lower sulfur trioxide contents (e.g. 35% by weight) in falling film evaporators or forced circulation evaporators with subsequent liquification at 30 to 35 °C. It is stabilized by the addition of thionyl chloride ($SOCl_2$) or oxalyl chloride and mostly processed directly by the manufacturer due to its polymerization tendency.

- Widespread application mainly for sulfonation

Applications: 100% sulfur trioxide is utilized for the manufacture of chlorosulfonic acid, fluorosulfonic acid, thionyl chloride and amidosulfonic acid and for the sulfonation of organic compounds e.g. for the detergent industry (linear alkylbenzenesulfonates). It is increasingly used as a sulfonation agent instead of chlorosulfonic acid and oleum/sulfuric acid, because its use produces no corrosive hydrogen chloride and less waste acid.

The USA production in 1982 was 35×10^3 t of sulfur trioxide (calculated as sulfur).

1.6.5 Disulfur Dichloride

Manufacture: Disulfur dichloride is continuously manufactured by passing chlorine into liquid sulfur at ca. 240 °C and distillative purification.

Disulfur dichloride (S_2Cl_2): manufactured from sulfur and chlorine

$$Cl_2 + 2\,S \xrightarrow[FeCl_3,\ AlCl_3]{240\,°C} S_2Cl_2$$

The sulfur dichloride formed as a byproduct is converted into disulfur dichloride by adding excess sulfur.

Applications: Disulfur dichloride is utilized for the manufacture of sulfur dichloride, thionyl chloride and sulfur tetrafluoride and by its reaction with polyols to form additives for high pressure lubricating oils and cutting oils. In addition it is used as a catalyst in the chlorination of acetic acid. Solutions of sulfur in disulfur dichloride are utilized in large quantities for the room temperature vulcanization of rubbers.

Applications:

- Starting material for the manufacture of thionyl chloride, sulfur dichloride, sulfur tetrafluoride etc.
- In the manufacture of additives for high pressure lubricants and cutting oils

1.6.6 Sulfur Dichloride

Manufacture: Sulfur dichloride is produced by reacting liquid disulfur dichloride with gaseous chlorine at low temperatures in the presence of catalysts e.g. iodine:

Sulfur dichloride (SCl_2):

- Manufactured from S_2Cl_2 and chlorine
- Starting product for thionyl chloride manufacture. Other applications as for disulfur dichloride

$$S_2Cl_2 + Cl_2 \xrightarrow{I_2} 2\,SCl_2$$

Since it is very unstable, decomposing into its starting materials, it is only produced as an intermediate product.

Applications: Sulfur dichloride is utilized in the same way as disulfur dichloride e.g. for the manufacture of thionyl chloride and in sulfidizing and chlorination reactions.

1.6.7 Thionyl Chloride

Manufacture: Thionyl chloride is industrially produced by reacting sulfur dioxide or sulfur trioxide and chlorine with sulfur dichloride or disulfur dichloride.

Thionyl chloride ($SOCl_2$):

- Manufactured from SO_2 or SO_3 and Cl_2 with SCl_2 or S_2Cl_2 with an activated carbon catalyst

It is also produced by the reaction of phosphorus trichloride with sulfuryl chloride.

1. $SO_2 + SCl_2 + Cl_2 \rightarrow 2\,SOCl_2$
2. $SO_3 + 2\,SCl_2 + Cl_2 \rightarrow 3\,SOCl_2$
3. $2\,SO_2 + S_2Cl_2 + 3\,Cl_3 \rightarrow 4\,SOCl_2$
4. $SO_3 + S_2Cl_2 + 2\,Cl_2 \rightarrow 3\,SOCl_2$
5. $SO_2Cl_2 + PCl_3 \rightarrow SOCl_2 + POCl_3$

The unreacted sulfur dioxide is processed by a gas phase reaction with chlorine on an activated carbon catalyst and the sulfuryl chloride formed returned to the thionyl chloride synthesis. The unreacted sulfur dichloride is reacted with sulfur to disulfur dichloride or with oxygen to thionyl chloride. Pure thionyl chloride is obtained by fractional distillation.

- Most important chlorination agent in the chemical industry
- 1979 consumption in the Western World: 46×10^3 t

Applications: It is estimated that the Western World consumption of thionyl chloride in 1979 was 46×10^3 t. Thionyl chloride is used as a chlorination agent e.g. in the manufacture of organic intermediates, plant protection agents and pesticides and in the syntheses of pharmaceuticals, dyes and pigments. It is also utilized for the dehydration of metal chloride hydrates and the chlorination of metal oxides. Recently it has been used as a nonaqueous electrolyte in galvanic cells.

The acidic reaction gases (e.g. hydrogen chloride, sulfur dioxide) produced during chlorination reactions with thionyl chloride (e.g. selective chlorination of alcohols and carboxylic acids) can be largely absorbed by alkali scrubbing.

1.6.8 Sulfuryl Chloride

Sulfuryl chloride (SO_2Cl_2):

- Manufacture from SO_2 and chlorine on activated carbon

Manufacture: Industrially sulfuryl chloride is almost exclusively produced by reacting sulfur dioxide with chlorine on an activated carbon catalyst in a well-cooled tubular converter. In a recent process this reaction is carried out in sulfuryl chloride with dimethylformamide or biuret as a catalyst with yields $> 99\%$.

Applications: Sulfuryl chloride is often used as a chlorinating and sulfochlorinating agent in organic chemistry (e.g. for selective side-chain chlorination of aromatic compounds) especially in the manufacture of organic intermediates for dyes, pharmaceuticals, plant protection agents and disinfectants. For environmental reasons many consumers have switched over to sulfuryl chloride as the byproduct, sulfur dioxide, can be reprocessed with chlorine.

- Chlorination and sulfochlorination agent

1.6.9 Chlorosulfonic Acid

Manufacture: Chlorosulfonic acid is produced by the reaction of liquid sulfur trioxide or oleum partly dissolved in chlorosulfonic acid with hydrogen chloride. Gas phase reactions between an excess of dry hydrogen chloride and sulfur trioxide from sulfuric acid plants are also used to produce chlorosulfonic acid.

Chlorosulfonic acid ($ClSO_3H$):

- Manufactured from SO_3 and HCl

Applications: Chlorosulfonic acid is mainly utilized as a mild sulfonating and chlorosulfonating agent in organic synthesis, but can also be used for sulfation.

- Condensation and sulfonation agent
- Also utilized for chlorosulfonation and sulfation

The current USA capacity is ca. 52×10^3 t/a and 38.5×10^3 t were produced in West Germany in 1985.

1.6.10 Fluorosulfonic Acid

Manufacture: Fluorosulfonic acid is produced by feeding liquid hydrogen fluoride with cooling into a solution of sulfur trioxide in fluorosulfonic acid.

Fluorosulfonic acid (FSO_3H):

- Manufactured from SO_3 and HF in fluorosulfonic acid

Applications: Fluorosulfonic acid is utilized as a fluorination agent in inorganic and organic chemistry in the production of sulfofluorides and sulfonic acids and as a catalyst for alkylation and polymerization reactions. It is also used in the polishing of lead crystal glass. No statistics are available regarding its consumption and production figures.

- Applications:
  - fluorination agent
  - catalyst
  - agent for polishing lead crystal glass

1.6.11 Sulfurous Acid Salts

Of the salts of sulfurous acid only the sodium salts and calcium hydrogen sulfite are industrially important.

Sodium hydrogen sulfite ($NaHSO_3$):

- Manufactured from SO_2 + NaOH or Na_2SO_3

Sodium hydrogen sulfite is manufactured by reacting sulfur dioxide with sodium hydroxide or a sodium sulfite suspension.

It is mainly used as a bleach. The 1985 USA consumption was ca. 66×10^3 t.

Sodium disulfite ($Na_2S_2O_5$):

- Manufactured from SO_2 + NaOH in a saturated solution of $NaHSO_3$
- Utilized in:
  - photographic industry
  - paper industry
  - textile industry
  - leather industry

Sodium disulfite is produced by reacting sulfur dioxide with 50 to 70% sodium hydroxide in saturated sodium hydrogen sulfite solution. Sodium disulfite is isolated from the resulting suspension by centrifugation.

Applications: Sodium disulfite is used in the photographic, paper, textile and leather industries as well as in water treatment.

Sodium sulfite (Na_2SO_3):

- Manufactured from SO_2 + NaOH in saturated Na_2SO_3 solution
- Utilized in:
  - photographic industry
  - paper industry
  - textile industry
  - food industry

Sodium sulfite is obtained upon reacting sulfur dioxide-containing gases with sodium hydroxide in a saturated sodium sulfite solution at 60 to 80 °C. The salt is separated by centrifugation and dried in a pneumatic-conveyor drier. It is marketed as its anhydrous salt.

Applications: Sodium sulfite is utilized as a reducing agent, in the manufacture of sodium thiosulfate, as an oxidation-prevention agent for developer solutions in the photographic industry, as an antichlorination agent in the paper and textile industries, for the preservation of food and for the treatment of boiler water. The 1984 production in the USA was 111×10^3 t.

$Ca(HSO_3)_2$ from SO_2 + $CaCO_3$

- Utilized in the production of sulfite cellulose

Calcium hydrogen sulfite solutions are produced by the reaction of sulfur dioxide-containing gases with limestone.

Applications: These solutions are used in the production of sulfite cellulose.

1.6.12 Sodium Thiosulfate, Ammonium Thiosulfate

Sodium thiosulfate ($Na_2S_2O_3$):

- Manufactured from S + Na_2SO_3 or S + $NaHSO_3$

Manufacture: Sodium thiosulfate is produced by reacting sulfur with sodium sulfite suspensions or sodium hydrogen sulfite solutions (sodium hydroxide and sulfur dioxide) at 50 to 100 °C in agitator vessels:

$$Na_2SO_3 + S \rightarrow Na_2S_2O_3$$

$$2\,NaOH + SO_2 + S \rightarrow Na_2S_2O_3 + H_2O$$

or by reduction of sulfur dioxide (e.g. from flue gas desulfurization) with sodium sulfide in the presence of sodium carbonate:

$$2\,Na_2S + Na_2CO_3 + 4\,SO_2 \rightarrow 3\,Na_2S_2O_3 + CO_2$$

The product is usually formed in aqueous solution and is purified by crystallization upon cooling. It usually crystallizes as $Na_2S_2O_3 \cdot 5\,H_2O$.

Applications: Ca. 90% of the sodium thiosulfate produced is utilized as a fixing salt in the photographic industry. It is also used as an antichlorination agent in bleaching plants, in paper fabrication and in flue gas desulfurization. The 1987 USA capacity was 28×10^3 t/a. Demand for sodium thiosulfate is decreasing in favor of ammonium thiosulfate.

- Utilized in the photographic industry as well as in bleaching plants and paper manufacture

Manufacture: Ammonium thiosulfate is produced by reacting sulfur with ammonium sulfite in aqueous alkaline ammonia solution at 80 to 110 °C:

Ammonium thiosulfate [$(NH_4)_2S_2O_3$]:

- Manufactured from $(NH_4)_2SO_3$ + S

$$2\,NH_3 + SO_2 + H_2O \longrightarrow (NH_4)_2SO_3$$
$$(NH_4)_2SO_3 + S \longrightarrow (NH_4)_2S_2O_3$$

Upon cooling the purified solution to ca. 20 °C, anhydrous $(NH_4)_2S_2O_3$ crystallizes out.

Applications: Ammonium thiosulfate is almost exclusively used as a fixing salt in the photographic industry. Compared with sodium thiosulfate it has the advantages of shorter fixing and washing times, ca. 50% higher fixing bath yields and easier silver recovery from the exhausted fixing baths.

- Utilized as a fixing salt in the photographic industry

1.6.13 Sodium Dithionite and Sodium Hydroxymethanesulfinate

Sodium dithionite has almost completely replaced zinc dithionite for environmental reasons. Dithionite is produced by the reduction of sulfurous acid either in an unbound form or in the form of hydrogen sulfite. The following reducing agents are utilized in the industrial production of dithionite: zinc dust, sodium amalgam, sodium formate and sodium tetrahydroborate. The current worldwide demand is 250 to 300×10^3 t/a. The zinc dust process is the most important manufacturing process followed by the formate process.

Sodium dithionite ($Na_2S_2O_4$):

- Manufactured using
  - the zinc dust process producing ca. 40% of world production

- Zinc dust process: liquid or gaseous sulfur dioxide reacts with zinc dust to zinc dithionite in an aqueous suspension at ca. 40 °C:

$$Zn + 2\,SO_2 \longrightarrow ZnS_2O_4$$

Addition of sodium carbonate or sodium hydroxide precipitates the zinc as its carbonate or its hydroxide:

$$ZnS_2O_4 + 2\,NaOH \longrightarrow Zn(OH)_2 + Na_2S_2O_4$$

After vacuum condensation of the filtrate and addition of sodium chloride, anhydrous sodium dithionite precipitates out.

  - formate process accounting for ca. 30% of world production

- Formate process: Sulfur dioxide under pressure (2 to 3 bar) is reacted with a solution of sodium formate in 80% aqueous methanol, the pH being maintained at 4 to 5 by adding sodium hydroxide.

$$HCOONa + 2\,SO_2 + NaOH \longrightarrow Na_2S_2O_4 + CO_2 + H_2O$$

Anhydrous sodium dithionite is formed as a finely crystalline anhydrous precipitate.

  - Amalgam process $NaHSO_3 + NaHg_x$ accounting for ca. 20% of world production

- Amalgam process: cooled aqueous solutions of sodium hydrogen sulfite are reduced with sodium amalgam (from sodium chloride electrolysis). The sodium dithionite formed is recovered by crystallization.

  - $NaBH_4$ process accounts for ca. 10% of world production

- Sodium tetrahydroborate process: reaction of sulfur dioxide (also sulfur dioxide-containing combustion gases) with a strongly alkaline aqueous solution of sodium tetrahydroborate yields sodium dithionite:

$$NaBH_4 + 8\,NaOH + 8\,SO_2 \longrightarrow 4\,Na_2S_2O_4 + NaBO_2 + 6\,H_2O$$

Applications: Its reducing properties make sodium dithionite important in the dying and printing of textiles. It is also used as a bleaching agent in the textile and paper industries.

- Utilized in the textile and paper industries

Dithionites are also starting materials for the manufacture of *sodium hydroxymethanesulfinate* in which they are reacted with formaldehyde in the presence of sodium hydroxide. In a process variant using zinc dithionite as a starting material, it reacts with formaldehyde in aqueous solution forming zinc salts of hydroxymethanesulfinic acid and hydroxymethanesulfonic acid. The latter is reduced to zinc hydroxymethanesulfinate by the addition of zinc at 90 °C. Subsequent addition of sodium hydroxide or sodium carbonate yields the sodium compound.

Sodium hydroxymethanesulfinate ($HOCH_2SO_2Na$):

- Manufactured from dithionites, formaldehyde and NaOH

Applications: Sodium hydroxymethanesulfinate is used in direct and discharge printing for converting vat dyes into their leucoforms. The current worldwide demand is ca. 15 to 18×10^3 t/a.

- Application: direct and discharge printing

1.6.14 Hydrogen Sulfide

Hydrogen sulfide is present in large quantities in crude oil and particularly in natural gas. This is processed to elemental sulfur (Sec. 1.6.1.3.2). In addition it is found in small quantities dissolved in mineral sources and is formed as a byproduct in a large number of chemical reactions.

Hydrogen sulfide (H_2S) found in:

- Crude oil
- Natural gas
- Mineral sources

Industrially hydrogen sulfide is mainly manufactured from pure sulfur and pure hydrogen at ca. 350 °C in the presence of catalysts such as cobalt molybdeum oxide on γ-Al_2O_3.

Industrially manufactured from pure S and pure H_2 at ca. 350 °C in the presence of Co/Mo-oxide catalysts

Applications: Hydrogen sulfide is the starting material for the manufacture of sodium hydrogen sulfide, sodium sulfide and organic sulfur compounds, such as thiophenes or thiols.

Hydrogen sulfide also plays an important role in the enrichment of heavy water (HDO, D_2O) which is used, for example, as a moderator in nuclear technology (Sec. 6.4.4). This application exploits the temperature dependence of the following isotope exchange equilibrium (Girdler sulfide process):

$$HDS_{gas} + H_2O_{liquid} \rightleftarrows H_2S_{gas} + HDO_{liquid}$$

Applications:

- Manufacture of NaHS and Na_2S
- Manufacture of sulfur-containing organic compounds
- In the production of heavy water

1.6.15 Sodium Sulfide

Currently most sodium sulfide is still manufactured using the classical reduction of sodium sulfate with coal:

$$Na_2SO_4 + 2\,C \xrightarrow{730\text{ to }790\,^\circ C} Na_2S + 2\,CO_2 \qquad \Delta H = +204\text{ kJ/mol}$$

Sodium sulfide (Na_2S):

- Manufacture from
  - $Na_2SO_4 + 2C$ in pot furnaces, continuous furnaces, rotary plate furnaces
  - sodium polysulfide + $NaHg_x$

The reaction is carried out in pot furnaces (low brick-lined hearths with lids), continuous furnaces or rotary plate furnaces. The melt produced in the furnace is mashed with water, the undissolved material removed and the solution concentrated to 60 to 62% sodium sulfide by evaporation in cast iron vessels. This concentrated solution solidifies at ca. 90 °C.

The production of sodium sulfide by the reaction of sodium polysulfide solutions (from sodium sulfide and sulfur) with sodium amalgam (from chloralkali electrolysis) is also industrially important. In this reaction sulfur is reduced to sodium sulfide:

$$Na_2S + 3\,S \longrightarrow Na_2S_4$$

$$Na_2S_4 + 6\,NaHg_x \longrightarrow 4\,Na_2S + 6\,Hg_x$$

Sodium sulfide is also formed by the reaction of hydrogen sulfide with sodium hydroxide and the reaction of barium sulfide with sodium carbonate.

- Applications:
  - hair-removal agent
  - in ore flotation
  - in the textile industry

Applications: Sodium sulfide is mainly used in tanning as a hair-removing agent (so-called asher). It is also utilized in ore flotation, dying with sulfur dyes in the textile industry and in the chemical industry e.g. for the precipitation of heavy metal ions.

The 1986 capacity for sodium sulfide production in Western Europe was 111×10^3 t/a. The 1985 West German production of sodium sulfide, sodium hydrogen sulfide and sodium polysulfide was 11.3×10^3 t (as sulfur).

1.6.16 Sodium Hydrogen Sulfide

Sodium hydrogen sulfide is produced by reacting hydrogen sulfide with either sodium sulfide or sodium hydroxide.

Applications: Sodium hydrogen sulfide is utilized in the form of a 45% solution and in solid form (with ca. 71% sodium hydrogen sulfide content) for hair removal in the leather industry. It is also used in ore flotation, the dye industry and the synthesis of organosulfur compounds. The 1985 USA production of sodium hydrogen sulfide was ca. 56×10^3 t.

Sodium hydrogen sulfide (NaHS):

- Manufactured from
  - $H_2S + NaOH$
  - $H_2S + Na_2S$
- 1983 USA application spectrum:

– ore flotation	30%
– leather industry	20%
– dyestuffs	20%
– chemicals	10%
– others	20%

1.6.17 Carbon Disulfide

Manufacture: Currently carbon disulfide is still manufactured in large quantities by the reaction of charcoal with liquid sulfur at ca. 720 to 750 °C in brick-lined retorts or in electrical furnaces (electrothermal process). The use of methane as a carbon source in this process is gaining in importance particularly in North America, Japan and Europe, the reaction then operating at 650 to 750 °C:

Carbon sulfide (CS_2):

- Manufactured from
  - carbon + sulfur, particularly in countries without methane or natural gas in retort or electrothermal processes
  - CH_4 or natural gas + sulfur at 650 to 750 °C

$$CH_4 + 2\,S_2 \longrightarrow CS_2 + 2\,H_2S$$

Carbon disulfide is purified by distillation. The hydrogen sulfide formed is processed to elemental sulfur in Claus plants.

The 1984 production of carbon disulfide in the Western World was 530×10^3 t of which more than 70% was in North America and Western Europe.

1984 carbon disulfide production in Western World:

530×10^3 t of which > 70% was in North America and Western Europe

Applications: Carbon disulfide is mainly used in the viscose industry for fiber production. This accounted for 35% of the total USA consumption of 145×10^3 t in 1985.

1985 USA application spectrum for CS_2:

• Viscose industry (rayon)	35%
• CCl_4 manufacture	30%
• Cellophane production	10%
• Rubber chemicals	10%
• Miscellaneous	15%

Smaller quantities are utilized in the manufacture of cellophane from viscose or as a starting material in the production of carbon tetrachloride. In addition carbon disulfide is used in the production of vulcanization accelerators, flotation agents, corrosion inhibitors, plant protection agents and intermediates for pharmaceuticals.

References for Chapter 1.6: Sulfur and Sulfur Compounds

General Information:

Ullmanns Encyklopädie der technischen Chemie, 4th Ed., Vol. 21, pp 1–64, pp 75–178, Verlag Chemie, Weinheim-Deerfield Beach-Basel (1982).

U. H. F. Sander, H. Fischer, U. Rothe and R. Kola: *Sulfur, sulfur dioxide and sulfuric acid,* The British Sulphur Corporation Ltd., Verlag Chemie, Weinheim-Deerfield Beach-Basel (1984).

Sulfur:

R. Kiyoura: "International trends in sulfur – overview for 1985–1986", *Chemical Economy and Engineering Review* **18,** pp 14–21 (May 1986).

"Recovered sulfur", *Sulphur* **184,** pp 13–16 (May/June 1986).

Minerals Yearbook Vol. 1, US Department of the Interior, Washington, pp 925–946 (1987).

Chemical Economics Handbook, Stanford Research Institute, Menlo Park, California, USA: "Sulfur – marketing research report" (Feb. 1983).

Sulfuric Acid:

"Is the single-contact process still a viable option?", *Sulphur* **186,** pp 23–29 (Sept./Oct. 1986).

"Sulfuric acid: the senior chemical citizen stumbles", *Chemical Business,* pp 2–5 (June 1985).

"Cogeneration and energy recovery in sulfuric acid plants", *Sulphur, Supplement,* pp 1–20 (Sept./Oct. 1984).

"Sulfur – tightening supplies for the long term", *Chemical Week,* pp 26–32 (March 27, 1985).

World sulfur and sulfuric acid atlas, 4th Ed., The British Sulphur Corporation Ltd., London (1985).

"The changing sulfuric acid industry", *Chem, Week,* pp 40–43 (Feb. 1982).

A. Bhattacharya, "Production of sulfuric acid and cement from phosphogypsum using the OSW process", *Chem. Age India* **27,** pp 1009–1014 (1976).

R. A. Bauer and B. P. Vidon, "The Ugine Kuhlmann pressure process", *Chem. Eng. Prog.* **74,** pp 68–69 (1978).

B. van der Berkhof, "Sulfuric acid as a byproduct of the European non-ferrous metals industry", *Sulphur* **146,** pp 24–28 (Jan./Feb. 1980).

Statistical Supplement, The British Sulphur Corporation Ltd. (1986).

Chemical Economics Handbook, Stanford Research Institute, Menlo Park, California, USA: "Sulfuric acid – marketing research report" (Feb. 1983).

Chemical Economics Handbook, Stanford Research Institute, Menlo Park, California, USA: "Sulfuric acid – report" (Feb. 1987).

Various Sulfur Chemicals:

Chemical Economics Handbook, Stanford Research Institute, Menlo Park, California, USA: "Sulfur dioxide, chemical profile" (Jan. 1, 1986).

Chemical Economics Handbook, Stanford Research Institute, Menlo Park, California, USA: "Miscellaneous sulfur chemicals – data summary" (April 1987).

1.7 Halogens and Halogen Compounds

1.7.1 Fluorine and Fluorine Compounds

1.7.1.1 Economic Importance and Applications

The world consumption of fluorine can be adduced from the consumption of fluorspar, since to date the contribution of hexafluorosilicic acid from the production of phosphorus-containing fertilizers has been minor (ca. 15% in the USA in 1984). Table 1.7.1-1 gives the estimated fluorspar consumption for 1983 (1984 for the World and USA) in 10^3 t.

1984 Fluorspar consumption in 10^6 t:

- World: 4.5
- USA: 0.682

Table 1.7.1-1. Estimated Fluorspar Consumption in 1983 (World, USA in 1984) in 10^3 t.

World	USA	Mexico	Canada	France	W. Germany	Italy	Spain	Great Britain
4500	682	26	73	180	290	170	127	165

Acid-grade fluorspar, so-called "acidspar", is mainly mined for the chemical industry and is used in the manufacture of hydrogen fluoride. The main applications of hydrogen fluoride are in the production of chlorofluorohydrocarbons, synthetic cryolite and aluminum fluoride. The last two compounds, which can also be manufactured from hexafluorosilicic acid, are needed for the production of primary aluminum pig.

Utilized in the manufacture of:

- Steel
- Aluminum
- Fluorochemicals

Metallurgical grade fluorspar, so-called "metspar", with a CaF_2-content of 60 to 85% is used in steel manufacture. Ceramic grade fluorspar has a quality in between.

1.7.1.2 Raw Materials

The raw materials for the manufacture of industrial fluorochemicals are:

- Fluorspar – CaF_2
- Fluorapatite – $3Ca_3(PO_4)_2 \cdot CaF_2$ with 2 to 4% by weight fluorine
- Natural cryolite in small quantities – Na_3AlF_6. The extraction of cryolite in Greenland was stopped in 1963, at the moment the demand is being supplied from existing stocks.

Fluorine sources:

- Calcium fluoride
- Fluorapatite
- Natural cryolite (minor)

Fluorine content in the Earth's crust: 0.09%

The estimated fluorine content in the Earth's crust is 0.09%.

World reserves of fluorspar: 123×10^6 t

1.7.1.2.1 Fluorspar

Statistics over the world reserves of fluorspar fluctuate very strongly as a result of uncertainty about the definition of economic workability. Table 1.7.1-2 gives the 1984 estimates of fluorspar reserves in several countries in 10^6 t.

Table 1.7.1-2. 1984 Estimates of Fluorspar Reserves in Several Countries in 10^6 t CaF_2.

World	P.R. China	Mexico	S. Africa	France	Italy	Gt. Britain	Spain	USA
122.800	22.750	21.650	37.150	8.870	7.260	2.640	8.740	0.840

The world capacity for fluorspar extraction is ca. 5.9×10^6 t/a CaF_2. The largest producers in 1984 were Mexico (15% world share), Mongolia (15%), the P.R. China (14%), the USSR (13%) and South Africa (7%). The production in the USA is under 180×10^3 t/a. About 50% of this production is acidspar.

Extraction of fluorspar:

- By mining

Beneficiation:

- Grinding
- Flotation
- Drying
  [Briquetting (for the steel industry)]

Extraction of fluorspar. Fluorspar is extracted by both open-cast and underground mining and is found together with other minerals such as heavy spar ($BaSO_4$), galena (PbS), zinc blende (ZnS) and quartzite (SiO_2). As a result complex beneficiation is necessary. The raw ore has a fluorspar content of 30 to 60%.

Processing is carried out in modern plants by grinding the ore to < 0.1 mm (crushers, ball-mills) followed by multistage flotation. (Flotation: separation of the different constituents of the finely ground ores by utilizing the different wettabilities of the components.) Ca. 90% of the calcium fluoride present in the ore is obtained as acidspar with a content of 96.5 to 97.5%, the impurities being silica, calcium carbonate and barium sulfate.

Acidspar: 96.5–97.5% calcium fluoride, rest: silica, calcium carbonate, barium sulfate

The fine calcium fluoride thereby produced contains ca. 10% of adherent moisture, which has to be removed before use. Metspar for the steel industry has to be briquetted, if a coarse product is not obtained during processing.

1.7.1.2.2 Fluorapatite

The hexafluorosilicic acid solutions produced as a by-product in the extraction of phosphoric acid by fluorapatite digestion with sulfuric acid in the presence of silica (Sec. 1.5.1.2.1) are important raw materials for the manufacture of fluorochemicals. The reserves of available fluorine from fluorapatite are estimated to be 360×10^6 t CaF_2. The industrial utilization of this hexafluorosilicic acid is, however, still relatively limited. The reasons are:

Hexafluorosilicic acid from fluorapatite: currently a little used source of fluorine

- Aqueous solutions with hexafluorosilicic acid contents up to 30% are produced which, when used, mean that large quantities of water have to be transported.
- Up to now only the large-scale production of fluorides from hexafluorosilicic acid has been successful, that of hydrofluoric acid unsuccessful.
- Some of the impurities in hexafluorosilicic acid, in particular the phosphorus compounds present, make further processing difficult.

Nevertheless, it is expected that this source of fluorine will increase in importance in the future.

1.7.1.3 Fluorine and Inorganic Fluorides

1.7.1.3.1 Fluorine

Despite the fact that it is now feasible to produce fluorine using purely chemical processes, industrial manufacture is exclusively electrochemical. Potassium fluoride-hydrofluoric acid adducts, particularly with a KF:HF molar ratio of 1:2.0 to 2.2, are electrolyzed at temperatures of ca. 70 to 130 °C. Potassium fluoride provides the necessary melt conductivity.

Fluorine:

From the electrolysis of molten KF/HF-mixtures at 8 to 12 V with carbon anodes

$$2HF \xrightarrow{e^-} H_2 + F_2$$

Description of the fluorine cell

The cathodes and cell vessel are usually made of steel, the anodes of degraphitized carbon. The cathodic and anodic compartments are not separated by a diaphragm, but by "steel shirts" suspended from the lid into the melt. Fig. 1.7.1-1 shows a cross section through an industrial fluorine cell. The hydrofluoric acid consumed during the electrolysis is continuously replenished.

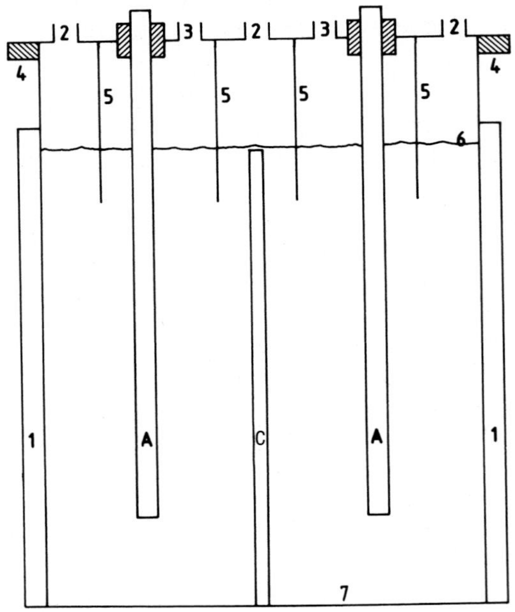

Fig. 1.7.1-1. Schematic cross-section through an industrial fluorine cell. 1) cooling jacket, 2) hydrogen outlet, 3) fluorine outlet, 4) vessel-lid seal, 5) steel shirt, 6) melt level, 7) vessel, at same potential as the cathode, C cathode, A anodes.

Technical data:

- Voltage: 8 to 12 V
  (standard electrode potential for F_2: 2.85 V)
- Current: 4 to 15 kA
- Current density: 0.1 to 0.15 A/cm^2
- Yield: 90 to 95% (relative to current consumed)

The lifetime of the anodes is, for example, 40 to 80×10^6 Ah in the type developed by the US Atomic Energy Commission. Since the voltage applied is well above the theoretical value, considerable amounts of heat, ca. 35 MJ/kg fluorine, have to be dissipated. On the other hand, the cooling water temperature must be maintained above the melting point of the electrolyte to prevent its solidification.

The fluorine and hydrogen produced contain ca. 10% (by volume) of hydrogen fluoride. The hydrogen fluoride content is strongly reduced by cooling below −100 °C. The residual hydrogen fluoride in the hydrogen is removed by an alkaline scrubber and that in the fluorine reduced, if necessary, by passing over sodium fluoride.

The fluorine thereby obtained is directly processed further (to uranium hexafluoride, sulfur hexafluoride), compressed in pressure cylinders or liquified (b.p. −188 °C).

The 1986 fluorine capacity in the Western World was estimated to be 2.4×10^3 t/a, of which 50% was in the USA, 35% in Europe and 15% in Japan. 55% of the fluorine produced is needed for the production of uranium hexafluoride, 40% for sulfur hexafluoride manufacture and 2.4% for carbon tetrafluoride manufacture. The manufacture of fluorographite for batteries consumes ca. 100 t/a of fluorine.

Fluorine capacity in the Western World: 2.4×10^3 t/a,

of which 55% is for UF_6
40% is for SF_6

1.7.1.3.2 Hydrogen Fluoride

Economic Importance

Hydrogen fluoride is a key product in the manufacture of fluorochemicals. Table 1.7.1-3 gives the estimated hydrogen fluoride capacities for 1986 for a few important industrialized countries.

1986 hydrogen fluoride capacities:

- World: 1.150×10^6 t/a
- USA : 0.306×10^6 t/a

Table 1.7.1-3. Estimated Hydrogen Fluoride Capacities in 1986 in 10^3 t/a.

World	USA	Canada	Mexico	W. Europe	E. Europe	USSR	Africa	Australia	Asia	Japan
1150	306	35	120	362.5	140	100	6	10	150	135

The US consumption of hydrogen fluoride in 1985 was ca. 260×10^3 t.

Manufacture

Industrially, hydrogen fluoride is manufactured by the reaction of sulfuric acid with acidspar (see Sec. 1.7.1.2.1).

$$CaF_2 + H_2SO_4 \xrightarrow{200-350\,°C} 2HF_g + CaSO_4$$

$$\Delta H_{25\,°C} = +59 \text{ kJ} \cdot \text{mol}^{-1}$$

Hydrogen fluoride from reaction of fluorspar with sulfuric acid in externally heated rotary furnaces

This reaction is nowadays generally carried out in rotary furnaces ca. 20 m in length and up to 3 m in diameter. Such furnaces produce more than 45 t hydrogen fluoride per day.

A ca. 5 to 10% excess of sulfuric acid is required due to the simultaneous reaction of the sulfuric acid with the steel of the rotary furnace and the impurities in the fluorspar: silica and oxidic and sulfidic minerals. Since water is formed in some of these side reactions e.g.

$$2CaF_2 + SiO_2 + 2H_2SO_4 \longrightarrow SiF_4 + 2CaSO_4 + 2H_2O$$

sulfur trioxide in the form of oleum is necessary to remove this.

As the reaction of fluorspar with sulfuric acid is endothermic, energy has to be supplied. This is accomplished by heating the rotary furnace either directly (ATO process in which the mixture of hydrogen fluoride and combustion gases formed is reacted directly in a fluidized bed reactor with calcined aluminum hydroxide to aluminum fluoride) or more commonly, indirectly.

In the Bayer process (see schema in Fig. 1.7.1-2) the required heat of reaction is supplied at several stages of the process:

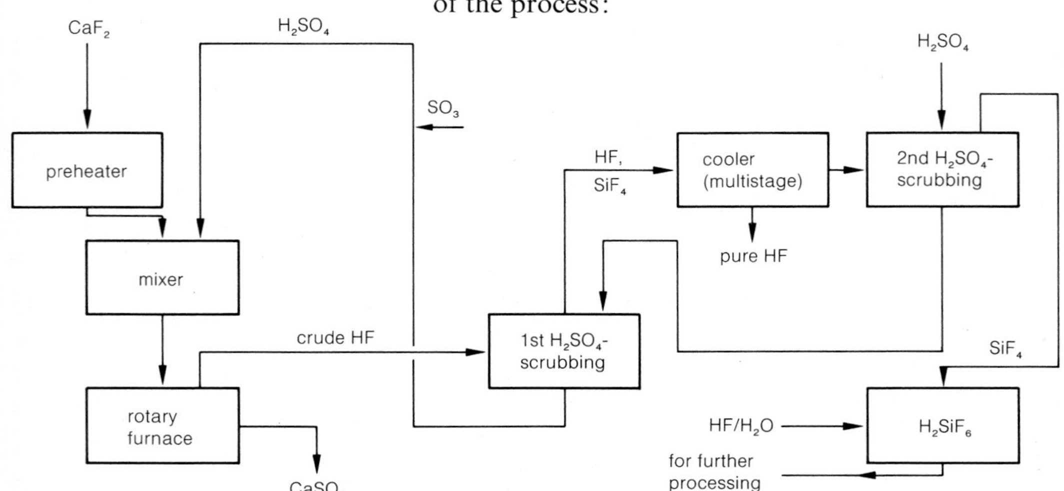

Fig. 1.7.1-2. Flow sheet of the Bayer process for the manufacture of hydrogen fluoride.

- Preheating of fluorspar with combustion gases
- Preheating of the sulfuric acid by heat exchange with the raw hydrogen fluoride gas
- Addition of sulfur trioxide as oleum and
- Heating of the rotary furnace

Fluorspar heated to 500 °C and preheated sulfuric acid are fed into a mixer, whereupon the reaction begins. It is completed in a rotary furnace.

The raw hydrogen fluoride, which still contains air, SiF_4, H_2, SO_2, CO_2 and dust, is scrubbed with sulfuric acid. (The scrubbing acid thereby produced is, after concentration by the addition of oleum, reused in the process.) After multistage cooling pure hydrogen fluoride is obtained (b.p. +19.9 °C). The hydrogen fluoride can be further purified by distillation.

The remaining gas is scrubbed with sulfuric acid to remove the residual hydrogen fluoride. The silicon tetrafluoride still present is reacted with water to hexafluorosilicic acid either in the absence of hydrofluoric acid with silicon dioxide precipitation:

$$3\,SiF_4 + 2\,H_2O \longrightarrow 2\,H_2SiF_6 + SiO_2$$

or with added hydrofluoric acid without silicon dioxide precipitation:

$$SiF_4 + 2\,HF \xrightarrow{H_2O} H_2SiF_6$$

These hexafluorosilicic acid solutions are then further utilized (see e.g. Sec. 1.7.1.3.3).

A much operated variant of the Bayer process is that of Buss AG in which the prereaction is carried out in a continuously operating kneader. In a further variant developed by Du Pont, the necessary heat of reaction is provided by producing sulfuric acid in situ by the reaction of sulfur trioxide with water.

3.8 t anhydrite, $CaSO_4$, are produced per t hydrogen fluoride. The anhydrite leaving the rotary tube still contains residual sulfuric acid, which is neutralized with lime. After processing, this anhydrite can be utilized in e.g. the manufacture of cement floors, in mining as tunnel infill or in the cement industry as a

Anhydrite byproduct utilized after processing in:

- Cement floors
- Mining
- Cement production

setting regulator. It has to be dumped, if no use can be found for it.

(The process developed by the Swedish firm Nynäs Petroleum for the manufacture of high specific surface area silica, Fluosil®, by the high temperature pyrohydrolytic decomposition of silicon tetrafluoride

$$SiF_4 + 2H_2O \longrightarrow SiO_2 + 4HF$$

has had only limited importance.)

Applications

Hydrogen fluoride used for:

- Inorganic fluorides
- Fluorohalogenhydrocarbons
- Glass etching and polishing
- Pickling stainless steels
- Semiconductor manufacture

Hydrogen fluoride is mainly utilized:

- In the manufacture of inorganic fluorides such as aluminum fluoride, boron trifluoride, uranium tetrafluoride, ammonium hydrogen fluoride
- In the manufacture of organic fluoro compounds, particularly fluorohalogenhydrocarbons
- For etching and polishing in the glass industry
- As a catalyst in alkylation reactions
- For pickling stainless steels
- In the manufacture of semiconductors

Hydrogen fluoride is usually commercially available in steel containers either liquified or in aqueous solutions containing $> 70\%$ HF. Steel is attacked by hydrogen fluoride solutions containing $< 60\%$ HF and hence needs to be coated for such solutions.

1.7.1.3.3 Aluminum Fluoride

Aluminum fluoride is used in addition to cryolite (Sec. 1.7.1.3.4) as a flux in the aluminum industry.

The estimated 1982 capacities for aluminum fluoride in 10^3 t/a are shown in Table 1.7.1-4.

Aluminum fluoride:

used as flux in aluminum industry

1982 capacities in 10^3 t/a:

- World: 630
- N. America: 185

Table 1.7.1-4. Estimated 1982 Capacities for Aluminum Fluoride in 10^3 t/a.

World	North America	South America	West Europe	East Europe	Asia	Japan	Africa
630	185	6	168	140	111	49	20

The East European capacity for aluminum fluoride was probably considerably higher in 1986 than in 1982.

Aluminum Fluoride Manufacture from Hydrogen Fluoride

In the Lurgi process aluminum hydroxide is first calcined at 300 to 400 °C and then reacted with hydrogen fluoride in a fluidized bed reactor at 400 to 600 °C.

Manufacture of aluminum fluoride:

- From aluminum hydroxide and hydrogen fluoride

$$2\,Al(OH)_3 \xrightarrow{300\text{ to }400\,^\circ C} Al_2O_3 + 3\,H_2O$$

$$Al_2O_3 + 6\,HF \xrightarrow{400\text{ to }600\,^\circ C} 2\,AlF_3 + 3\,H_2O\,.$$

(Formerly, aqueous hydrofluoric acid was reacted with aluminum oxide hydrate and the aluminum fluoride hydrate, which crystallized, was calcined in a rotary furnace.)

In the ATO process calcined aluminum hydroxide is reacted with a hydrogen fluoride-combustion gas mixture as already mentioned in Sec. 1.7.1.3.2.

Aluminum Fluoride Manufacture from Hexafluorosilicic Acid

Chemie Linz AG process

In this process a hexafluorosilicic acid solution is reacted with aluminum hydroxide at 100 °C, whereupon silica precipitates. After separating off the silica, the aluminum fluoride crystallizes as the trihydrate. Heating at above 500 °C then yields anhydrous aluminum fluoride.

- From aluminum hydroxide and hexafluorosilicic acid

$$2Al(OH)_3 + H_2SiF_6 \xrightarrow{100\,^\circ C} 2AlF_3 + SiO_2 + 4H_2O\,.$$

This process is operated in Sweden, Rumania, Tunisia and Japan.

UKF process

In the UKF process hexafluorosilicic acid is reacted with ammonia to ammonium fluoride and silica. After silica removal, the ammonium fluoride solution is reacted with double its molar quantity of aluminum oxide hydrate to a mixture of ammonium aluminum hexafluoride and aluminum oxide hydrate. After separation, this is converted at 500 °C into aluminum fluoride, ammonia and water.

$$H_2SiF_6 + 6NH_3 + 2H_2O \xrightarrow{H_2O} 6NH_4F + SiO_2$$

$$6NH_4F + Al_2O_3 \xrightarrow{100\,°C} [(NH_4)_3AlF_6 + 0.5Al_2O_3] + 3NH_3 + 1.5H_2O$$

$$2[(NH_4)_3AlF_6 + 0.5Al_2O_3] \xrightarrow{500\,°C} 4AlF_3 + 6NH_3 + 3H_2O$$

This process is currently not operated commercially.

1.7.1.3.4 Sodium Aluminum Hexafluoride, Cryolite

Statistics concerning the 1982 production capacities for cryolite are given in Table 1.7.1-5.

1982 cryolite capacities in 10^3 t/a:

- World: 398
- N. America: 125

Table 1.7.1-5. Estimated 1982 Capacities for Cryolite in 10^3 t/a.

World	North America	South America	West Europe	East Europe	Asia	Japan
398	125	6	114	81	72	50

A variety of processes have been described for its manufacture.

Manufacture of cryolite from hydrogen fluoride or, preferably, hexafluorosilicic acid

As in the case of aluminum fluoride manufacture hydrofluoric acid, or nowadays preferably hexafluorosilicic acid, are used as starting materials. In the latter case, the ammonium fluoride solution produced from hexafluorosilicic acid is, for example, reacted with sodium aluminate.

$$6NH_4F + 3NaOH + Al(OH)_3 \longrightarrow Na_3AlF_6 + 6NH_3 + 6H_2O\,.$$

In variants of this process, the ammonium fluoride solution is first reacted with sodium hydroxide to ammonia and sodium fluoride and then with aluminum fluoride to cryolite. After separation, the cryolite is calcined at 500 to 700 °C.

In the manufacture of cryolite, as in the manufacture of aluminum fluoride, three things need particularly to be taken into consideration:

- The high purity required of fluxes used in the electrolytic manufacture of aluminum. Phosphate (< 0.1%), silica (< 0.5%) and iron interfere. In the case of hexafluorosilicic acid from apatite digestion, phosphate impurities are to be expected and appropriate purification steps have to be carried out.
- The slight solubility of both aluminum fluoride and cryolite in water. In their production the fluoride concentration in the effluent must be carefully monitored and measures taken to ensure that the maximum permitted levels are not exceeded.
- The presence of fluoride residues in the silica by-product obtained when hexafluorosilicic acid is used as the source of fluorine. The silica must be disposed of appropriately.

Cryolite is, as is aluminum fluoride, mainly utilized in the electrolytic manufacture of aluminum. Other applications of cryolite are as: a flux in the manufacture of porcelain enamels, an additive in the manufacture of abrasives and an aid in the recasting of light metals.

Applications for cryolite:

- As a flux in aluminum manufacture
- Manufacture of porcelain enamels
- Additive in the manufacture of abrasives
- Aid in the manufacture of light metals

1.7.1.3.5 Alkali Fluorides

Sodium fluoride and potassium and ammonium hydrogen fluorides are particularly commercially important.

NaF, KHF_2, $NH_4F \cdot HF$.

They are manufactured by reacting either hydrogen fluoride or hexafluorosilicic acid with the corresponding alkali hydroxides. Ammonium hydrogen fluoride is mainly produced by the reaction of anhydrous ammonia and hydrogen fluoride in the melt.

Sodium fluoride is mainly used in water fluoridation, in the manufacture of organofluoro compounds by fluorine displacement and as a preservative.

Potassium hydrogen fluoride is used as a frosting agent in the glass industry and as a starting material for the manufacture of elemental fluorine (Sec. 1.7.1.3.1).

Ammonium hydrogen fluoride is used for the dissolution of silicate minerals in the extraction of crude oil and for the pretreatment of aluminum prior to anodization.

1.7.1.3.6 Hexafluorosilicates

Sodium and potassium hexafluorosilicates are obtained by reacting alkali salts (chlorides) with hexafluorosilicic acid and separating off the poorly soluble alkali hexafluorosilicates.

Magnesium, zinc and copper hexafluorosilicates, which are freely soluble in water, are manufactured from hexafluorosilicic acid and the appropriate oxides and are recovered by evaporating the solutions.

Hexafluorosilicates mainly used in:

- Wood protection
- Water fluoridation

The hexafluorosilicates are mainly used as preservatives in wood protection (particularly magnesium hexafluorosilicate). Sodium hexafluorosilicate is used in water fluoridation.

1.7.1.3.7 Uranium Hexafluoride

Uranium hexafluoride is the key compound in the separation of the uranium isotopes ^{235}U and ^{238}U. In its manufacture uranium(IV) oxide is first reacted with hydrogen fluoride to uranium tetrafluoride, which is then reacted with elemental fluorine to uranium hexafluoride:

Uranium hexafluoride:

see Chapter 6, nuclear fuel cycle

$$UO_2 + 4HF \longrightarrow UF_4 + 2H_2O$$

$$UF_4 + F_2 \longrightarrow UF_6 .$$

A detailed description of the process is given in Chapter 6.

1.7.1.3.8 Boron Trifluoride and Tetrafluoroboric Acid

The manufacture of boron trifluoride proceeds either discontinuously by reacting borates with fluorspar and oleum or continuously by reacting, for example, hydrogen fluoride and boric acid in the presence of sulfuric acid to bind the water formed:

Boron trifluoride:

from boric acid or borates, hydrofluoric acid, fluorosulfonic acid or fluorspar in the presence of sulfuric acid

$$Na_2B_4O_7 + 6CaF_2 + 7SO_3 \xrightarrow{H_2SO_4} 4BF_3 + 6CaSO_4 + Na_2SO_4$$

$$H_3BO_3 + 3HF \xrightarrow{H_2SO_4} BF_3 + 3H_2O\ .$$

The reaction of boric acid with fluorosulfonic acid also yields boron trifluoride:

$$3HSO_3F + H_3BO_3 \longrightarrow BF_3 + 3H_2SO_4\ .$$

Pure boron trifluoride is marketed in pressurized gas cylinders. It is utilized as its complexes or addition compounds e.g. with ethers, alcohols, carboxylic acids etc. or as a pure substance particularly in organic chemistry as a Friedel-Crafts catalyst.

Boron trifluoride:

utilized mainly as a Friedel-Crafts catalyst

Tetrafluoroboric acid, which is produced as a ca. 50% aqueous solution from boric acid and hydrofluoric acid, is also industrially important.

Tetrafluoroboric acid, fluoroborates:

used in galvanic metal deposition, as fluxes and flame retardants

$$4HF + H_3BO_3 \longrightarrow HBF_4 + 3H_2O\ .$$

Alkali, ammonium and transition metal fluoroborates can be produced from the acid. These fluoroborates are utilized in the galvanic deposition of metals, as fluxes, as flame retardants etc.

1.7.1.3.9 Sulfur Hexafluoride

Sulfur hexafluoride (sublimation temperature −63.9 °C) is manufactured from sulfur and elemental fluorine in a strongly exothermic reaction at several 100 °C:

Sulfur hexafluoride:

from sulfur and fluorine

$$S + 3F_2 \longrightarrow SF_6$$

Lower sulfur fluorides are formed as byproducts together with the extremely poisonous disulfur decafluoride, which is destroyed by subsequent pyrolysis:

$$S_2F_{10} \xrightarrow{400\,°C} SF_4 + SF_6 \,.$$

The lower sulfur fluorides and traces of hydrogen fluoride still present in the raw sulfur hexafluoride are removed by scrubbing with aqueous alkali. Oxygen, nitrogen and carbon fluorides (contained in the fluorine) are removed by distillation under pressure.

Worldwide production:

several 1000 t/a

Several thousand tons are produced annually worldwide.

Applications:

- In switching installations
- In magnesium casting
- In multiple glazing of windows

Sulfur hexafluoride is used as a protective gas in high voltage installations [due to its high dielectic constant (1.00204), high electrical breakdown resistance, nontoxicity], for the inhibition of magnesium-melt ignition during casting (addition of several 1/10th's % to the air), in multiple glazing of windows, particularly for thermal and noise insulation.

1.7.1.4 Organofluoro Compounds by Electrochemical Fluorination

Electrochemical fluorination:

Manufacture of functional perfluoro compounds by electrolysis of the corresponding nonfluorinated compounds in liquid hydrogen fluoride

When no special measures are taken, reactions of organic compounds with elemental fluorine generally lead to their decomposition and the formation of lower carbon fluorides. Functional perfluoroaliphatic compounds are manufactured using electrochemical fluorination, a fluorination technique discovered by Simons in about 1941. In this technique a few per cent of e.g. aliphatic carboxyl or sulfonyl fluorides are dissolved in liquid hydrogen fluoride and electrolyzed using nickel anodes and nickel or iron cathodes (the organic compounds providing sufficient conductivity in the fluorohydrocarbon). Hydrogen is produced cathodically and the perfluorocarboxyl or sulfonyl fluorides, corresponding to the organic compound utilized, anodically, e.g.

$$C_nH_{2n+1}SO_2F + 2n + 1\ HF \rightarrow C_nF_{2n+1}SO_2F + 2n + 1\ H_2$$

n e.g. = 8

Perfluoro compounds are, in general, insoluble in fluorohydrocarbons and can be drawn off as the denser phase. Gaseous trifluoromethanesulfonyl fluoride (b.p. −21.7 °C) is driven off together with the hydrogen. The yields decrease with increasing chain length. In the case of perfluorooctanesulfonyl fluoride, for example, it is 30 to 35%, with respect to the octanesulfonyl fluoride used.

Electrolysis is carried out at voltages of 5 to 8 V, exceptionally low current densities (0.01 to 0.02 A/cm^2) and temperatures of 0 to 20 °C. Intensive cooling is necessary to remove the large quantities of heat produced. The hydrogen emerging from the cell is saturated with fluorohydrocarbons and therefore has to be passed through a low temperature cooler.

The process was first utilized commercially by the 3-M Corp. in 1951. A flow sheet of an electrofluorination plant is shown in Fig. 1.7.1-3.

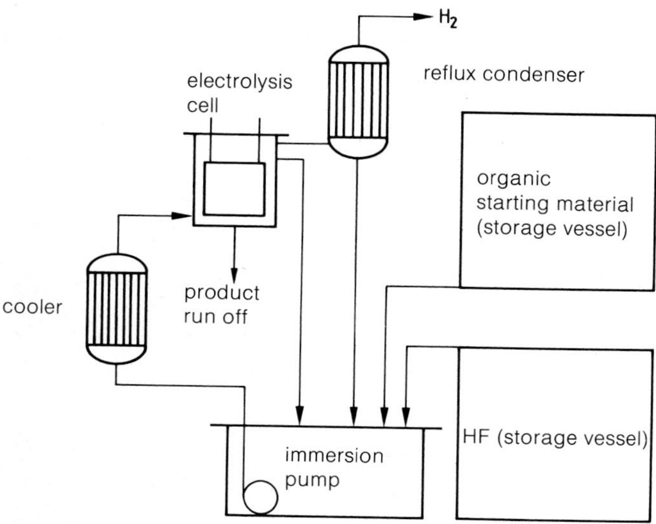

Fig. 1.7.1-3. Flow sheet of an electrofluorination plant.

Applications for perfluoroalkanesulfonic and carboxylic acids and their derivatives:

- Herbicides
- Flame retardants
- Textile finishing
- Tensides
- Emulsifiers
- Fire extinguishing agents
- Catalysts

Electrofluorination products are mainly utilized in the manufacture of

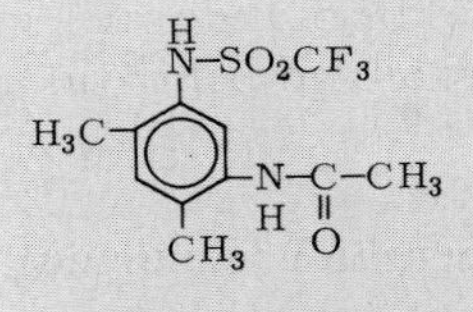

Mefluidide

- Herbicides (e.g. Mefluidide)
- Flame retardants such as $C_4F_9SO_3K$
- Tensides such as $C_8F_{17}SO_3K$ or $C_8F_{17}SO_3^-$-$N(C_2H_5)_4^+$
- Grease and dirt repellants for textiles
- Fire extinguishing agents
- Emulsifiers for tetrafluoroethene polymerization
- Leveling and flow control agents for paints

References for Chapter 1.7.1: Fluorine and Fluorine Compounds

Raw Materials:

L. Pelham, "Sources and Availability of Raw Materials for Fluorine Chemistry", *J. Fluorine Chem.* **30**, pp 1–17 (1985).

J. V. Smith, "Halogen and phosphorus storage in the Earth", *Nature* **289**, pp 762–765 (1981).

G. Bayer und H. G. Wiedemann, "Fluorrohstoffe – Vorkommen, Verwendung, Probleme" (Fluorine raw materials – occurrence, applications, problems), *Chemie in unserer Zeit* **19**, pp 33–41 (1985).

Commercial Information:

Chemical Profile, "Hydrofluoric Acid", Schnell Publishing Company, Inc. (Jan. 1, 1986).

Resumes des Centenaire de la Decouverte du Fluor, Paris, 25–29 August 1986, p. 193.

Inf. Chim. **No. 214,** Special, pp 193–200 (June 1981).

K.-H. Schmidt, "Flußsäuremarkt im Umbruch" (Hydrogen fluoride market in reorganization), *Chem. Ind.* **XXXII,** pp 453–457 (1980).

Electrofluorination:

E. Hollitzer und P. Sartori, "Die elektrochemische Fluorierung – ein Ueberblick" (Electrochemical fluorination – a survey), *Chem. Ing.-Tech.* **58**, pp 31–38 (1986).

Reviews:

H. C. Fielding and B. E. Lee, "Hydrofluoric Acid, Inorganic Fluorides and Fluorine", in: *The Modern Inorganic Chemicals Industry* (R. Thompson, ed.), The Chemical Society, Burlington House, London, pp 149–167 (1977).

Kirk-Othmer: *Encyclopedia of Chemical Technology*, 3rd Ed., Vol. 10, pp 630–828. John Wiley & Sons, New York (1980).

Winnacker-Küchler, *Chemische Technologie*, 4th Ed., Vol. 2, pp 526–549, Anorganische Technologie I (Inorganic technology I). Carl Hanser Verlag, München, 1982.

1.7.2 Chloralkali Electrolysis, Chlorine and Sodium Hydroxide

1.7.2.1 Economic Importance

Up to the beginning of the 1980's, the chlorine production of a region was an indicator of the strength of its chemical industry, since the evolution of chlorine production paralleled that of the total chemical production. Chlorine production appears to have reached the zenith of its growth or is even in decline. Chlorine is being displaced from its largest application fields such as plastics, solvents and pulp bleaching for a number of reasons, some ecological. The world chlorine production capacity fell ca. 4% between 1980 and 1986 and a further fall is expected in the USA (despite the > 90% utilization of the effective USA capacity and ca. 90% utilization of the European capacity in the summer of 1986). The chlorine capacities on Jan. 1, 1986 in 10^6 t/a are given in Table 1.7.2-1.

Chlorine:

worldwide capacity decline of 4% between 1980 and 1986

Table 1.7.2-1. Chlorine Capacities on 1st Jan. 1986 in 10^6 t/a.

World*	EEC	W. Europe excl. EEC	North America	USA**	South America	Africa	Near East	Asia/ Oceania	Japan	Eastern Europe	USSR
38.372	9.345	1.409	14.195	12.140	1.296	0.382	0.520	5.445	3.248	5.780	3.400

* excluding P.R. China
** 11.231×10^6 t/a expected on 1.1.1987

The capacity of the planned projects on Jan. 1 1986 was 2.005×10^6 t/a, of which 36% was in Eastern Europe.

1.1 t of sodium hydroxide and 28 kg of hydrogen are produced per t of chlorine.

Balancing the chlorine and sodium hydroxide turnovers with demand has been and is a difficult problem, first one product and then the other being dominant in the past.

The hydrogen turnover is not, in general, a problem. The excess of supply over demand for chemical synthesis can be burnt in power stations. The quantity of hydrogen produced by electrolysis is only a few % of the total production.

For utilization of chlorine and sodium hydroxide see Sec. 1.7.2.4.

1986 chlorine capacities in 10^6 t/a:

- World: 38.3
- USA: 12.1

Byproducts per t chlorine:

- Sodium hydroxide: 1.1 t
- Hydrogen: 28 kg

Sodium hydroxide is currently in excess on the world market

1.7.2.2 Starting Materials

97% of chlorine produced by the electrolysis of aqueous NaCl solutions

Chlorine is almost exclusively produced by the electrolysis of aqueous sodium chloride solutions. Only a small quantity is produced by the electrolysis (or oxidation) of hydrochloric acid (or hydrogen chloride) (Sec. 1.7.3). Small quantities of chlorine are also produced as byproducts in the electrochemical manufacture of metals such as e.g. sodium.

Availability of NaCl as a raw material is unlimited

Sodium chloride: The availability of sodium chloride as a raw material for the electrolytical manufacture of chlorine and sodium hydroxide is unlimited. It is either extracted from natural deposits (up to ca. 70%) or from seawater. Just in the USA, the economically workable deposits of sodium chloride are estimated to be greater than 55×10^{12} t. Extraction is either carried out by mining or leaching (i.e. dissolution of subterranean salt deposits by injection of freshwater and pumping out the brine).

Extraction of NaCl:

- From natural deposits, either by mining or leaching
- From seawater

While some of the salt produced by mining contains 99% sodium chloride, some contains only 95 to 98%, the rest being clays, anhydrite, quartz, dolomite, fluorspar and mica. In the latter case the salt is concentrated to 98 to 99% sodium chloride by sieving and gravitational separation. The latter utilizes the differences in specific gravity between sodium chloride and anhydrite and clays in a slurry of magnetite in a saturated sodium chloride solution.

In leaching, the insoluble components are left behind underground.

Evaporated salt produced by precipitating out the impurities from brine:

purity > 99.95%

Evaporated salt can be obtained from brine by, e.g., multistage vacuum evaporation in Monel metal vessels with copper-nickel or titanium coiled-tube evaporators. Pretreatment of the brine is necessary to attain an adequate purity. Calcium, magnesium and sulfate ions, in particular, have to be removed:

- Mg^{2+} as $Mg(OH)_2$ by the addition of $Ca(OH)_2$,
- Ca^{2+} as $CaCO_3$ by the addition of sodium carbonate (or CO_2 from flue gases),
- SO_4^{2-} by evaporating to the point at which sodium sulfate is about to precipitate out.

Extraction of salt from seawater by evaporation or electrodialysis

Extraction of salt from seawater occurs almost exclusively by solar evaporation in salt meadows, except in

Japan where it is not possible for climatic reasons. Here electrodialysis is used to concentrate the seawater, accounting for more than 50% of Japan's salt requirements.

Seawater is evaporated as follows: concentration of the seawater in the first pond; transportation to the next evaporator zone, in which calcium sulfate precipitates out; and finally crystallization of sodium chloride in a further evaporator zone. The residual brine is enriched in potassium and magnesium salts. The salt obtained is too impure to be used for electrolysis. Washing in special units is sufficient to increase the sodium chloride content to > 99%. 1 m^3 of seawater yields ca. 23 kg of sodium chloride.

Depending on the electrolysis process utilized: mercury, diaphragm or membrane, different additional purification steps are required. In the mercury process solid salt is utilized, which is dissolved in water. If evaporated salt is used in this process, the purification can be carried out in a small branch loop. When mined salt is utilized, care has to be taken during dissolution to settle out the insoluble impurities. Soluble impurities are removed by precipitating SO_4^{2-} with Ba^{2+}, precipitating Mg^{2+} and Fe^{3+} as hydroxides by the addition of NaOH and precipitating Ca^{2+} as carbonate with sodium carbonate (see the production of evaporated salt).

Salt purification dependent upon electrolysis process to be utilized:

Mercury, Diaphragm, Membrane } process

Purification for the mercury process:

multistage precipitation with Ba^{2+}, NaOH, Na_2CO_3

Solid salt is also required in the membrane process, but with a much higher purity, particularly as regards multivalent ions. The Ca^{2+}-content is thus reduced to below 0.1 ppm (ca. 3 ppm Ca^{2+} in the mercury process) with the aid of ion exchangers such as Lewatit® TP 208 (Sec. 1.7.2.3.3).

Purification for the membrane process:

additional purification by ion exchangers

Brine is generally used in the diaphragm process. Multivalent ions such as Ca^{2+}, Mg^{2+}, Fe^{3+}, Al^{3+} and silica, which block the diaphragm, are precipitated out by the addition of sodium hydroxide and sodium carbonate.

Purification for the diaphragm process:

precipitation with Na_2CO_3

Economic importance of sodium chloride:

The USA utilization spectrum for sodium chloride in 1982 is given in Table 1.7.2-2.

Table 1.7.2-2. 1982 Consumption of Sodium Chloride in the Different Industrial Sectors in the USA in 10^6 t.

Total	Electrolysis & Na_2CO_3	Other chemicals	Food	Animal nutrition	Other industries	Salting of roads	Various
34.8	16.8	0.78	2.1	1.52	1.59	8.2	3.8

1982 NaCl production in 10^6 t:

- World: 168.7
- USA: 34.2

The 1982 production of salt in the world and in several selected countries is given in Table 1.7.2-3.

Table 1.7.2-3. 1982 Sodium Chloride Production in 10^6 t.

World	North America	USA	West Germany	USSR	P.R. China	India	Australia
168.7	50.3	34.2	11.5	15.4	16	10	5.6

1.7.2.3 Manufacturing Processes

Industrially operated electrolysis processes:

- Mercury process
- Diaphragm process
- Membrane process (7%)

Three electrolysis processes are operated for the manufacture of chlorine, sodium hydroxide and hydrogen from aqueous salt solutions:

- Mercury process
- Diaphragm process
- Membrane process

The relatively recently developed membrane process has experienced a strong advance of late: accounting for only 1% of the worldwide chlorine production in 1978 this had risen to ca. 7% by the end of 1985. The current capacity of the Japanese membrane process plants is 2.2×10^6 t/a. This is due to statutory closure of the mercury process plants. The largest plant outside Japan is currently the Akzo plant at Botlek (NL) with a capacity of 250×10^3 t/a. It is generally assumed that this process will eventually displace the other two processes.

1.7.2.3.1 Mercury Process

The amalgam cells consist of slightly inclined elongated steel troughs over the bottoms of which flows a thin mercury layer, which acts as the cathode and absorbs the sodium. Chlorine is produced at horizontal anodes which are adjustable in height and are incorporated into the cell lids and it is drawn off upwards through gas extraction slits. The amalgam emerging from the ends of the cells is converted in "decomposers" by hydrolysis with water on graphite into mercury, 50% sodium hydroxide solution and hydrogen in a strongly exothermic reaction (see Fig. 1.7.2-1, 1.7.2-2 and 1.7.2-3).

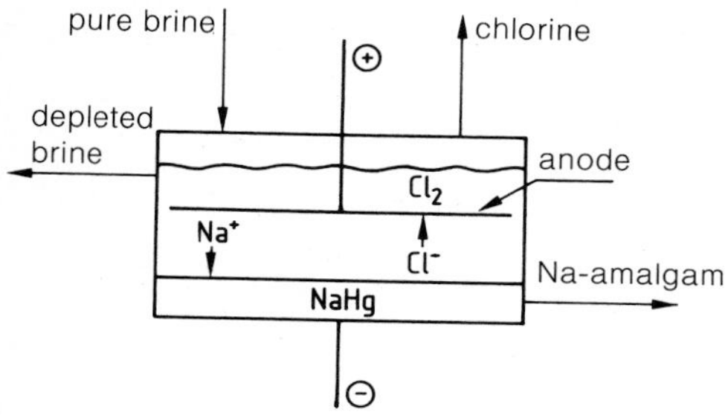

Fig. 1.7.2-1. Schematic representation of the electrolysis of aqueous salt solutions by the mercury process.

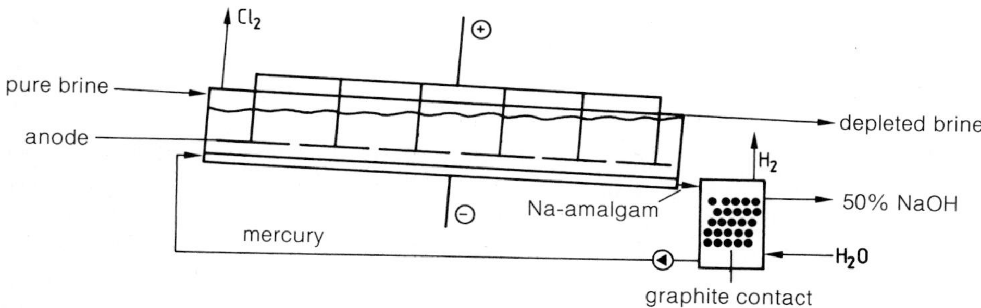

Fig. 1.7.2-2. Mercury process electroysis cell.

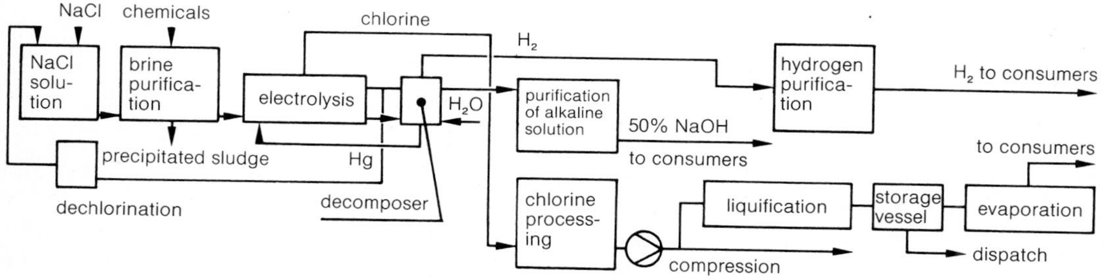

Fig. 1.7.2-3. Flow sheet of the mercury process.

Description of the mercury cells:

- Cathode surface area: 10 to 30 m^2
- Mercury layer thickness: 3 mm
- Sodium concentration in the mercury: 0.2 to 0.4% (by weight)
- 50 to 180 individual anodes per cell
- Cathode-anode separation: 3 mm
- Anode material: graphite or, preferably, titanium coated with platinum group metal compounds (so-called dimensionally stable anodes ®DSA)
- Brine throughput per cell: 3 to 20 m^3/h

A salt solution with a NaCl content of ca. 310 g/L is electrolyzed at ca. 80 °C, during which the NaCl content falls to 260 to 280 g/L. This is then reconcentrated with solid salt and recycled.

During electrolysis the following reactions take place:

Anodic reaction:

Mercury process:

electrolysis of aqueous NaCl solution on Hg-cathode and graphite- or titanium-anodes; separate decomposition of the Na-amalgam formed

$$Cl^- \rightarrow 0.5\,Cl_2 + e^-; \text{ Deposition voltage ca. } 1.24\text{ V}$$

Cathodic reaction:

$$x\text{Hg} + Na^+ + e^- \rightarrow NaHg_x; \text{ Deposition voltage ca. } -1.66\text{ V}$$

Typical side reactions are:

anodic

$$Cl_2 + 2NaOH \longrightarrow NaOCl + NaCl + H_2O$$

cathodic

$$Cl_2 + 2e^- \longrightarrow 2Cl^-$$

$$ClO^- + 2H^+ + 2e^- \longrightarrow H_2O + Cl^-$$

The electrochemical yield amounts to 94 to 97%, the energy consumption to ca. 3300 kWh/t chlorine, the effective cell voltage to ca. 4.2 V and current densities to 8 to 15 kA/m^2.

The cathodically formed amalgam is decomposed with water:

$$NaHg_x + H_2O \longrightarrow 0.5\,H_2 + NaOH + x\,Hg\,.$$

The electrical energy stored in the amalgam is thereby converted to heat.

Capacities of large-scale plants:

- 50 to 300×10^3 t/a chlorine
- 56 to 340×10^3 t/a sodium hydroxide

Mercury process

capacity of large-scale plants:

- up to 300×10^3 t/a chlorine
- up to 340×10^3 t/a sodium hydroxide

In modern units the height of the anodes is computer controlled. Chemical and physical processes are used to reduce the mercury concentration in the effluent, exhaust gases and products to the ppb level.

1.7.2.3.2 Diaphragm Process

Industrial diaphragm cells consist of a box in which the anode plates are mounted vertically parallel to one another. The cathodes are flat hollow steel mesh structures covered with asbestos fibers, optionally impregnated with fluoroorganic resins, and fit between the anodes (see Fig. 1.7.2-4, 1.7.2-5 and 1.7.2-6).

- Monopolar electrode arrangement: cell anode surface areas of up to 50 m^2 (activated titanium). Cathodes and anodes are all connected with one another
- Bipolar electrode arrangement: electrode surface area of up to ca. 35 m^2. Cathodes and anodes connected back to back.

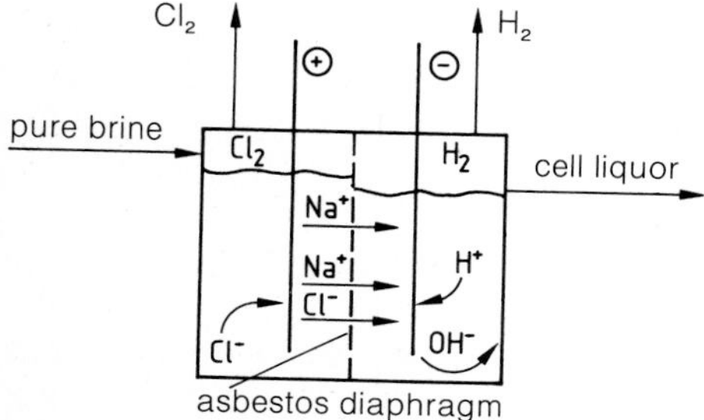

Fig. 1.7.2-4. Schematic representation of the electrolysis of aqueous salt solutions by the diaphragm process.

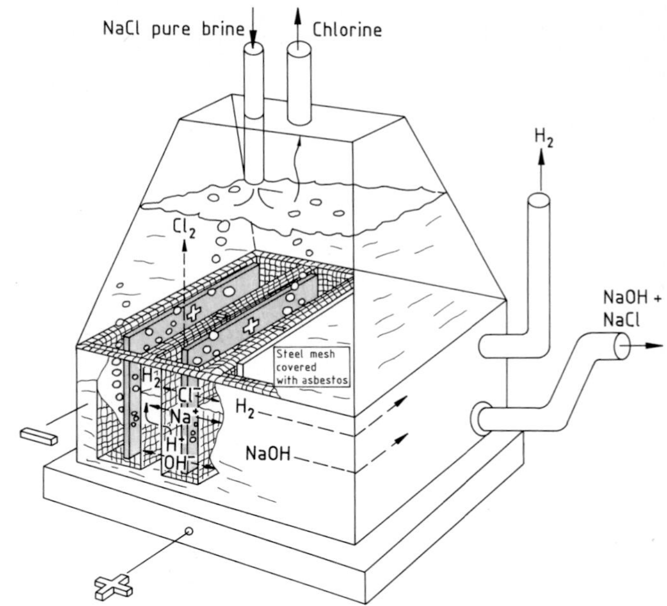

Fig. 1.7.2-5. Typical electrolysis cell for the diaphragm process.

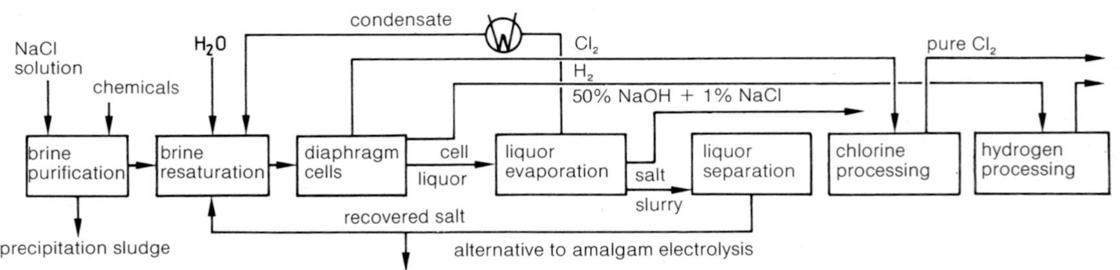

Fig. 1.7.2-6. Flow sheet of the diaphragm process.

The salt solution enters into the anode chamber and then passes through the diaphragm into the cathode chamber. The chlorine produced at the anode is drawn off upwards and hydrogen and sodium hydroxide mixed with residual salt are produced at the cathode.

Diaphragm process:

Electrolysis of aqueous NaCl solution on titanium anodes and steel cathodes. Anodic and cathodic chambers separated by an asbestos diaphragm. Brine passes through the diaphragm into the cathodic chamber. Dilute NaCl-containing sodium hydroxide formed

The asbestos diaphragm has a number of functions:

- It hinders the mixing of hydrogen and chlorine. The tangled fiber structure of the asbestos allows liquids to pass through, but not fine gas bubbles (the 4% of chlorine which dissolves in the brine does, however, pass into the cathode chamber where it is reduced thereby reducing the yield),

- It hinders to a large extent the back-diffusion of cathodically-formed OH^- ions to the anode. The flow rate of the brine into the anode chamber is regulated to limit this and also the hydrostatic pressure in the chamber.

Upon electrolysis, the NaCl content of an initially saturated salt solution falls to 170 g/L. The anodic reactions are the same as in the mercury process. However, hydrogen is produced at the steel cathode:

$$H_2O + e^- \longrightarrow 0.5\,H_2 + OH^- .$$

The catholyte solution flowing out of the cathode chamber contains ca. 12% NaOH and 15% NaCl (by weight).

Evaporation of sodium hydroxide:

The catholyte solution is evaporated to 50% by weight of NaOH, whereupon all the salt bar a residual 1% by weight precipitates. This salt is very pure and can be further utilized for concentrating diluted brine or, in the case of combined plants, in the mercury process.

The evaporation takes place in multi(up to four)stage forced circulation evaporators. 5 t of water have to be evaporated per t of 50% sodium hydroxide solution. A further purification of this salt-containing sodium hydroxide is possible, but very expensive.

Capacity of large-scale units:

- 360×10^3 t/a of chlorine corresponding to
- Ca. 410×10^3 t/a of sodium hydroxide at a specific current density of 2.2 to 2.7 kA/m^2.

Diaphragm process

capacity of large-scale unit:

- up to 360×10^3 t/a chlorine
- up to 410×10^3 t/a sodium hydroxide

The electrical energy consumption is ca. 20% less than that in the mercury process.

1.7.2.3.3 Membrane Process

In the membrane process the cathode and anode chambers are separated by a water-impermeable ion-conducting membrane (Fig. 1.7.2-7). The development of a membrane which is stable under the conditions of electrolysis, i.e. high salt concentration, high pH-gradient between the chambers and the presence of the strong oxidizing agents: chlorine and hypochlorite, has been a problem.

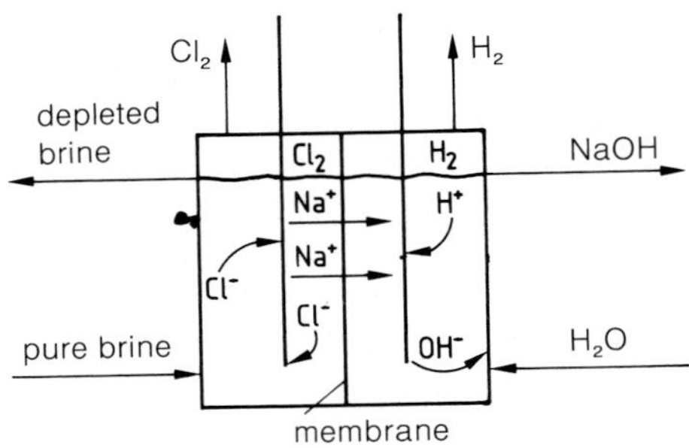

Fig. 1.7.2-7. Schematic representation of the electrolysis of aqueous salt solutions by the membrane process.

A number of firms (Du Pont, Asahi Chemical, Asahi Glass, Tokuyama Soda etc.) have developed membranes which fulfill these requirements. These membranes consist of a poly(perfluoroethene) backbone with side-chains which contain sulfonic acid or carboxylic acid groups:

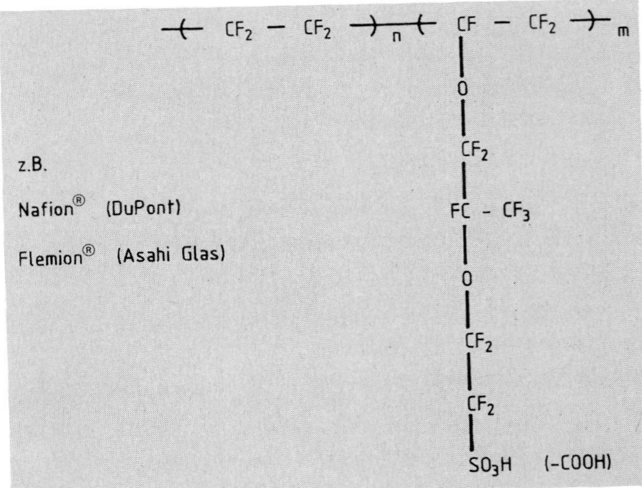

Currently carboxylic acid or two-layer membranes are generally used. The latter consist of a sulfonic acid membrane upon which a relatively thin layer of carboxylic acid membrane has been laminated on the cathodic side. These membranes are often strengthened with mesh.

Two-layer membranes benefit both from the lower electrical resistance of the sulfonic acid membranes compared with carboxylic acid membranes and from the higher repulsion by the carboxylic acid membranes of hydroxyl ions traveling in the direction of the anode. The repulsion of carboxylic acid membranes is much higher due to the small or absent solvatization shell of carboxylic acid groups which allows little or no OH^- ion penetration compared with that of sulfonic acid groups with their relatively large solvatization shells.

There are a number of manufacturers of membrane cells and several options are available:

- Monopolar or bipolar construction
- Cells with or without ("zero-gap") gaps between the membrane and the electrodes

The choice between mono- or bipolar cells is often dependent upon local technical factors. Zero-gap cells in which the membrane is in contact with the electrodes are in principle preferable, due to their lower electrical resistance. However, the cells with gaps between the membrane and the electrodes are currently (1986) more highly developed. Furthermore, the choice of electrodes also influences the electrical energy consumption as the overvoltage of hydrogen is much lower on nickel cathodes than on steel cathodes.

The operation of membrane cells: The same processes take place on the anodes and cathodes as in the diaphragm cells. Activated titanium is used for the anodes and stainless steel or nickel is preferred for the cathodes. No water transport takes place in the absence of current, but upon the application of current solvatization water is transported by the current-carrying Na^+ ions as they travel from the anode to the cathode chambers.

Membrane process:

cathodes and anodes separated by an ion-conducting membrane. Titanium anodes, stainless steel or nickel cathodes. Na^+ ions travel from the anode to the cathode chamber. Very pure 20 to 35% NaOH produced

The brine has to be much purer than for the mercury process. Ca^{2+} content must be below 0.1 ppm, otherwise $Ca(OH)_2$ precipitates in the membrane which rapidly leads to its destruction (Sec. 1.7.2.2).

The concentration of virtually chloride-free sodium hydroxide in the cathode chamber is between 20 and 35% by weight, depending upon the type of membrane used. With the newest membrane types the current yield with respect to sodium hydroxide is over 95%. This nonquantitative current yield is due to the passage of hydroxide ions into the anode chamber which causes chlorate formation. Since the brine is recycled, as with the mercury process, appropriate measures have to be taken to limit its chlorate concentration, e.g. by feeding in hydrogen chloride.

The bipolar membrane cell shown in Fig. 1.7.2-8 consists of individual modules in which the cathodic side of one module is pressed against the anodic side of another module. The individual modules are manufactured in two half shells into which electrodes consisting of narrow horizontal steel strips have been welded one above another like a blind. The membranes are mounted between these two half shells with special seals.

The pure brine is fed into the anode chamber and dilute sodium hydroxide or water fed into the cathode chamber. The depleted brine and chlorine are run off

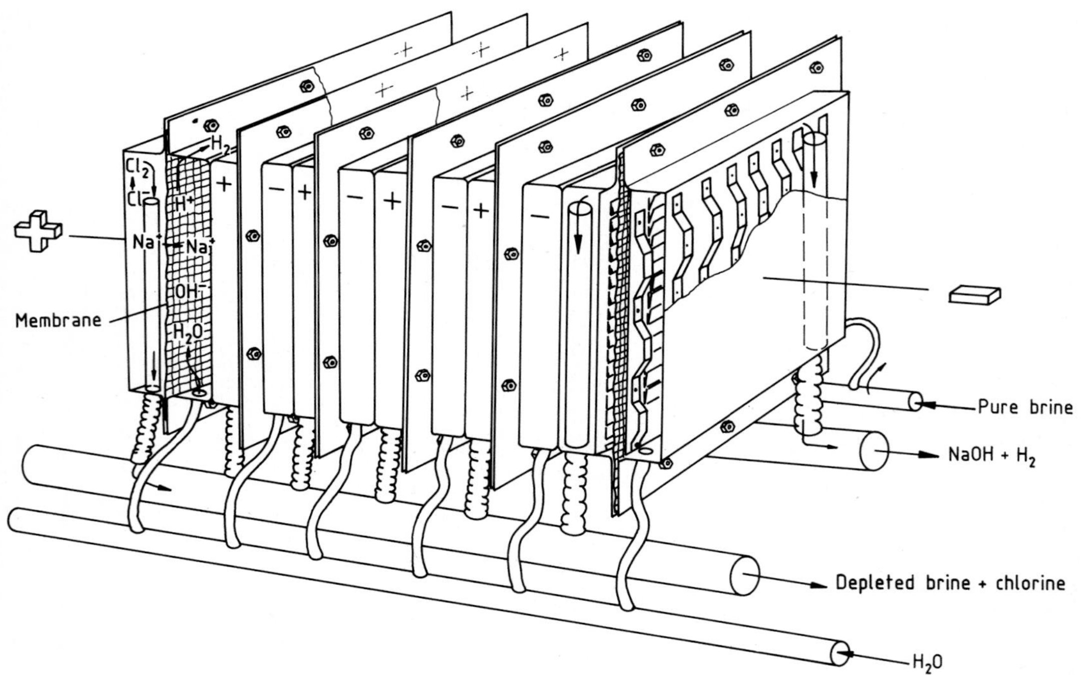

Fig. 1.7.2-8. Membrane process electrolysis cell (bipolar Hoechst-Uhde membrane cell).

from the anode chambers using overflow pipes and the sodium hydroxide and hydrogen from the cathode chambers likewise. An electrical energy consumption of 2100 to 2450 kWh/t NaOH produced has been cited for a current density of 3 kA/m^2.

1.7.2.3.4 Evaluation of the Mercury, Diaphragm and Membrane Processes

Mercury process:

Advantages:

- Pure 50% sodium hydroxide solution (without evaporation)
- Pure chlorine gas

Disadvantages:

- Higher voltage than with the diaphragm process and hence 10 to 15% higher electrical energy consumption
- More stringent brine purification requirements
- Stringent mercury contamination avoidance measures required

Diaphragm process:

Advantages:

- Utilization of less pure brine
- Lower voltage than in the mercury process

Disadvantages:

- Sodium hydroxide produced is both dilute and chloride-contaminated, evaporation required
- Chlorine gas contains oxygen
- Rigorous measures required to avoid asbestos emission

The economics of the two processes are comparable. A decision for one or other process is determined, amongst other things, by: the quality of the available salt, the cost of thermal and electrical energy and the statutory limitations in the particular locality.

Economics of the mercury and diaphragm process similar

Membrane process:

Advantages:

- Pure sodium hydroxide
- Electrical energy consumption only about 77% of that of the mercury process
- No mercury or asbestos used

Disadvantages:

- Sodium hydroxide content only ca. 33% by weight
- Chlorine gas contains oxygen
- Very high purity brine required
- Present high cost and short lifetime of the membranes

Economics of the membrane process superior to those of the mercury and diaphragm processes

1.7.2.4 Applications for Chlorine and Sodium Hydroxide

1.7.2.4.1 Chlorine

More than 70% of USA chlorine consumption utilized in the production of organic chemicals

A summary of the quantities of chlorine used in the USA in 1982 for different applications is given in Table 1.7.2-4.

Table 1.7.2-4. Quantities of Chlorine Consumed in the USA in 1982 for Different Applications in 10^6 t.

Total	All organic chemicals	Only C_2 chemicals	Paper & pulp	Inorganic chemicals	$TiCl_4$	Water treatment	Cleaning & sanitation
8.580	5.922	3.304*	1.240	0.714	0.357	0.400	0.304

* mainly vinyl chloride

An increase in the utilization of chlorine in the production of organic chemicals is expected in 1987. The USA utilization spectrum does not hold for other regions.

1.7.2.4.2 Sodium Hydroxide

The utilization spectrum of the 10.4×10^6 t of sodium hydroxide consumed in the USA in 1985 is given in Table 1.7.2-5 in %.

Table 1.7.2-5. Utilization Spectrum of the USA Consumption of Sodium Hydroxide in 1985 in %.

Organic chemicals	Inorganic chemicals	Paper & pulp	Export	Soap & cleaning agents	Oil industry	Textiles	Bauxite digestion	Miscellaneous
30	20	20	10	5	5	4	3	3

References for Chapter 1.7.2: Choralkali Electrolysis, Chlorine, Sodium Hydroxide

Commercial Information:

"Chloralkalies profit from plant shutdowns", *Chem. Eng. News*, pp 11–19 (Sept. 8, 1986).

"New entries vie in chlorine technology", *Chem. Week*, pp 22–23 (Aug. 13, 1986).

"Chlorine: tight market, tight profits", *Chem. Week*, pp 18–20 (Sept. 10, 1986).

"Japan abandons mercury cells", *Chem Week*, pp 11–12 (July 23, 1986).

"Strong chlorine demand hits caustic soda", *Eur. Chem. News*, **21,** p. 8 (July 1986).

"Clore – un recul confirme"(Chlorine – a setback confirmed), *Informations Chimie*, **n°272,** pp 175–198 (May 1986).

Chemical Economics Handbook, Stanford Research Institute, Menlo Park, California, USA, "Chlorine/Sodium Hydroxide" (July 1984).

Chemical Profile: "Caustic Soda", Schnell Publishing Company, Inc. (May 5, 1986).

Technical Information:

F. Y. Masuda, "Recent developments of the chlorine industry in Japan", *J. Appl. Electrochem.*, **16,** pp 317–331 (1986).

V. H. Thomas and E. J. Rudd, "Energy saving advances in the chloralkali industry", in: *Modern Chloralkali Technology*, Vol. II, Ed. C. Jackson, Ellis Horwood Ltd., Chichester (GB) (1983).

R. D. Varjian, "Energy analysis of the diaphragm chloralkali cell", Lectures in Electrochemical Engineering, *AIChE Symposium Series*, pp 219–226 (1981).

V. H. Thomas and R. D. Penny, "Review of mercury cathode chlorine technology", Lectures in Electrochemical Engineering, *AIChE Symposium Series*, pp 227–233 (1981).

K. Hannesen und D. Bergner, "Eine Membranzelle bipolarer Konstruktion zur Erzeugung von Chlor und Natronlauge" (A membrane cell of bipolar construction for the production of chlorine and caustic soda), *Chem. Ing.-Tech.*, **58,** pp 236–237 (1986).

Sodium Chloride:

Kirk-Othmer: *Encyclopedia of Chemical Technology*, 3rd Ed., Vol. 21, pp 205-223, John Wiley & Sons, New York-Chichester-Brisbane-Toronto (1983).

"Sodium Chloride", *Chemical Economics Handbook*, Stanford Research Institute, Menlo Park, California, USA (June 1984).

Reviews:

Kirk-Othmer: *Encyclopedia of Chemical Technology*, 3rd Ed., Vol. 1, pp 799–865, John Wiley & Sons, New York-Chichester-Brisbane-Toronto (1978).

R. W. Purcell: "The Chloralkali industry" and A. Campell: "Chlorine and chlorination", in: *The Modern Inorganic Chemicals Industry*. Ed. R. Thompson, The Chemical Society, Burlington House, London (1977), pp 106–133 and pp 134–148.

F. R. Minz und R. Schliebs, "Moderne Verfahren der Großchemie: Chlor und Natronlauge" (Modern processes in the heavy chemicals industry: chlorine and caustic soda), *Chemie in unserer Zeit* **12,** pp 135–141 (1978).

Winnacker-Küchler, *Chemische Technologie*, 4th Ed., Vol. 2, pp 379-480, Anorganische Technologie I (Inorganic technology I), Carl Hanser Verlag, München (1982).

1.7.3 Hydrochloric Acid – Hydrogen Chloride

1.7.3.1 Manufacture of Hydrogen Chloride

HCl:

from $NaCl + H_2SO_4$ (of little importance)

Hydrogen chloride is produced:

- As a byproduct in the reaction of sodium chloride with sulfuric acid to sodium sulfate:

$$NaCl + H_2SO_4 \longrightarrow NaHSO_4 + HCl$$
$$NaHSO_4 + NaCl \longrightarrow Na_2SO_4 + HCl$$
$$2NaCl + H_2SO_4 \longrightarrow Na_2SO_4 + 2HCl$$

Sulfur dioxide and aerial oxygen can be used instead of sulfuric acid (Hargreaves process). Ca. 40 to 50×10^3 t/a are produced in the USA using this process (1 to 2% of the total production).

HCl:

from $H_2 + Cl_2$ (ca. 8% of the USA production)

- By the direct reaction of hydrogen with chlorine

$$H_2 + Cl_2 \longrightarrow 2HCl$$

This process is strongly exothermic (flame temperature $> 2000\,^\circ C$) and is especially used when particularly pure hydrogen chloride (hydrochloric acid) is required e.g. in the food sector. It places considerable demands on the construction materials of the plant, particularly that of the burner for which quartz or graphite are preferred. The synthesis furnace and the adjoining cooler can be constructed of steel when dry chlorine and dry hydrogen are used.

220 to 240×10^3 t of hydrogen chloride have been produced in the USA in recent years using this process, ca. 8% of the total production.

HCl:

byproduct in the production of organic and inorganic chemicals (ca. 90% in the USA)

- As byproducts in chlorinations, halogen exchange reactions and in the production of organic and inorganic chemicals.

  Typical examples are:

  - Manufacture of aliphatic and aromatic chlorohydrocarbons,
  - Manufacture of isocyanates by the reaction of amines with phosgene,

- Manufacture of fluorohydrocarbons by the reaction of chlorohydrocarbons with hydrogen fluoride,
- Manufacture of phosphoric acid and phosphorous acid esters (Sec. 1.5.2.2),
- Manufacture of high surface area silica by the flame hydrolysis of silicon tetrachloride (Sec. 5.8.4.1.1).

1.7.3.2 Economic Importance of Hydrogen Chloride and Hydrochloric Acid

The production and consumption statistics for hydrogen chloride and hydrochloric acid are relatively uncertain, since in many cases it is recycled either directly (e.g. in the oxychlorination process) or indirectly (see below). Table 1.7.3-1 gives an estimate of the 1985 hydrochloric acid and hydrogen chloride production.

USA hydrochloric acid production in 1985:

ca. 3×10^6 t (as HCl)

Excess HCl is chemically or electrochemically converted into H_2 + Cl_2 or is disposed of by deep well injection

Table 1.7.3-1. Hydrochloric Acid and Hydrogen Chloride Production in 10^6 t/a (as HCl).

USA in 1985	Canada in 1981	W. Europe in 1978	W. Germany in 1978	Japan in 1980
ca. 3.00	0.186	2.000	0.880	1.000

Hydrochloric acid is marketed in the USA as muriatic acid with 20° Baumé (31.45% by weight HCl) and 22° Baumé (35.21% by weight HCl). There are diverse applications for hydrochloric acid e.g. for:

- Pickling of metals (250 to 300×10^3 t/a)
- Acid treatment of oil and gas wells (250 to 300×10^3 t/a)
- Neutralization in inorganic and organic chemistry
- Hydrolysis of proteins and carbohydrates
- Regeneration of ion exchangers, which are utilized in the purification of glucose and high-fructose corn syrup (HFCS) (250 to 300×10^3 t/a)
- Chlorine dioxide manufacture (see Sec. 1.7.4.2.5).

The figures in brackets refer to the estimated USA consumption. Hydrogen chloride is utilized, e.g., in the manufacture of:

- Methyl and ethyl chloride
- Chlorosulfonic acid

Since the amount of hydrogen chloride byproduct often exceeds demand, it is necessary to convert it partly into chlorine. This is particularly necessary when it is a byproduct at sites at which there is no further use for it. This is carried out both by electrolysis (Sec. 1.7.3.3) and by a modified Deacon process (Sec. 1.7.3.4). In the USA hydrochloric acid is also disposed of by deep well injection.

1.7.3.3 Electrolysis of Aqueous Hydrochloric Acid Solutions

The decomposition voltage for the electrolysis of hydrochloric acid is ca. 2V.

$$2HCl \xrightarrow{e^-} Cl_2 + H_2$$

Electrolysis of 23% hydrochloric acid with graphite electrodes yields $H_2 + Cl_2$ and ca. 20% acid. Anodic and cathodic chambers are separated by a PVC-cloth diaphragm

The electrolysis cell is similar to the membrane cell, with PVC-cloth acting as the diaphragm. The bipolar electrodes are graphite. Small quantities of platinum group compounds may be added to the cathode compartment to reduce the overvoltage. Fig. 1.7.3-1 shows a schema of hydrochloric acid electrolysis.

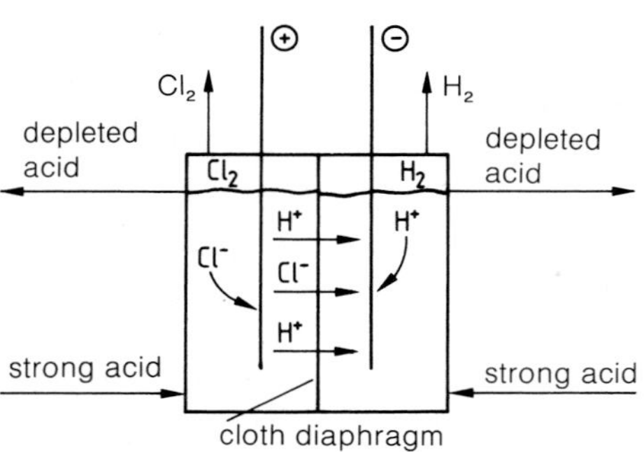

Fig. 1.7.3-1. Schema of hydrochloric acid electrolysis.

Ca. 23% by weight hydrochloric acid is fed into both the cathode and anode chambers, part of the electrolyte diffusing into the cathode chamber. The depleted acid leaving the chambers has a concentration of 17 to 20%. Hydrogen chloride from the production of organic chemicals (see above) is adiabatically absorbed in this depleted acid, the resulting heat being used to evaporate part of the water together with steam distillable organic impurities. The purity of the hydrochloric acid used is important and postpurification with activated carbon can be necessary.

The electrode gap is ca. 6 mm. The yield is 97 to 98% at a current density of 4000 A/m^2. Large-scale plants produce ca. 70×10^3 t/a of chlorine.

Capacity of large-scale units: up to ca. 70×10^3 t/a

Ca. 350×10^3 t/a of chlorine are currently produced using this process. Its economics are similar to those of the mercury process.

Total production: 350×10^3 t/a chlorine

1.7.3.4 Nonelectrolytic Processes for the Manufacture of Chlorine from Hydrogen Chloride

The nonelectrolytic processes for the manufacture of chlorine from hydrogen chloride (Deacon, air oxidation of hydrogen chloride; Weldon, manganese dioxide oxidation of hydrogen chloride) which marked the beginning of industrial chlorine chemistry, are almost insignificant nowadays.

However, Du Pont in Corpus Christi (USA) brought a modified Deacon process on stream in 1975, the

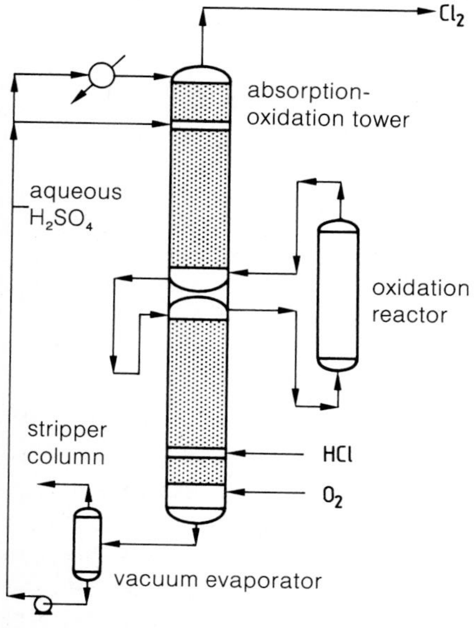

Fig. 1.7.3-2. Schema of the Kel-Chlor® process.

Kellog Kel-Chlor® process. In this process the hydrogen chloride produced as a byproduct in the manufacture of fluorohydrocarbons is oxidized. Fig. 1.7.3-2 shows a schema of this process.

Kel-Chlor® process: oxidation of HCl with O_2 to Cl_2 with nitrogen-oxygen compounds as catalysts. Capacity: 260×10^3 t/a

Hydrogen chloride is oxidized with oxygen under pressure in the presence of catalytically acting nitrogen-oxygen compounds, with concentrated sulfuric acid removing the water formed. The capacity of the plant amounts to 260×10^3 t/a of chlorine.

To date no further plants have been built using this process.

References for Chapter 1.7.3: Hydrochloric Acid — Hydrogen Chloride

Commercial Information:

C. J. Verbanic, "Hydrochloric, the acid nobody knows", *Chem. Business,* pp 34–37 (June 1986).

Chemical Economics Handbook, Stanford Research Institute, Menlo Park, California, USA, "Hydrochloric acid" (Oct. 1982).

Reviews:

Kirk-Othmer: *Encyclopedia of Chemical Technology,* 3rd Ed., Vol. 12, pp 983–1015, John Wiley & Sons, New York-Chichester-Brisbane-Toronto (1980).

F. R. Minz und R. Schliebs, "Moderne Verfahren der Großchemie: Chlor und Natronlauge" (Modern processes in the heavy chemicals industry: chlorine and caustic soda), *Chemie in unserer Zeit* **12,** pp 142–145 (1978).

Ullmanns Encyklopädie der technischen Chemie, 4th Ed.. Vol. 8, pp 355–359, Verlag Chemie, Weinheim-Deerfield Beach-Basel (1975).

Ullmanns Encyklopädie der technischen Chemie, 4th Ed., Vol. 20, pp 363–374, Verlag Chemie, Weinheim-Deerfield Beach-Basel (1981).

Kel-Chlor® Process:

"Chlorine recovery from hydrogen chloride", in: *Encyclopedia of Chemical Processing and Design.* Ed.: J. McKetta, Vol. 8, pp 95–114, Marcel Dekker, Inc., New York-Basel (1978).

1.7.4 Chlorine-Oxygen Compounds

1.7.4.1 Economic Importance

Sodium hypochlorite USA consumption in 1982 in households, swimming pools and water treatment: 210×10^3 t

Hypochlorite: Sodium hypochlorite is generally used in situ, so production and consumption statistics are difficult to come by. 210×10^3 t of sodium hypochlorite were consumed in 1982 in the USA: in the household sector, in swimming pools for disinfection and in water treatment. Atochem quoted a 1980 capacity in France of 50×10^3 t/a and ICI a capacity in Great Britain of 250×10^3 t/a.

The 1986 USA capacity for calcium hypochlorite, $Ca(OCl)_2$, was 106.5×10^3 t/a. $Ca(OCl)_2$ is mainly utilized in the swimming pool, bleaching and disinfection sectors, but is becoming increasingly supplanted by chlorinated isocyanurates.

Calcium hypochlorite:
1986 USA capacity: 106.5×10^3 t/a

Sodium chlorite: The 1980 capacity of Atochem in France was 50×10^3 t/a, the capacity of Aragonensas in Spain is 1.5×10^3 t/a and that of Olin in the USA is ca. 5×10^3 t/a.

Sodium chlorite capacities in 10^3 t/a:
Atochem (France): 50
Aragonensas (Spain): 1.5
Olin (USA): ca. 5

Sodium and potassium chlorate:

Table 1.7.4-1. 1984 Capacities of Sodium Chlorate in 10^3 t/a.

USA	Canada	Western Europe	Japan
355	398	376	75

Sodium chlorate:
1984 capacities in 10^3 t/a:
- USA: 355
- Canada: 398

A 1985 capacity for potassium chlorate in Sweden of 14×10^3 t/a has been quoted. It is not produced in the USA.

Perchlorates and perchloric acids: The consumption of perchloric acid (70%) is very low. The USA consumption is estimated to be ca. 450 t/a. The most important perchlorate is ammonium perchlorate. Its USA capacity is ca. 22×10^3 t/a.

Ammonium perchlorate:
most important perchlorate
USA capacity: ca. 22×10^3 t/a

Chlorine dioxide: The consumption of chlorine dioxide can be adduced from the consumption of sodium chlorate for the manufacture of chlorine dioxide, assuming a 90% yield of chlorine dioxide. This does not take into account the relatively small quantities of chlorine dioxide produced from sodium chlorite. The USA consumption of chlorine dioxide in 1984 was 180×10^3 t, 215×10^3 t being predicted for 1988.

Chlorine dioxide consumption in the USA:
1984: ca. 180×10^3 t
1988: 215×10^3 t (est.)

1.7.4.2 Manufacture of Chlorine-Oxygen Compounds

1.7.4.2.1 Hypochlorite

In Solution

Solutions of sodium and calcium hypochlorite in mixtures with sodium and calcium chloride can be easily obtained by reacting aqueous sodium hydroxide or calcium hydroxide slurries with chlorine:

Hypochlorite from bases and chlorine with cooling

$$2NaOH + Cl_2 \longrightarrow NaOCl + NaCl + H_2O$$

$$Ca(OH)_2 + Cl_2 \longrightarrow Ca(OCl)Cl + H_2O$$

The solutions ("bleaching solutions") contain approximately equimolar quantities of chloride and hypochlorite ions.

Content given as "available chlorine"

Sodium hypochlorite solutions contain 12 to 15% and calcium hypochlorite solutions 3 to 3.8% of available chlorine. Available chlorine is the quantity of chlorine produced upon adding hydrochloric acid relative to the weight of the product:

$$NaOCl + 2HCl \longrightarrow NaCl + Cl_2 + H_2O\ .$$

The reaction of sodium hydroxide with chlorine is strongly exothermic (heat of reaction = 103 kJ/mol). Production can be carried out continuously and is monitored by redox potential measurements. Since hypochlorite is easily converted to chlorate at higher temperatures, the reaction temperature must be kept below 40 °C. Coolers constructed from titanium are used. In general, the chlorination is carried out in such a way that a slight excess of alkali is retained so as to increase the stability of the solution. The same holds for the manufacture of calcium hypochlorite solutions.

Bleaching solutions are generally further used in situ, because they are easily decomposed by light or traces of heavy metals.

Direct manufacture of hypochlorite solutions by electrolysis of sodium chloride solutions in diaphragmless cells for small consumers

The manufacture described above assumes the availability of chlorine. Direct electrolytic manufacture of hypochlorite solutions is appropriate for special applications (cooling of power stations with seawater, effluent treatment etc.), when chlorine is unavailable.

This is achieved by electrolysis of seawater or brine in cells without diaphragms or membranes. The anodes are of activated titanium, the cathodes of titanium. The 40 to 60% current yield is relatively poor, due to the hydrogen produced at the cathode reducing the hypochlorite formed. The electrolysis cells are technically uncomplicated and small. The hypochlorite solutions obtained contain several grams of hypochlorite per L.

Solid Hypochlorite

Bleaching powder (chloride of lime), which was first used industrially at the beginning of the nineteenth century, was for over a century the only transportable form of chlorine from which chlorine could be made available by acidification with hydrochloric acid. It contains 36% of available chlorine. Since the transportation of liquid chlorine became technically feasible at the beginning of this century, the manufacture of bleaching powder has steadily decreased in importance. It is manufactured by reacting moist calcium hydroxide with chlorine. This reaction is fairly slow.

Bleaching powder (chloride of lime):

formerly only transportable form of chlorine. Importance now markedly reduced

Calcium hypochlorite: There are many processes for the manufacture of calcium hypochlorite ("high percentage bleaching powder"). The oldest is the Griesheim Elektron process ("Perchloron process") in which a calcium hydroxide suspension is chlorinated to the extent that the calcium chloride formed mostly dissolves, but not the calcium hypochlorite.

Calcium hypochlorite:

high percentage bleaching powder. Produced e.g. by chlorination of calcium hydroxide suspensions and separation of the calcium hypochlorite as the dihydrate

$$2Ca(OH)_2 + 2Cl_2 \longrightarrow Ca(OCl)_2 \cdot 2H_2O + CaCl_2$$

The calcium hypochlorite, which precipitates as the dihydrate, is filtered off and dried.

In the Olin process, a calcium hydroxide suspension in a sodium hypochlorite solution is chlorinated and a triple-salt precipitates out upon cooling to $-15\,^{\circ}C$:

$$Ca(OH)_2 + 2NaOCl + Cl_2 + 11H_2O \longrightarrow Ca(OCl)_2 \cdot NaOCl \cdot NaCl \cdot 12H_2O$$

This triple-salt reacts with a bleaching powder suspension to form calcium hypochlorite dihydrate which is filtered off and dried.

$$Ca(OCl)_2 \cdot NaOCl \cdot NaCl \cdot 12H_2O + Ca(OCl)Cl \longrightarrow 2Ca(OCl)_2 \cdot 2H_2O + 2NaCl + 10H_2O$$

The sodium chloride byproduct can be utilized in chloralkali electrolysis.

Other manufacturing processes, such as the ICI, Thann and Pennwalt processes, are modifications of this process.

In the PPG process, chlorine is converted in a carbon dioxide stream with sodium carbonate to chlorine monoxide and hypochlorous acid, which are dissolved in water. Reaction with a calcium hydroxide slurry yields calcium hypochlorite:

$$Ca(OH)_2 + 2HOCl \longrightarrow Ca(OCl)_2 + 2H_2O$$

The available chlorine content of technical grade calcium hypochlorite is 70 to 74%. The material reacts vigorously with oxidizable materials and decomposes exothermically upon ignition.

"Chlorinated trisodium phosphate" $[Na_3PO_4 \cdot 11H_2O]_4 \cdot NaOCl$ produced from trisodium phosphate and sodium hypochlorite solutions

Solid sodium hypochlorite is not a commercial product, because it is too chemically unstable. "Chlorinated trisodium phosphate", $[Na_3PO_4 \cdot 11H_2O]_4 \cdot NaOCl$, on the other hand, is industrially important in the cleaning agent sector. This has an available chlorine content of ca. 3.65%. It is obtained by reacting a sodium phosphate solution with a sodium hypochlorite solution at 75 to 80 °C in the appropriate molar concentrations.

In the USA lithium hypochlorite is also manufactured in small quantities.

1.7.4.2.2 Chlorites

Sodium chlorite:

from chlorine dioxide, sodium hydroxide and hydrogen peroxide as a reducing agent

Only sodium chlorite is industrially significant. It is manufactured by reacting chlorine dioxide (Sec. 1.7.4.2.5) with sodium hydroxide and a reducing agent, usually hydrogen peroxide (other reducing agents have technical disadvantages):

$$2ClO_2 + 2NaOH + H_2O_2 \longrightarrow 2NaClO_2 + 2H_2O + O_2$$

A clear excess of hydrogen peroxide is necessary, since part of it decomposes in the alkaline solution.

Due to its easily initiated exothermic self-decomposition, solid sodium chlorite is either marketed as the monohydrate or as mixtures with sodium chloride or sodium nitrate.

1.7.4.2.3 Chlorates

Sodium and potassium chlorate are used industrially. The latter is produced by metathesis with potassium chloride:

Potassium chlorate:

from sodium chlorate by metathesis

$$NaClO_3 + KCl \longrightarrow KClO_3 + NaCl$$

Sodium chlorate is industrially produced electrochemically from sodium chloride in a diaphragmless cell:

Sodium chlorate:

from sodium chloride by electrolysis in diaphragmless cells with anodes of activated titanium and steel cathodes. The electrochemical reaction follows a slow chemical reaction

$$NaCl + 3H_2O \longrightarrow NaClO_3 + 3H_2$$

(The formerly operated purely chemical process is no longer of any importance.) Steel cathodes and anodes of platinum or titanium activated by a ruthenium oxide/titanium oxide mixture are used in the chlorate cells, the latter supplanting the graphite anodes previously used. The electrode separation is ca. 3 to 5 mm. The electrolysis is carried out at ca. 80 °C, at a voltage of 3.0 to 3.5 V and an energy consumption of 4700 to 5400 kWh/t sodium chlorate, the latter accounting for ca. 50% of the manufacturing costs (assuming an electricity price of ca. 4 US c/kWh). The cathodically formed hydrogen is contaminated with small amounts of chlorine and hence has to be appropriately processed e.g. by stripping with water. There is a large number of manufacturers of chlorate cells.

The chemical and electrochemical processes in the cells are very complex. The following sequence of reactions is now assumed to take place:

- Anodically produced chlorine reacts with water to hypochlorous acid and with the cathodically produced hydroxide ions to hypochlorite anions.

$$Cl_2 + H_2O \longrightarrow HOCl + HCl$$

$$OH^- + Cl_2 \longrightarrow OCl^- + HCl$$

- Hypochlorous acid and hypochlorite anions react to form chlorate anions as follows:

$$2HOCl + ClO^- \longrightarrow ClO_3^- + 2HCl$$

As this reaction is relatively slow, the electrolysis recycling unit has a large dead volume to allow for the completion of the reaction.

The most important side reaction is the anodic electrochemical oxidation of the hypochlorite anion to chlorate anions, which can be approximately represented as follows:

$$6ClO^- + 3H_2O \xrightarrow{-6e^-} 2ClO_3^- + 1.5O_2 + 6H^+ + 4Cl^-$$

The simultaneous formation of oxygen reduces the electricity yield to a third less than that of the purely chemical formation of chlorate. Every attempt is therefore made to suppress the electrochemical oxidation of hypochlorite by maintaining certain process parameters for the concentrations, temperature (60 to 75 °C), pH value (6.9), flow conditions and residence time in the electrolysis cell. Modern units have an electricity yield $> 93\%$.

Another important side reaction is the cathodic reduction of hypochlorite anions:

$$ClO^- + H_2O \xrightarrow{+2e^-} Cl^- + 2OH^-$$

This is, to a large extent, suppressed by the addition of ca. 3 g of chromate/L electrolyte. This coats the cathode with a layer of hydrated chromium oxide, which strongly hinders the diffusion of hypochlorite anions to it. The sodium chlorate produced during the electrolysis is worked up as crystals from solutions containing ca. 600 g/L sodium chlorate and 100 g/L sodium chloride by evaporation, the sodium chloride precipitating out first. Technical grade sodium chlorate is 99.5% pure. It decomposes above 265 °C forming the perchlorate. To an increasing extent, sodium chlorate is being supplied in solution, in some cases without separating off the sodium chloride, to save energy, but at the cost of strongly increased transport charges.

In the so-called "Munich process" for the manufacture of chlorine dioxide (Sec. 1.7.4.2.5), the chlorate-containing electrolyte is directly reacted with hydrochloric acid.

1.7.4.2.4 Perchlorates and Perchloric Acid

Sodium, potassium and ammonium perchlorate are particularly important industrially. Potassium and ammonium perchlorate are produced by metathesis with sodium perchlorate, which is itself manufactured from sodium chlorate by electrochemical synthesis:

Sodium perchlorate:

produced by electrochemical oxidation of sodium chlorate in diaphragmless cells on lead oxide or platinum anodes

Potassium and ammonium perchlorate:

from sodium perchlorate by metathesis

$$NaClO_3 + H_2O \longrightarrow NaClO_4 + H_2$$

Oxygen is produced anodically as a byproduct. The cells and cathodes of the diaphragmless cells consist of steel, the anodes of platinum or β-lead dioxide on graphite. The cell voltages are about 4.75 V (lead dioxide) or 6 V (platinum). The energy consumption is 2.5 to 3 kWh/kg sodium perchlorate. When platinum anodes are used chromate can be added to suppress the cathodic reduction. Some platinum dissolves, particularly at higher temperatures and at sodium chlorate concentrations below 100 g/L. When lead dioxide anodes are used, addition of sodium fluoride effects an improvement in the electricity yield.

The perchlorate formed is recovered by crystallization.

Perchloric acid is produced using the Merck process in which chlorine dissolved in cold (ca. 0 °C) perchloric acid is electrochemically oxidized:

Perchloric acid:

by electrochemical oxidation of chlorine dissolved in perchloric acid

$$Cl_2 + 8H_2O \longrightarrow 2HClO_4 + 7H_2$$

Diaphragm cells (with plastic cloth diaphragms) are used with platinum anodes and silver cathodes. A silver cooler is utilized to remove the heat produced. The cell voltage is 4.4 V, the electricity yield 60%. Part of the electrolyte is continuously taken off and concentrated to ca. 70% perchloric acid.

1.7.4.2.5 Chlorine Dioxide

Of the oxides of chlorine only chlorine dioxide has achieved industrial significance. It is a gas at room temperature. As a result of its explosive properties, it can only be utilized in situ and even then has to be diluted with inert gases (nitrogen, carbon dioxide) to 10 to 15% (by volume).

Manufacture of larger quantities of chlorine dioxide: by reaction of sodium chlorate with hydrochloric or sulfuric acid and separation or reduction of the simultaneously formed chlorine

The gaseous, explosive chlorine dioxide has to be immediately diluted with inert gases

Sodium chlorite is used for small quantities. When large quantities are required, sodium chlorate is used as the starting material and is reacted with hydrochloric acid:

$$NaClO_3 + 2HCl \longrightarrow ClO_2 + 0.5\,Cl_2 + NaCl + H_2O$$

The most important side reaction is the augmented formation of chlorine:

$$NaClO_3 + 6HCl \longrightarrow 3Cl_2 + NaCl + 3H_2O$$

Sulfuric acid and sodium chloride can be used instead of hydrochloric acid. The simultaneous addition of sulfur dioxide reduces the chlorine to chloride, so that the otherwise necessary separation of chlorine from chlorine dioxide, by e.g. stripping with water, is no longer necessary (chlorine dioxide is much more soluble in water than chlorine).

In modern units, 90% of the theoretical yield of chlorine dioxide is obtained.

A number of the industrial processes follow the reaction scheme described above. The ER processes from Erco and the SVP processes from Hooker are mainly operated in the USA and Canada. Sodium sulfate disposal problems arise when these processes are operated with sulfuric acid or sulfur dioxide.

These problems do not arise in the Kesting (Munich) process in which the solution produced by the electrolytic manufacture of sodium chlorate is immediately reacted with hydrochloric acid and the chlorine dioxide-chlorine mixture separated from the solution in the column by blowing. After separation the chlorine is reacted to hydrogen chloride with hydrogen produced by electrolysis and is fed back into the process. The only starting material for chlorine dioxide manufacture is thus the chlorine, which is present in the chlorine dioxide obtained and which is fed back into the process as hydrochloric acid.

Manufacture of smaller quantities of chlorine dioxide:

reaction of sodium chlorite with chlorine

Sodium chlorite as starting material: Small quantities of chlorine dioxide are produced by reacting sodium chlorite with chlorine:

$$2NaClO_2 + Cl_2 \longrightarrow 2NaCl + 2ClO_2$$

1.7.4.3 Applications of Chlorine-Oxygen Compounds

Hypochlorite: Sodium hypochlorite (bleaching solution) is utilized for the bleaching and decolorization of pulp and textiles, for disinfection, e.g. in swimming baths, and for the manufacture of hydrazine (Sec. 1.4.2).

Applications of

- Hypochlorites:
  - bleaching
  - disinfection
  - neutralization of poison gases
  - hydrazine manufacture

Calcium hypochlorite and bleaching powder are utilized for disinfection, for the treatment of cooling water, in swimming baths and for rendering poison gases of the "2,2′-dichloroethylsulfide"-type harmless.

"Chlorinated trisodium phosphate" is a component of household and industrial cleaning agents, particularly in the USA.

Sodium chlorite is used for the small-scale manufacture of chlorine dioxide, mainly at water treatment plants. In the USA sodium chlorite is five times dearer than sodium chlorate.

- Sodium chlorite:
  - manufacture of chlorine dioxide for small-scale users
- Sodium chlorate:
  - manufacture of chlorine dioxide and perchlorates
  - in uranium extraction
  - as a herbicide
- Potassium chlorate:
  - e.g. match manufacture

Chlorate: More than 88% of the sodium chlorate produced in the USA is converted into chlorine dioxide for pulp bleaching. It is further utilized (figures in brackets refer to the % consumption in the USA) as a starting material for the manufacture of other chlorine-oxygen compounds (7%), for the oxidation of U^{4+} in the extraction of uranium (2%) and as a herbicide (2%). Potassium chlorate is used in the manufacture of fireworks and matches. In the USA, 9 to 10×10^3 t/a of potassium chlorate is consumed in the manufacture of matches.

Perchlorates are mainly utilized in fireworks and, especially ammonium perchlorate, as an oxidation agent in rocket fuel.

- Ammonium perchlorate:
  - oxidizing agent in rocket fuels

Chlorine dioxide is utilized as a bleaching agent for wood pulp. Its main advantage over chlorine is that no chlorolignin is formed. Chlorine dioxide can also be utilized in the treatment of potable water, less chlorinated hydrocarbons being formed than with chlorine treatment.

- Chlorine dioxide:
  - pulp bleaching
  - water treatment

References for Chapter 1.7.4: Chlorine-Oxygen Compounds

Commercial Information:

"Chemical makers (of bleaches) play hard-nosed poker in the pool", *Chem. Week,* pp 10–11 (March 7, 1984).

Chemical Profile: "Calcium hypochlorite", Schnell Publishing Company, Inc. (July 1, 1986).

Chemical Economics Handbook, Stanford Research Institute, Menlo Park, California, USA, "Sodium chlorate" (Feb. 1985).

"Chemical Profile: Sodium chlorate", *Chemical Marketing Reporter* (June 18, 1984).

"Sodium chlorate from Canada continues to swamp the States", *Chemical Marketing Reporter,* p. 26 (Feb. 17, 1986).

"Kaliumchlorat" (Potassium chlorate), *Europa Chemie,* p. 542 (1984).

Technical Information:

J. E. Colmam, "Electrolytic production of sodium chlorate", Lectures in Electrochemical Engineering, *AIChE Symposium Series,* pp 244–263 (1981).

W. Roeske, "Chlordioxid für die Wasseraufbereitung – Eigenschaften und Anwendung" (Chlorine dioxide for water treatment – properties and utilization), *bbr (Brunnenbau, Bau von Wasserwerken, Rohrleitungsbau)* **36,** pp 39–41 (1985).

D. Christ und W. Roeske, "Chlordioxid für die Wasseraufbereitung – Herstellung, Dosierung und Messung" (Chlorine dioxide for water treatment – production, dosing and measurement), *bbr* **36,** pp 457–460 (1985).

P. Wintzer, "Entwicklung und Trend der Chlordioxid-Bleiche mit integrierter Chlorat-Elektrolyse für die Zellstoff-Industrie" (Developments and trends in chlorine dioxide-bleaching with integrated chlorate electrolysis for the wood pulp industry), *Chem. Ing.-Tech.* **52,** pp 392–398 (1980).

Reviews:

Kirk-Othmer: *Encyclopedia of Chemical Technology,* 3rd Ed., Vol. 5, pp 580–667, John Wiley & Sons, New York-Chichester-Brisbane-Toronto (1979).

Ullmanns Encyklopädie der technischen Chemie, 4th Ed., Vol. 9, pp 542–572, Verlag Chemie, Weinheim-Deerfield Beach-Basel (1975).

1.7.5 Bromine and Bromine Compounds

1.7.5.1 Natural Deposits and Economic Importance

Bromine, as bromide ions, occurs in:

- Seawater
- Natural brines
- Salt deposits

Bromine mainly occurs in nature as soluble bromides. The concentration of bromine in the earth's crust is $1.6 \times 10^{-4}\%$.

It occurs in:

- Seawater: 0.065 g/L
- Dead Sea: 4 g/L
- Natural brines: 3 to 4 g/L
- Salt deposits: 0.005 to 0.45% (by weight)

The most important bromine deposits are in Arkansas (USA) and the Dead Sea (Israel/Jordan).

The residual solutions produced in the potash industry contain up to 6 g of bromide/L.

Bromine reserves virtually unlimited

The world reserves of bromine are virtually unlimited. The bromine content of the Dead Sea alone is estimated to be 10^9 t.

Production:

The bromine production of various countries is given in Table 1.7.5-1.

Table 1.7.5-1. Bromine Production in Various Countries in 10^3 t.

USA (1985)	Israel (1985)	Great Britain (1980)	USSR (1985)	France (1980)	Japan (1980)	West Germany (1980)	Italy (1985)
159	97	27	15	14	12	4	1

1985 bromine production in 10^3 t:

- USA: 159
- Israel: 97

The capacity of the firms producing bromine in the USA: Great Lakes Chemicals, Ethyl Corporation (which acquired the Dow Chemical bromine plants on Jan 1, 1987), Morton Thiokol and Arkansas Chemical was 288×10^3 t in 1985. The most important inorganic bromine compound is calcium bromide.

1.7.5.2 Manufacture of Bromine and Bromine Compounds

Almost all bromine compounds are manufactured either directly or indirectly from elemental bromine. Its production is therefore of key importance.

Elemental bromine:

starting material for inorganic and organic bromine compounds

1.7.5.2.1 Bromine

Bromine is manufactured from:

- Seawater
- Bromide-enriched starting materials

In both cases the bromide ion is oxidized with elemental chlorine:

Bromine manufacture:

from bromides by chlorine oxidation

$$2Br^- + Cl_2 \longrightarrow 2Cl^- + Br_2$$

and the bromine formed taken off as a gas.

Manufacture of Bromine from Seawater ("Cold Debromination")

For bromine extraction the alkaline seawater has to be acidified, because bromine disproportionates in alkaline solution:

In "cold debromination" (manufacture from seawater), the bromine formed, after acidifying the seawater and introducing chlorine, is expelled with air, absorbed in a sodium carbonate solution and finally liberated by acidification

$$3\,Br_2 + 6\,OH^- \longrightarrow 5\,Br^- + BrO_3^- + 3\,H_2O$$

forming bromide and bromate.

The seawater is therefore acidified with sulfuric acid to a pH of 3.5, 130 g of 100% sulfuric acid being necessary per t seawater. The slight excess of chlorine necessary to oxidize the bromide is fed in at the same time as the sulfuric acid. The bromine formed is expelled by air in so-called "blow-out" towers. The bromine (and possibly chlorine or bromine chloride)-containing air is fed into absorption towers in which it is brought into contact with a sodium carbonate solution, whereupon the bromine is disproportionated into bromide and bromate according to the above equation. The bromine in the absorption solution is then converted into elemental bromine with sulfuric acid and expelled with steam:

$$5\,NaBr + NaBrO_3 + 3\,H_2SO_4 \longrightarrow 3\,Br_2 + 3\,Na_2SO_4 + 3\,H_2O$$

Plants in Great Britain and Japan operate this process. Preconditions for its economic operation are high seawater temperatures and the possibility of disposing of the debrominated solutions so that they do not mix with the freshwater entering the plant.

Manufacture of Bromine from Bromide-enriched Starting Materials – Brines, Residual Solutions from the Potash Industry – ("Hot Debromination")

The bromine formed in the extraction of bromine from brines is expelled with steam ("hot debromination", the industrially more important process)

In this process bromine is also liberated by chlorine oxidation, but it is expelled with steam rather than air. After the steam-bromine mixture is condensed and separated in a settler, the bromine is purified by multistep distillation. The plants are designed to ensure the recycling of bromine-containing aqueous solutions and bromine- or chlorine-containing gases into the reaction tower. The debrominated salt-containing effluent is neutralized before further processing. Its heat content is transferred in countercurrent to the incoming bromine-containing brines. More than 95% of the bromine in the starting solution is extracted.

Due to the high corrosiveness of moist bromine, materials such as glass, tantalum, titanium, ceramics and poly(tetrafluoroethene) have to be utilized in the plants. 0.5 to 0.65 kg of chlorine and 11 kg of steam are consumed per kg of bromine produced.

Most of the bromine produced is extracted using this process.

1.7.5.2.2 Hydrogen Bromide

Hydrogen bromide is manufactured by the combustion of bromine and hydrogen at ca. 500 °C, or the reaction of these two elements over platinum catalysts at 350 °C:

Hydrogen bromide:

- By reacting hydrogen with bromine
- As a byproduct in the bromination of organic compounds

$$H_2 + Br_2 \longrightarrow 2HBr \qquad \Delta H = -103\,kJ$$

Bromine-free hydrogen bromide can be obtained either by passing the reacted gases over hot activated carbon or by using an excess of hydrogen.

Hydrogen bromide is a byproduct in the bromination of organic chemicals:

$$-\overset{|}{\underset{|}{C}}-H + Br_2 \longrightarrow -\overset{|}{\underset{|}{C}}-Br + HBr$$

Hydrogen bromide dissolves in water forming hydrobromic acid. At 25 °C and atmospheric pressure 193 g of hydrogen bromide dissolve in 100 g of water.

1.7.5.2.3 Alkali Bromides, Calcium Bromide, Zinc Bromide

Bromides are mainly produced by the reaction of the corresponding hydroxide, carbonate etc. with hydrobromic acid:

Alkali, alkaline earth and zinc bromides produced by reacting hydroxides and carbonates with hydrobromic acid

$$NaOH + HBr \longrightarrow NaBr + H_2O$$

The Dead Sea Works (Israel) plan a unit with a capacity of 20 to 30 × 10^3 t/a for the extraction of $CaBr_2$ from the waters of the Dead Sea. A pilot plant is already operating successfully.

Ammonium bromide is directly manufactured from bromine by reducing it with ammonia:

$$8NH_3 + 3Br_2 \longrightarrow 6NH_4Br + N_2$$

1.7.5.2.4 Alkali Bromates

Bromates:

mainly by the electrochemical oxidation of bromides

Alkali bromates are produced either by passing bromine into alkali hydroxide solutions, whereupon a mixture of bromide and bromate is initially formed by disproportionation (see Sec. 1.7.5.2.1) and from which the more poorly soluble bromate is separated, or by electrochemical oxidation of bromide [in an analogous process to the manufacture of sodium chlorate (Sec. 1.7.4.2.3)]:

$$NaBr + 3H_2O \longrightarrow NaBrO_3 + 3H_2$$

1.7.5.3 Applications for Bromine and Bromine Compounds

Utilization of bromine in organic chemical manufacture:

- Fuel additives
- Flame retardants
- Plant protection agents
- Fire extinguishing agents
- Tear gases
- Inhalation anaesthetics
- Dyes

Typical examples of the utilization of bromine in organic chemical manufacture are:

- Fuel additives such as 1,2-dibromomethane
  The percentage of total USA bromine consumption utilized in this sector decreased from 80% in 1970 to 50% in 1979. A ban on leaded fuel would reduce its consumption in this sector to zero.
- Flame retardants such as brominated diphenyl ether, hexabromocyclododecatriene, tetrabromobisphenol-A etc.
  There are toxicological reservations about the use of decabromophenyloxide, which have led to restrictions on their use in West Germany.
- Plant protection agents such as 1,2-dibromoethane, methyl bromide and bromine-containing pyrethroids.
  There are considerable toxicological reservations about the use of brominated alkanes.
- Fire extinguishing agents such as bromodifluoromethane, bromotrifluoromethane and bromochlorodifluoromethane

- Tear gases such as bromoacetophenone and bromoacetone
- Inhalation anaesthetics such as 2-bromo-2-chloro-1,1,1-trifluoroethane
- Dyes, e.g. brominated anthraquinones

Hydrogen bromide is mainly used for the manufacture of bromides.

Hydrogen bromide:

for manufacture of bromides

80% of the sodium bromide consumed in the USA is used in the extraction of crude oil in clear drilling, completing and packing fluids. Calcium bromide and zinc bromide which provide still denser solutions (1.7 kg/L for a 53% by weight aqueous solution of $CaBr_2$ and 2.3 kg/L for an aqueous solution containing 55% by weight of $ZnBr_2$ and 20% by weight of $CaBr_2$) are used for the same application. Up to 150 m^3 of packing fluid are used per well. Its consumption is presently decreasing, due to the current decline in the number of wells coming on stream.

Sodium, calcium and zinc bromides:

mainly used in the form of aqueous solutions in the crude oil/natural gas sector as clear drilling, completing and packing fluids

Lithium bromide, as a 55% by weight aqueous solution, is utilized as a drying agent for air, e.g. in air conditioning units. Ammonium bromide is utilized e.g. in photography and as a flame retardant. Potassium bromide is also used in photography.

Lithium bromide:

- Drying agent in air conditioning units

Ammonium bromide:

- Photography
- Flame retardant

Alkali bromates (heat and impact sensitive) are utilized e.g. in the treatment of flour or in hair-setting lotions.

Potassium bromide:

- Photography

Table 1.7.5.2 gives the USA bromine consumption spectrum in 1986.

Alkali bromates:

- Flour treatment

Table 1.7.5-2. USA Bromine Consumption Spectrum in 1986.

Flame retardants	Crude oil/gas	1,2-dibromo-ethane	Agriculture	Export	Various
30	28	20	10	4	8

The agricultural consumption of bromine is mainly for bromomethane and the miscellaneous consumption is principally for disinfection purposes.

References for Chapter 1.7.5: Bromine and Bromine Compounds

Commercial Information:

Chemical Profile: "Bromine", Schnell Publishing Company, Inc. (Jan. 1, 1986).

"Bromine outlets & developments", *Manufacturing Chemist,* p. 69 and 71 (Dec. 1985).

"Bromine, Israel's rising star", *Chem. Eng.,* p. 12C and 12D (June 10, 1985).

"Drilling markets falter for sodium bromide", *Chem. Week,* pp 11 – 12 (April 9, 1986).

"Dead Sea Works goes it alone", *Eur. Chem. News., Fertilizers & Agrochemicals Supplement,* p. 22 (Aug. 1986).

"A good bromine deal for Ethyl and Dow", *Chem. Week,* p. 22 (Sept. 24, 1986).

"Flame retardant use restricted", *Manufacturing Chemist,* p. 13 (Oct 1986).

Reviews:

Ullmann's Encyclopedia of Industrial Chemistry, 5th Ed., Vol. A4, pp 391-403 and pp 422-429, Verlag Chemie, Weinheim-Deerfield Beach-Basel (1987).

Kirk-Othmer: *Encyclopedia of Chemical Technology,* 3rd Ed., Vol. 4, pp 226 – 263, John Wiley & Sons, New York-Chichester-Brisbane-Toronto (1978).

R. B. McDonald and W. R. Merriman: "The Bromine and Bromine-Chemicals Industry", in: *The Modern Inorganic Chemicals Industry,* Ed.: R. Thompson, The Chemical Society, Burlington House, London, pp 168 – 182 (1977).

1.7.6 Iodine and Iodine Compounds

1.7.6.1 Economic Importance

Production of iodine from:

- Brines
- Residual solutions from the Chilean niter industry

Iodine is extracted from:

- Brines, which often accrue in crude oil or natural gas production
- Residual solutions from the Chilean niter industry

The extraction of iodine from ashed algae or kelp is currently not very significant. However, the reserves, just in the Pacific region, are estimated to be 10×10^3 t/a, but at current iodine prices extraction is uneconomical.

The estimated world iodine reserves are given in Table 1.7.6-1.

Table 1.7.6-1. Estimated World Iodine Reserves in 10^3 t.

World	USA	Chile	Japan	West Germany	Indonesia	Various countries
2608	240	363	1814	5	5	181

The USA iodine consumption in 1984 was distributed among the different classes of product as shown in Table 1.7.6-2.

Table 1.7.6-2. USA Iodine Consumption in 1984 Distributed among the Different Product Classes in t.

Total	Iodine	Potassium iodide	Sodium iodide	Ethylenediamine dihydroiodide	Various inorganic compounds	Various organic compounds
2502	73	581	55	516	583	698

In the USA, Dow Chemical had an iodine capacity of ca. 220 t/a (this plant produced iodine as a byproduct from bromine manufacture, its capacity being therefore dependent upon bromine demand). This plant was shut down in May 1986. The Woodward Iodine Operations (a subsidiary of the Japanese firm Ise) has a capacity of 800 to 900 t/a. A second USA producer is North American Brine Resources, which is also Japanese owned. In 1985 the USA imported 2268 t of iodine. The worldwide consumption of iodine is ca. 12×10^3 t/a.

Worldwide iodine consumption in 1984: ca. 12×10^3 t/a.

1.7.6.2 Manufacture of Iodine and Iodine Compounds

1.7.6.2.1 Iodine

From Brines

Worldwide, 80% of the iodine extracted currently comes from brines. The brines contain between 30 and more than 100 ppm of iodine (as iodide). (The deposits in the USA are mainly in Michigan and Oklahoma.) The extraction is similar to that of bromine. The brines are mixed with hydrochloric or sulfuric acid and oxidized with excess chlorine. The elemental iodine formed is blown-out with air and absorbed in a sulfuric acid-hydroiodic acid-water mixture in an absorber. Part of this is continuously taken off and the hydrogen iodide present oxidized with chlorine to iodine. The iodine is finally melted under sulfuric acid to dry the iodine and oxidize the organic impurities.

Iodine extraction from brines:

- Oxidation of iodide with chlorine
- Iodine blown out
- Absorption in hydroiodic acid/sulfuric acid
- Reduction with sulfuric acid to hydrogen iodide
- Part of mixture taken off and oxidized with chlorine to iodine

The following equations correspond to the individual steps:

$$2NaI + Cl_2 \longrightarrow I_2\uparrow + 2NaCl$$

$$I_2 + SO_2 + 2H_2O \xrightarrow{\text{in HI/H}_2\text{SO}_4} 2HI + H_2SO_4$$

$$2HI + Cl_2 \xrightarrow{\text{in H}_2\text{SO}_4} 2HCl + I_2$$

Alternatively:

- Absorption of iodine as polyiodide on anionic ion exchangers
- Desorption with alkali
- Worked up to iodine

In Japan another process is used in which the iodine formed by chlorine oxidation is absorbed as polyiodide on an anionic ion exchanger. Desorption with alkali yields concentrated iodide- and iodate-containing solutions which are worked up to elemental iodine. A process has been recently proposed in which hydrogen peroxide is used for oxidation in order to reduce the corrosion problems.

From Residual Solutions from Niter Manufacture

Iodine extraction from residual solutions from niter production:

reduction of sodium iodate with sodium bisulfite to iodine

The Chilean niter deposits contain up to 0.3% iodine in the form of calcium iodate. After dissolution and recrystallization of the niter, the supernatent liquor contains up to 9 g/L of sodium iodate. Iodine is liberated by reduction with sulfur dioxide (in the form of sodium bisulfite), further reduction being avoided by maintaining stoichiometry:

$$2NaIO_3 + 5SO_2 + 4H_2O \longrightarrow Na_2SO_4 + 4H_2SO_4 + I_2$$

The iodine is filtered off and purified by sublimation. Iodine dissolved in the supernatent liquor is extracted with kerosine and stripped with an aqueous acidic iodide solution. The aqueous now iodide-free solution is then reutilized for dissolving the niter.

1.7.6.2.2 Hydrogen Iodide

Hydrogen iodide is manufactured by the reaction of iodine and hydrogen on a platinum catalyst at 500 °C:

$$H_2 + I_2 \xrightarrow[500\,°C]{Pt} 2HI$$

Hydroiodic acid is also industrially produced by the reaction of hydrazine with iodine:

$$N_2H_4 + 2I_2 \longrightarrow 4HI + N_2$$

At 10 °C, 234 g of hydrogen iodide dissolve per 100 g of water. Acids with 47% by weight hydrogen iodide are commercially available.

1.7.6.2.3 Alkali Iodides

Alkali iodides are mainly produced by the neutralization of alkali hydroxide or carbonate solutions with hydroiodic acid:

$$MOH + HI \rightarrow MI + H_2O$$

$M = Na, K$

1.7.6.2.4 Alkali Iodates

Alkali iodates are produced from the corresponding chlorates by reaction with iodine in the presence of hot nitric acid:

$$2MClO_3 + I_2 \longrightarrow Cl_2 + 2MIO_3$$

M: Na, K

Electrochemical manufacture is also possible.

1.7.6.3 Applications for Iodine and Iodine Compounds

- For catalysts, e.g. titanium iodide in the stereospecific polymerization of butadiene
- For disinfection, e.g. nonionic tensides with small quantities of iodine
- For pharmaceutical purposes, e.g. medicines for the treatment of thyroid gland diseases; sodium iodide as additives in table salt in iodine-deficient areas such as Switzerland; potassium iodide as preventative medication (avoids absorption of radioactive iodine by the thyroid gland) in the event of nuclear accidents and as an X-ray contrast agent
- In the photographic industry (silver iodide)
- Iodine-containing activated carbon for the removal of traces of mercury e.g. from gas streams
- For the manufacture of very pure metals such as zirconium or titanium by way of their tetraiodides (van Arkel process)
- For the induction of rain-cloud formation (silver iodide)

Applications for iodine and iodine compounds:

- For catalysts
- For disinfection
- For pharmaceutical purposes
- In the photographic industry
- For inducing rain
- As animal nutrition supplements
- As stabilizers
- For dyes and inks

- For animal nutrition supplements, particularly for cattle
- For Polyamide 6.6 stabilization
- For dyes

The estimated applications spectrum of the ca. 3000 t of iodine consumed in the USA in 1986 is as follows:

- Disinfection: 23%
- Pharmaceuticals: 22%
- Catalysts: 17%
- Food and feedstuff additives: 13%
- Ink and colorants: 10%
- Agricultural chemicals: 9%
- Photography: 2%
- Nylon stabilizers: 2%
- Miscellaneous: 2%

References for Chapter 1.7.6: Iodine and Iodine Compounds

Commercial Information:

"Surplus of iodine now, but shortage coming", *Chem. Eng. News*, pp 12–13 (June 14, 1982).

"Outlook for iodine: a sellers' market", *Chem. Week*, pp 71–72 (Jan. 21, 1987).

Chemical Economics Handbook, Stanford Research Institute, Menlo Park, California, USA, "Iodine products" (May 1982) and "Manual of Current Indicators, Supplemental Data", p. 336 (Oct. 1986).

Technical Information:

R. P. Crozier, "Iodine: its markets and availability", *Mining Magazine*, pp 282–289 (April 1982).

H. H. Weetall and W. Hertl, "Iodine recovery from brine solutions", *Inorg. Chimica. Acta*, **104**, pp 119–123 (1985).

Reviews:

Kirk-Othmer: *Encyclopedia of Chemical Technology*, 3rd Ed., Vol. 13, pp 649–677, John Wiley & Sons, New York-Chichester-Brisbane-Toronto (1981).

Ullmanns Encyklopädie der technischen Chemie, 4th Ed., Vol. 13, pp 421–428, Verlag Chemie, Weinheim-Deerfield Beach-Basel (1977).

2 Mineral Fertilizers

The most important mineral fertilizers are those that contain phosphorus, nitrogen and potassium. Only these will be considered in the following sections. More of them are manufactured than any other chemical product except cement.

2.1 Phosphorus-Containing Fertilizers

2.1.1 Economic Importance

2.1.1.1 General Information

The worldwide consumption of phosphorus in mineral fertilizers for the fertilizer years between 1980 and 1985 is as follows (Table 2-1):

World consumption of mineral phosphate fertilizers in 1984/85:

34.0378×10^6 t P_2O_5

Table 2-1. World Consumption of Phosphorus in Mineral Fertilizers in 10^6 t P_2O_5.

Fertilizer Year	1980/81	1982/83	1983/84	1984/85
	31.346	30.631	32.864	34.078

The consumption in different regions of the world in 1984/85 is given in Table 2-2:

Table 2-2. 1984/85 Consumption of Phosphorus in Mineral Fertilizers in the Different Regions in 10^6 t P_2O_5.

North America	USA	Western Europe	Eastern Europe	Latin America	Africa	Asia	Oceania
4.977	4.210	5.638	10.008	2.455	1.128	8.727	1.145

The economic development in the individual regions is very different. In 1986, 20% of the USA phosphoric acid production capacity was idle.

2.1.1.2 Importance of Superphosphate

Superphosphate: a fertilizer strongly declining in importance (P_2O_5 content: only ca. 20%). Share of superphosphate in total phosphorus fertilizer consumption in the USA decreased from 90% in 1940 to 8% in 1977

The production of this fertilizer is decreasing, because at 19 to 20% P_2O_5 content a very high proportion of ballast has to be transported and in addition the biological availability of the phosphorus is lower than in other fertilizers. Since 1960 no further plants for superphosphate have been built in the industrialized countries. The proportion of superphosphate in the total consumption of mineral phosphorus fertilizers in the USA had fallen from 90% in 1940 to 8% in 1977 and in subsequent years has decreased still further. The capacities for superphosphate in the different regions in 1985 are given in Table 2-3.

Table 2-3. 1985 Capacities for Superphosphate in 10^6 t P_2O_5.

World	North America	Latin America	Western Europe	Eastern Europe	Africa	Asia	Oceania
7.750	0.255	0.414	1.168	1.786	0.532	1.597	1.993

There are no statistics for several countries including Canada, the USSR and the People's Republic of China in particular.

2.1.1.3 Importance of Triple Superphosphate

Triple superphosphate: importance still increasing (P_2O_5 content: ca. 40%)

estimated world capacity:

12.835×10^6 t/a P_2O_5

The production of triple superphosphate with a P_2O_5 content greater than 40% and biologically more available phosphorus is increasing. The worldwide increase in capacity is, however, much lower than for ammonium phosphates (part of the world capacity can also be used for the manufacture of ammonium phosphates). An estimate of the capacities in the different regions is given in Table 2-4.

Table 2-4. 1985 Capacities for Triple Superphosphate in 10^6 t P_2O_5.

World	N. America	S. America	W. Europe	E. Europe	Africa	Asia	Oceania
12.835	1.783	1.295	1.355	4.585	1.708	2.034	0.075

The USA production of triple superphosphate was 1.011×10^6 t P_2O_5 in 1984.

2.1.1.4 Importance of Ammonium Phosphates

Ammonium phosphates are utilized both as solid fertilizer and in solution (for nonfertilizer applications of ammonium phosphates see Sec. 1.5.1.2.2), 90% of the production being solid products in the USA [consisting of 75% diammonium phosphate (DAP), 15% monoammonium phosphate (MAP) and 10% of other ammonium phosphates].

Ammonium phosphate increasingly important as a solid fertilizer

Share of dissolved ammonium phosphates ("fluid ammonium phosphates") in the USA: 10%

Solid ammonium phosphates are particularly important fertilizers in the USA, where they accounted in 1983/84 for 39% of the phosphorus-containing fertilizers. A considerable and increasing proportion of the diammonium phosphate produced is used in the form of N,P,K-mixtures. Ammonium phosphate liquid fertilizers are also very important in the USA.

Table 2-5 shows the 1985 capacity for solid ammonium phosphates in 10^6 t/a P_2O_5.

1985 World capacity for solid ammonium phosphates:

22.68×10^6 t/a P_2O_5

Table 2-5. 1985 Capacities for Solid Ammonium Phosphates in 10^6 t/a P_2O_5.

World	North America	Western Europe	Eastern Europe	Latin America	Africa	Asia	Oceania
22.680	7.109	2.828	5.984	0.935	1.190	4.152	0.482

2.1.1.5 Importance of Nitrophosphates

The manufacture of this type of phosphorus fertilizer is largely restricted to Europe. The Worldwide capacity in 1985 was estimated to be 4.4×10^6 t/a P_2O_5, Western Europe accounting for 2×10^6 t/a P_2O_5 and Eastern Europe for 1.1×10^6 t/a P_2O_5.

Nitrophosphates: mainly in Western Europe (capacity 2×10^6 t/a P_2O_5) and Eastern Europe (capacity 1.1×10^6 t/a P_2O_5)

2.1.1.6 Importance and Manufacture of Thermal (Sinter, Melt) and Basic Slag (Thomas) Phosphates

These products, which only dissolve very slowly in the soil, are only produced and marketed in limited geographic areas. Their significance is very small in a worldwide context. Their production will be considered below.

"Rhenania Phosphates":

by sintering apatite, silica and sodium carbonate. P_2O_5 content ca. 29%. Produced in West Germany, Poland and East Germany

Sinter phosphates: "Rhenania phosphates" with a P_2O_5 content of ca. 29% are obtained by sintering apatite, silica and sodium carbonate or sodium hydroxide. Their annual production in West Germany is over 300×10^3 t (corresponding to ca. 90×10^3 t/a P_2O_5). Their production in West Germany was cut back in 1982. Poland and East Germany have similarly sized capacities.

Defluoridized apatites:

world capacity 1.5×10^6 t/a (mainly in USA and USSR). P_2O_5 content ca. 42%

Defluoridized apatites are obtained by sintering apatite in the presence of water vapor, silica and other substances. The world capacity (mainly in the USA and USSR) for this type of product, with a P_2O_5 content of ca. 42%, is estimated to be 1.5×10^6 t/a. Defluoridized apatite is also used as an animal nutrition supplement.

Melt phosphates:

- By melting apatite with Mg-compounds and silica
- P_2O_5-content ca. 21%
- Production mainly in East Asia
- Ca. 3% of worldwide phosphate fertilizer production

Melt phosphates: A product with a P_2O_5 content of ca. 21% is obtained by melting apatite with magnesium compounds and silica. These fertilizers are mainly manufactured in Eastern Asia (capacity in Japan and Korea 670×10^3 t/a).

Melt and sinter phosphates accounted for ca. 3% of the worldwide production of phosphate fertilizers in 1976, with its share decreasing relative to the other types.

Thomas phosphates:

- From the smelting of phosphorus-containing iron ores
- P_2O_5 content between 10 and 18%
- 1976 production in the EEC: ca. 700×10^3 t P_2O_5

Thomas phosphates: The slag resulting from the smelting of phosphorus-containing iron ores contains in addition to phosphorus mainly calcium, magnesium, iron and silica. The citrate-soluble P_2O_5 content of these "Thomas phosphates" varies between 10 and 18%. In 1976 ca. 700×10^3 t P_2O_5 of Thomas phosphates were produced in the EEC (main production in Great Britain, West Germany, France, Benelux). Future production depends mainly on the availability of phosphorus-containing iron ore.

2.1.2 Manufacture of Phosphorus-Containing Fertilizers

2.1.2.1 Superphosphate

Superphosphate (mixture of monocalcium phosphate and calcium sulfate):

- From apatite and sulfuric acid

The manufacture of superphosphate from apatite and sulfuric acid proceeds according to the idealized equation:

$$2\,Ca_5[(PO_4)_3(F)] + 7\,H_2SO_4 \longrightarrow 3\,Ca(H_2PO_4)_2 + 7\,CaSO_4 + 2\,HF$$

The reaction takes place in two steps: anhydrite and phosphoric acid are formed in the first fast step. This phosphoric acid then reacts slowly – over a period of weeks – ("curing") with further apatite producing monocalcium phosphate hydrate. Part (ca. 10 to 40%) of the fluoride contained in the apatite is expelled in the form of gaseous silicon tetrafluoride, the rest remains in the superphosphate. Any carbonato-apatite and calcium carbonate present in the fluoro-apatite react forming carbon dioxide.

Reaction proceeds in two steps:

- Fast step: Formation of anhydrite and phosphoric acid
- Slow step ("curing"): Conversion of phosphoric acid into monocalcium phosphate

The industrial production of superphosphate proceeds in five stages:

Industrial manufacture proceeds in five process stages

- Grinding of the apatite
- Reaction with sulfuric acid
- Solidification and crushing of the primary reaction product
- "Curing" – completion of the reaction
- Comminution and possible granulation of the end product

The grinding of the apatite ($\geq 33.5\%\ P_2O_5$), if necessary after prior crushing, yields a material with a particle size of e.g. 90% < 150 μm. The reaction with the ca. 70% sulfuric acid is nowadays mainly carried out continuously in mixing units with stirrers or in a stainless steel conical mixing funnel without stirrer developed by TVA. On average 60 kg sulfuric acid is necessary per 100 kg apatite.

The initially liquid digestion mixture solidifies within 5 to 20 mins and then has to be crushed. A large number of apparatuses have been developed for this primary size reduction. For example, chamber systems in which the material solidifies and is then mechanically removed. In modern continuous plants the reaction mixture is placed on long rubber conveyer belts upon which it solidifies. During solidification gaseous fluoro compounds are expelled and have to be sucked off and absorbed.

A large number of apparatuses have been developed for crushing the primary reaction product.

To complete the afterreaction, the product has to "cure" for several weeks. Then it can be crushed and delivered. Since it is dusty and has a tendency to cake, it is often granulated. This can take place both before and after the curing.

P_2O_5 content of superphosphate:

- West Germany: 18% } (citrate-soluble)
- USA: 20 to 21% }

In West Germany superphosphate is marketed with 18% citrate-soluble P_2O_5, whereas in the USA the P_2O_5 content is 20 to 21%.

2.1.2.2 Triple Superphosphate

Triple superphosphate (monocalcium phosphate):

- From apatite and phosphoric acid
- Process analogous to superphosphate manufacture
- P_2O_5 content ca. 47% (citrate soluble)
- $CaO:P_2O_5$ molar ratio 0.92 to 0.95

The manufacture of triple superphosphate proceeds according to the equation:

$$2\,Ca_5[(PO_4)_3(F)] + 14\,H_3PO_4 \longrightarrow 10\,Ca(H_2PO_4)_2 + 2\,HF$$

Industrially its production is very similar to that for superphosphate. Apatite with a P_2O_5 content $\geq 31\%$, ground to e.g. 70% < 74 μm, and wet-process acid with 52 to 54% P_2O_5 are used as the starting materials. The molar ratio of CaO to P_2O_5 in the final product should be between 0.92 and 0.95 and the P_2O_5 content ca. 47%. The solidification of the reaction mixture occurs faster than in the manufacture of superphosphate.

In the USA, triple superphosphate is often used as a granulate, which can be produced using various methods. In the Dorr-Oliver process, for example, the digestion mixture, while still liquid, is deposited upon circulating triple superphosphate granules, sieved and dried. The ratio of separated to recirculated granules is 1:12. In general, however, triple superphosphate is first prepared and then granulated with water and steam.

2.1.2.3 Ammonium Phosphates

Solid ammonium phosphate fertilizers:

- Monoammonium phosphate (MAP)
- Diammonium phosphate (DAP) and mixtures
- Content ca. 85%

The following ammonium phosphates are used as fertilizers:

Monoammonium phosphate $NH_4H_2PO_4$ (MAP)

Diammonium phosphate $(NH_4)_2HPO_4$ (DAP)

Ammonium polyphosphate $[NH_4PO_3]_n$ (APP)

either separately or as mixtures. Triammonium phosphate is not a commercial product because of its high ammonia vapor pressure.

Mono- and diammonium phosphate are used as solid fertilizers, whereas ammonium polyphosphate is mainly employed in solution as a liquid fertilizer, since unlike the orthophosphates, it is very soluble and is more difficult to granulate than the orthophosphates. It also, as a result of its complexing properties, keeps impurities (iron, aluminum, magnesium etc.) in solution.

Liquid ammonium phosphate fertilizers:

- Ammonium polyphosphate (APP)

Properties of APP:

- Good water solubility
- Sequestration of impurities as water-soluble complexes
- Difficult to granulate

Ammonium phosphate fertilizers are relatively impure (content ca. 85%), due to their being prepared with nonpurified wet-process acid. Commercial monoammonium phosphate contains 11 to 13% N and 48 to 53% P_2O_5 (theoretically 12.2% N, 61.7% P_2O_5) and commercial diammonium phosphate 16 to 18% N and 46 to 48% P_2O_5 (theoretically 21.2% N, 53.7% P_2O_5).

Solid fertilizers: In the manufacture of solid ammonium phosphate fertilizers there are two main problems:

- Manufacture of storable, noncaking products
- Least possible energy consumption during manufacture

Manufacture of solid ammonium phosphate fertilizers:

- Neutralization of phosphoric acid with ammonia and processing into a granulate or a powder
- Utilization of the heat of neutralization for drying the final products

To achieve these aims the manufacturing steps: neutralization of phosphoric acid with ammonia in an exothermic reaction and production of solid materials (e.g. by granulation or prilling) are linked with one another. Several of the many processes for the manufacture of ammonium phosphates will be discussed below.

In the TVA granulation process, slurries of ammonium phosphates, with a deficiency or excess of ammonia with respect to monoammonium phosphate, are produced by the reaction of ammonia with phosphoric acid. These slurries are granulated by adding the quantities of acid or ammonia necessary for a stoichiometric product and recycling the fine fraction. The granulate is then dried in, for example, rotary driers, a process which requires considerable energy.

In the TVA pipecross reactor process, anhydrous ammonia alone, or diluted with an equal quantity of water, is reacted with phosphoric acid. The reaction product is immediately added to recycled particles in the adjacent granulation unit. At this point, further ammonia or acid can be added to the granulating material. The process is designed to utilize most of the heat of neutralization in the drying of the granules.

Nongranular (powdery) monoammonium phosphate is obtained in the Swift process by reacting liquid ammonia with phosphoric acid containing 50% P_2O_5 in an impeller-stirred reactor. The reaction products (finely divided monoammonium phosphate and steam, temperature ca. 126 °C) are fed in at the top of a tower in which the steam is driven out by an air current from below and the solid ammonium phosphate sinks to the bottom. The wet-process acid in this process must not be deslimed beforehand.

The manufacture of ammonium phosphate can also be combined with the production of mixed fertilizers: e.g. ammonium phosphate with ammonium sulfate (partial substitution of sulfuric acid for phosphoric acid) or ammonium nitrate. The ammonium phosphates can also be converted into mixed fertilizers during granulation by the addition of potassium salts, urea etc.

Liquid ammonium phosphate fertilizers: Ammonium polyphosphates can be manufactured from either phosphoric acids containing high or low concentrations of polyphosphoric acid, or from orthophosphoric acid solutions.

Manufacture of liquid ammonium phosphate fertilizers (APP):

- From polyphosphoric acid and ammonia – problems: scaling of the pipe reactor walls with Al- and Mg-phosphates
- By reacting orthophosphoric acid with ammonia at 300 °C yielding polyphosphates by water elimination

Initially, polyphosphoric acid was used which was produced by the combustion of white phosphorus (Sec. 1.5.1.2.1), solutions with 11% N and 37% P_2O_5 being obtained. These "furnace acids" are now too costly for fertilizers, due to increased energy costs, and the polyphosphoric acids now used are manufactured from wet-process acid.

If polyphosphoric acids with a high content of P_2O_5 (40 to 50% of total P_2O_5 as polyphosphoric acids) in the polymer form are used, the reaction with ammonia has to be carried out with external cooling to avoid hydrolysis to orthophosphates. The energy intensive nature of polyphosphate production has resulted in the current use of acids with about 20 to 30% of the P_2O_5 in the polymer form. In a tube reactor developed by TVA, the acid is reacted with gaseous ammonia in an exothermic reaction at 230 to 240 °C, whereupon a considerable proportion of the orthophosphates is condensed to polyphosphates with the elimination of water. The resulting melt is absorbed in an aqueous solution of the end product and the required quantities of water and, if necessary, ammonia added at the same time. Solutions with 11% N and 37% P_2O_5 are usually produced, with 60 to 68% polyphosphate.

Solutions of liquid fertilizers contain:

- Ca. 11% N and 37% P_2O_5
- 60 to 68% polyphosphate

In 1985 in the USA, there were more than 135 TVA tube reactor plants operating, each producing 25 t of product per hour, 20 of which were mobile. A problem with these processes is the scaling of the walls of the tube reactors with aluminum and magnesium phosphates, which results in enforced down time for scale removal.

In a process developed by the firm Swift, orthophosphoric acid is used as the starting material. Preheated acid is reacted in a special jet reactor with gaseous ammonia at high temperatures. The proportion of polyphosphate in the end product is 60% at a temperature of 300 °C.

In addition to these soluble fertilizers, suspensions of ammonium phosphate, which may also contain other fertilizer substances, are very important, especially in the USA.

In the USA ammonium phosphate suspensions also marketed as fertilizers

2.1.2.4 Nitrophosphates

The digestion of apatite with nitric acid is of major commercial importance, in addition to the digestion of apatite with sulfuric acid (to phosphoric acid or superphosphate) and with phosphoric acid (to triple-superphosphate). In contrast the digestion with hydrochloric acid is of no importance. The exothermic reaction with nitric acid proceeds, in principle, as follows:

Manufacture of nitrophosphates:

Digestion of apatite with nitric acid to phosphoric acid and calcium nitrate. Soluble hygroscopic calcium nitrate has to be converted to other compounds or (partially) separated

$$Ca_5[(PO_4)_3(F)] + 10\ HNO_3 \xrightarrow{45\ \text{to}\ 80\,^\circ C} 5\ Ca(NO_3)_2 + HF + 3\ H_3PO_4$$

$$\Delta H = -290\ \text{kJ/mol}$$

50 to 60% nitric acid (up to a 20% excess) is used. The fluoride remains largely in the reaction mixture. Part

of the nitric acid is reduced to nitrogen oxides, which have to be removed from the tail gases.

Unlike the digestion with sulfuric acid, calcium is present in a soluble form. Since calcium nitrate is very hygroscopic, it mostly has to be either converted or separated (and then further utilized). [In a Lonza AG process the calcium nitrate is left in the mixture. The digestion mixture is dehydrated up to the point that not more than 2 to 3 moles of water are present per mole calcium nitrate. Then the not yet solidified mass is granulated and coated with basic slag (Thomas meal), calcium cyanamide etc.]

The conversion can occur:

Addition of ammonia and carbon dioxide to the digestion mixture:

formation of calcium hydrogen phosphate, ammonium nitrate and calcium carbonate (Carbonitric process)

- Either by adding ammonia and carbon dioxide to the reaction mixture, whereupon ammonium nitrate and (only citrate soluble) calcium hydrogen phosphate (dicalcium phosphate) are produced in addition to calcium carbonate (Carbonitric process):

$$3H_3PO_4 + 5Ca(NO_3)_2 + 10NH_3 + 2CO_2 + 2H_2O \longrightarrow 10NH_4NO_3 + 2CaCO_3 + 3CaHPO_4$$

Addition of sulfate (as sulfuric acid or ammonium or potassium sulfate). Calcium sulfate formed in addition to calcium hydrogen phosphate (Sulfonitric process)

- Or by adding sulfate as sulfuric acid, ammonium sulfate or potassium sulfate, whereupon calcium is converted in an analogous reaction to calcium sulfate. The free phosphoric acid forms calcium hydrogen phosphate upon neutralization (Sulfonitric process).

(Partial) separation of calcium nitrate from the digestion mixture by crystallizing out at low temperatures as the tetrahydrate. The calcium nitrate separated can be converted with ammonia/carbon dioxide to a mixture of calcium carbonate and ammonium nitrate

An example of calcium separation is the Odda process, in which calcium nitrate crystallizes out as the tetrahydrate upon cooling, the amount depending upon the temperature achieved. The calcium-depleted mother liquor is then neutralized with ammonia. The calcium nitrate separated can, for example, be converted into calcium carbonate and ammonium nitrate by reacting with ammonia and carbon dioxide and then, after filtering off the calcium carbonate, worked up to ammonium nitrate:

$$Ca(NO_3)_2 \cdot 4H_2O + 2NH_3 + CO_2 \longrightarrow CaCO_3 + 2NH_4NO_3 + 3H_2O$$

The calcium carbonate is then generally mixed with a 97% ammonium nitrate melt to lime ammonium nitrate ("calnitro") (Sec. 2.2.2.2).

2.2 Nitrogen-Containing Fertilizers

2.2.1 Economic Importance

General information: The consumption of mineral nitrogen-containing fertilizers is given in Tables 2-6 and 2-7.

1984/85 consumption of nitrogen fertilizers:

- Worldwide: 70.136×10^6 t
- USA: 10.436×10^6 t

Table 2-6. World Consumption of Nitrogen Fertilizers in 10^6 t N.

Fertilizer year	1981/82	1982/83	1983/84	1984/85
Consumption	60.498	61.079	66.961	70.136

Table 2-7. Consumption of Nitrogen Fertilizers in the Different Regions in the Fertilizer Year 1984/85 in 10^6 t N.

N. America	USA	W. Europe	E. Europe	S. America	Africa	Asia	Oceania
11.733	10.436	11.017	15.332	3.140	1.920	26.638	0.356

In the USA considerable, and year by year increasing, quantities of liquid nitrogen fertilizers are used. The liquid fertilizers are mainly ammonia, ammonia solutions, ammonium salts and urea and their mixtures. The proportions of liquid and solid fertilizers in the USA for 1982 and estimated for 1990 are given in Table 2-8.

Table 2-8. Proportions of Liquid and Solid Nitrogen Fertilizers Produced from Ammonia in the USA for the Fertilizer Years 1982 and 1990 in %.

	1982	1990*
liquid fertilizers	57	64 to 65
solid fertilizers	43	35 to 36

* estimated

2.2.1.1 Importance of Ammonium Sulfate

1983/1984 ammonium sulfate capacities:

- Worldwide: 5.980×10^6 t/a N
- North America: 0.736×10^6 t/a N

The estimated capacities for ammonium sulfate in the fertilizer year 1983/84 are shown in Table 2-9.

Table 2-9. Estimated Ammonium Sulfate Capacities in the Fertilizer Year 1983/84 in 10^6 t/a N.

World	N. America	S. America	W. Europe	E. Europe	Africa	Asia	Middle East	Oceania
5.980	0.736	0.545	1.364	1.269	0.092	1.842	0.072	0.060

Ammonium sulfate share of worldwide nitrogen fertilizer consumption: 4.7%

The world consumption of ammonium sulfate in the fertilizer year 1982/83 was 2.853×10^6 t N, which corresponds to 47.7% of the available capacity and ca. 4.7% of the worldwide consumption of nitrogen fertilizers. Ammonium sulphate is largely produced in North America and Europe as a coproduct (e.g. in the manufacture of polyamide) and as a byproduct (e.g. in the steel industry), whereas it is specifically produced in Asia and South America. Ammonium sulfate is particularly used in the less developed countries.

The future development of ammonium sulfate production depends strongly upon the situation regarding (textile-)fiber production. Ammonium sulfate production is growing more slowly than that of the other nitrogen fertilizers. There are indications that in the future ammonium sulfate may become important as a sulfur fertilizer (e.g. for maize in the USA).

2.2.1.2 Importance of Ammonium Nitrate

Ammonium nitrate consumption:

- World in 1983/84: 14×10^6 t
- For explosives in USA in 1982: 1.65×10^6 t

Ammonium nitrate capacity in USA in 1982: 8.4×10^6 t/a with 83% utilization

Pure ammonium nitrate, ammonium nitrate-sulfate mixtures, and calcium carbonate-ammonium nitrate mixtures ("calnitro", see Sec. 2.2.2.2) are included in this class of substances. The world consumption of ammonium nitrate in 1983/84 was ca. 14×10^6 t N. Although in absolute terms ammonium nitrate production is still rising, in relative terms it is losing ground to urea in the total nitrogen fertilizer production. The USA capacity for ammonium nitrate was 8.4×10^6 t/a in 1982 and 7.0×10^6 t was produced, of which 1.65×10^6 t was used for explosives.

2.2.1.3 Importance of Urea

Urea is the nitrogen compound with the largest production volume in the world. Its utilization is increasing steadily, it being the preferred nitrogen fertilizer worldwide. (Urea is also used in much smaller quantities for the manufacture of industrial products. The following statistics relate to total urea manufacture.) The production and consumption of urea for 1983/84 are shown in Table 2-10:

Urea is the most important nitrogen fertilizer

1983/84 Worldwide consumption:

21.8×10^6 t N or ca. 32.8% of worldwide nitrogen fertilizer consumption

Table 2-10. Production and Consumption of Urea in the Different Regions in 1983/84 in 10^6 t N.

	World	N. America	S. America	W. Europe	E. Europe	Africa	Asia	Oceania
Production	23.044	2.302	1.015	1.829	6.438	0.619	10.743	0.099
Consumption	21.842	1.532	1.118	1.176	3.822	0.803	13.318	0.073

In 1983/84 urea made up 32.8% of the world nitrogen fertilizer consumption. In the USA in 1982 the urea consumption was distributed as follows: solid fertilizers, 42.5%; liquid fertilizers, 37.6%; animal nutrition supplements, 13.2%; plastics and glues, 6%.

2.2.2 Manufacture of Nitrogen-Containing Fertilizers

2.2.2.1 Ammonium Sulfate

Ammonium sulfate is produced as a byproduct in various processes both inside and outside the chemical industry (see below). In addition, it is directly produced by ammonia neutralization of sulfuric acid, concentration of the solution and crystallization of the product:

$$2NH_3 + H_2SO_4 \longrightarrow (NH_4)_2SO_4$$

The reaction is carried out in special reactors ("saturators") in which the small crystals have a sufficient residence time to grow into the coarse crystals required by agriculture.

Ammonium sulfate manufactured from ammonia and sulfuric acid in:

- "Saturators"
- Reaction towers

If sulfuric acid concentrations above 70% and gaseous ammonia are used, the heat of neutralization is sufficient to evaporate off the water. In a process operated in Japan, the sulfuric acid is pumped in at the head of a tower and the ammonia gas fed in from below. Solid ammonium sulfate can be taken from the bottom and does not need additional drying.

In 1980 there were 60 plants in the USA for the manufacture of ammonium sulfate. Only in 10 of these is ammonia reacted with sulfuric acid. This accounted for 24% of the total production of ammonium sulfate.

Ammonium sulfate produced as byproducts from:

- Metallurgy
- Desulfurization of flue gases
- Coking plant ammonia
- Waste acids from the Organic Chemical Industry
- Caprolactam manufacture

Ammonium sulfate is produced as a byproduct in:

- The working up of the sulfur dioxide formed during the oxidation of sulfide ores.
- The working up of the sulfur dioxide removed from power station flue gases during desulfurization. In the large coal-fired power station at Mannheim (FRG), commissioned in 1984, the sulfur dioxide is washed out of the cooled flue gases with ammonia as ammonium sulfite. Then the solution is intensively treated with air to convert most of the sulfite to sulfate. The nitrogen monoxide present in the sulfur dioxide-free flue gases is oxidized with ozone to nitrogen dioxide and then scrubbed with aqueous ammonia. Further oxidation yields ammonium nitrate. The product streams obtained can be mixed and worked up to ammonium sulfate-nitrate mixed fertilizers. The main problem in this process is the occurrence of aerosols that are difficult to separate. Desulfurization of all flue gases would yield ca. 40×10^3 t/a of ammonium sulfate.
- The working up of the ammonia produced as a byproduct during coke manufacture to ammonium sulfate. 16% of the ammonium sulfate manufactured in the USA in 1980 was produced in this way.
- The working up of the sulfuric acid often obtained as a byproduct in organic processes such as nitration, alkylation etc. to ammonium sulfate. The organic impurities from the syntheses can, however, interfere very strongly (e.g. through foaming or influencing crystal shape).

"Classic" caprolactam process:

- 4.4 t ammonium sulfate per t caprolactam

- The manufacture of caprolactam (starting product for polyamide 6) by the Beckmann rearrangement of cyclohexanone oxime in fuming sulfuric acid:

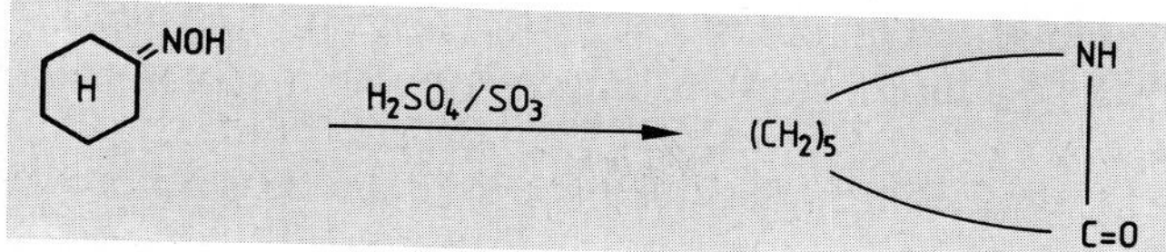

This is the most important source of ammonium sulfate as a byproduct, 47% of the ammonium sulfate manufactured in the USA in 1980 having been produced in this way. In the "classic" caprolactam process in which the oxime is formed by reaction with hydroxylamine sulfate (produced by the Raschig process) and is then rearranged in sulfuric acid (Sec. 1.4.3.2.1 and 1.4.3.2.2), 4.4 t ammonium sulfate are obtained per t caprolactam. Modern processes yield substantially less byproduct or no byproduct at all.

Modern caprolactam processes:

- Much less ammonium sulfate byproduct

2.2.2.2 Ammonium Nitrate

Ammonium nitrate is manufactured by the neutralization of nitric acid with gaseous ammonia in a strongly exothermic reaction:

Ammonium nitrate: manufactured from nitric acid and gaseous ammonia e.g. in circulating systems

$$NH_3 + HNO_3 \longrightarrow NH_4NO_3$$

in units in which the reaction components are circulated to ensure rapid mixing. In some cases these reactions are carried out under pressure to enable the steam formed to be used for the preheating of the ammonia and the acid. In this way reactor temperatures of up to 180 °C can be obtained.

The thermal sensitivity of ammonium nitrate means that the volume of the reactors must be kept as small as possible. The acid used must be free of impurities (particularly chlorides) and the reaction must be so controlled that excess acid is avoided.

Thermal sensitivity of ammonium nitrate means:

- Use of small reactors
- Use of pure (chloride-free) nitric acid
- Avoidance of excess acid

If nitric acid concentrations above 50% are used, it is possible, with suitable process design, to evaporate all the water introduced with the acid without supplying additional energy.

The ammonium nitrate melt leaving the reactor generally has a water content of 3 to 5%, in some plant types even as low as 0.5%. The melt is fed into the head of so-called "prilling towers" (up to 60 m high),

The ammonium nitrate melt is prilled or granulated

in which the melt is so dispersed that droplets are formed which, upon cooling with air fed in from below, solidify as they descend. If the melt only contains 0.5% water the resulting prills can be used directly, otherwise they have to be dried further. Granulation can also be used instead of prilling.

The (hygroscopic) ammonium nitrate prills have to be posttreated to avoid caking

Since ammonium nitrate is hygroscopic, it has to be posttreated. In view of its strong oxidizing power, only inorganic substances such as attapulgite, kieselguhr or clay can be used.

In West Germany for safety reasons, only ammonium nitrate-mixtures, with e.g. calcium carbonate, may be used as fertilizers

In West Germany, the use of pure ammonium nitrate as a fertilizer is forbidden on safety grounds. Mixtures, particularly with calcium carbonate, are used instead. Currently mixtures with N-contents of up to 28% are allowed.

(For the manufacture of ammonium nitrate from calcium nitrate see Sec. 2.1.2.4).

2.2.2.3 Urea

Urea:
produced from carbon dioxide and ammonia. Urea formed in two steps:

- 1st step: Formation of ammonium carbamate (exothermic, quantitative)
- 2nd step: Conversion of carbamate into urea (endothermic, ca. 70% yield)

Industrially, urea is only produced from ammonia and carbon dioxide. Since carbon dioxide is a byproduct in the production of hydrogen from natural gas or crude oil (in the case of natural gas only in 90% of the required amount) for use in the synthesis of ammonia, a urea plant is often coupled with an ammonia synthesis plant.

In the first step ammonium carbamate is formed from ammonia and carbon dioxide:

$$CO_{2\,gas} + 2\,NH_{3\,liq.} \rightarrow NH_2COONH_{4\,liq.}$$

$$\Delta H = -117\ kJ/mol$$

At high pressures this reaction is quantitative. Ammonium carbamate is in equilibrium with urea and water:

$$NH_2COONH_{4\,liq.} \rightleftharpoons H_2NCONH_{2\,liq.} + H_2O_{liq.}$$

$$\Delta H = 15.5\ kJ/mol$$

Ca. 70% of the carbon dioxide is converted into urea at an ammonia to carbon dioxide molar ratio of 4:1, a temperature of 200 °C and a pressure of 250 bar.

The unconverted ammonium carbamate and the unreacted ammonia have to be removed from the reaction mixture obtained: an aqueous solution of urea, ammonium carbamate and ammonia. The many industrial urea processes particularly differ in the way in which the separation and the recycling of ammonia and carbon dioxide are carried out, the minimizing of the energy consumption of these large plants (up to 1700 t urea per day) being crucial.

In the reaction mixture consisting of an aqueous solution of urea, carbamate and ammonia:

- Ammonia has to be expelled and
- Carbamate thermally decomposed

In the first plants constructed, which used the "once through" process, the mixture of ammonia and carbon dioxide resulting from pressure release and thermal decomposition of ammonium carbamate was worked up to ammonium nitrate or sulfate, the carbon dioxide being vented. Modern plants operate with total recycling of carbon dioxide and ammonia acheiving ammonia yields of 98.6 to 99.5%. Since considerable product loss occurs in the subsequent prilling of the urea, the real reaction yields are even higher.

"Once through" processes, in which part of the carbon dioxide is vented, are obsolete

The current processes comprise:

- Solution recycling and
- Stripping processes

Current types of processes:

- Solution recycling and
- Stripping processes

total recycling of ammonia and carbon dioxide

In the former, the synthesis solution pressure is released bit by bit after it leaves the reactor and the ammonia and carbon dioxide released are dissolved under their respective pressures in an absorber in water or countercurrently in the urea mother liquor (Fig. 2-1).

Solution recycling processes:

see Fig. 2-1 and 2-2

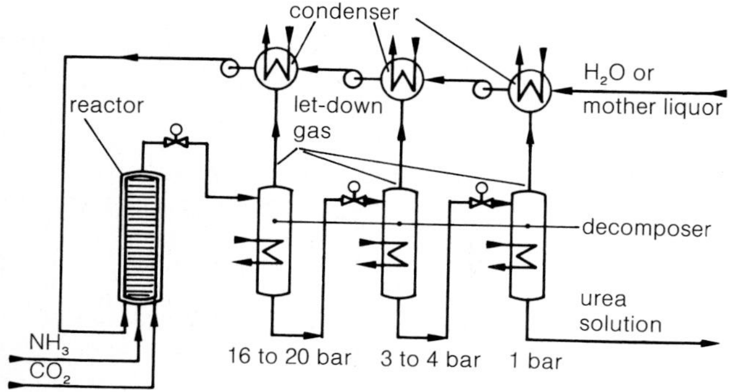

Fig. 2-1. Flow sheet of a conventional solution circulation process for urea manufacture. [Taken from D. Fromm and D. Lütow, "Modern Processes in the Heavy Chemicals Industry: Urea", *Chemie in unserer Zeit* **13,** pp 78-81 (1979).]

The absorption solution pressure is increased from stage to stage until the synthesis pressure is regained. A concentrated carbamate solution containing little water is obtained by this countercurrent pressure absorption and the equilibrium is thereby shifted towards urea formation.

Recycling processes in which the first decomposition stage is operated at 60 to 80 bar require even less heat (Mitsui-Toatsu, Montedison processes) (Fig. 2-2, see below).

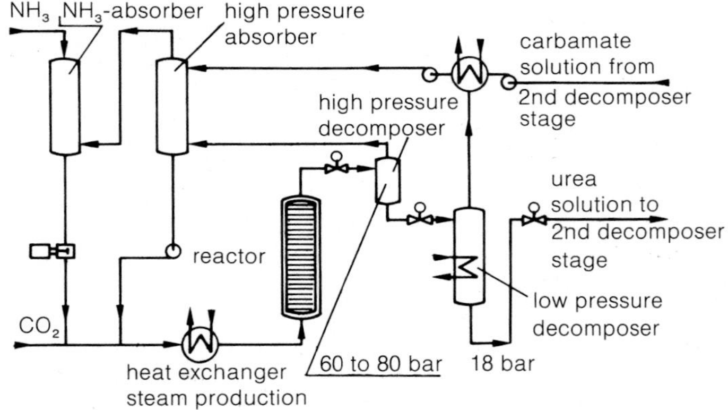

Fig. 2-2. Flow sheet for urea manufacture by a solution circulation process with a high pressure depressurization stage. [Taken from D. Fromm and D. Lütow, "Modern Processes in the Heavy Chemicals Industry: Urea", *Chemie in unserer Zeit* **13,** pp 78-81 (1979).]

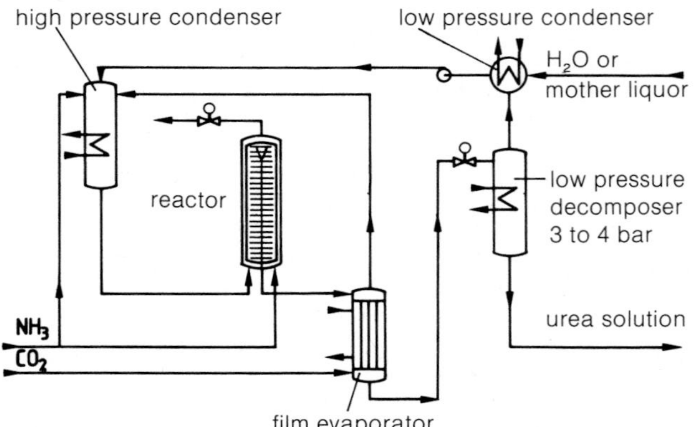

Fig. 2-3. Flow sheet for urea manufacture by the stripping process (with CO_2-stripping). [Taken from D. Fromm and D. Lütow, "Modern Processes in the Heavy Chemicals Industry: Urea", *Chemie in unserer Zeit* **13,** pp 78-81 (1979).]

In the stripping processes, the solution, upon leaving the reactor, is fed into the head of a film evaporator in which the solution flowing downwards is contacted countercurrently with all the carbon dioxide required in the process. The gases which leave by the head of the film evaporator are partly condensed in the pressurized solution coming from the low pressure decomposer, together with part of the freshly supplied ammonia. This mixture of liquid and gas is introduced into the reactor. Since ca. 85% of the carbamate is decomposed in the stripper, a single adjoining low pressure decomposer is sufficient to complete the dissociation. The heat produced in the high pressure condenser is converted into steam and used for operating the low pressure decomposer (Stamicarbon Process).

Stripping process:

see Fig. 2-3

Instead of stripping with carbon dioxide, ammonia can be used (Snam Progetti Process). The stripping processes are energetically somewhat more favorable than the recycling processes with high pressure decomposers.

Stripping processes energetically somewhat more favorable than recycling processes

The 72 to 77% urea solutions obtained with these processes are, after prepurification e.g. with activated carbon to remove oil, either vacuum evaporated until crystallization takes place or evaporated in film evaporators to a urea melt (m.p. 132.7 °C). These urea melts or molten crystals are generally prilled (granulated urea is also produced).

The urea solution, after ammonia and carbamate removal, is worked up by:

- Dehydration in a film evaporator, or
- Crystallization

Urea for fertilizers is mainly prilled and the prills posttreated to avoid caking

Urea readily cakes upon storage. Its behavior can be improved by posttreating with e.g. kieselguhr or formaldehyde. Sulfur-coated urea is used as a controlled release fertilizer, particularly in the USA (20 to 30×10^3 t/a).

Urea forms biuret upon heating:

Important: low biuret content in urea

$$2H_2NCONH_2 \longrightarrow H_2NCONHCONH_2 + NH_3$$

which is harmful to some plants. The biuret content must therefore be kept low (commercial urea contains 0.3 to 1%).

So-called "controlled release fertilizers" have been developed, since urea is metabolized relatively quickly. Examples are crotonylidene urea (I) (from urea and

Controlled release fertilizers:

Condensation products of urea with aldehydes such as acetaldehyde, isobutyraldehyde and in particular formaldehyde

acetaldehyde) and isobutylidene urea IBDH (II) (using isobutyraldehyde). The most important are, however, the urea-formaldehyde products with a production capacity in the USA of 145×10^3 t/a in 1980. They are mainly used in nonagricultural applications.

2.3 Potassium-Containing Fertilizers

2.3.1 Occurrence of Potassium Salts

Potassium salt deposits in many parts of the world with largest deposits in the USSR and Canada

World reserves: $> 21 \times 10^9$ t K_2O

Potassium salts are found in numerous deposits in many parts of the world. By far the largest are in the USSR and Canada.

The 1983 estimates of the available reserves of potassium salts are given in Table 2-11.

Table 2-11. Estimated (1983) World Reserves of Potassium Salts in 10^6 t K_2O.

World	North America	Canada	Latin America	Western Europe	Eastern Europe	USSR	Africa	Asia	Oceania
> 20772	> 14633	14310	100	940	3800	3000	126	> 1160	14

The Canadian deposits are found in the provinces of Saskatchewan, New Brunswick and Manitoba, those in the USA in the states of Utah, North Dakota, New Mexico and California.

Deposits consist of salt mixtures including sodium chloride, potassium chloride, magnesium salts, calcium sulfate

The salt deposits consist of salt mixtures which mainly contain the following components:

- Sodium chloride (rock salt, halite)
- Potassium chloride (sylvine)
- Potassium magnesium chloride ($KMgCl_3 \cdot 6H_2O$, carnallite)

- Magnesium sulfate ($MgSO_4 \cdot H_2O$, kieserite)
- Potassium magnesium chloride sulfate [$(KCl \cdot MgSO_4)_4 \cdot 11\,H_2O$, kainite]
- Calcium sulfate (anhydrite)

2.3.2 Economic Importance of Potassium-Containing Fertilizers

Table 2-12 gives the production and consumption statistics for potassium-containing fertilizers in the different regions for the fertilizer year 1984/85.

1984/85 potassium-containing fertilizer production in 10^6 t K_2O:

- World: 28.9
- USSR: 9.8
- North America: 8.7

Table 2-12. Production and Consumption Figures for Potassium-Containing Fertilizers in the Different Regions for the Fertilizer Year 1984/85 in 10^6 t K_2O.

	World	North America	USA	Western Europe	Eastern Europe	USSR	Latin America	Africa	Asia	Oceania
Production	28.9	8.7	1.4	5.5	13.2	9.8	–	–	1.5	–
Consumption	25.9	5.4	5.0	5.3	9.5	6.2	1.6	0.4	3.3	0.3

The largest producer country is the USSR with 9.8×10^6 t/a K_2O.

Worldwide more than 90% of the potassium-containing fertilizer used is potassium chloride. Potassium sulfate, potassium magnesium sulfate and potassium nitrate are, however, also used as fertilizers in particular for plants which have a poor tolerance for chloride ions (e.g. tobacco, spinach, cucumbers etc.) or where magnesium-containing fertilizer is also required.

$>$ 90% of all potassium fertilizers consist of potassium chloride, remainder of potassium sulfate, potassium magnesium sulfate or potassium nitrate

The worldwide consumption of potassium sulfate in 1982 was 1.16×10^6 t K_2O, of which 0.56×10^6 t was consumed in Europe, the region with the highest consumption. The USA consumption in 1984 was 0.328×10^6 t and 0.280×10^6 t in 1985. West Germany and Belgium together provide half of the world production.

The world production of potassium nitrate in 1983 was ca. 0.40×10^6 t, of which ca. 80% was used as fertilizer and 20% was for industrial purposes.

Haifa Chemicals (Israel) with 0.22×10^6 t/a and Vestac (USA) with 0.09×10^6 t/a are the two most important producers of potassium nitrate fertilizer. In Chile, there is a 0.03×10^6 t/a capacity for potassium sodium nitrate manufacture.

2.3.3 Production of Potassium-Containing Fertilizers

2.3.3.1 Potassium Chloride

Extraction of potassium salts:

- By mining
- By leaching
- From potassium-containing waters

The extraction of potassium salts occurs mostly by mining (currently in West Germany to a depth of ca. 1200 m), but leaching processes (solution mining, with one plant each in Canada and Utah/USA) and direct extraction from lakes (Dead Sea; Great Salt Lake, Utah; Searles Lake, California; Lake McLoed, Australia) are also used.

Four types of separation process — used separately or in combination:

- Solution process
- Flotation
- Electrostatic fractionation
- Gravitational separation

The solid salt has to be more or less strongly ground, depending upon the degree of fusion, before it can be further processed. Potassium chloride can be separated from the salt mixtures by:

- Thermal dissolution processes
- Flotation
- Electrostatic fractionation
- Gravitational (dense medium) separation

Solution processes exploit the different solubilities of the salts

These processes are also, in some cases, combined with one another. In the thermal dissolution process, the differences in solubility of the various constituents of the raw salt in water are utilized to attain high percentage potassium chloride. The particular process used depends upon the type of salt mixture, but energy consumption, disposal of byproducts etc. are also important.

Flotation processes:

- 50% of sylvinite so processed
- Precondition, coarsely intergrown minerals

Flotation or a combination of flotation with solution processes are currently used for processing 50% of the sylvinite. The availability of coarsely intergrown minerals (predominantly in Canada, USA, USSR) is an important precondition for the use of flotation processes. One plant in West Germany also uses this process.

Electrostatic process:

- Opposite charging of components
- Separation of components using free fall in an electric field
- Lower energy consumption
- Ca. 10^7 t/a crude salt worked up worldwide

In electrostatic fractionation, the different triboelectric chargeabilities of the various components of a salt mixture (in particular sylvinite, kieserite and rock salt) are utilized.

The process consists of the following steps:

- Dry grinding of the crude salt
- Conditioning
- Heating and drying
- Triboelectric charging and adjustment of the relative humidity of the air
- (Multistage) separation

Grinding reduces the crude salt to < ca. 1 to 2 mm in size, depending upon composition. Fines and dust interfere with this process. Conditioning entails the addition of organic compounds, mostly carboxylic acids, in quantities of 50 to 200 g/t raw salt. The salt is then dried e.g. by passing hot air over it. The triboelectric charging takes place in so-called "fluidized bed salt warmers" in which the various salt components become oppositely charged. The residence time of the salt in this unit is less than 1 min.

The separation of the charged salt mixture takes place (multistep, with partial recirculation) in so-called "plate separators". In the 2 m high and up to 10 m wide unit, particles fall unhindered, are deflected sideways by an electric field of 4 to 5 kV/cm and separated according to the sign of the charge. The electrodes rotate at 10 to 30 rpm against brushes which remove the deposited particles. Energy consumption is low. The throughput of a separation unit is several thousand tons per day. 10^7 t/a are separated worldwide using this process.

Only coarsely intergrown minerals can be separated by dense medium separation. It has found only limited commercial application (Canada). Salt solutions containing ferrosilicon serve as the gravitational liquids.

Dense medium separation: only of minor importance

The potassium chloride produced for use as a fertilizer or in fertilizer mixtures is mostly granulated or compacted.

Much of the salt mined together with potassium chloride and produced as a byproduct in the course of its separation can not be exploited commercially.

Solid residues can to some extent be disposed of in mines, however, there are limits to this procedure in that the bulk volume of the material being returned is much greater than the volume of salt originally present

Disposal of residues:

- Deposition in mines
- Tipping
- In solution: injection under pressure in geologically impermeable strata
- Discharge into rivers

in the bed. A further possibility is tipping, which is not without its problems due to the solubility of the salt in water. If the geological conditions permit, dissolved residues can be injected under pressure into impermeable subterranean strata. Discharge into rivers is only possible to a limited extent and has encountered considerable resistance in some cases (salt problems in the rivers Werra, Weser and Rhine in West Germany for example).

2.3.3.2 Potassium Sulfate

Potassium sulfate production:

- From potassium chloride either with sulfuric acid or a sulfur dioxide, air and water mixture (Hargreaves process)
- By metathesis of potassium chloride with e.g. magnesium sulfate

Potassium sulfate is produced from potassium chloride. This is reacted with sulfuric acid or with a mixture of sulfur dioxide, air and water (Hargreaves process):

$$2KCl + H_2SO_4 \longrightarrow K_2SO_4 + 2HCl$$

$$2KCl + SO_2 + 0.5O_2 + H_2O \longrightarrow K_2SO_4 + 2HCl$$

Worldwide production of potassium sulfate from potassium chloride and sulfuric acid: 0.7×10^6 t/a K_2O

There are corrosion problems in the reaction with sulfuric acid, in which the reaction mixture has to be heated in an oven to 700 °C. Therefore the reaction of potassium chloride with sulfuric acid is carried out with direct heating in a fluidized bed. Ca. 0.7×10^6 t/a (as K_2O) of potassium sulfate is produced worldwide using this process. The hydrogen chloride byproduct is also further utilized.

Potassium sulfate can also be produced by metathesis with the sulfates of other metals, particularly magnesium sulfate. This reaction takes place in two steps:

$$2KCl + 2MgSO_4 + 6H_2O \xrightarrow{H_2O} K_2SO_4 \cdot MgSO_4 \cdot 6H_2O + MgCl_2$$

The double salt is separated and reacted with additional potassium chloride:

$$2KCl + K_2SO_4 \cdot MgSO_4 \cdot 6H_2O \xrightarrow{H_2O} 2K_2SO_4 + MgCl_2$$

Production of potassium sulfate from potassium chloride and magnesium sulfate worldwide: $> 1 \times 10^6$ t/a (as K_2O)

Over 1.0×10^6 t/a (as K_2O) are produced using this process.

2.3.3.3 Potassium Nitrate

This is produced by reacting potassium chloride with nitric acid:

$$4KCl + 4HNO_3 + O_2 \longrightarrow 4KNO_3 + 2Cl_2 + 2H_2O$$

The process is in fact much more complex, than this equation indicates.

Potassium nitrate can also be obtained by metathesis reactions with other nitrates (e.g. sodium, calcium or ammonium), in analogy with the production of potassium sulfate, as follows:

$$NaNO_3 + KCl \longrightarrow KNO_3 + NaCl$$

$$Ca(NO_3)_2 + K_2SO_4 + 2H_2O \longrightarrow CaSO_4 \cdot 2H_2O + 2KNO_3$$

Potassium nitrate:

- From potassium chloride and nitric acid
- From potassium chloride by metathesis with e.g. sodium nitrate

References for Chapter 2: Mineral Fertilizers

Reviews:

Kirk-Othmer: *Encyclopedia of Chemical Technology,* 3rd. Ed., Vol. 10, pp 31–125, John Wiley & Sons, New York, Chichester, Brisbane, Toronto (1980).

Ullmanns Encyklopädie der technischen Chemie. 4th Ed., Vol. 10, pp 201–256, Verlag Chemie, Weinheim-Deerfield Beach-Basel (1975).

Concerning the Situation of the Fertilizer Industry in 1983:

"ECN Chemscope", *Eur. Chem. News* (Feb. 1983).

Phosphor-Containing Fertilizers:

Commercial information:

"Whither DAP", *Phosphorus and Potassium* **140**, pp 22–24 (Nov./Dec. 1985).

"World Phosphate Fertilizer Statistics 1984/85", *Phosphorus and Potassium* **141**, pp 41–42 (Jan./Feb. 1986).

"Fertilizer producers gird for harder times", *Chem. Week*, pp 8–9 (June 18, 1986).

Chemical Economics Handbook, Stanford Research Institute, Menlo Park, California, USA:

- Triple Superphosphate (Nov. 1985)
- Normal Superphosphate (Nov. 1985)
- Ammonium phosphates (Nov. 1986).

Technical information:

"Manufacture of Nitrophosphates", *Chem. Eng. News,* pp 50–51 (Sept. 14, 1981).

"Improved techniques for ammonium phosphate production", *Phosphorus and Potassium* **144**, pp 27–33 (July/Aug. 1986).

Nitrogen-Containing Fertilizers:

Commercial information:

Nitrogen, **159**, pp 16–19 and 35–36 (Jan./Feb. 1986).

Eur. Chem. News, pp 13–14 (Nov. 25, 1985).

Chemical Economics Handbook, Stanford Research Institute, Menlo Park, California, USA:

- Nitrogen Chemicals (July 1983)
- Explosives (June 1984).

Technical information:

H. Gomaa and J. LeBlanc, "Environmental Considerations for the Nitrogen Fertilizer Industry", *Fert. News,* **30**, pp 23–27 (June 1985).

Ullmann's Encyclopedia of Industrial Chemistry, 5th Ed., Vol. A2, pp 243–256, VCH Verlagsgesellschaft, Weinheim-Deerfield Beach-Basel (1985).

D. Fromme und D. Lützow, "Moderne Verfahren der Großchemie: Harnstoff" (Modern processes in the heavy chemicals industry: urea), *Chemie in unserer Zeit,* **13**, pp 78–81 (1979).

Potassium-Containing Fertilizers:

Commercial information:

Phosphorus and Potassium, **136**, p. 7 (March/April 1985).
Phosphorus and Potassium, **141**, pp 41–42 (Jan/Feb 1986).
Chemical Profile, Potassium sulfate, Schnell Publishing Company, Inc., (Jan. 1, 1986).
"Potash", *Chemical Economics Handbook,* Stanford Research Institute, Menlo Park, California, USA (Nov. 1983).

Technical information:

Kirk-Othmer: *Encyclopedia of Chemical Technology,* 3rd. Ed., Vol. 18, pp 920–950, John Wiley & Sons, New York, Chichester, Brisbane, Toronto, Singapore (1982).
Ullmanns Encyklopädie der technischen Chemie. 4th Ed., Vol. 13, pp 447–496, Verlag Chemie, Weinheim-Deerfield Beach-Basel (1977).
Ullmanns Encyklopädie der technischen Chemie. 4th Ed., Vol. 20, pp 340–354, Verlag Chemie, Weinheim-Deerfield Beach-Basel (1981).
R. Bock, "Elektrostatische Trennung der Komponenten von Kali-Rohsalzen" (Electrostatic separation of the components of crude potassium salts), *Chem.-Ing.-Techn.* **53**, pp 916–924 (1981).
A. Sigwald und G. Fricke, "Die elektrostatische Aufbereitung von Kali-Rohsalzen" (The electrostatic working up of crude potassium salts), *Chem.-Ing.-Techn.* **55**, pp 39–45 (1983).

3 Metals and their Compounds

3.1 Alkali and Alkaline Earth Metals and their Compounds

3.1.1 Alkali Metals and their Compounds

3.1.1.1 General Information

Of the alkali metals: lithium, sodium, potassium, rubidium and caesium, elemental sodium and its compounds are the most important industrially, particularly the mineral and industrial heavy chemicals: sodium chloride, sodium carbonate, sodium hydroxide, sodium sulfate etc. In second place is potassium, which as its salts (chloride, sulfate, nitrate, phosphate) is an important component of mineral fertilizers. Lithium and its compounds have a much lower but steadily increasing importance. Caesium and rubidium are only utilized in very small quantities for special applications.

Commercial importance:

- Alkali metals: $Na \gg Li > K > Cs > Rb$
- Alkali metal compounds: $Na > K \gg Li \gg Cs > Rb$

3.1.1.2 Lithium and its Compounds

3.1.1.2.1 Natural Deposits and Commercial Importance

The most important mineral raw material for the industrial extraction of lithium is spodumene ($LiAlSi_2O_6$), which is mainly found together with lepidolite, petalite and amblygonite (all with Li contents of 4 to 7%) in pegmatite rock. The main producer countries are the USA (North Carolina), the USSR, China, Australia, Zimbabwe and Brazil (probably with Austria and Zaire in the future).

The working up of brines from North and South American salt lakes [USA (Utah, Nevada), Chile, Bolivia] is becoming increasingly important. In this process lithium precipitates out (preferably in the form

Lithium content:

- In the Earth's crust: 65 ppm
- In seawater: 0.17 ppm

Industrially important Li minerals:

- Spodumene $LiAlSi_2O_6$
- Lepidolite (lithium mica) $KLi_2Al(F, OH)_2Si_4O_{10}$
- Petalite $LiAlSi_4O_{10}$
- Amblygonite $LiAl(F, OH)PO_4$

of the poorly soluble lithium carbonate) as a byproduct in the production of borax, potassium salts, magnesium chloride and sodium chloride (1000 t of lithium had already been extracted from the Chilean "Salar de Atacama" in 1985).

USA is the largest Li producer and consumer

The certain world reserves amount to 7.3×10^6 t lithium (of which 60% is in salt lakes). 7400 t lithium (corresponding to 39×10^3 t of lithium carbonate) was produced in 1984 of which ca. 70% was in the USA.

Li consumption will increase 5 to 10%/a up to 1995

Main Li applications in:

- Aluminum electrolysis
- Glass and ceramic industry
- Al-Li alloys
- Batteries
- Nuclear technology

Lithium consumption doubled in the period 1975–1985 and furthermore will increase disproportionately in the future. It amounted to 5.1×10^3 t in the Western World in 1985, mainly for the applications given in the marginal notes. The most important consumers are the USA (with ca. 40% of world consumption), Western Europe, Japan and the USSR.

3.1.1.2.2 Metallic Lithium

Metallic lithium:

by electrolysis of a molten LiCl/KCl mixture (ca. 1:1)

The proportion of elemental lithium in the total production of lithium and lithium compounds is ca. 10% worldwide. It is exclusively manufactured by the electrolysis of molten mixtures of lithium chloride (45 to 55%) and potassium chloride at 400 to 460 °C in steel cells with a graphite anode and a steel cathode. The cell voltage is 6.0 to 6.5 V. The lithium formed collects on the surface of the molten salt electrolyte.

Elemental lithium is utilized for the manufacture of lithium hydride and lithium amide and for the synthesis of organolithium compounds (e.g. butyl- and phenyl-lithium) which are utilized as reducing agents in organic chemistry and as catalysts in polymerization reactions e.g. in the manufacture of cis-1,4-polyisoprene (steric rubber). In metallurgy, lithium is used in extremely light and strong Al-Li alloys (containing 2 to 3% Li) for space applications. Its utilization in batteries with high power densities and long term stability is gaining in importance (for watches, computers, heart pacemakers and cars). Finally the $^{6}_{3}Li$ isotope with a natural occurrence of 7.5% is important in nuclear weapons technology and as a breeding material for future (nuclear-)fusion reactors.

3.1.1.2.3 Lithium Compounds

Lithium carbonate: Lithium carbonate is the industrially most important lithium compound and is the starting material for the production of most of the other lithium salts. It is produced during the processing of lithium minerals and brines. Enriched and calcined lithium ore (spodumene) is e.g. digested with sulfuric acid at 200 to 250 °C, the lithium sulfate formed being extracted with water, purified by treatment with calcium hydroxide and then precipitated as lithium carbonate by treating with sodium carbonate. Lithium-containing brines are evaporated in solar ponds, purified and converted with sodium carbonate into lithium carbonate.

Most important Li compounds:
Li_2CO_3, LiCl, LiOH, LiH

Lithium carbonate is utilized in large quantities in the manufacture of aluminum by melt electrolysis (ca. 25% of the total lithium consumption). Lithium carbonate is also used as a flux in the glass, enamel and ceramic industries, which accounts for a further ca. 25% of lithium consumption. Lithium-containing glasses with eucryptite ($LiAlSiO_4$) crystallization are generally thermal shock resistant, due to their low thermal expansion coefficients. In psychiatry high purity lithium carbonate is utilized for the treatment of manic-depressive complaints.

Li_2CO_3 applications:

- Aluminum manufacture
- Glass, enamel and ceramic industries
- Medicine
- Intermediate in the manufacture of other Li salts

Lithium hydroxide: Lithium hydroxide is produced by the reaction of lithium carbonate with calcium hydroxide:

$$Li_2CO_3 + Ca(OH)_2 \longrightarrow CaCO_3 + 2\,LiOH$$

Lithium hydroxide monohydrate is industrially important in the manufacture of greases e.g. on the basis of lithium stearate. In the USA more than 60% of all greases are produced with the help of lithium-soaps (ca. 25% of the total lithium consumption).

Lithium hydride: Lithium hydride is manufactured by the reaction of metallic lithium with hydrogen:

$$2\,Li + H_2 \xrightarrow{700\,^\circ C} 2\,LiH\,.$$

It has industrial importance as a hydrogen source, a drying agent and a reducing agent in organic synthesis, particularly in the form of its derivatives lithium aluminum hydride and lithium borohydride.

Lithium chloride: Lithium chloride is utilized for air-drying, as an electrolyte in dry batteries and in the manufacture of metallic lithium.

3.1.1.3 Sodium and its Compounds

3.1.1.3.1 General Information

Industrially important Na compounds:

- $NaCl$, $NaOH$, Na_2CO_3
- Na silicates, Na aluminosilicates
- Na_2SO_4, Na_2SO_3, Na_2S
- NaF, Na_3AlF_6

The industrially most important sodium compounds are sodium chloride, sodium hydroxide (Sec. 1.7.2) and sodium carbonate. In addition a multitude of sodium salts such as sodium fluoride (Sec. 1.7.1), sodium-bromine and -iodine compounds (Sec. 1.7.5/1.7.6), sodium-sulfur compounds (Sec. 1.6), sodium-chlorine-oxygen compounds (Sec. 1.7.4), sodium silicates and sodium aluminosilicates (Sec. 5.1.2/5.1.3), sodium sulfate, sodium hydrogen sulfate and sodium hydrogen carbonate are utilized in almost all sectors of industrial inorganic chemistry.

3.1.1.3.2 Metallic Sodium

Metallic sodium:

Consumption dependent upon the (diminishing) production of Pb-containing antiknock agents

Economic Importance: The production of elemental sodium is closely coupled to the utilization of lead-containing antiknock agents for Otto motor fuels, which is diminishing due to antipollution legislation. In the USA, by far the most important producer country, the average annual decrease in production in the period 1975 to 1984 was 6.8%. Only 86×10^3 t was produced in 1984. The largest producer worldwide is Du Pont (Niagara Falls, USA) with a capacity of 51×10^3 t/a.

1986 USA capacity: 76×10^3 t/a

Metallic sodium manufacture:

By melt electrolysis of NaCl (in the presence of $CaCl_2$ and $BaCl_2$) at ca. 600 °C

Electricity consumption: 10 kWh/kg Na

Manufacture: Metallic sodium is currently almost exclusively produced by the electrolysis of molten, specially purified sodium chloride (modified Downs process). Older processes on the basis of thermochemical processes or the electrolysis of molten sodium hydroxide in the Castner Cell are no longer important. Ternary salt mixtures of sodium, calcium and barium chlorides, which melt at ca. 600 °C (m.p. of pure NaCl: 808 °C), are currently used in the sodium chloride melt electrolysis process. Graphite anodes are used with cathodes and diaphragms (wire gauze) of steel. The process is energy intensive, 10 kWh of electricity being consumed per kg sodium produced. The cell voltage is ca. 7 V and the electricity yield 85 to 90%. The sodium deposited at the cathode is cooled to 120 °C, freed from impurities (Ca, CaO, Na_2O) and marketed at a purity of 99.8%. The chlorine produced simultaneously is purified of salt dust and liquified (share in world chlorine production ca. 1 to 2%).

Applications: The most important application sector for metallic sodium (ca. 30% of USA consumption) is in the synthesis of anti-knock agents such as tetramethyl- and tetraethyl lead from lead-sodium alloys and alkyl chlorides:

$$4PbNa + 4C_2H_5Cl \rightarrow (C_2H_5)_4Pb + 3Pb + 4NaCl\,.$$

The utilization of sodium in the manufacture of difficultly reducible metals such as uranium, thorium, zirconium, tantalum and particularly titanium (from titanium tetrachloride) has slightly increased in the USA to ca. 27% of total consumption. It is also utilized in smaller quantities as a reducing agent in organic chemistry e.g. in the manufacture of pharmaceuticals, dyes and herbicides. Sodium is also the starting material for the manufacture of different sodium compounds such as sodium peroxide, sodium amide, sodium alcoholates (for organic chemical synthesis), sodium azide (for explosives), sodium hydride and sodium borohydride (reducing agents) and organo-sodium compounds (catalysts e.g. in polymerization reactions). Finally sodium is used as a coolant in fast breeder reactors (Sec. 6.4.5), due to its high thermal conductivity and low neutron absorption.

As in the case of lithium, sodium-based batteries are promising alternatives in the high power battery sector. Secondary sodium-sulfur batteries operating at 300 °C and weighing only a quarter of an equivalent lead-acid accumulator have already been evaluated.

Na applications:

- In the manufacture of
  - antiknock agents
  - difficultly reducible metals
  - NaH, $NaBH_4$, Na_2O_2 etc.
- As a reducing agent and catalyst in organic synthesis
- As a coolant in nuclear reactors
- Sodium-sulfur batteries

3.1.1.3.3 Sodium Carbonate

General Information

Sodium carbonate (soda ash) is a heavy chemical product comparable in importance to sodium hydroxide.

It is produced both from natural deposits (trona) and synthetically. The development of processes for the synthetic manufacture of sodium carbonate is closely involved in the history of industrial inorganic chemistry (LeBlanc process, Solvay process and the technical developments which resulted from them). The applications of sodium carbonate in e.g. the manufacture of glass and for cleaning purposes have been known since ancient times.

Development of industrial sodium carbonate manufacturing processes such as the Leblanc (1790) and Solvay (1865) processes stimulated the development of the inorganic chemical industry

Economic Importance

1985 world production of sodium carbonate: 28.7×10^6 t

The world production of sodium carbonate doubled between 1960 and 1975, but this growth had petered out by 1985 at which time the production at 28.7×10^6 t was similar to that for sodium hydroxide. Table 3.1.1-1 gives the regional distribution of sodium carbonate production in 1985.

Table 3.1.1-1. 1985 Production of Sodium Carbonate by Region in 10^6 t.

World	North America	South America	Western Europe	Eastern Europe	Asia/ Oceania	Africa
28.7	8.6	0.3	5.2	9.5	4.8	0.3

USA is the largest producer and exporter of sodium carbonate (100% from natural deposits)

The country with the largest production is the USA with 7.7×10^6 t. It is also the most important sodium carbonate exporter with 1.5×10^6 t in 1985. Then come the USSR, China, Bulgaria and West Germany.

Since the shutting down of the last synthetic sodium carbonate manufacturing unit in the USA in 1985, USA production has been exclusively derived from natural sodium carbonate deposits. Other producers of "natural" sodium carbonate are Kenya, Mexico and the USSR. In 1985 30% of the world production was from natural deposits. This figure is expected to rise still further due to the low energy requirements and low pollution of "natural" sodium carbonate production together with innovative developments in the recovery of sodium carbonate from brines (solution mining). Total sodium carbonate consumption will rise still further, despite much lower consumption in the glass industry due to increased glass recycling and increased use of plastic containers. Increased sodium carbonate demand is particularly forecast for the plate glass and glass fiber industries.

Sodium Carbonate Manufacture from Natural Deposits

The most important raw material for the industrial production of "natural" sodium carbonate is trona ($Na_2CO_3 \cdot NaHCO_3 \cdot 2H_2O$). It is mined in large quantities from the largest trona deposit worldwide in Wyoming (USA) with mineable reserves of ca. 45×10^9 t and worked up to pure sodium carbonate using two processes. In the *monohydrate process* the mineral is first calcined, then dissolved in water, the insoluble part being filtered off and the filtrate evaporated. The sodium carbonate monohydrate that precipitates out is calcined to pure sodium carbonate. In the *sesquicarbonate process* trona is first dissolved in water, then freed from impurities and lastly calcined.

Most important "natural" sodium carbonate deposits:

- Mineral (trona):
  - Wyoming/USA (ca. 45×10^9 t)
- In salt lakes:
  - California/USA, East Africa (ca. 10^9 t)

Working up to pure sodium carbonate:

- Monohydrate process
- Sesquicarbonate process

Sodium carbonate deposits in the Californian salt lakes are estimated to be greater than 600×10^6 t, with a further 300×10^6 t in the East African salt lakes. The salt concentrates are worked up by first carbonating, whereupon sodium hydrogen carbonate is precipitated. This yields sodium carbonate, carbon dioxide and water upon heating, the former being further treated as in the monohydrate process.

Synthetic Sodium Carbonate Manufacture

The industrial Solvay process (ammonia-sodium carbonate process) is based on the precipitation of the relatively poorly soluble sodium hydrogen carbonate from an aqueous sodium chloride solution according to:

$$NaCl + NH_4HCO_3 \longrightarrow NaHCO_3 + NH_4Cl.$$

Solvay process:

Sodium carbonate from NaCl and $CaCO_3$ with the help of NH_3

Process steps:

- Preparation of a concentrated NaCl-solution
- Pyrolysis of $CaCO_3$ to $CaO + CO_2$
- Saturation of NaCl-solution with NH_3
- Precipitation of $NaHCO_3$ by saturating with CO_2
- Separation of $NaHCO_3$
- Calcination (CO_2-splitting off) to Na_2CO_3
- Preparation of milk of lime
- NH_3-recovery

This reaction is carried out by passing gaseous ammonia into a concentrated sodium chloride solution and then saturating it with carbon dioxide. The precipitated sodium hydrogen carbonate is separated off and then calcined to sodium carbonate in e.g. a rotary kiln:

$$2NaHCO_3 \longrightarrow Na_2CO_3 + CO_2 + H_2O.$$

Solvay process:

- Energy intensive
- Waste disposal problem ($CaCl_2$ brine)
- More expensive than production of "natural" Na_2CO_3

Carbon dioxide and the ammonia recovered by treating the ammonium chloride formed with milk of lime are returned to the process-cycle. Ammonia is only therefore a reaction-aid. The overall reaction is:

$$CaCO_3 + 2NaCl \longrightarrow Na_2CO_3 + CaCl_2$$

with calcium chloride-containing brine, which is difficult to dispose of, as a byproduct. A modified Solvay process (new Asahi process) with reduced energy consumption and improved calcium chloride recovery possibilities has therefore been recently developed.

Leblanc process:

Industrially unimportant since the 1920's

The Leblanc process introduced at the end of the eighteenth century also uses sodium chloride as a starting material. It is reacted with sulfuric acid to sodium sulfate, the sodium sulfate reduced with coal to sodium sulfide and this reacted with calcium carbonate to sodium carbonate. This process has not been industrially important since the 1920's.

Applications

Applications of Na_2CO_3:

- Glass manufacture
- Inorganic Na compounds
- Pulp and paper industries
- Soap and detergent production
- Smaller quantities utilized in a multitude of application areas

A large part of the sodium carbonate consumed (e.g. ca. 50% in the USA) is utilized in the glass industry, of which ca. two thirds is used in the production of plate glass. Sodium carbonate is used here both as a raw material and as a flux for the glass melts (Sec. 5.1.2.2). A further 23% is required for the manufacture of chemicals of which about a third is used in the production of sodium phosphates, mainly sodium tripolyphosphate. The rest is used in the production of silicates (sodium metasilicate pentahydrate and sodium orthosilicate), sodium chromate and dichromate, sodium hydrogen carbonate, sodium nitrate, organic chemicals etc. About 5% is utilized in the pulp and paper industries and a further 5% in the manufacture of soaps and detergents. Small amounts of sodium carbonate are used in almost all sectors of industry (dressing of ores, metallurgy, leather industry, ceramic and enamel manufacture, textile industry etc.). Sodium carbonate is thus one of the most versatile chemical products.

3.1.1.3.4 Sodium Hydrogen Carbonate

$NaHCO_3$ (nahcolite):

Very large deposits in the USA (ca. 26×10^9 t)

Natural deposits: Sodium hydrogen carbonate is found in nature as the mineral nahcolite together with dawsonite ($AlNa(OH)_2CO_3$) e.g. in the huge oil shale deposits in the USA (Wyoming, Utah and Colorado with reserves of ca. 26×10^9 t as sodium carbonate). The alkaline salt lakes in the USA, Mexico and Kenya also contain sodium hydrogen carbonate.

Economic importance: Sodium hydrogen carbonate demand is much smaller than that for sodium carbonate. The 1985 world capacity was ca. 1.5×10^6 t/a. The 1984 USA production was 320×10^3 t (4% of the sodium carbonate production). Up to 1989 an annual growth of 2% is expected. This may be higher due to the utilization of nahcolite, sodium sesquicarbonate and sodium hydrogen carbonate in the desulfurization of power station flue gases.

1985 world capacity: ca. 1.5×10^6 t/a

$NaHCO_3$ production:

- USA: 320×10^3 t
- FRG: 50×10^3 t

Manufacture: In the Solvay process sodium hydrogen carbonate precipitates out as an intermediate product. This is insufficiently pure for immediate use. Pure sodium hydrogen carbonate is produced by the reaction of cooled filtered solutions of calcined sodium carbonate with pure carbon dioxide as follows:

$NaHCO_3$ manufacture:

from both synthetic and "natural" sodium carbonate; high purity requirements (up to food grade)

$$Na_2CO_3 + H_2O + CO_2 \rightarrow 2\ NaHCO_3$$

with sodium hydrogen carbonate precipitating out. This has to be carefully dried, e.g. in disc driers, to avoid a back reaction. The manufacture of sodium hydrogen carbonate is normally integrated into the manufacture of synthetic sodium carbonate. In the USA it is produced from "natural" sodium carbonate.

Applications: The most important market for sodium hydrogen carbonate is the food industry (e.g. baking powder manufacture) which accounted for 30% of the USA consumption in 1985. 20% was utilized in the manufacture of animal feedstuffs. Other consumers are the rubber, chemical, pharmaceutical, textile, leather and paper industries. The residue is distributed in a multitude of application areas (neutralizing agents, manufacture of soaps and detergents etc.).

USA application spectrum for $NaHCO_3$ in 1985:

- Food industry: 30%
- Animal feedstuffs: 20%
- Rubber and chemicals sectors: 15%
- Pharmaceuticals: 11%
- Textile, leather and paper industries: 10%
- Fire extinquisher powder: 9%
- For soaps, detergents, neutralizing agents etc.: 5%

with in the future: flue gas desulfurization

3.1.1.3.5 Sodium Sulfate

Sodium sulfate is found in many natural deposits and the world reserves at current consumption levels are sufficient for many hundreds of years. In addition to

Na_2SO_4 production:

- From natural deposits
- As a byproduct in chemical and metallurgical processes
- From NaCl + H_2SO_4

its production from natural deposits, it is produced in large quantities as a byproduct e.g. in the production of potassium salts, sodium chloride and borax as well as in chemical and metal-production processes. Its dedicated manufacture, e.g. from sodium chloride and sulfuric acid, has diminished in importance.

Economic Importance

1985 Na_2SO_4 production: 4.2×10^6 t

The world production of sodium sulfate (anhydrous or as Glauber's salt, $Na_2SO_4 \cdot 10H_2O$) has remained constant since 1975 at ca. 4.5×10^6 t, of which 2.0×10^6 t is from natural deposits. Table 3.1.1-2 gives the 1985 sodium sulfate production figures for the major producing countries.

Table 3.1.1-2. 1985 Sodium Sulfate Production for the Major Producing Countries in 10^6 t.

World	USA	USSR	Spain	Canada	Mexico	Japan	Belgium
4.2	0.8	0.6	0.5	0.4	0.4	0.3	0.3

Production from Natural Deposits and as a Byproduct

Natural Na_2SO_4:

- From minerals (thenardite, glauberite, mirabilite)
- From salt lakes, salt brines, salt deposits (e.g. potassium salt deposits)

The production of pure sodium sulfate or Glauber's salt from natural minerals, such as thenardite Na_2SO_4 or glauberite $Na_2SO_4 \cdot CaSO_4$, is still important in Spain, Canada, USA and the USSR, but is decreasing in importance relative to other processes. The production of sodium sulfate from brines (USA, USSR, Canada) or as a byproduct in the production of sodium chloride, sodium carbonate, borax, potassium- and lithium-salts is more important. Sodium sulfate decahydrate is formed during the working-up of potassium salts e.g. in the reaction of kieserite ($MgSO_4 \cdot H_2O$) with sodium chloride. This Glauber's salt is converted into anhydrous sodium sulfate by heating in a suspended particle drier, spray drier, fluidized bed drier or evaporator crystallizer. In the USA in 1986 ca. 50% of the total sodium sulphate production capacity of 1×10^6 t was from natural sources.

Sodium sulfate is produced in large quantities as a byproduct in various chemical and metallurgical processes e.g. in the production of sodium dichromate, vitamin C, formic acid, resorcinol, and the manufacture of viscose fibers.

Manufacture of Synthetic Sodium Sulfate

Sodium sulfate is directly produced by the reaction of sodium chloride with sulfuric acid at high temperatures (Mannheim process, Sec. 1.7.3.1) or by the Hargreaves process in which sulfuric acid is replaced by sulfur dioxide, air (oxygen) and steam:

$$4\,NaCl + 2\,SO_2 + O_2 + 2\,H_2O \xrightarrow{\text{ca. } 650\,^\circ C} 2\,Na_2SO_4 + 4\,HCl\,.$$

In Europe at the end of the 1970's, approximately half of the sodium sulfate was produced using this process.

Applications

The main consumers of sodium sulfate are the cellulose, detergent and glass industries. In West Germany up to two thirds is utilized in the manufacture of detergents (diluent), in the USA only 39%. In North America its utilization in the Kraft or sulfate paper process is more important (in 1980: 50% of the Na_2SO_4 consumption in the USA and 90% in Canada). Its utilization in the digestion of wood in which it is reduced to sodium sulfide, the actual active component, is declining (1960: 75 kg/t paper, 1982: 30 to 37 kg/t and less). This is a result of improved recycling processes and the substitution of sodium sulfate by sodium sulfide, sodium hydroxide and sulfur. In the manufacture of glass (mainly float glass) sodium sulfate is utilized for clarification and in small amounts can be used instead of sodium carbonate (Sec. 5.1.1.4.1). Small quantities of sodium sulfate are utilized in dye manufacture, in dyeing, in electroplating and in the manufacture of animal feedstuffs and chemicals (e.g. sodium sulfide). Glauber's salt can be used as a heat-storage medium.

Na_2SO_4 consumption:

- Pulp and paper industry
- Detergent industry
- Glass industry
- Chemical industry

3.1.1.3.6 Sodium Hydrogen Sulfate

Sodium hydrogen sulfate is manufactured by reacting sodium chloride with sulfuric acid in heated cast iron retorts (Sec. 1.7.3.1). The liquid product is solidified in refrigeration units. It can also be manufactured in liquid form directly from sodium sulfate and sulfuric

$NaHSO_4$ manufacture:

- $NaCl + H_2SO_4 \rightarrow NaHSO_4 + HCl$
- $Na_2SO_4 + H_2SO_4 \rightarrow 2\,NaHSO_4$
- $Na_2Cr_2O_7 + 2\,H_2SO_4 \rightarrow 2\,CrO_3 + 2\,NaHSO_4 + H_2O$

acid and is produced as a byproduct in the manufacture of chromic acid, but this is contaminated with Cr(III) and Cr(VI) compounds.

Sodium hydrogen sulfate is used as a cleaning agent, due to its water solubility and the acidity of its aqueous solutions, and as a flux.

3.1.1.3.7 Sodium Borates

Natural Deposits and Economic Importance

Boron occurrence:

- In the Earth's surface: 10 ppm
- In seawater: 4.6 ppm

Most important mineral sources of boron:

- Sodium borates (mainly as borax, $Na_2B_4O_7 \cdot 10H_2O$)
- Sodium calcium borates
- Calcium borates
- Salt lakes, brines

Sodium borates are the most important industrial boron compounds. They are found in nature as e.g. tincal or borax($Na_2B_4O_7 \cdot 10H_2O$), kernite ($Na_2B_4O_7 \cdot 4H_2O$) and tincalconite ($Na_2B_4O_7 \cdot 5H_2O$). Other important boron minerals are the sodium calcium borates ulexite ($NaCaB_5O_9 \cdot 8H_2O$) and probertite ($NaCaB_5O_9 \cdot 5H_2O$) and the calcium borate colemanite ($Ca_2B_6O_{11} \cdot 5H_2O$). The largest ore deposits are found in the USA (Boron, California) and Turkey (Kirka district, Hisarcik). Less rich deposits are found in Argentina, Peru, China and the USSR. The world reserves were estimated in 1983 to be 270×10^6 t (as B_2O_3).

World reserves: ca. 270×10^6 t (as B_2O_3)

(the largest deposits are thought to be in Turkey)

The 1984 world production was 1.01×10^6 t (as B_2O_3) of which 60% (mainly borax) was in the USA and 30% (mainly colemanite/ulexite) in Turkey. Ca. 9% of the USA production arises from the processing of salt-rich brines from Searle's Lake in California (main product: potassium salts). The world boron consumption has in recent years increased at an average of 2% annually and is expected to increase 0.5 to 3.5% annually to 1990. Above average growth in demand is expected in the glass fiber and metallic glass sectors.

1984 world production: 1.0×10^6 t (as B_2O_3) with

- 60% in USA
- 30% in Turkey

Extraction

Manufacture of pure sodium tetraborate ($Na_2B_4O_7 \cdot 5H_2O$ or $Na_2B_4O_7 \cdot 10H_2O$) by working up crude mineral borax or Ca borates.

Pure sodium tetraborates are produced from crushed crude sodium borate minerals (tincal, kernite) by dissolution with heating in a weak borax-containing mother liquor, separating off the impurities (clays) and selective crystallization. Either the penta- or decahydrate is formed during vacuum crystallization depending upon the temperature (above or below 60.8 °C).

Anhydrous borax is formed by the calcining of water-containing sodium borates initially in rotary kilns then in standing furnaces, it being produced as a liquid which is poured e.g. into molds. Borax is also manufactured from calcium borates by heating with a sodium carbonate/sodium hydrogen carbonate/sodium hydroxide solution, whereupon the calcium precipitates out and the sodium borate crystallizes out.

Borax is also produced from brines by fractional crystallization in yields of up to 91% (Kerr-McGee Corp., USA) partly directly, partly after prior carbonation to precipitate excess sodium as sodium hydrogen carbonate.

Borax manufacture from brines by fractional crystallization with yields $\leq 91\%$

The other sodium borates such as sodium metaborate ($NaBO_2 \cdot 4H_2O$) and disodium octaborate ($Na_2B_8O_{13} \cdot 4H_2O$) are of minor importance.

Applications

Borax-decahydrate, -pentahydrate, anhydrous borax and boric acid are the most industrially and commercially important boron compounds. They are mainly used directly, but are also the starting materials for the manufacture of other important boron compounds e.g. inorganic borates and perborates, refractory boron derivatives (boron carbide, boron nitride, borides), elemental boron, ferroboron, boron halides, fluoroborates, boron hydrides and organic boron compounds. Crude ore concentrates such as calcium and calcium sodium borates are also directly utilized in e.g. the manufacture of textiles, insulating glass fibers and steel, but are less important.

In the USA 80% of the boron compound consumption (as B_2O_3) is as sodium borates

Sodium borates are mainly utilized in the glass, ceramic, enamel and porcelain industries (e.g. in borosilicate glasses in which the 12 to 15% boron trioxide is used both as a flux and to reduce the thermal expansion coefficient of the glass; glass wool with 5 to 7% boron trioxide for insulation purposes; glass fibers with 8 to 9% boron trioxide). It is also used for the manufacture of sodium perborate (detergent and cleaning agent), fertilizers (boron is a necessary trace element for plant growth) and as a corrosion protection agent in antifreezes. It is also utilized for metallurgical purposes (in fluxes and welding and soldering

Main application sectors for sodium borates:

- Glass, ceramic and enamel industries
- Detergents
- Fertilizers
- Flame retardants
- Corrosion protection agents
- Metallurgy

and in the future:

- Metallic glasses

agents) and as a flame retardant in cellulose materials. The consumption of boron for metallic glasses for transformer coils is low, but there is potential for considerable expansion.

In the USA, the most important boron consumer, utilization of boron compounds in the glass and ceramic industries predominates (ca. 57% of consumption). In Western Europe the glass and detergent industries, each with 42%, are the most important consumers of boron.

3.1.1.4 Potassium and its Compounds

3.1.1.4.1 General Information

Potassium:

- Indispensible plant nutrient; ca. 95% of total K_2O production utilized in the fertilizer sector (1984/85 ca. 28.9×10^6 t)
- Seventh most frequent element in Earth's crust

Industrially most important K compounds:

- KOH, K_2CO_3, $KHCO_3$, K phosphates
- $KMnO_4$, KCN, KI, KIO_3, KBr, $KBrO_3$

By far the most important potassium compounds are the fertilizer salts: potassium sulfate, potassium magnesium sulfate, potassium nitrate and particularly potassium chloride (Sec. 2.3). Only 5 to 6% of all potassium compounds, on the basis of K_2O, are consumed outside the fertilizer industry. The industrially most important potassium compounds are potassium hydroxide and potassium carbonate (potash).

3.1.1.4.2 Metallic Potassium

Metallic K has only limited industrial importance

Manufactured from:

- KCl + Na or
- KF + CaC_2

The industrial manufacture of elemental potassium is insignificant, the 1982 capacity in the Western World amounting to ca. 200 t/a (0.1% of the sodium production). It is mainly manufactured by the reduction of molten potassium chloride with sodium at 760 to 880 °C, the resulting potassium/sodium alloy being purified by fractional distillation. Potassium is also produced in small quantities by the reaction of potassium fluoride with calcium carbide:

$$2\,KF + CaC_2 \xrightarrow{1000-1100\,^\circ C} CaF_2 + 2\,C + 2\,K$$

(Greisheim process). The formerly favored melt electrolysis of potassium hydroxide or potassium chloride is no longer operated. Metallic potassium is utilized in the manufacture of potassium peroxide, KO_2 (oxygen generator) and low melting potassium/sodium alloys (reducing and drying agent, heat transfer medium).

3.1.1.4.3 Potassium Hydroxide

Economic Importance: The 1985 world production of potassium hydroxide was ca. 0.7×10^6 t, of which 0.2×10^6 t was manufactured in the USA mainly using the mercury process. The increase in demand in the next few years is estimated to be ca. 3%/a.

Manufacture: Potassium hydroxide is almost exclusively manufactured by the electrolysis of potassium chloride solutions using the mercury process and the increasingly important membrane process. The technology of potassium chloride electrolysis is similar to that of sodium chloride electrolysis, with a slightly lower cell voltage. In the case of the mercury process a very pure 45 to 50% potassium hydroxide is produced compared to a ca. 30% solution with the membrane process. These solutions are concentrated by evaporation, solid ca. 90% potassium hydroxide being obtained mostly by vacuum evaporation. Byproducts of the electrolysis are chlorine and hydrogen. The manufacture of potassium hydroxide by the reaction of potassium carbonate with slaked lime is no longer operated.

KOH manufacture:

by electrolysis of KCl solutions (mercury and membrane processes)

KOH in two forms:

- Ca. 50% (directly from mercury process)
- Ca. 90% (caustic alkali; by vacuum evaporation)

Applications: Potassium hydroxide is utilized in the manufacture of other potassium compounds [potassium carbonate, potassium phosphates (e.g. tetrapotassium diphosphate), potassium permanganate, potassium bromate, potassium iodate, potassium cyanide etc.], of dyes, special soaps and battery liquids. It is also utilized in photography, in glass manufacture and as a drying and absorption agent. However, its use is decreasing in many of these application areas.

1984 USA application spectrum for KOH:

- K_2CO_3: 25%
- Liquid fertilizers: 15%
- Soaps: 10%
- K phosphates: 9%
  ($K_4P_2O_7$ for liquid detergents)
- Synthetic rubber: 5%
- Crop protection agents: 3%
- $KMnO_4$: 2%
- Other chemicals: 26%
- Export: 5%

3.1.1.4.4 Potassium Carbonate

Manufacture: Potassium carbonate (potash) was formerly produced by the ashing of wood and other plant raw materials. Since the middle of the last century, the saline residues from the rock salt industry and the salt deposits have been the raw materials for potassium carbonate production. The currently most important process is the carbonation of electrolytically produced potassium hydroxide. 50% potassium hydroxide solutions (e.g. from the mercury process) are saturated with carbon dioxide, the solution partially evaporated

K_2CO_3 manufacture:

by carbonation of KOH. Part of the $K_2CO_3 \cdot 1.5H_2O$ resulting from carbonation, is calcined in rotary kilns at 250 to 350 °C and the potassium carbonate hydrate $K_2CO_3 \cdot 1.5H_2O$ which precipitates out separated off. After drying, the product is either marketed as potash hydrate or is calcined to anhydrous potash in a rotary kiln at temperatures of 250 to 350 °C. Anhydrous potassium carbonate is also produced in a fluidized bed process, in which potassium hydroxide is reacted with CO_2-gas in countercurrent in a fluidized bed reactor.

In the USSR part of K_2CO_3 manufacture from alkali aluminosilicates e.g. nepheline $KNa_3[AlSiO_4]_4$

In the USSR, potassium carbonate is also produced from deposits of alkali aluminosilicates (e.g. nepheline) together with aluminum oxide, cement and sodium carbonate.

Main applications of K_2CO_3:

- Glass manufacture
- Soaps, detergents
- Enamel
- Food industry
- For other K compounds

Applications: Potassium carbonate is utilized in the glass industry (special glasses, crystal glass, CRT-tubes for televisions), in the manufacture of soap and enamel, in the food industry and in pigment manufacture. It is also the starting product for other potassium compounds, e.g. potassium hydrogen carbonate (raising agent in the food industry, constituent of fire extinguisher powders) and potassium silicate (detergent). It is also used in the synthesis of many organic chemicals and pharmaceutical products.

References for Chapter 3.1.1: Alkali Metals and their Compounds

General Information:

A. F. Trotman-Dickenson: *Comprehensive Inorganic Chemistry.* 1st Ed., Vol. 1, pp 331–529, pp 665–669, pp 693-694, pp 880–892, Pergamon Press, Oxford (1973).

Kirk-Othmer: *Encyclopedia of Chemical Technology,* 3rd Ed., Vol. 1, pp. 866–882 (1978); Vol. 4, pp. 67–110 (1978); Vol. 14, pp. 448–476 (1981); Vol. 18, pp. 912–947 (1982); Vol. 21, pp. 194–255 (1983), John Wiley & Sons, Inc., New York-Chichester-Brisbane-Toronto.

Ullmanns Encyklopädie der technischen Chemie, 4th Ed., Vol. 8, pp 663–665, pp 669–672, pp 675–676 (1974); Vol. 13, pp 441–494 (1977); Vol. 16, pp 261–277 (1978); Vol. 17, pp 143–159, pp 159–177, pp 211–229 (1979), Verlag Chemie, Weinheim-Deerfield Beach-Basel.

Winnacker-Küchler, *Chemische Technologie*, 4th Ed., Vol. 4, pp 326–338, pp 342–344, Carl Hansen Verlag, München (1986).

Specific Information:

Lithium and its Compounds:

Minerals Yearbook, Vol. I, US department of the Interior, Washington, pp 645–651 (1987).

Mineral Commodity Summaries, Bureau of Mines, Washington, 1986, p. 92.

World Mineral Statistics 1980–84, London, 1986, p. 157.

J. E. Lloyd: "Lithium Chemicals", *Spec. Publ. Royal Soc. Chem.* **40,** pp 98–122 (1986).

S. Buckley: "Lithium – recession delays market lift off", *Industrial Minerals,* New York, pp 25–34 (1983).

R. Bauer: "Chemie – wie es nicht im Lehrbuch steht" (Chemistry – as it is not written in textbooks), *Chemie in unserer Zeit* **19,** pp 167–173 (1985).

Sodium and its Compounds:

Metallic Sodium:

Minerals Yearbook, Vol. I, US Department of the Interior, Washington, pp 835–838 (1978–1979).

"Sodium", *Chemical Economics Handbook,* Stanford Research Institute, Menlo Park, California, USA (Oct. 1986).

"Chemical Profile: Sodium", *Chem. Mark. Rep.,* p. 7 (June 23, 1986).

Sodium Carbonate, Sodium Hydrogen Carbonate:

Minerals Yearbook, Vol. I, US Department of the Interior, Washington, pp 835–842 (1978–1979), p. 772 (1981).
"Sodium carbonate and sodium hydrogen carbonate", *Chemical Economics Handbook*, Stanford Research Institute, Menlo Park, California, USA (June 1986).
"Product Review: Sodium carbonate and sodium hydrogen carbonate", *Chemical Economics Handbook*, Stanford Research Institute, Menlo Park, California, USA (Aug. 1984).
Mineral Commodity Summaries, Bureau of Mines, Washington, 1986, pp 144–145.

Sodium Sulfate, Sodium Hydrogen Sulfate:

"Chemical Profile: Sodium Sulfate", *Chem. Mark. Rep.*, p.9 (Nov. 17 1980).
Minerals Yearbook, Vol. I, US Department of the Interior, Washington, pp 883–890 (1987).
"Sodium sulfate", *Chemical Economics Handbook*, Stanford Research Institute, Menlo Park, California, USA (June 1986).
"Sodium sulfate – salient statistics", *Chemical Economics Handbook*, Stanford Research Institute, Menlo Park, California, USA (May 1982).

Sodium Borates:

Minerals Yearbook, Vol. I, US Department of the Interior, Washington, pp 187–194 (1987).
Ullmann's Encyclopedia of Industrial Chemistry, 5th Ed., Vol. A4, pp 263–280, Verlag Chemie, Weinheim-Deerfield Beach-Basel (1986).
Mineral Commodity Summaries, Bureau of Mines, Washington, pp 24–25 (1986).
"Boron minerals and chemicals – Marketing Research Report", *Chemical Economics Handbook*, Stanford Research Institute, Menlo Park, California, USA (May 1986).
J. Farmer, "Structural chemistry in the borate industry", *Chemistry and Industry*, pp 145–150 (March 6, 1981).

Potassium and its Compounds:

Minerals Yearbook, Vol. I, US Department of the Interior, Washington, pp 775–786 (1987).
"Chemical Profile: Caustic potash", *Chem. Mark. Rep.*, (Oct. 1, 1984).
"Potassium hydroxide – salient stastistics", *Chemical Economics Handbook*, Stanford Research Institute, Menlo Park, California, USA (April 1982).
"Potassium hydroxide", *Chemical Economics Handbook*, Stanford Research Institute, Menlo Park, California, USA (Oct. 1986).

3.1.2 Alkaline Earth Metals and their Compounds

3.1.2.1 General Information

The most important of the alkaline earth metal, beryllium, magnesium, calcium, strontium and barium, is magnesium (more important than sodium). It is followed at two orders of magnitude by calcium. The production of strontium and barium is insignificant, that of beryllium only a few hundred t/a. The calcium compounds are the most important of the alkaline earth compounds. This is mainly due to the huge industrial and economic importance of calcium carbonate (limestone) [not only in the chemical industry, but also in the building industry (Sec. 5.3.2) and in the metallurgical industry] and other calcium minerals such as calcium phosphate (apatite, Sec. 2.1), calcium magnesium carbonate (dolomite), complex calcium silicates (e.g. cement, Sec. 5.3.3), calcium sulfates (gypsum, anhydrite, Sec. 5.3.4) and calcium fluoride (fluorspar, Sec. 1.7.1).

Economic Importance:

- Alkaline earth metals: $Mg \gg Ca > Be > Ba > Sr$
- Alkaline earth compounds: $Ca \gg Mg \gg Ba \gg Sr > Be$

3.1.2.2 Beryllium and its Compounds

Beryllium content in the Earth's crust: 2 ppm

Most important minerals:

- Beryl $Al_2Be_3Si_6O_{18}$
- Bertrandite $Be_4(OH)_2Si_2O_7$

in Brazil, India, USSR, Argentina, USA, South Africa

1985 Beryllium production:

- World: ca. 350 t
- USA: 210 t

Economic Importance: Beryllium is a relatively rare element. The commercially important minerals are beryl $3\,BeO \cdot Al_2O_3 \cdot 6\,SiO_2$ and bertrandite $4\,BeO \cdot 2\,SiO_2 \cdot H_2O$, with beryllium contents of 4 and 0.2 to 0.3% respectively. They are mined primarily in the USA, USSR and Brazil and worked up to beryllium compounds such as $Be(OH)_2$ or BeO. The worldwide workable reserves of beryllium are estimated at 0.2×10^6 t, that in the USA at 73×10^3 t (as Be). The 1985 world production was ca. 400 t beryllium, of which 240 t was produced in the USA.

Beryllium manufacture:

- BeF_2 + Mg
- Melt electrolysis of $BeCl_2$

Manufacture: Metallic beryllium is manufactured either by the reduction of beryllium fluoride with magnesium in graphite crucibles at elevated temperatures or, less commonly, by melt electrolysis of beryllium chloride.

Beryllium applications:

- Alloys: 70 to 80%
- Be metal, Be based alloys: 15 to 20%
- BeO: 6 to 8%
- Other products: rest

Applications: Beryllium is an industrial special metal and belongs together with aluminum and magnesium to the light metals. It is mainly (70 to 80%) utilized in the manufacture of beryllium/copper alloys with 0.5 to 2.5% beryllium (hardenable beryllium bronzes) e.g. for electrical equipment. Beryllium is an interesting moderator- and reflector-material, due to its interesting nuclear physical properties. Beryllium metal and beryllium-based alloys are also utilized for aerospace applications, mainly due to their high elasticity-weight ratio, their high tensile stress and their high thermal conductivity.

BeO:

for oxidic ceramics

The 1986 USA consumption of metallic beryllium was ca. 300 t. Beryllium oxide is utilized for oxide ceramic materials, e.g. base materials for printed circuit boards, due to its high melting point and high chemical stability.

The demand for beryllium and beryllium oxide, particularly in the aerospace and electronics industries, as well as in defence, is currently increasing at a rate of ca. 3 to 5% annually. A further expansion in its utilization is hindered by its high price, its complex processing and the toxicity of beryllium dust and vapor.

3.1.2.3 Magnesium and its Compounds

3.1.2.3.1 Natural Deposits

Magnesium is the eighth most frequent element in the Earth's crust at a content of 2.1% and the third most frequent element in seawater. Magnesium from seawater is an inexhaustible source of raw materials, one km^3 containing 1.3×10^6 t of Mg (0.13%) and there being an estimated 10^{18} km^3 of seawater. The industrially most important magnesium minerals are magnesium carbonate ($MgCO_3$, magnesite), calcium magnesium carbonate ($CaCO_3 \cdot MgCO_3$, dolomite), magnesium chloride, e.g. as a double salt with potassium chloride (carnallite, $KCl \cdot MgCl_2 \cdot 6H_2O$), magnesium sulfate (kieserite, $MgSO_4 \cdot H_2O$) and silicates such as asbestos (Sec. 5.2.2) or olivine [$(Mg, Fe)_2SiO_4$].

Main deposits of Mg:

- Seawater
- Dolomite
- Magnesite rock
- Brines and salt deposits

3.1.2.3.2 Metallic Magnesium

Economic Importance: The 1985 world production of magnesium was ca. 325×10^3 t to which must be added ca. 70×10^3 t of remelted magnesium (secondary magnesium) mainly from recycled aluminum waste. The main producing countries are the USA, the USSR and Norway with 43%, 26% and 17% respectively in 1985. The consumption of metallic magnesium has increased substantially in recent years, due mainly to increasing demand for die-cast magnesium alloy products and its utilization in steel manufacture for steel desulfurization. Its demand is expected to increase disproportionately in the future, a consumption of 400×10^3 t being estimated for 1990. To satisfy this increasing demand two new electrolysis plants are planned in Canada with a total capacity of 100×10^3 t/a.

1985 world Mg production: 325×10^3 t

World Mg consumption:

- 1985: 282×10^3 t
- 1990: 400×10^3 t (estimated)

1986 USA Mg consumption: 226×10^3 t

Main consumers: USA, USSR, West Germany

Manufacture: Magnesium is manufactured either electrolytically from magnesium chloride (from seawater or salt lakes and salt deposits) or by thermal reduction of ores such as dolomite with ferrosilicon. 40% of the magnesium producers in the Western World producing 80% of the world production use electrolysis processes.

Mg manufacture by two processes:

- Melt electrolysis of $MgCl_2$
- Silicothermal reduction of dolomite

energy requirements equally high for both processes

In the melt electrolysis manufacture of magnesium, favored by the largest magnesium manufacturers worldwide, mixtures of anhydrous magnesium chlor-

Melt electrolysis utilized in the USA, USSR and Norway (ca. 80% of the world capacity)

ide with alkali and alkaline earth chlorides (the electrolyte usually containing 8 to 24% $MgCl_2$) are electrolyzed at 700 to 800 °C with a decomposition voltage of 5 to 7 V on the iron cathodes. The chlorine byproduct produced at the graphite anodes is not usually marketed directly, being used for the production of magnesium chloride. The magnesium deposited collects at the surface of the melt and is sucked off. The electricity yield is 80 to 90% and the power loading of a cell up to 250 kA. The average specific electricity consumption is 18 to 20 kWh/kg magnesium, with only 12 to 14 kWh/kg magnesium in the process developed by Norsk Hydro. This is similar to that for the modern production of aluminum. In the Dow process, a hydrated magnesium chloride ($MgCl_2 \cdot 1.5$ to $2H_2O$) is utilized instead of the anhydrous salt.

Specific energy consumption similar to that for Al production

In the silicothermal production of magnesium mainly operated in Brazil, calcined dolomite is reacted with 70 to 85% ferrosilicon at ca. 1200 °C under vacuum as follows:

The silothermal process is utilized in Brazil

$$2\,(CaO \cdot MgO) + Si(Fe) \rightarrow 2\,Mg + \underbrace{Ca_2SiO_4 + Fe}_{\text{slag}}\,.$$

The magnesium vapor initially formed is precipitated in a condensation vessel. In a variant of this process (magnetherm process) aluminum oxide is added to obtain an electrically conducting liquid $CaO\text{-}SiO_2\text{-}Al_2O_3$-slag and the required energy (ca. 18 kWh/kg Mg) supplied by resistive heating.

The magnetherm process is preferred in Italy, France and Yugoslavia

The crude magnesium obtained from electrolysis or thermal reduction has to be purified (refined) before further processing. This is carried out by mixing salt melts (alkali and alkaline earth chlorides or fluorides) with the liquid metal. The purest magnesium is manufactured by distillation.

Applications: Most of the magnesium is utilized in the manufacture of alloys, primarily magnesium/aluminum alloys (over 99% of all magnesium alloys contain aluminum). These can be split into extremely light "real" magnesium alloys with magnesium as the main component and admixtures of up to 10% of aluminum, manganese, zinc, silicon, beryllium etc. (used e.g.

Application spectrum for Mg in the Western World in 1985:

- Al alloys with < 10% Mg: 51%
- Mg-based alloys with < 10% Al, Mn, Zn, Si, Be: 20%
- Reduction agent, Grignard reagents, cathode protection etc.: 12%
- Steel desulfurization: 9%
- Cast iron metallurgy: 5%
- Miscellaneous: 3%

in the motor industry) and aluminum alloys with up to 11% magnesium. These can be subdivided on the basis of their properties into casting alloys and wrought alloys. Magnesium is also utilized in the thermal reduction of metal chlorides and other compounds e.g. in the manufacture of titanium, uranium, zirconium, hafnium and beryllium. In the iron and steel industry magnesium is used in increasing quantities as a desulfurization and deoxidification agent and for the manufacture of cast iron nodules. Magnesium powder is utilized in pyrotechnical applications and magnesium chips in organic synthesis for the preparation of Grignard reagents. Magnesium is also utilized in the synthesis of the antiknock agents: tetramethyl- and tetraethyl-lead.

Mg is, with a density of 1.74 g/cm^3, the lightest construction metal

3.1.2.3.3 Magnesium Carbonate

Natural Deposits and Economic Importance: Natural magnesium carbonate (magnesite) is the industrially most important magnesium mineral. It is found in large quantities in both crystalline and amorphous forms in exploitable deposits in the People's Republic of China, the USSR, North Korea, Brazil and Australia. Its estimated world reserves are ca. 11×10^9 t to which can be added the very much larger deposits of the double carbonate dolomite which is also utilized in the manufacture of magnesium and its compounds. 12×10^6 t of raw magnesite were mined in 1985, the main producer countries being the USSR (with an 18% share), China (18%), North Korea (17%), Austria (9%) and Greece (8%).

$MgCO_3$ (dolomite):

- In quantity the most important Mg-mineral
- Main mining countries: China, USSR and North Korea (> 50% share)

Manufacture: Magnesium carbonate is produced by mining (mainly open-pit), and is processed by gravitational separation, flotation or magnetic separation. Magnesium carbonate-hydrate $MgCO_3 \cdot 3H_2O$ and basic magnesium carbonate ($4MgCO_3 \cdot Mg(OH)_2 \cdot 4H_2O$, hydromagnesite, magnesia alba) are also synthesized in small amounts: in precipitation reactions by reacting magnesium salt solutions (e.g. residual brines in the potash industry) with ammonium carbonate and hydrogen carbonate, by carbonating precipitated magnesium hydroxide under pressure or by calcining or partially calcining dolomite ($CaO \cdot MgO$ or $CaCO_3 \cdot MgO$) and then reacting with carbon dioxide to soluble magnesium hydrogen carbonate,

Synthetic $MgCO_3$ from Mg-salts and CO_2

which is converted upon heating into basic magnesium hydrogen carbonate.

Natural $MgCO_3$ almost exclusively utilized for MgO manufacture

Applications: Little natural unburnt magnesite is utilized directly. It is mainly converted by calcination (above 550 °C) into different, mainly refractory, magnesia qualities. This technology is of little importance in the USA, since ca. 90% of the industrially produced magnesium compounds originate from the magnesium content of seawater and brines. Synthetic, in particular basic, magnesium carbonate is utilized in the manufacture of thermal insulating materials; as a filler for paper, plastics and rubber; in the paint and printing ink industry; as an additive in table salt (to reduce caking) and in pharmaceuticals (e.g. antacids).

3.1.2.3.4 Magnesium Oxide

1984 MgO world production: ca. 6.5×10^6 t

Economic Importance: Magnesium oxide is the most important industrial magnesium compound. Its 1984 world production amounted to ca. 6.5×10^6 t. This has stagnated since the mid-1970's mainly due to the decline in steel production, the steel industry being the main consumer of magnesium oxide. About one third of the world demand is currently met by magnesium oxide from seawater and brines, but this is decreasing in importance because its energy requirements are three times higher than those of the magnesite process (a number of plants in Japan and the USA have already reduced or stopped production).

Raw materials for MgO manufacture:

- Magnesite and dolomite: 70%
- $Mg(OH)_2$ from seawater and $MgCl_2 \cdot 6H_2O$ from brines: 30%

Manufacture: The raw materials for the production of magnesium oxide are either natural magnesium carbonate (magnesite) which is calcined, or magnesium chloride or hydroxide from seawater and brines (salt deposits). Dolomite can also be utilized as a starting material.

MgO manufacture by calcination of $MgCO_3$, $Mg(OH)_2$ and $MgCl_2 \cdot 6H_2O$

MgO types:

- Chemical (caustic) magnesia
- Dead-burnt (sintered) magnesia
- Fused magnesia

In the production of magnesium oxide from seawater magnesium hydroxide is precipitated with the aid of quick and slaked lime, or still better with dolomite. The separated hydroxide is then calcined. The different magnesia qualities differ in the type and temperature of the process used to calcine magnesium carbonate or hydroxide. More reactive caustic magnesia ("chemical magnesia") is burnt at 600 to 900 °C, dead-burnt sintered magnesia at 1600 to 2000 °C. Fused magnesia is melted at 2800 to 3000 °C in an electric arc furnace.

Applications: The most important magnesium oxide product is sintered magnesia (m.p. 2800 °C) for the refractory industry (lining of metallurgical furnaces, raw materials for refractory bricks). There are three types of refractory bricks: (basic) magnesite, magnesite-chrome and chrome-magnesite bricks. Magnesia-containing refractory liners for furnaces are particularly important in steel production (LD-process). Magnesium oxide is also utilized in heat storage materials. Reactive chemical magnesia has gained in importance in recent years. In 1984 caustic calcined magnesia made up 40% of the USA magnesium oxide consumption (worldwide only ca. 20%). It is utilized in the manufacture of fertilizer and animal feedstuffs, building materials (Sorel cement, lightweight building panels), chemical and pharmaceutical products (e.g. vulcanization accelerators) and in water treatment. Fused magnesia is utilized as a refractory material for high temperature crucibles and as an insulating material in the electrical heating industry.

Main application for MgO: refractory industry and in steel works

3.1.2.3.5 Magnesium Chloride

Economic Importance: Since the production of magnesium chloride is closely linked with the manufacture of metallic magnesium, it follows developments in this sector. In 1984 ca. 10^6 t of magnesium chloride was manufactured in the Western World.

$MgCl_2$ is predominantly manufactured by Mg producers

1984 production in Western World: 10^6 t

Manufacture: Magnesium chloride is found in large quantities in nature (Sec. 3.1.2.3.1). It is produced from the residual brines in the potash industry, from seawater and brines, or from magnesium carbonate and oxide.

$MgCl_2$-manufacture from:

- Seawater
- Salt lakes
- Underground brines
- Residual brines of the potash-industry
- $MgCO_3$ or MgO

In the Dow Chemical process milk of lime is added to seawater, the precipitated hydroxide then being reacted with hydrogen chloride (the calcium precipitating as the sulfate due to the simultaneous addition of sulfuric acid) and the magnesium chloride solution evaporated at ca. 200 °C. The resulting product has ca. 1.5 to 2 molecules of hydration water.

In the Norsk-Hydro process burnt magnesite is utilized instead of milk of lime, the precipitated magnesium hydroxide being calcined and then chlorinated in the presence of coal at 1000 to 1200 °C to anhydrous magnesium chloride:

$$2\,MgO + 2\,Cl_2 + C \rightarrow 2\,MgCl_2 + CO_2\ (\text{or } CO)\,.$$

Anhydrous magnesium chloride is also produced by the careful dehydration of the hexahydrate in a stream of hydrogen chloride at temperatures $> 230\,°C$, or by directly reacting magnesite with chlorine and carbon monoxide at high temperatures (MPLC-process).

Over 80% of the $MgCl_2$ is utilized for the electrolytic manufacture of Mg. Remainder mainly used in the construction industry

Applications: Most of the magnesium chloride, either anhydrous or sesquihydrate (Dow Chemical process), is utilized in the electrolytic manufacture of magnesium. The hexahydrate mixed with magnesium oxide (forming magnesium oxychlorides) is utilized in the manufacture of Sorel cement and lightweight building panels. Magnesium chloride is also utilized in the granulation of fertilizers, in the oil and sugar industries, as a dust binder (in mining, road building) and as a thawing agent.

3.1.2.3.6 Magnesium Sulfate

$MgSO_4$ products:

- Anhydrous $MgSO_4$
- Kieserite $MgSO_4 \cdot H_2O$
- Epsom salt $MgSO_4 \cdot 7\,H_2O$
- Potassium magnesium sulfate

Economic Importance: The world production of magnesium sulfate products including potassium magnesium sulfate (fertilizer) was 3.4×10^6 t in the mid-1970's of which 2.3×10^6 t was kieserite, $MgSO_4 \cdot H_2O$ (a byproduct of the potash industry). The main producer countries are: the USA, West Germany and East Germany.

Manufacture: Magnesium sulfate is widely distributed in nature, e.g. in salt deposits as kieserite, as epsom salt $MgSO_4 \cdot 7\,H_2O$, in the form of double salts such as kainite $4\,KCl \cdot 4\,MgSO_4 \cdot 11\,H_2O$ or langbeinite $K_2SO_4 \cdot 2\,MgSO_4$, and in brines. Large quantities of kieserite, epsom salt or anhydrous magnesium sulfate are produced in the processing of potassium salts. Magnesium sulfate is also produced by reacting magnesium carbonate or seawater with sulfuric acid.

$MgSO_4$ – most important byproduct in the potash industry

Main application for $MgSO_4$ products (kieserite, potassium magnesium sulfate):

in fertilizer sector

Applications: Magnesium sulfates are utilized in the potash industry for the manufacture of potassium sulfate (from potassium chloride), sodium sulfate and potash magnesia (K/Mg sulfate). Magnesium sulfate, particularly as kieserite, is also used as a fertilizer (ca. 80% of total consumption). It is also utilized in the textile industry, in the cellulose industry and in the manufacture of building materials, refractory materials, animal feedstuffs and motor oil additives.

3.1.2.4 Calcium and its Compounds

3.1.2.4.1 Natural Deposits

Calcium is the fifth most frequent element in the Earth's crust at ca. 4%. The most important minerals are calcium carbonate (limestone, chalk, marble, shell limestone, calcite etc.), calcium magnesium carbonate (dolomite, $CaCO_3 \cdot MgCO_3$), calcium sulfate (gypsum, $CaSO_4 \cdot 2H_2O$, and anhydrite, $CaSO_4$), calcium phosphate [apatite, $Ca_5(PO_4)_3(F)$], calcium fluoride (fluorspar, CaF_2) and calcium aluminum silicates (e.g. anorthite, $CaO \cdot Al_2O_3 \cdot 2SiO_2$). The reserves of limestone are almost unlimited, ca. 7% of the earth's crust consisting of this mineral. It is the most important starting material for the manufacture of calcium and its compounds.

Limestone and chalk are together with coal, oil, iron ore, sulfur and sodium chloride the most important raw materials for modern industry

$CaCO_3$-containing rocks are widely distributed, making it an almost inexhaustible raw material source

3.1.2.4.2 Metallic Calcium

The 1985 world production of metallic calcium was ca. 2×10^3 t, of which more than 50% was produced in the USA. It is currently manufactured by the thermal reduction of calcium oxide with aluminum in vacuum at 1200 °C:

1985 world Ca production: ca. 2×10^3 t with > 50% in USA

Manufacture:

by aluminothermal reduction of CaO

$$6\,CaO + 2\,Al \rightarrow 3\,CaO \cdot Al_2O_3 + 3\,Ca_{gas}$$

the resulting calcium vapor being condensed. The melt electrolysis of calcium chloride is no longer operated.

Calcium is utilized as a reducing agent in the manufacture of special metals such as zirconium, thorium, uranium and the rare earths and also in the manufacture of $SmCo_5$ permanent magnets. It is used as a refining agent in metallurgy (steel, copper, magnesium, tantalum, lead), is contained in the form of lead/calcium alloys in maintenance-free batteries and is the starting material for the synthesis of calcium hydride (hydrogen source).

Applications:

- In manufacture of Zr, Th, U, rare earth metals
- Refining agent in metallurgy
- In maintenance-free batteries
- In manufacture of magnetic material $SmCo_5$

3.1.2.4.3 Calcium Carbonate

Largest limestone producers:

- USA (with 25% share)
- USSR
- Japan
- West Germany

Economic Importance: It is estimated that in 1983 ca. 2.4×10^9 t of limestone and other calcium carbonate-containing minerals were mined in the Western World of which 570×10^6 t was in the USA. In the rock and earth extraction industry limestone is in second place, only the mining of sand and pebbles being more important. The 1984 USA production of synthetic (precipitated) calcium carbonate was 160×10^3 t.

Production of $CaCO_3$:

- Coarse-grained $CaCO_3$
  - by mining
- Fine-grained $CaCO_3$
  - by carbonating milk of lime

Mining and Manufacture: Calcium carbonate-containing minerals are mainly extracted by open cast mining. High grade limestone is also extracted in underground workings. A comparatively small quantity of calcium carbonate is produced synthetically e.g. by carbonating milk of lime.

USA application spectrum of natural $CaCO_3$ in 1983:

- Construction industry: 74%
- Portland cement: 16%
- Quicklime manufacture: 4%
- Fertilizer: 4%
- Steel industry: 1%

Applications of fine particulate $CaCO_3$ as a white filler in the:

- Paint industry
- Paper industry

Applications: Natural calcium carbonate is mainly utilized in the construction industry e.g. in road building and in the cement industry (1 t limestone/1 t cement). It is also utilized in the manufacture of quicklime, as a flux and refining agent (desulfurization agent) and as a sintering aid and slag-former in metallurgy e.g. in the production of pig iron. Significant quantities of limestone are used in agriculture, in glass manufacture, as a desulfurization agent in the purification of flue gases and in fine- and coarse-ground form as a filler. Precipitated calcium carbonate is an important filler e.g. in paper manufacture in which 75% of this calcium carbonate is utilized. It is also used in the rubber and plastics industries and in the manufacture of paint, adhesives and printing inks.

3.1.2.4.4 Calcium Oxide and Calcium Hydroxide

Economic Importance: Calcium oxide (quicklime) and calcium hydroxide (slaked lime, lime hydrate) are together with calcium carbonate the most important calcium compounds. They are utilized in almost all sectors of industry. The 1985 world production (including burnt dolomite) was 112×10^6 t. Table 3.1.2-1 gives the 1985 production figures for the main producer countries.

Table 3.1.2-1. 1985 Calcium Oxide and Calcium Hydroxide Production in the Main Producer Countries in 10^6 t.

World	USSR	USA	Japan	West Germany	Brazil	Poland	Mexico	Czechoslovakia	East Germany	France
112	29.5	12.2	7.4	7.0	5.0	4.1	3.6	3.2	3.1	3.0

The main product is quicklime. It is produced in a multitude of plants, large and small, throughout the world.

Breakdown of USA production in 1985:

- Quicklime: 83%
- Lime hydrate: 15%
- Burnt dolomite: 2%

Manufacture: Calcium oxide is produced by calcining limestone at ca. 1000 to 1200 °C. Various shaft furnaces or rotary kilns [particularly in the USA (e.g. with a capacity of 1000 t/d)] are utilized as burning units. Calcium hydroxide is produced by the slow addition of water (slaking) to calcium oxide, the process being strongly exothermic.

Manufacture of burnt lime in shaft or rotary furnaces

Manufacture of CaO strongly endothermic, of $Ca(OH)_2$ exothermic

In the so-called dry slaking process just sufficient water is added to yield an easily usable hydrate.

Applications: Much of the preferably softly burnt lime is utilized in metallurgy e.g. in the iron and steel industry for the removal of phosphorus and sulfur from molten metal. Further amounts are utilized in the chemical industry for the manufacture of calcium compounds (calcium carbide, calcium cyanamide) or sodium carbonate using the Solvay process and as a neutralizing and precipitating agent. In environmental technology lime is utilized in water and effluent treatment and in increasing quantities for flue gas desulfurization. Other consumers are the sugar industry, agriculture and the refractory industry (dead-burnt dolomite). Finally large quantities of quicklime and lime hydrate are utilized in the construction industry (manufacture of mortar, of building materials e.g. sand-lime bricks, ground consolidation in road building). In the USA and Japan its use in metallurgy predominates, whereas in West Germany and Italy its utilization in the construction industry predominates.

USA application spectrum for CaO in 1985:

- Metallurgy (of which 34% is for steel manufacture and the rest for bauxite processing, copper flotation, Mg manufacture): 39%
- Water and effluent treatment: 16%
- Chemicals, sugar, glass: 11%
- Construction industry: 9%
- Flue gas desulfurization: 8%
- Paper manufacture: 7%
- Refractory industry: 3%

3.1.2.4.5 Calcium Chloride

Economic Importance: Calcium chloride is produced in large quantities as a byproduct in different chemical processes e.g. in the Solvay process or propylene oxide

$CaCl_2$ produced in large quantities as a byproduct in chemical processes

manufacture in the chlorohydrin process. Only a small quantity is worked up to pure calcium chloride. The 1985 USA production of calcium chloride was 725×10^3 t (as the dihydrate) for the most part (> 80%) from natural sources (brines, minerals).

Manufacture: Synthetic water-containing calcium chloride is produced as a 30 to 45% solution or as flakes with a 75% calcium chloride content from the calcium chloride-containing spent solution of the Solvay process (Sec. 3.1.1.3.3) by vacuum and then atmospheric pressure evaporation. Calcium chloride is also manufactured from waste acid by reaction with limestone. Anhydrous calcium chloride is obtained e.g. by evaporation in fluidized bed driers or spray towers.

USA application spectrum for $CaCl_2$ in 1984:

- Road deicing agent: 35%
- Dust binder: 20%
- Technical purposes: 20%
- Crude oil production: 12%
- Concrete manufacture: 5%

Applications: Calcium chloride is utilized as a drying agent, a dust binder (road consolidation, in mines), a cooling-, defrosting- and antifreeze-agent, e.g. road deicing at temperatures below those at which sodium chloride is effective. It is also utilized as a concrete additive and in various chemical and metallurgical processes. The addition of calcium chloride to crude oil drilling-muds (for conditioning the drill cleaning agent) is gaining in importance. In the USA it is predominantly used for road deicing.

3.1.2.4.6 Calcium Carbide

CaC_2 production is declining, mainly due to the decline in acetylene chemistry (replacement by petrochemical feedstocks)

1982 world production: 6.1×10^6 t

Economic Importance: The world production of calcium carbide has strongly decreased since the mid-1950's, due to the preferred use of petrochemical feedstocks in the organic chemical industry (substitution of acetylene by ethene). In 1965 it was still ca. 10×10^6 t, but in 1982 was only 6.1×10^6 t. The largest producers with a total world share > 50% are China, East Germany, the USSR and Japan. The 1985 USA production was 0.2×10^6 t.

CaC_2 manufacture is very energy intensive; raw materials purity requirements stringent

Manufacture: Calcium carbide is industrially manufactured by reacting highly purified calcium oxide with coke in an electrical arc reduction furnace at 2000 to 2200 °C:

$$CaO + 3\,C \rightarrow CaC_2 + CO, \qquad \Delta H = 464\ \text{kJ/mol}\,.$$

The ca. 80% calcium carbide formed (balance being mainly CaO) is initially obtained as a liquid, but is removed as blocks. Carbide furnaces are mostly operated with three-phase current at a power of up to 70 MW using Söderberg (carbon-)electrodes dipped deeply into the reaction mixture. The electricity price in addition to the raw materials costs (lime and coal) is crucial for the economics of carbide production, because the process is very energy intensive (2800 to 3100 kWh/t).

Applications: Calcium carbide is mainly utilized for the:

Applications for CaC_2:

- Acetylene manufacture
- Azotization to calcium cyanamide
- Desulfurization of iron during steel manufacture

- Production of acetylene:

$$CaC_2 + 2H_2O \longrightarrow C_2H_2 + Ca(OH)_2$$

  which is mainly utilized in welding (autogenous technology) and in the manufacture of special cast iron.

- Desulfurization and deoxidization of raw iron and steel, in increasing quantities.
- Calcium cyanamide (nitro-lime) production:

$$CaC_2 + N_2 \longrightarrow CaCN_2 + C$$

  Recently, an inexpensive process has been developed in Japan for the manufacture of calcium hydride from calcium carbide and hydrogen under pressure.

3.1.2.5 Strontium and its Compounds

Strontium mainly occurs naturally as its sulfate (celestine, $SrSO_4$) and its carbonate (strontianite, $SrCO_3$). In 1985 137×10^3 t of strontium ores were mined worldwide, mainly in Mexico, Spain, Turkey and Great Britain. The largest manufacturer of strontium compounds is Kali-Chemie AG (West Germany) with a market share > 50%.

Strontium only important for special applications. USA application spectrum in 1984:

- Manufacture of special glasses (color television, computer monitors): 53%
- Pyrotechnical and signalling uses: 14%
- Magnetic materials: 11%
- Pigments and fillers: 8%
- Electrolytic zinc manufacture: 6%

The most important strontium compound is strontium carbonate. In 1985 ca. 100×10^3 t of strontium carbonate was consumed in the manufacture of CRT-screen glass for color televisions and computer monitor screens, ferritic magnetic materials and electrolytic zinc (precipitation of lead and cadmium) etc. Strontium nitrate is important for pyrotechnical and signalling applications. Metallic strontium has almost no industrial significance. The USA turnover of strontium compounds in 1985 was ca. 24×10^3 t.

3.1.2.6 Barium and its Compounds

3.1.2.6.1 Natural Deposits and Economic Importance

Raw material for manufacture of barium and its compounds is heavy spar ($BaSO_4$)

The most important barium mineral is heavy spar ($BaSO_4$, barite). It is mainly mined in China, the USA, India and the USSR. Its worldwide extraction is closely coupled to natural gas and crude oil exploration and was 6.0×10^6 t in 1985. The certain world reserves are ca. 450×10^6 t.

Applications for natural $BaSO_4$:

- Oil and natural gas exploration: 88 to 90%
- Barium chemicals: 7 to 8%
- Fillers, glass manufacture: 3 to 5%

Of the heavy spar extracted only ca. 7 to 8% is worked up to barium chemicals (barium carbonate, barium sulfide, synthetic barium sulfate, barium chloride, barium hydroxide etc.). Ca. 90% is utilized in crude oil and natural gas exploration as a suspension in drilling-muds. Just in the USA ca. 2.5×10^6 t was utilized for this purpose in 1984, an average of 30 t being used per drilling. A small part of the heavy spar, ca. 3 to 5%, is utilized as a filler in the paint, varnish, plastics, and rubber industries and in glass manufacture. Metallic barium is hardly utilized industrially and is only manufactured in large quantities by Degussa (ca. 30 t/a). It is mainly used as a getter material in the manufacture of valves.

3.1.2.6.2 Barium Carbonate

$BaCO_3$ most important Ba compound after natural $BaSO_4$

Economic Importance: Barium carbonate is produced almost exclusively synthetically. The worldwide demand was ca. 200×10^3 t/a in 1985 of which Kali-Chemie produced more than 50%. The USA capacity in 1985 was only 18×10^3 t.

$BaCO_3$ manufacture:

$BaSO_4 \rightarrow BaS \rightarrow BaCO_3$

Manufacture: In barium carbonate manufacture barium sulfate is first reduced with coal to barium sulfide at 1000 to 1200 °C in a rotary kiln. This is then reacted in solution either with carbon dioxide or sodium carbonate, whereupon barium carbonate precipitates:

$$BaSO_4 + 4\,C \rightarrow 4\,BaS + 4\,CO$$

$$BaS + CO_2 + H_2O \rightarrow BaCO_3 + H_2S \text{ or}$$

$$BaS + Na_2CO_3 \rightarrow BaCO_3 + Na_2S$$

Applications: The most important utilization sector for barium carbonate is the manufacture of clay tiles and ceramic products, the addition of barium carbonate hindering the bleading of salts (Na-, Ca-sulfate). In the USA this accounts for 31% of the demand, a further 29% being utilized in the glass industry for the manufacture of special optical glasses and CRT-screens. Barium carbonate is also used in the manufacture of special ceramics (barium ferrite and barium titanate), in the oil industry, in photographic papers and in the manufacture of other barium compounds.

Applications:

- Tile and ceramic industry
- Glass industry
- Oil industry
- Photographic papers

3.1.2.6.3 Barium Sulfide

Barium sulfide is the most important intermediate in the manufacture of barium compounds. It is also used in the manufacture of lithopone ($ZnS + BaSO_4$) and hydrogen sulfide. The USA production in 1984 was 69×10^3 t.

BaS:

- Most important intermediate for Ba compound manufacture
- Manufacture of H_2S

3.1.2.6.4 Barium Sulfate

Synthetic (precipitated) barium sulfate is manufactured by reacting barium sulfide solution with sodium sulfate. Synthetic barium sulfate is utilized as a filler and a white pigment in the manufacture of paper (barytes white) and in the paint, rubber and plastics industries. Coprecipitation of barium sulfate with zinc sulfide yields the white pigment lithopone, which has declined in importance due to the development of titanium dioxide pigments. Ca. 50 to 55×10^3 t/a of synthetic barium sulfate is currently produced in Western Europe.

Synthetic $BaSO_4$:

- Manufactured from $BaS + Na_2SO_4$
- Applications:
  - Filler
  - White pigment; significance of lithopone ($BaSO_4$ + ZnS) has strongly declined

References for Chapter 3.1.2: Alkaline Earth Metals and their Compounds

General Information:

A. F. Trotman-Dickenson: *Comprehensive Inorganic Chemistry.* 1st Ed., Vol. 1, pp 531–664, Pergamon Press, Oxford (1973).

Kirk-Othmer: *Encyclopedia of Chemical Technology,* 2nd Ed., Vol. 19, pp 48–54 (1969); 3rd Ed., Vol. 3, pp 457–479, pp 803–829 (1978); Vol. 4, pp 412–448, pp 505–519 (1979); Vol. 14, pp 343–382, pp 570–646 (1981), John Wiley & Sons, Inc., New York-Chichester-Brisbane-Toronto.

Ullmann's Encyclopedia of Industrial Chemistry, 5th Ed., Vol. A3, pp 325–341 (1985); Vol. A4, pp 11–33, pp 515–553 (1985), Verlag Chemie, Weinheim-Deerfield Beach-Basel.

Ullmanns Encyklopädie der technischen Chemie, 4th Ed., Vol. 8, pp 301–311, pp 442–460 (1974); Vol. 13, pp 484, pp 497–508 (1977); Vol. 16, pp 319–329, pp 331–356 (1978), Verlag Chemie, Weinheim-Deerfield Beach-Basel.

Specific Information:

Beryllium and its Compounds:

Minerals Yearbook, Vol. I, US Department of the Interior, Washington, pp 177–181 (1987).

"Beryllium minerals", *Chemical Economics Handbook,* Stanford Research Institute, Menlo Park, California, USA (June 1986).

Mineral Commodity Summaries, Bureau of Mines, Washington 1986, pp 20–21.

Magnesium and its Compounds:

Metallic Magnesium:

Minerals Yearbook, Vol. I, US Department of the Interior, Washington, pp 653–660 (1987).

"Magnesium – salient statistics", *Chemical Economics Handbook,* Stanford Research Institute, Menlo Park, California, USA (June 1986).

Mineral Commodity Summaries, Bureau of Mines, Washington, pp 94–97 (1986).

Winnacker-Küchler, *Chemische Technologie,* 4th Ed., Vol. 4, pp 301–320, Carl Hanser Verlag, München (1986).

Magnesium Compounds:

"Markets: Magnesium oxide makes a comeback", *Chemical Week,* p. 43 (Jan. 7, 1981).

Minerals Yearbook, Vol. I, US Department of the Interior, Washington, pp 661–667 (1987).

B. M. Coope: "Magnesium markets", *Industrial Minerals,* pp 57–98 (1983).

T. P. Whaley: "Magnesium compounds of industrial significance", *Spec. Publ. – Royal Soc. Chem.* **40,** pp 123–169 (1981).

J. H. Canterford: "Magnesia – An important industrial mineral", *Mineral Processing and Extractive Metallurgy Review,* Vol. 2, pp 57–104 (1985).

Mineral Commodity Summaries, Bureau of Mines, Washington, pp 96–97 (1986).

World Mineral Statistics 1979–1983, Brit. Geol. Survey, London, p. 150 (1985).

"Magnesium – salient statistics", *Chemical Economics Handbook,* Stanford Research Institute, Menlo Park, California, USA (Sept. 1982).

Calcium and its Compounds:

"Calcium chloride: salient statistics", *Chemical Economics Handbook,* Stanford Research Institute, Menlo Park, California, USA (Dec. 1986).

"Calcium carbide: salient statistics", *Chemical Economics Handbook,* Stanford Research Institute, Menlo Park, California, USA (June 1986).

"Limestone and lime – product review", *Chemical Economics Handbook,* Stanford Research Institute, Menlo Park, California, USA (June 1986).

Minerals Yearbook, Vol. I, US Department of the Interior, Washington, pp 209–213, pp 635–644 (1987).

Industrial Statistics Yearbook 1982, Vol. II, United Nations, New York, pp 40–42, p. 405, pp 525–526 (1985).

"Aragonite, a mineral mined from the sea", *The Glass Industry,* p. 15, (Dec. 1980).

Mineral Commodity Summaries, Bureau of Mines, Washington, pp 86–87 (1981).

Strontium and its Compounds:

Mineral Commodity Summaries, Bureau of Mines, Washington, pp. 152–153 (1986).

Minerals Yearbook, Vol. I, US Department of the Interior, Washington, pp 1027–1030 (1984).

Barium and its Compounds:

World Mineral Statistics 1975–1979, Institute of Geological Sciences, London, p. 23 (1981).

Mineral Commodity Summaries, Bureau of Mines, Washington, pp 16–17 (1986).

Europa Chemie **12,** 182 (1986).

Minerals Yearbook, Vol. I, US Department of the Interior, Washington, pp 158–161 (1987).

"Chemical profile: barium carbonate", *Chem. Mark. Rep.* (April 1, 1986).

"Barite – salient statistics", *Chemical Economics Handbook,* Stanford Research Institute, Menlo Park, California, USA (Feb. 1982).

"Barium Chemicals – USA Data Summary", *Chemical Economics Handbook,* Stanford Research Institute, Menlo Park, California, USA (April 1986).

3.2 Aluminum and its Compounds

3.2.1 General Information

In quantity aluminum is the most important nonferrous metal, more important than copper, zinc or lead.

Al is by far the most important nonferrous metal

The most important aluminum compounds apart from aluminum silicate-based ceramics and clay products (Sec. 5.5) are aluminum hydroxide and aluminum oxide which are raw materials in the manufacture of aluminum. Other industrially important aluminum compounds are aluminum sulfate, aluminum chloride, sodium aluminate, aluminum fluoride and cryolite (Sec. 1.7.1).

Industrially important Al compounds:

$Al(OH)_3$, Al_2O_3, $Al_2(SO_4)_3$, $AlCl_3$, $NaAlO_2$, AlF_3, Na_3AlF_6

3.2.2 Occurrence

Aluminum is with 8% the third most abundant element in the Earth's crust. In nature it is mainly present as oxides with Na_2O, K_2O, CaO, SiO_2 and other oxides e.g. in felspars (e.g. in anorthite $CaO \cdot Al_2O_3 \cdot 2SiO_2$ and orthoclase $K_2O \cdot Al_2O_3 \cdot 6SiO_2$) and their weathering products, clay minerals (e.g. montmorillonite $Al_2O_3 \cdot 4SiO_2 \cdot H_2O$ and kaolinite $Al_2O_3 \cdot 2SiO_2 \cdot 2H_2O$) and in different micas (e.g. muscovite, margarite). The most important minerals for aluminum manufacture are the bauxites (laterite), weathering products from clays. They contain between 40 and 65% aluminum oxide and are a mixture of different aluminum oxide hydrates and hydroxides [e.g. boehmite, γ-AlO(OH), diaspore, α-AlO(OH), gibbsite, α-$Al(OH)_3$] with aluminum silicate, iron and titanium oxides etc.

Al is the third most abundant element in the Earth's crust (after oxygen and silicon). It is found in:

- Felspars
- Clays
- Mica
- Bauxites

Most important raw materials for Al manufacture are the bauxites:

Mono- and trihydrates of Al_2O_3 contaminated with SiO_2, Fe_2O_3 and TiO_2

The quantities of nonbauxitic raw materials (nepheline and alunite ores) used for the manufacture of primary aluminum pig (e.g. in the USSR) are small.

More than 90% of the bauxites mined is processed to aluminum oxide, the rest being utilized in the manufacture of refractory materials, grinding materials and aluminum chemicals.

Bauxite utilized in the manufacture of:

- Al_2O_3 (90%)
- Refractory materials and grinding materials
- Al chemicals

1985 world bauxite extraction: 85.1×10^6 t

- 38% in Australia
- 15% in Guinea
- 8% in Brazil

The known world reserves of bauxite are 23×10^9 t of which ca. 26% is in Guinea and 21% in Australia. The world extraction of bauxite is closely connected with the production of primary aluminum pig and was 85.1×10^6 t in 1985 (see Table 3.2-1). The USA is the largest consumer with 10.5×10^6 t in 1984.

Table 3.2-1. Estimated Bauxite Extraction in 1985 in 10^6 t.

World	Australia	Guinea	Brazil	Jamaica	Surinam
85.1	32.4	13.1	6.7	6.2	4.6

3.2.3 Aluminum Metal

3.2.3.1 Economic Importance

1985 worldwide production of primary aluminum pig:

15.3×10^6 t

The production of primary aluminum pig has stagnated since the beginning of the 1980's at ca. 16×10^6 t/a. Table 3.2-2 gives the 1985 production statistics for the most important producer countries.

Table 3.2-2. 1985 Primary Aluminum Pig Production of Important Producer Countries in 10^6 t.

World	USA	USSR	Canada	Australia	W. Germany	Norway	Brazil	P.R. China	Spain
15.3	3.5	2.2	1.3	0.85	0.75	0.72	0.54	0.41	0.37

Increasing energy costs, intenser competition and increased utilization of recycled aluminum from scrap and waste disposal (4.4×10^6 t in 1984 in the Western World) have in recent years led to capacity reduction (USA, Japan, France) and site relocation. Capacity is being steadily expanded in the bauxite producing countries e.g. Australia and Brazil and in countries with cheap electricity e.g. Venezuela and Norway.

3.2.3.2 Manufacture

Industrial Al manufacture currently almost exclusively by the Hall-Heroult process

The industrial manufacture of aluminum is based on the Hall-Heroult process developed in 1886. In this process aluminum oxide produced using the Bayer process (Sec. 3.2.4.2) is dissolved in a cryolite (Na_3AlF_6) melt and electrolyzed at 940 to 980 °C with direct current. Molten metallic aluminum is deposited at the carbon cathode and is drawn off. The oxygen formed at the carbon anode (presintered or Söderberg electrodes) burns the anodes to carbon dioxide or carbon monoxide according to:

$$2\,Al_2O_3 \rightarrow 4\,Al + 3\,O_2$$

$$3\,C + 3\,O_2 \rightarrow 3\,CO_2\ (\text{or } CO)\,.$$

The electrolyte consists of a complex mixture of ca. 75% cryolite, 2 to 9% aluminum oxide (optimally 2 to 5%) and various additives such as aluminum fluoride, fluorspar, lithium carbonate and magnesium fluoride. These additives serve to reduce the melting point, increase the conductivity and electricity yield and reduce the fluorine emission. The electricity yield is ca. 85 to 95%, the cell voltage up to 5 V, the anode consumption ca. 0.5 kg/kg aluminum and the power rating of a unit with many (100 to 200) individual cells connected in series is 50 to 300 kA. The specific power requirement has been reduced in recent years to 13 to 16 kWh/kg aluminum by various process improvements. The fluorine emission and thereby also the fluorine consumption has also been considerably reduced. Ca. 99% of the hydrogen fluoride-containing gases from the electrolysis cells are removed by adsorption and chemisorption on aluminum oxide and recycled.

Electrolyte:

- Na_3AlF_6 (synthetic cryolite): ca. 75%
- Al_2O_3: 2 to 9%
- AlF_3: 5 to 15%
- CaF_2: 2 to 6%
- LiF (Li added as Li_2CO_3): 2 to 5%
- MgF_2 (seldom added): 2 to 3%

Melt additives lead to:

- Increased conductivity
- Reduced F emission
- Increased electricity yield

Al production is energy intensive, energy costs making up ca. 25% of the Al manufacturing costs in the Western Industrialized Countries

F emission, the old problem of the Al industry, to a very large extent solved

The liquid pure aluminum (99.5 to 99.9% purity) from the electrolysis furnace is after appropriate refining remelted and cast. The purest aluminum (99.99 to 99.999%) is manufactured by three layer melt electrolysis with the aid of fluorine-containing fluxes or by fractional crystallization.

In addition to the electrolysis of clay, other minor manufacturing processes are operated such as carbothermal reduction of aluminous ores with carbon in the USSR and the electrolysis of aluminum chloride (Alcoa process). The latter has not been operated since 1982 due to technical and economic problems.

Al manufacture from $AlCl_3$ using the Alcoa-process currently insignificant

3.2.3.3 Aluminum Applications

Table 3.2-3 gives a summary of the world consumption of primary and recycled aluminum.

Table 3.2-3. 1984 World and Regional Aluminum Consumption in 10^6 t.

Primary aluminum pig		Total consumption including recycled aluminum							
World	USA	World	N. & S. America	USA	Western Europe	Eastern Europe	Asia	Africa	Australia & Oceania
15.9	4.5	21.2	7.3	6.3	5.1	4.0	4.2	0.3	0.3

The largest consumers are the USA, the USSR, Japan and West Germany.

1985 USA aluminum application spectrum:

- Vehicles and aircraft 20%
- Containers and packaging 27%
- Construction industry 20%
- Electronics 9%
- Office and household equipment 7%
- Machinery and equipment 5%
- Iron and steel industry (alloying metal) 4%
- Export 8%

Aluminum is preponderantly used in the form of alloys, since these have a higher strength than pure aluminum while having a better or similar corrosion resistance. The most important alloying elements are silicon, zinc, manganese, magnesium, copper and lithium. Usually only a few percent of these are alloyed with aluminum either solely or, preponderantly, in combinations. There are two types: wrought and cast alloys. The first industrially important aluminum alloy was the age-hardenable duralumin (Al/Cu/Mg). Better strengths can now be achieved with Al/Zn/Mg/Cu-alloys or more recently with metal matrix composites, a combination of aluminum with fibers, whiskers or oxide-, carbide- or nitride-particles. The main application fields are in vehicles and aircraft, containers, packaging, the construction industry and electronics.

3.2.4 Aluminum Oxide and Aluminum Hydroxide

3.2.4.1 Economic Importance

1985 worldwide production of hydrated alumina:

31.9×10^6 t as Al_2O_3.

The worldwide production of aluminum hydroxide [hydrated alumina, $Al(OH)_3$] is strongly ($> 90\%$) dependent upon the manufacture of aluminum and was ca. 32×10^6 t in 1985. Table 3.2-4 gives the 1985 hydrated alumina production statistics for the most important producer countries.

Table 3.2-4. 1985 Hydrated Alumina Production for the Most Important Producer Countries in 10^6 t.

World	Australia	USA	USSR	West Germany	Jamaica	Japan	Surinam	Canada
31.9	8.8	3.5	3.5	1.7	1.6	1.0	1.0	1.0

The production figures for aluminum oxide (alumina, Al_2O_3) are somewhat lower because not all the hydrated alumina is calcined.

3.2.4.2 Manufacture

The industrial manufacture of aluminum hydroxide and aluminum oxide currently proceeds almost exclusively by the Bayer process, which was developed by the Austrian Joseph Bayer in 1887. In this process (see Fig. 3.2-1) ground bauxite is digested in autoclaves or more recently continuous tube reactors with 3 to 6 mol/L sodium hydroxide solution at temperatures of 120 to 250 °C optionally with calcium oxide (to remove carbon dioxide). Aluminium oxide hydrate is thus dissolved as sodium aluminate. The iron-containing very finely divided so-called "red mud" is then separated and washed in thickening and fitration units. Ca. 1 to 2 t of red mud is produced for every ton of alumina.

$Al(OH)_3$ and Al_2O_3 are almost exclusively industrially produced using the Bayer process by digestion of bauxite with sodium hydroxide

Fe content of bauxite precipitates as a red mud

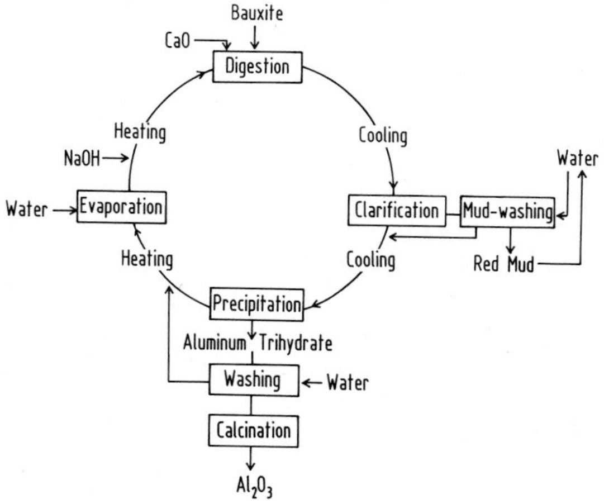

Fig. 3.2-1. Bayer process flow sheet for the production of alumina hydrate and alumina.

The filtrate consisting of a solution with a sodium oxide/aluminum oxide ratio of 1.5 to 1.8 is cooled with stirring and then seeded with large quantities of aluminum hydroxide, whereupon a large part of the dissolved hydroxide precipitates out as hydrargillite, α-$Al(OH)_3$:

α-$Al(OH)_3$ (hydrargillite) precipitated out from the sodium aluminate solution by seeding. This is mostly calcined to α-Al_2O_3

$$Al(OH)_4^- + H_2O \longrightarrow \alpha\text{-}Al(OH)_3 + OH^-$$

After aluminum hydroxide separation the aluminum hydroxide-deficient alkali is partly evaporated and then returned to the bauxite digestion.

The aluminum hydroxide obtained is mainly calcined to aluminum oxide (α-Al_2O_3) in a rotary or more energy savingly (VAW-Lurgi process) in a fluidized bed kiln at 1200 to 1300 °C. A small quantity of the hydroxide is carefully dried and used as such.

Active aluminas:

- By precipitation reactions from Al-salt solutions
- By careful calcination of $Al(OH)_3$

Active aluminas (various oxides and oxide hydrates) with high specific surface areas, good adsorption capacities, catalytic properties and high chemical reactivity are either produced by precipitation processes from aluminum salt solutions, e.g. through thermal posttreatment of aluminum hydroxide gels, or by the calcination of α-aluminum hydroxide under specific conditions (low temperatures, very rapid heating).

Electrocorundum from:

- Pure bauxite
- Alumina

by reduction melting

Electrocorundum is obtained by reduction melting of the purest possible bauxite or alumina at ca. 2000 °C with coke or anthracite in an arc reduction furnace, the impurity oxides in the raw material (iron, titanium, silicon) being thereby reduced and mostly removed.

3.2.4.3 Applications

Main applications of Al_2O_3:

- Electrolytic production of Al
- Ca. 10% is utilized in the refractory, grinding, glass, enamel and plastics industries

Most of the α-aluminum oxide is utilized for aluminum manufacture (ca. 0.5 t aluminum/t Al_2O_3). The rest ($< 10\%$) is utilized for the manufacture of refractory, grinding, ceramic, glass and enamel products and electrocorundum. It is also used as a filler in polymers (polyesters, epoxy resins).

Main application of $Al(OH)_3$:

- Alumina manufacture.

Less important applications:

- Al chemicals
- Active aluminas
- Flame retardant for plastics

Apart from its further processing to metallic aluminum, α-aluminum hydroxide is mainly used in the manufacture of aluminum compounds (aluminum fluoride, synthetic cryolite, sodium aluminate, aluminum sulfate) and active aluminas. In a finely divided form it is utilized as a flame retardant and/or filler in carpets, plastics, foam rubber, wall elements and toothpaste.

Applications for active alumina:

- Adsorption agent for water
- Adsorption agent for gases
- Catalyst(-carrier)

In 1983 ca. 3% of the USA consumption of α-aluminum hydroxide (ca. 22×10^3 t) was processed to active alumina, which is used as an adsorption agent for water and gases, as catalysts or catalyst carriers e.g. in the Claus process (Sec. 1.6.1.2.2) and as a filter material. Electrocorundum (with an Al_2O_3 content of

≥ 80%) is an important grinding and polishing agent due to its high hardness and is also utilized in the refractory industry and in the manufacture of hard concrete materials.

Applications for electrocorundum:

- Grinding and polishing agents
- Refractory industry
- Hard concrete sector

3.2.5 Aluminum Sulfate

Economic Importance: Aluminum sulfate is the most important aluminum compound after aluminum oxide and aluminum hydroxide with a worldwide production in 1982 of 2.9×10^6 t. The most important producers are the USA (with 1.1×10^6 t in 1984 on the basis of 17% Al_2O_3), Western Europe (with 0.9×10^6 t) and Japan (with 0.8×10^6 t on the basis of 14% Al_2O_3).

1982 worldwide aluminum sulfate production: 2.9×10^6 t

Manufacture: Aluminum sulfate is industrially produced by the reaction of aluminum hydroxide or other aluminum raw materials such as bauxite or kaolin with sulfuric acid at ca. 170 °C in a pressure-resistant agitator vessel. After appropriate concentration the melt obtained contains ca. 57% aluminum sulfate (ca. 13 molecules of hydration water/mole). Double salts of aluminum sulfate with potassium, ammonium or sodium sulfate (alums) e.g. potassium alum $KAl(SO_4)_2 \cdot 12H_2O$ have been largely supplanted by aluminum sulfate.

Aluminum sulfate [with 57% $Al_2(SO_4)_3$] industrially manufactured from $Al(OH)_3$ or bauxite and H_2SO_4

Applications: Ca. 50% of the aluminum sulfate produced is utilized in the paper and pulp industry (paper glues, pH adjustment etc.). Almost an equal amount is used in water purification as a flocculant. Aluminum sulfate is also the starting material for other aluminum compounds e.g. aluminum hydroxide gels.

$Al_2(SO_4)_3$ applications:

- Paper and pulp industry
- Water purification

3.2.6 Aluminum Chloride

Economic Importance: Aluminum chloride is almost exclusively utilized in its anhydrous form. In 1984 ca. 30×10^3 t was produced in the USA and ca. 50×10^3 t in Western Europe.

Manufacture: Anhydrous aluminum chloride is currently manufactured predominantly by the chlorination of liquid aluminum in ceramic-lined reaction vessels at 600 to 750 °C. Its manufacture by reductive

Anhydrous $AlCl_3$:

from Al metal and chlorine

$AlCl_3 \cdot 6H_2O$:
from $Al(OH)_3$ and HCl

chlorination (carbochlorination) of aluminum oxide or clay is much less energy intensive and has therefore been preferred in some recent plants. Hydrated aluminum chloride is manufactured from aluminum hydroxide and hydrochloric acid or hydrogen chloride.

Main application of $AlCl_3$:
catalyst in organic chemistry. Its utilization in the petrochemical industry is currently decreasing (e.g. in USA)

Potential utilization of $AlCl_3$:
electrolytic manufacture of Al: Alcoa process (Sec. 3.2.3.2)

Applications: Anhydrous aluminum chloride is mainly used as a catalyst in organic chemistry e.g. in Friedel-Crafts reactions. In the USA in 1960 ca. 40 to 50% was utilized in this sector (mainly in the manufacture of ethyl benzene, ethyl chloride, dye precursors, detergents and polymers). Its significance in the petrochemical industry has strongly decreased with the advent of zeolite-based catalysts.

Aluminum chloride hexahydrate is used in pharmaceuticals and cosmetics as a flocculant and for impregnating textiles.

3.2.7 Sodium Aluminate

$NaAlO_2$:

- Intermediate in the digestion of bauxite with NaOH
- Manufactured from $Al(OH)_3$ and NaOH

In 1983 the USA sodium aluminate production was ca. 45×10^3 t with 25×10^3 t in Japan. Sodium aluminate (theoretical formula: $NaAlO_2$) is manufactured by dissolving hydrated alumina in 50% sodium hydroxide solution. It is utilized in water purification, in the paper industry, for the aftertreatment of titanium dioxide-pigments, in the manufacture of aluminum-containing catalysts and aluminosilicates and in the construction industry.

References for Chapter 3.2: Aluminum and its Compounds

General Information:

Ullmann's Encyclopedia of Industrial Chemistry, 5th Ed., Vol. A1, pp 459–480, pp 527–540, pp 557–594, Verlag Chemie, Weinheim-Deerfield Beach-Basel (1984).

A. F. Trotmann-Dickenson: *Comprehensive Inorganic Chemistry.* 1st Ed., Vol. 1, pp 993–1058, Pergamon Press, Oxford (1973).

Kirk-Othmer: *Encyclopedia of Chemical Technology,* 3rd Ed., Vol. 2, pp 129–252, John Wiley & Sons Inc., New York-Chichester-Brisbane-Toronto (1978).

Specific Information:

Aluminum metal:

Minerals Yearbook, Vol. I, US Department of the Interior, Washington, pp 109–132 (1985).

Chemical Economics Handbook, Stanford Research Institute, Menlo Park, California, USA, "Aluminum and aluminum raw materials – product review" (Sept. 1985).

"World Mineral Statistics 1979–1983", Brit. Geol. Survey, London, p. 5 (1985).

"Metallstatistik 1974 – 1984" (Metal statistics 1974 – 1984), Metallgesellschaft AG, Frankfurt a.M., pp 14 – 19, pp 73 – 74 (1985).
P. Kratzer: "Die Lage der Aluminiumindustrie" (The state of the aluminum industry), *Metall* **38,** p. 569 (1984).
A. S. Russell: "Pitfalls and pleasures in new aluminum process development", *J. Metals*, p. 132 (1981).

Aluminum Compounds:

Minerals Yearbook, Vol. I, US Department of the Interior, Washington, pp 170 – 175 (1985).
Chemical Economics Handbook, Stanford Research Institute, Menlo Park, California, USA: "Aluminum and aluminum raw materials – product review" (Sept. 1985).
"Metallstatistik 1974 – 1984" (Metal statistics 1974 – 1984), Metallgesellschaft AG, Frankfurt a.M., pp 12 – 13 (1985).
K. A. Evans, N. Brown: "Speciality inorganic aluminum compounds", *Spec. Publ. – R. Soc. Chem.* **40,** pp 164 – 194 (1981).
"Mineral Commodity Summaries", Bureau of Mines, Washington, pp 18 – 19 (1986).
"Industrial Statistics Yearbook 1982", Vol. II, United Nations, New York, pp 397 – 399 (1985).
"Bauxite", *Proc. Bauxite Symp.*, Los Angeles (1984), Ed. L. Jacob, Soc. Min. Eng., New York.
P. V. Bonsignore, "Alumina trihydrate as a flame retardant for polyurethane foams", *J. Cellular Plastics*, pp 220 – 225 (1981).
H. S. Katz and I. V. Milewski, *Handbook of Fillers and Reinforcement for Plastics,* pp 172 – 173, pp 237 – 247, van Nostrand Reinhold Co., New York (1978).

3.3 Chromium Compounds and Chromium

3.3.1 Chromium Compounds

3.3.1.1 Economic Importance

The following chromium compounds are economically important:

- Chromates and dichromates
- Chromium(VI) oxide ("chromic acid")
- Chromium(III) oxide
- Basic chromium(III) sulfate (chrome tanning agents)
- Chromium(VI) and chromium(III) compounds and chromium(IV) oxide (see Sec. 5.9: Pigments).

Chromium compounds:

Chromates, dichromates, chromium(VI) oxide, chromium(III) oxide, basic chromium(III) sulfate (chrome tanning agents), chrome pigments (Sec. 5.9.3.2 and 5.9.3.3)

A flow sheet summarizing the production pathways from chrome ore to the important chromium products is given in Fig. 3.3-1 (see p. 256).

The most important manufacturers of chromium chemicals, with capacities of $Na_2Cr_2O_7 \cdot 2H_2O$ in 10^3 t/a in brackets, are:

- American Chrome & Chemicals, USA (41 in 1986)
- Bayer AG in West Germany, Spain, South Africa, Argentina and Brazil (120 – 130 in FRG and Spain in 1982)

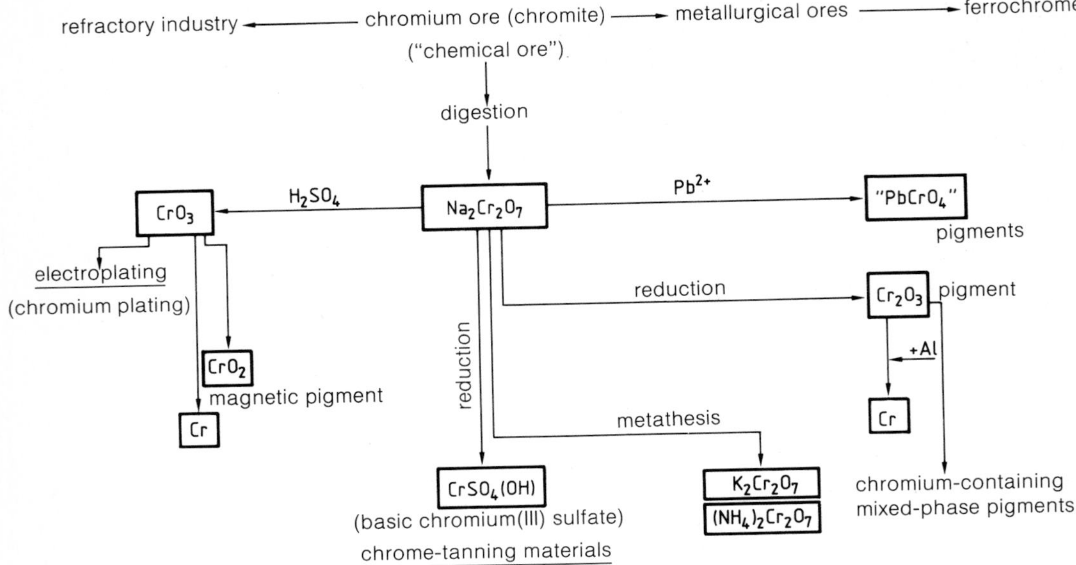

Fig. 3.3-1. Production pathways to the important chromium products.

- British Chrome & Chemicals Ltd. in Great Britain (70 in 1982)
- Diamond Shamrock Corp., USA (109 in 1986)*
- Kromsan Bilesikleri Sanayi, Turkey (22 in 1985)
- Nippon Chemical Industries, Japan (36 in 1982)
- Nippon Denko KK, Japan (41 in 1982)
- Stoppani, Italy (45 in 1982)

* Chrome activities taken over by Occidental Chemical Corp. in 1986.

Production of chromium compounds concentrated in West Germany, Brazil, Argentina, Italy, South Africa, Great Britain, Turkey, USA, USSR, Rumania, Poland and P.R. China

There are small plants in Mexico, India and Pakistan. Ca. 30% of the World capacity is concentrated in the USSR, the People's Republic of China, Rumania and Poland. Estimated production figures for sodium dichromate in the different regions (based on an average over the period 1981 – 1985) are given in Table 3.3-1.

Table 3.3-1. Estimated Production of Sodium Dichromate (averaged over the period 1981 – 1985) in 10^3 t/a.

World	USA	EEC	Japan	Other Western countries	Communist countries
670	150	195	45	80	200

The worldwide capacities were not fully utilized in 1986. The USA consumption of sodium dichromate in 1985 was 109×10^3 t.

USA consumption in 1985:

- Sodium dichromate: 109×10^3 t
- Chromium(VI) oxide: 39×10^3 t

About a third of the chromium compounds manufactured worldwide are used for chrome tanning.

The USA chromium(VI) oxide capacity in 1986 was 48.5×10^3 t/a [American Chrome & Chemicals Corp.: 12 (extendable to 20) $\times 10^3$ t/a; Diamond Shamrock Corp.: 36.5×10^3 t/a]. The USA consumption in 1985 was 39×10^3 t.

3.3.1.2 Raw Material: Chromite

The raw material for all chromium chemicals is chrome iron ore, so-called chromite, which is a spinel. It contains magnesium and aluminum as minor constituents and has the approximate composition (Fe,Mg)-$[Cr,Al,Fe(III)]_2O_4$. Calcium, silicon dioxide and, in some cases, vanadium are also present in small quantities.

Raw material: chromite

Main economically workable reserves in Southern Africa, deposits are very large

The known economically workable reserves of chromite were estimated to be 1.058×10^9 t in 1984, with a further 5.77×10^9 t not being economically workable at the present time. The USA does not possess any workable reserves, but has deposits of 10×10^6 t in Alaska, California, Oregon and Washington. The economically workable reserves are not expected to run out in the near future and are summarized for different countries in Table 3.3-2.

World chromite reserves estimated in 1984:

1.058×10^9 t

Table 3.3-2. Estimated (1984) Economically Workable Reserves of Chromite in Different Countries in 10^6 t.

World	South Africa	USSR	Zimbabwe	Finland	India	Philippines	Brazil	Madagascar	Albania	Turkey	Iran	Cuba	Rest
1058	828	129	17	17	14	14	8	7	6	5	2	2	9

The extraction of chromite in the different countries in 1984 in 10^3 t is given in Table 3.3-3.

World chromite extraction in 1984: 9.586×10^6 t

Table 3.3-3. 1984 Chromite Extraction in the Different Countries in 10^3 t.

World	South Africa	USSR	Albania	Turkey	Zimbabwe	Finland	India	Brazil	Philippines	Greece	Others
9586	3407	2520	905	487	477	446	435	300	257	75	277

Utilization of chromite in the USA (Japan):

- Steel industry: 78% (92%)
- For refractory materials: 7% (5%)
- Chemical industry: 15% (3%)

The utilization of chromite depends upon the concentration of its minor components, such as magnesium and aluminum. Chromite was utilized in 1981 [USA% (Japan%)]:

- In the steel industry for the production of ferrochrome: 78% (92%). Ores are needed with a high chromium : iron ratio (Cr : Fe > 2.5 : 1) together with a high magnesium content.
- In the refractory materials industry: 7% (5%). Ores are required with a high content of chromium(III) oxide and aluminum oxide (> 60%) and as much magnesium and as little iron as possible.
- In the chemical industry: 15% (3%). Ores are required with as high chromium(III) oxide and iron(III) oxide contents as possible and as low aluminum, magnesium and silicon dioxide contents as possible. A typical composition (Rustenburg, S. Africa) is:
  44.5, Cr_2O_3; 26.4, FeO; 3.5, SiO_2; 10.6, MgO; 14.4, Al_2O_3 (in %).

Such ores are predominantly found in Southern Africa and the world's dependence thereon can therefore not be overlooked.

3.3.1.3 Manufacture of Chromium Compounds

3.3.1.3.1 Chromite Roasting to Alkali Chromates

Roasting of chromite:

alkaline oxidative roasting in rotary kilns or annular hearth furnaces. Chromium(III) converted to chromium(VI) and the sodium chromate leached out. Chromium yield: 75–85%

Only alkaline oxidative roasting is suitable for the large scale conversion of chromite to chromium chemicals. This leads to alkali(sodium) chromates.

$$4\,FeCr_2O_4 + 8\,Na_2CO_3 + 7\,O_2 \xrightarrow{1000-1100\,^\circ C} 8\,Na_2CrO_4 + 2\,Fe_2O_3 + 8\,CO_2$$

Sodium carbonate can be partially replaced by sodium hydroxide. (In some regions it is usual to add calcium carbonate to the roasting mixture, despite the fact that this procedure is extremely suspect due to the formation of toxic calcium chromate.) The accompanying elements aluminum, silicon and vanadium are converted during the roasting into sodium aluminate, silicate and vanadate, respectively, and consume corresponding quantities of alkali.

In this roasting, chromium is easily separated from the accompanying elements iron, aluminum, magnesium and silicon, but removal of vanadium can require a postpurification step.

Prior to roasting, the ore is finely ground. The chrome ore concentrate, which has been enriched using a gravitational process (e.g. spiral classifiers, settling machines), is delivered with a particle size of ca. 0.5 mm. Ball-milling coupled with air classification yields a particle size < 0.1 mm. The sodium carbonate is also ground. Since the sodium chromate formed melts at 792 °C, an inert diluent is necessary which can absorb sodium chromate like a sponge and therefore enable air to come into contact with the reaction mixture.

A particularly suitable diluent is the iron oxide left after the roasting, but calcined pyrites can also be used. A typical roasting mixture consists, for example, of:

- 100 parts of ore
- 60 to 75 parts of sodium carbonate
- 50 to 200 parts of diluent

The chrome ore digestion is currently carried out in rotary kilns or annular hearth furnaces. Fig. 3.3-2 gives a flow chart of a rotary kiln plant with units coupled at either end of it.

The brick-lined slowly rotating rotary kiln has an inclination of 2 to 5%. The roasting mixture is fed in countercurrently to the heating gases (heating materials: heavy oil, natural gas, powdered lignite). Since the air employed has to oxidize chromium(III) to chromium(VI) and iron(II) to iron(III) and provide oxygen for the burning of the heating materials, the air input is so regulated that the exhaust gases contain 12 to 16 vol.% of oxygen. The average residence time of the roasting mixture in the kiln is ca. 4 h.

After reaction, the mixture has a content of up to 30% sodium chromate with up to 45% in annular hearth furnaces. The chromium yields are 75 to 85%.

10% of the reaction mixture is entrained as dust in the exhaust gases emerging from the head of the kiln. This is removed by e.g. electrostatic gas purification units. The heat content of these gases can be used for steam production.

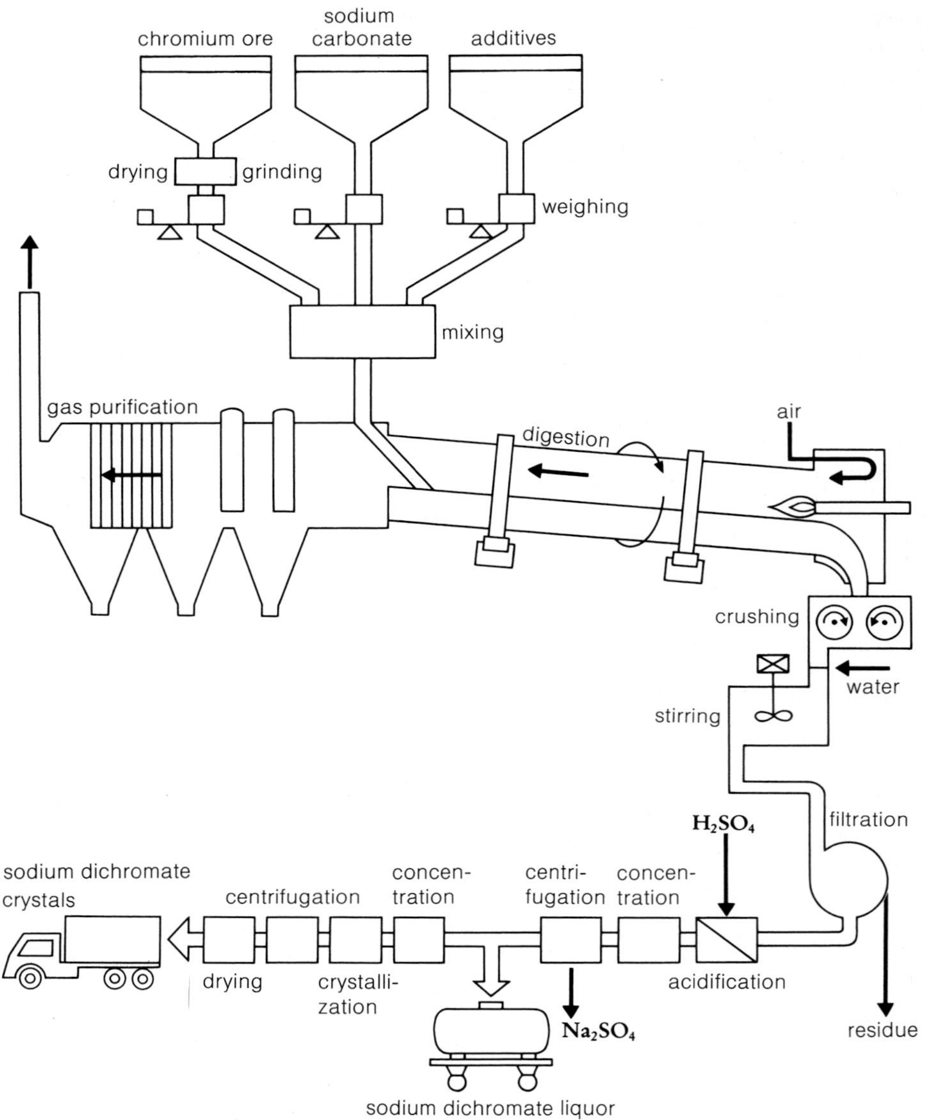

Fig. 3.3-2. Flow chart of a rotary kiln plant for the manufacture of sodium dichromate.

The digested material emerging from the furnace head can, for example, be ground in wet-grinding mills and leached with water. By appropriate adjustment of the conditions only sodium dichromate and part of the sodium vanadate dissolve and the iron, aluminum, silicon and magnesium left behind as insoluble oxides or hydroxides are filtered off with rotary or belt filters. This residue is washed countercurrently in a series of stages. A chromate solution of 500 g/L is thus obtained and the residue contains less than 1/10 % by weight of chromium(VI). This residue is either dried and returned as a diluent to the digestion mixture, or treated in the form of a sludge dispersion with reducing agents such as iron(II) sulfate or sulfur dioxide to remove chromium(VI), dehydrated and dumped in landfill in an orderly manner.

3.3.1.3.2 Alkali Dichromates

The alkali dichromates have a much higher chromium content than monochromates and are also easier to produce pure. They are therefore almost exclusively used in practice.

Commercially most important chromium(VI) compound:

sodium dichromate dihydrate.

Sodium chromate converted to dichromate with

- Sulfuric acid
- Carbon dioxide

Sodium Dichromate Dihydrate

Sodium dichromate dihydrate is produced from the chromate by reaction with sulfuric acid or carbon dioxide.

Sulfuric acid process

The reaction between sodium chromate and sulfuric acid proceeds as follows:

$$2Na_2CrO_4 + H_2SO_4 \longrightarrow Na_2Cr_2O_7 + Na_2SO_4 + H_2O \,.$$

The discontinuous or continuous addition of sulfuric acid to sodium chromate solution up to a pH of 3 is technically simple. The resulting solution of sodium dichromate dihydrate is then evaporated to 70 % by weight, whereupon almost all the sodium sulfate precipitates out and is separated off. The dichromate solution obtained is partially utilized as such. Further evaporation to ca. 1600 g dichromate/L and cooling to 30 to 40 °C leads to crystallization of the dichromate. These processes are either carried out stepwise or continuously. Separation is carried out in a skimmer or push type centrifuge.

Conversion of chromate into dichromate with sulfuric acid:

technically simple process but has dichromate-containing sodium sulfate as a byproduct

The "yellow", i.e. dichromate-containing, sodium sulfate byproduct is problematical in that it can not be readily used and has to be freed from chromium in a further process step before it can be used.

Carbon dioxide process

Conversion of chromate to dichromate with carbon dioxide:

technically complex process, but no difficult to utilize byproducts

$$2Na_2CrO_4 + H_2O + 2CO_2 \rightleftharpoons Na_2Cr_2O_7 + 2NaHCO_3$$

Under normal conditions, the reaction equilibrium lies mainly on the left side of this equation. To promote the formation of dichromate, the reaction has to be carried out under 7 to 15 bar pressure in a series of autoclaves (if necessary in a multistage process) with sodium chromate solution concentrations of 800 to 900 g/L and carbon dioxide (added either as a gas or a liquid) in countercurrrent. The reaction is exothermic, cooling being necessary. The sodium hydrogen carbonate precipitate is filtered off either rapidly after pressure release to avoid back-reaction, or under pressure.

The chromate-containing sodium hydrogen carbonate is either calcined at 200 to 300 °C to sodium carbonate:

$$2NaHCO_3 \longrightarrow Na_2CO_3 + CO_2 + H_2O$$

or converted with sodium hydroxide to sodium carbonate

$$NaHCO_3 + NaOH \rightarrow Na_2CO_3 + H_2O$$

and returned to the digestion. In the latter case, sodium hydroxide can substitute almost all the alkali utilized for the digestion. The further processing of the sodium dichromate solution has already been described in the section: *Sulfuric acid process*.

Potassium and Ammonium Dichromate

Potassium and ammonium dichromate obtained by metathesis of sodium dichromate with potassium or ammonium salts

Potassium and ammonium dichromate are obtained by metathesis of sodium dichromate with potassium or ammonium salts e.g.:

$$Na_2Cr_2O_7 + 2KCl \longrightarrow K_2Cr_2O_7 + 2NaCl\ .$$

For example, solutions with 400 to 450 g sodium dichromate/L can be mixed with ca. 200 g potassium chloride/L while heating to boiling point. The precipitated sodium chloride is then filtered off from the hot solution and the potassium dichromate precipitates out upon cooling.

Ammonium dichromate decomposes upon impact or heating above 180 °C:

$$(NH_4)_2Cr_2O_7 \longrightarrow N_2 + Cr_2O_3 + 4H_2O .$$

3.3.1.3.3 Chromium(VI) Oxide ("Chromic Acid")

$$Na_2Cr_2O_7 \cdot 2H_2O + 2H_2SO_4 \longrightarrow 2NaHSO_4 + 2CrO_3 + 3H_2O .$$

Chromium(VI) oxide, "chromic acid", from sodium dichromate and sulfuric acid at ca. 200 °C

This reaction can be carried out both in concentrated aqueous solution and, preferably continuously, in the molten state. In the latter, sodium dichromate is mixed with sulfuric acid in a twin screw and the mixture fed into an externally heated rotary kiln. In sequence, the water evaporates, the sodium hydrogen sulfate melts (at 170 °C) and then the chromium(VI) oxide melts (at 198 °C). Temperature control is critical, because chromium(VI) oxide decomposes at slightly higher temperatures. After the reaction is completed, the liquid mixture is separated in a settler, the chromium(VI) oxide being drawn off from below and the sodium hydrogen sulfate from above, and converted with cooling rollers into solid materials. Electrochemical processes for manufacturing chromium(VI) oxide from sodium dichromate are also known.

3.3.1.3.4 Chromium(III) Oxide

Chromium(III) oxide is obtained by reducing sodium dichromate with organic materials, charcoal or sulfur in an exothermic reaction:

Chromium(III) oxide obtained from sodium dichromate by reduction with carbon, sulfur or ammonium salts. Detailed information in Sec. 5.9.3.2.

$$2Na_2Cr_2O_7 + 3C \longrightarrow 2Cr_2O_3 + 2Na_2CO_3 + CO_2$$

$$Na_2Cr_2O_7 + S \longrightarrow Cr_2O_3 + Na_2SO_4 .$$

These reactions are carried out continuously in kilns. The reaction product is leached with water to remove soluble products, filtered, dried and ground. Very pure chromium(III) oxide (99%) of pigment quality is thereby obtained.

Chromium(III) can also be obtained by reacting sodium dichromate with ammonium salts such as ammonium chloride or sulfate:

$$Na_2Cr_2O_7 + 2NH_4Cl \longrightarrow 2NaCl + Cr_2O_3 + 4H_2O + N_2 .$$

Reduction with ammonium salts yields chromium(III) oxide with a particularly low sulfur content

The chromium(III) oxide thus obtained has a low sulfur content, particularly when an understoichiometric amount of ammonium salt is used, and can be used for the aluminothermic manufacture of chromium metal (Sec. 3.3.2.2.1).

3.3.1.3.5 Basic Chromium(III) Salts (Chrome-Tanning Agents)

Chrome-tanning agents:

crosslink the collagen carboxyl groups in animal hides. Basic chromium(III) sulfate or chromium(III) salts of organic acids are used

Chrome-tanning is the crosslinking of the collagen carboxyl groups in animal hides with the help of chromium(III) compounds. This results in a material with increased temperature stability and reduced swelling capacity.

Basic chromium(III) salts with nonstoichiometric compositions are used in tanning, particularly oligomers of basic chromium(III) sulfate with the approximate composition $Cr(OH)SO_4$. Basic chromium sulfates are marketed both in aqueous solution and as a spray dried powder.

During tanning further buffering is necessary. This is accomplished by adding bases such as sodium carbonate, dolomite or magnesium oxide. If sulfate is partially exchanged for organic anions such as glutarate, the chromium content of the tanning batch is particularly strongly consumed during the tanning process, which is ecologically desirable. Particularly high quality leather is obtained when the formate is used, formate acting by way of complex formation.

Basic chromium sulfates can be manufactured in different ways:

- As a byproduct in the dichromate oxidation of impurities in montan waxes, or of anthracene to anthraquinone:

$$Na_2Cr_2O_7 + 3\ H_2SO_4 + C_{14}H_{10} \longrightarrow Na_2SO_4 + 2\ Cr(OH)SO_4 + 3\ H_2O + C_{14}H_8O_2$$

Manufacture of basic chromium sulfates:

- From solutions obtained in the dichromate oxidation of organic materials such as montan waxes or anthracene

- In the reduction of dichromate with molasses:

$$8\,Na_2Cr_2O_7 + 24\,H_2SO_4 + C_{12}H_{22}O_{11} \longrightarrow 16\,Cr(OH)SO_4 + 8\,Na_2SO_4 + 12\,CO_2 + 22\,H_2O\ .$$

- By the reduction of dichromate with molasses

- In the reduction of dichromate with sulfur dioxide:

$$Na_2Cr_2O_7 + 3\,SO_2 + H_2O \longrightarrow Na_2SO_4 + 2\,Cr(OH)SO_4\ .$$

- By the reduction of dichromate with sulfur dioxide

In each case the basic chromium sulfates are formed together with sodium sulfate, which is desirable in that it acts as a buffer during the tanning process.

3.3.1.4 Applications for Chromium Compounds

Dichromate: the USA consumption spectrum in 1985 was as follows:

- Manufacture of chromium(VI) oxide ("chromic acid"): 35%
- Manufacture of pigments: 19%
- Manufacture of chrome tanning agents: 11%
- Manufacture of oil drilling mud: 5%
- Water treatment (corrosion prevention): 4%
- Metal treatment: 3%
- Export: 15%
- Miscellaneous: 8%

Pigment and chrome tanning agent manufacture is currently declining due to environmental objections to lead chromate pigments and increasing imports of shoes and leather goods respectively.

Chromium(VI) oxide: the USA consumption spectrum in 1985 was as follows:

- Manufacture of wood protection agents: 44%
- In chromium plating: 39%
- Export: 11%
- Miscellaneous (incl. CrO_2 manufacture): 6%

The manufacture of wood protection agents using chromium(VI) oxide is increasing strongly due to restrictions upon the use of pentachlorophenol and creosote, whereas its consumption in electroplating is decreasing due to smaller cars, increasing car imports and the replacement of metal with plastic in cars.

Chromium(III) oxide: is used in the manufacture of metallic chromium by aluminothermic reduction, as a pigment and as a polishing powder.

Basic chromium sulfate, $Cr(OH)SO_4$: is used for chrome tanning of leather.

3.3.2 Chromium Metal

3.3.2.1 Economic Importance

Chromium metal used for:

- Turbine blades
- Iron-free chromium alloys
- Cermets

Apart from the utilization of large quantities of ferrochrome (produced by reducing chromite with coal) in the manufacture of alloys containing iron and chromium, smaller quantities of chromium metal are used in, for example, the manufacture of turbine blades, the production of iron-free alloys and cermets (metal ceramics e.g. 23% by weight aluminum oxide, 77% by weight chromium).

Chromium metal capacities in Western Industrialized Countries:

15 to 20×10^3 t/a

Recent statistics on chromium production capacities are unavailable. It is produced in the USA, the USSR, Japan, West Germany and other countries. The total capacity in the western industrialized countries is probably 15 to 20×10^3 t/a.

3.3.2.2 Manufacture of Chromium Metal

3.3.2.2.1 Chemical Reduction

Chromium(III) oxide (Sec. 3.3.1.3.4) is always the starting material in the chemical reduction process, for which it has to be as pure as possible particularly with regard to sulfur content (This is to avoid the formation of nickel sulfide at grain boundaries in nickel alloys). The reduction can be carried out with silicon and in particular aluminum and carbon.

The reaction:

Metallic chromium:

- By reduction of chromium(III) oxide with aluminum

$$Cr_2O_3 + 2Al \longrightarrow Al_2O_3 + 2Cr$$

is insufficiently exothermic to be self-sustaining. Strong oxidizing agents such as potassium dichromate, chromium(VI) oxide, potassium perchlorate or barium peroxide are therefore added. The reaction is carried out batchwise in vessels lined with refractory materials. The metal obtained in ca. 85 to 90% yield has a purity of 97 to 99% (by weight). The main impurities are silicon, aluminum and iron. Operators of this process include Gesellschaft für Elektrometallurgie and Shieldalloy.

Reduction with carbon is achieved by slowly reacting briquettes of chromium(III) oxide and carbon at 1275 to 1400 °C in a vacuum of 40 Pa (0.3 mm Hg).

- By reduction of chromium(III) oxide with carbon

$$Cr_2O_3 + 3C \longrightarrow 2Cr + 3CO.$$

The US company Elkem operates this “Simplex process”.

3.3.2.2.2 Electrochemical Reduction of Chrome Alum

The US company Elkem produces chromium metal using the following process:

- Dissolution of ferrochrome in sulfuric acid/ammonium sulfate.
- Separation of iron as iron(III) alum.
- Crystallization of chrome alum, $NH_4Cr(SO_4)_2 \cdot 12H_2O$.
- Dissolution of chrome alum and the electrodeposition of metallic chromium.

- By electrolysis of chrome alum solutions (produced from ferrochrome)

Electrolysis is carried out in a diaphragm cell with a stainless steel cathode. A 3 to 6 mm thick layer of chromium is obtained, which is chipped off the cathode at the end of the electrolysis. The energy consumption amounts to 18.5 kWh/kg chromium. After degassing (hydrogen removal), the chromium is 99.3% pure (by weight).

3.3.2.2.3 Electrochemical Reduction of Chromium(VI) Oxide

- By electrolysis of acidic baths containing 300 g/L chromium(VI) oxide (chromium plating) with other ingredients

The electrochemical deposition of chromium upon metals is usually carried out in chromium(VI) oxide-containing baths to which ca. 1% sulfuric acid has been added. The material to be coated is used as the cathode, the anode usually being lead. A diaphragm is not necessary. The baths contain ca. 300 g/L of chromium(VI) oxide, in addition to a number of other components. The electricity yield is very poor, due to hydrogen being preferentially produced at the cathode. The energy consumption amounts to ca. 75 kWh/kg chromium. The chromium obtained is relatively pure.

This process is suitable both for the production of metallic chromium and for the coating of objects with thin, decorative or thicker hard chromium layers (chromium electroplating).

References for Chapter 3.3: Chromium Compounds and Chromium

Raw Materials:

Mineral Facts and Problems, US Bureau of Mines, Washington (1985).
Minerals Yearbook, US Department of the Interior, Washington (1984).
Mining Annual Review, US Bureau of Mines, Washington (1986).

Commercial Information:

Chemical Profile, Chromic Acid and Sodium Bichromate, Schnell Publishing Company, Inc. (April 1, 1986).
"Chromite and Inorganic Chromium Compounds", *Chemical Economics Handbook,* Stanford Research Institute, Menlo Park, California, USA (March 1983).

Reviews:

Kirk-Othmer: *Encyclopedia of Chemical Technology,* 3rd Ed., Vol. 6, John Wiley & Sons, New York-Chichester-Brisbane-Toronto, pp 54–120 (1979).
Ullmann's Encyclopedia of Industrial Chemistry, 5th Ed., Vol. A7, pp 67–97, VCH Verlagsgesellschaft mbH, Weinheim-Deerfield Beach-Basel (1987).
R. Schliebs, *Chemie in unserer Zeit* **14,** pp 13–17 (1980).

3.4 Silicon and its Inorganic Compounds

3.4.1 Elemental Silicon

3.4.1.1 General Information and Economic Importance

Silicon is in principle available in three different commercial forms:

- Silicon as a component of alloys in the form of ferrosilicon with Si-contents between 8 to 13% (FeSi 10) and 87 to 95% (FeSi 90) and calcium silicide with a Si-content of 60 to 65%
- Metallurgical grade (MG) silicon with purities of 98.5% (Si 98) to 99.7% (Si 99)
- Electronic grade (EG) silicon with a purity, which depending upon the application has to be > 99.999%

Commercial forms of silicon:

- Ferrosilicon
- Calcium silicide
- Metallurgical grade (MG) Si
- Electronic grade (EG) Si

The 1983 worldwide production of ferrosilicon was 3.4×10^6 t with the following main producers:

- USA 306×10^3 t
- USSR 789×10^3 t
- Norway 421×10^3 t

1983 worldwide production of ferrosilicon: 3.4×10^6 t

Consumer: the steel industry

The 1985 worldwide production of MG silicon amounted to ca. 480×10^3 t. The production plants are mainly sited in areas in which cheaper (hydro)electricity is available, in particular in the USA, Norway, France, Canada and the USSR. It is utilized for:

- Aluminum alloys: 320×10^3 t
- Silicones: 150×10^3 t
- Semiconductors and highly dispersed silicas: 20×10^3 t

1985 worldwide production of MG silicon: 480×10^3 t

Consumers of MG silicon:

- Aluminum industry
- Chemical industry

The current price of MG silicon is ca. 1.5 US $/kg and is largely dependent upon energy costs.

The worldwide production of electronic grade silicon for the semiconductor industry in 1986 was ca. 5000 t. The world's largest producer is Wacker Chemie accounting for ca. 40% of this. If the current efforts to reduce the manufacturing costs of solar cells are successful, the consumption of EG silicon is expected to rise in the 1990's.

1986 worldwide production of EG silicon: ca. 5000 t

Since EG silicon is the result of a multistep purification process (see Sec. 3.4.1.2.2), the prices rise with increasing purity. Thus in 1981 polycrystalline silicon obtained by the decomposition of $SiHCl_3$ cost 125 US $/kg, silicon single crystals ca. 500 US $/kg and the silicon wafers used in semiconductor technology ca. 1900 US $/kg.

3.4.1.2 Manufacture

3.4.1.2.1 Metallurgical Grade Silicon and Ferrosilicon

Manufacture of MG silicon: reduction of quartz with coke in arc reduction furnaces

MG silicon and ferrosilicon are manufactured by reducing quartzites with coke in an arc reduction furnace using the same process. The quartz types utilized in the manufacture of MG silicon should be as pure as possible (see Table 3.4-1), those for the production of ferrosilicon may be somewhat less pure (> 96% SiO_2).

Table 3.4-1. Quartz Specification for the Manufacture of MG Silicon.

Component	Content in %
SiO_2	98 minimum
Fe_2O_3	1.5 maximum
Al_2O_3	1.0 maximum
CaO	0.2 maximum
MgO	0.2 maximum

Arsenic-, phosphorus- and sulfur-contents are undesirable due to the formation of poisonous flue gases. High aluminum oxide contents lead to the formation of sticky slags, which can contaminate the final product.

Manufacture of ferrosilicon:

by addition of iron as turnings or shredded iron

In the production of ferrosilicon the iron in the charge is added in the form of turnings or shredded iron. A shaft electric furnace is used for silicon-contents above 45% in which the lower part is lined with carbon bricks. It operates with three phase electricity, which in the case of MG silicon is introduced by way of graphitized electrodes.

It is assumed that production proceeds in the following steps:

$$SiO_2 + C \longrightarrow SiO + CO$$
$$SiO + 2C \longrightarrow SiC + CO$$
$$2SiC + SiO_2 \longrightarrow 2Si + 2CO$$

Overall reaction

$SiO_2 + 2C \rightarrow Si + 2CO$

proceeds by way of intermediates (SiO,SiC)

The carbothermal reaction requires temperatures in excess of 2000 °C. After ca. 1 to 2 h the continuously operating furnace is tapped with an auxiliary electrode and the liquid silicon (m.p. 1413 °C) run off.

3.4.1.2.2 Electronic Grade (EG) Silicon

It has to be extremely pure so that the semiconductor properties of the silicon can be fully realized.

The presence of 1 ppb_a of phosphorus in EG silicon single crystals is sufficient to reduce its specific resis-

EG silicon: purity requirements especially for p- and n-doping elements extremely high: e.g. < 0.1 to 1 ppb_a boron and/or phosphorus

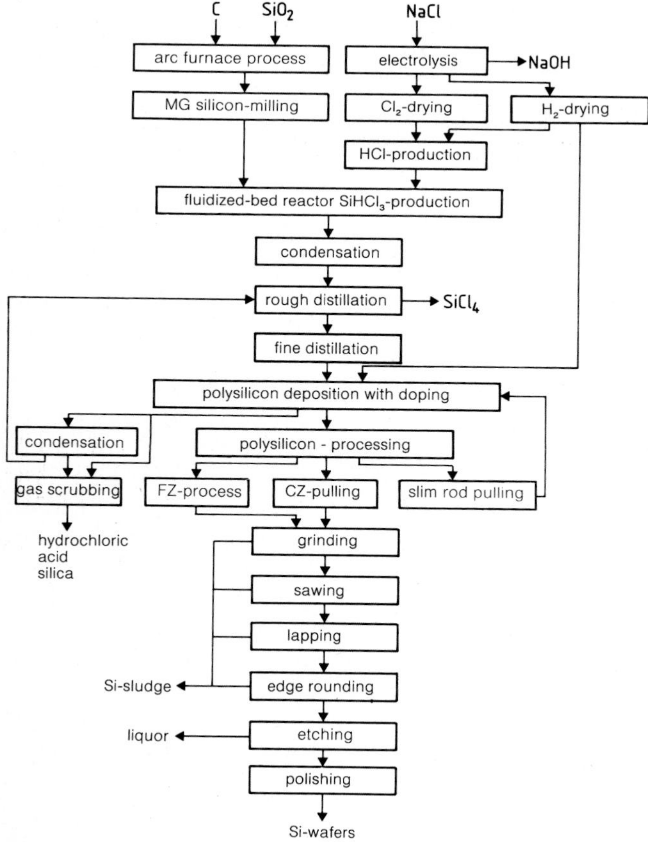

Fig. 3.4-1. Manufacture of EG silicon wafers.

tance from 150 kΩcm to 0.1 kΩcm. Thus the purity requirements for p- or n-doping elements have to be particularly stringent. The concentrations of boron and phosphorus for example should not exceed 0.1 to 1 ppb_a.

Purification of MG silicon by:

- Conversion into $SiHCl_3$:

  $Si + 3HCl \xrightarrow{300\,°C} SiHCl_3 + H_2$
- Distillation yielding highly purified $SiHCl_3$
- Decomposition of $SiHCl_3$ on Si-rods:

  $SiHCl_3 + H_2 \xrightarrow{1000\,°C} Si + 3HCl$

The process currently used worldwide for the production of ca. 80% of the EG silicon was developed by Siemens between 1953 and 1956 (Fig. 3.4-1). The starting material is MG silicon which is reacted with HCl in a fluidized bed reactor forming $SiHCl_3$ (trichlorosilane), which can be highly purified distillatively.

$$Si + 3HCl \xrightarrow{300\,°C} SiHCl_3 + H_2$$

By reversing the reaction at 1000 °C polycrystalline silicon (polysilicon) formed from $SiHCl_3$ deposits on thin pure Si-rods (slim rods):

$$SiHCl_3 + H_2 \xrightarrow{1000\,°C} Si + 3HCl$$

Since the BCl_3 present as traces in highly purified $SiHCl_3$ is hardly reduced under these conditions, a further purification takes place during this step. The silicon thus obtained contains 10^{13} boron atoms per cm^3 and is sufficiently pure for all current purposes.

Silicon for semiconductor components has to be doped with donors (P,As,Sb) or acceptors (B,Al)

Silicon for semiconductor components must be doped with well-defined quantities of electron donors (phosphorus, arsenic or antimony) or electron acceptors (boron, aluminum). This can be achieved by addition: before pulling from a crucible, during zone melting (introduction of PH_3) or by conversion of silicon into phosphorus by thermo-neutron bombardment.

The silicon single crystals required for the manufacture of semiconductor components (see Fig. 3.4-1) can be obtained in two ways: either by the more widely used crucible-free zone melting (float zone process) or by pulling single crystals from the melt using the Czochralski process (CZ-pulling).

3.4.1.3 Silicon Applications

Silicon in the form of ferrosilicon is used in large quantities as a deoxidizing agent in steel manufacture. Silicon steel alloys are utilized as dynamo and transformer plates due to their soft magnetic properties, as machine tool steels, as spring steels and as corrosion resistant casting steels for chemical plant. Its utilization in amounts of 2 to 25% in aluminum alloys is important, due to the resulting improvement in castability (engine blocks, cylinder heads).

Applications of MG Si:

- Deoxiding agent in steel production (ferrosilicon)
- Component of aluminum alloys
- Production of silicones

The utilization of MG silicon in the manufacture of silicones using the Rochow process is discussed in Chapter 4.

EG silicon has largely supplanted germanium in semiconductor technology and electronic power components (thyristors, rectifiers), accounting for 90% of the semiconductor market. There are various reasons for this. Silicon forms stable protective layers of SiO_2, which can be easily removed by etching. This has enabled masking and etching methods to be developed and refined to the extent that today integrated circuits containing up to a million components can be accomodated on a single silicon chip (microprocessors). In addition silicon components permit higher working temperatures (ca. 150 °C) than e.g. germanium components (ca. 75 °C).

EG silicon has largely displaced germanium in semiconductor technology due to the formation of more impervious SiO_2 protective layers, which can be etched as required (integrated circuits), and the higher working temperatures (150 °C) which can be used

Solar energy can be directly converted into electrical current (photovoltaic cells) using 300 μm thick 10 cm × 10 cm silicon wafers appropriately doped and provided with electrical contacts to take off the current. Single crystal wafers with efficiencies of up to 15% can be used or the cheaper unidirectionally solidified polycrystalline silicon with efficiencies of up to 12%.

Single or polycrystalline Si utilized in solar cells. Efficiences 15 to 12%

Single crystalline silicon solar cells have been used for years in space applications due to their reliability. For terrestial applications the cost of the resulting electricity needs to be reduced by a factor of 10 to be able to compete with conventional electricity production techniques. It is hoped to achieve this aim with rigorous rationalization of all process steps and the utilization of cheaper silicon grades (solar grade, possibly amorphous Si).

Up to 1986 solar cells with a capacity in excess of 60 MW had been installed, of which more than 48 MW was on the basis of crystalline silicon. In the consumer electronics sector lower efficiency (6 to 8%) amorphous cells are adequate, with ca. 1 μm thick amorphous, mostly hydrogen-containing Si-layers (α-Si:H), which are deposited on a suitable substrate at ca. 400 °C by glow discharge in silane (SiH_4) or disilane (Si_2H_6) (thin layer cells).

3.4.2 Inorganic Silicon Compounds

Most important inorganic silicon compounds:

- Silicon carbide
- Silicon nitride
- Metal silicides
- Silicon dioxide, hydrated silicas
- Silicates
- Silicon halides
- Silicic acid esters

The most important industrially utilized silicon compounds are the halides and their derivatives, silicon dioxide and silicas in their various forms, silicates in the form of glass, water glass, enamel frits, silicate fillers, zeolites, cement, silicon nitride and silicon carbide.

In this section only silicon halides and silicic acid esters will be discussed. The other classes of compounds will be discussed in the sections of this book appropriate to their applications i.e.:

Silicon carbide in:

- Section 5.5.7.2, "Nonoxide Ceramics": Manufacturing Processes for Silicon Carbide
- Section 5.2.3.7, "Silicon Carbide Fibers and Silicon Carbide-Coated Carbon Fibers"

Silicon nitride, Si_3N_4, in:

- Section 5.5.7.4, "Nonoxide Ceramics": Fine Silicon Nitride Ceramic Products

Metal silicides:

- Section 5.6.9, "Metallic Hard Materials": metal silicides

Silicon dioxide and silicas in:

- Section 5.8.3.1, "Natural Fillers": silicon-based
- Section 5.8.4.1, "Synthetic Fillers": silicas and silicates

Silicates and silicate products are utilized in many industrial sectors and are dealt with in:

- Section 5.1, "Silicate Products"
- Section 5.1.1, "Glass"
- Section 5.1.2, "Alkali Silicates"

- Section 5.1.3, “Zeolites”
- Section 5.2.3.1, “Textile Glass Fibers”
- Section 5.2.3.3, “Mineral Fiber Insulation Materials”
- Section 5.3.3, “Cement”
- Section 5.3.5, “Coarse Ceramic Products for the Construction Industry”
- Section 5.3.6, “Expanded Products”
- Section 5.5.4, “Clay Ceramic Products”
- Section 5.8.4.1, “Synthetic Fillers”: silicas and silicates

Silicon halides:

The most important industrially utilized silicon halides are silicon tetrachloride $SiCl_4$ and trichlorosilane $SiHCl_3$. Both are formed by the reaction of elemental silicon with HCl at temperatures above 300 °C, $SiCl_4$ being increasingly favored with increasing reaction temperature:

$$Si + 4HCl \longrightarrow SiCl_4 + 2H_2$$
$$Si + 3HCl \longrightarrow SiHCl_3 + H_2$$

The direct reaction of elemental silicon or ferrosilicon (> 90% Si) with chlorine to $SiCl_4$ is also used industrially.

Manufacture of $SiCl_4$:

- Reaction of Si (or ferrosilicon) with Cl_2
- Reaction of Si with HCl
- Reaction of SiO_2 with carbon and Cl_2
- Chlorination of SiC

Since the 1960’s an increasing number of processes starting from SiO_2 have been published in the patent literature:

$$SiO_2 + 2C + 2Cl_2 \longrightarrow SiCl_4 + 2CO$$

The reaction temperatures of ca. 1400 °C are achieved by heating the reactants at the center of the fluidized bed by burning carbon in oxygen or by resistive heating.

The manufacture of $SiCl_4$ from SiC also requires additional heat:

$$SiC + 2Cl_2 \longrightarrow SiCl_4 + C$$

Applications of $SiCl_4$:

- Synthesis of organo-functional Si-compounds
- Manufacture of EG silicon and highly dispersed SiO_2
- Siliconization of metals

$SiCl_4$ is the starting material for the synthesis of organo-functional silicon compounds. It is together with $SiHCl_3$, utilized in the manufacture of EG silicon for semiconductor applications, in the manufacture of highly dispersed SiO_2 and for the siliconization of metallic objects.

Tetrafluorosilane, SiF_4, in:

- Section 1.7.1.3.2, "Hydrogen Fluoride"

Hexafluorosilicic acid, H_2SiF_6, in:

- Section 1.7.1.3.2, "Hydrogen Fluoride"
- Section 1.7.1.3.3, "Aluminum Fluoride": AlF_3 *from hexafluorosilicic acid*

Hexafluorosilicates in:

- Section 1.7.1.3.6, "Hexafluorosilicates"

Silicic acid esters $Si(OR)_4$:

Silicic acid esters are produced by the reaction of $SiCl_4$ with the appropriate alcohols. The most important representative of this group is tetraethoxysilane (tetraethyl orthosilicate) $Si(OC_2H_5)_4$, which is used directly or after partial hydrolysis to ethylpolysilicates as a binder for ceramic pastes, for inorganic zinc dust paints (corrosion protection), for the surface treatment of glass and for the modification of silicones.

Other organic silicon compounds are treated in Chapter 4, "Silicones".

References for Chapter 3.4: Silicon and its Inorganic Compounds

General Information:

Winnacker-Küchler, *Chemische Technologie*. 3rd Ed., Vol. 6, pp. 488–491, Carl Hanser Verlag, München, 1973.
Ullmanns Encyklopädie der technischen Chemie, 4th Ed., Vol. 21, pp. 417–429, Verlag Chemie, Weinheim-Deerfield Beach-Basel, 1982.
G. Rath and J. Wagner, *Elektrowärme international* **34,** B 2 April, B 81–B 84 (1976).
United Nations Industrial Statistics, Yearbook 1983.
W. Lang, *Erzmetall* **39,** pp. 207–213 (1986).

Specific Information:
(particularly Electronic Grade Silicon)

W. Dietze, *Metall* **34,** pp. 676–677 (1980).
Siemens Zeitschrift **55 (Jan./Feb.),** pp. 1–81 (1981).
E. Spenke, *Z. Werkstofftech.* **10,** pp. 262–275 (1979).
Dynamit Nobel Brochures: *Silicon Products; Silicon Tetrachloride; Trichlorosilane.*
P. A. Munter, O. T. Aepli and R. A. Kossatz, Ind. Eng. Chem. **39,** pp. 427–431 (1947).

3.5 Manganese Compounds and Manganese

3.5.1 Manganese Compounds

3.5.1.1 Economic Importance

The following manganese compounds are industrially important:

- Manganese(II) salts such as oxide, sulfate, chloride and carbonate
- Manganese(II,III) oxide and manganese(III) oxide
- Manganese(IV) oxide
- Potassium permanganate ($KMnO_4$)

Industrially important manganese compounds:

- MnO, $MnSO_4$, $MnCl_2$, $MnCO_3$
- Mn_2O_3, Mn_3O_4
- MnO_2
- $KMnO_4$

Recent capacity and production statistics are only available for:

- Manganese(II) sulfate: The USA capacity in 1985 was 50×10^3 t/a, with Eastman Kodak accounting for 35×10^3 t/a from the manufacture of hydroquinone. The Belgian capacity (SEDEMA) is 78×10^3 t/a.
- Manganese(IV) oxide: The world capacity in 1985 was 180×10^3 t/a, of which Japanese manufacturers with their foreign subsidiaries accounted for almost 50%. The largest producer is SEDEMA in Belgium which with its US subsidiary Chemetals Corp. has a capacity of 50×10^3 t/a. Another US manufacturer is Foote Mineral Corporation with a capacity of 10×10^3 t/a. The 1985 world consumption of manganese(IV) oxide was 128×10^3 t.
- Potassium permanganate: The only West European manufacturer is Asturquimica SA in Spain. Western Europe imports it from the People's Republic of China, East Germany and Czechoslovakia. West Germany, the Benelux countries and Spain consume 1000, 700 and 300 t/a respectively.

Capacities:

- $MnSO_4$: USA 50×10^3 t/a; Belgium 78×10^3 t/a
- MnO_2: World 180×10^3 t/a; USA $\geq 20 \times 10^3$ t/a; W. Europe 87×10^3 t/a

3.5.1.2 Raw Materials

Manganese as its compounds is widely distributed in the Earth's crust, it being the 12th most common element with ca. 1000 ppm. In soil, rock and sediment its content varies between 200 and 4500 ppm. It is

Manganese is widely distributed in the Earth's crust (12th most frequent element). Manganese ores consist mainly of nonstoichiometric manganese(IV) oxides and manganese(II) carbonates and silicates

also contained in freshwater up to a concentration of several ppm, which has to be taken into account during the treatment of freshwater (Sec. 1.1.2.4).

Manganese ores are mainly (impure and nonstoichiometric) manganese(IV) oxide, manganese(II) carbonate and silicate. Furthermore, manganese is found in considerable quantities in the so-called manganese nodules on the seabed. The total quantity of these nodules is estimated to be 10^{12} t. Harvesting of these nodules would be particularly interesting for the copper, nickel and cobalt present, less so for the manganese content.

The most important of the considerable number of manganese ores are:

- Pyrolusite: manganese(IV) oxide
- Psilomelane: Ba-containing manganese(IV) oxide hydrate
- Cryptomelane: K-containing manganese(IV) oxide hydrate
- Manganite: manganese(III) oxide hydrate
- Hausmannite: mangenese(II,III) oxide
- Rhodochrosite: mangenese(II) carbonate

The most important deposits are sedimentary. The most important manganese ore-producing countries together with their certain and probable reserves (in 10^6 t manganese) are listed in Table 3.5-1. In 1984, 23.08×10^6 t of ore were mined worldwide with a manganese content of 4.85×10^6 t.

Most important manganese ore-producing countries:

- South Africa
- USSR

World reserves (certain and probable):

3597×10^6 t

Table 3.5-1. Important Manganese Ore Producing-Countries and their Reserves (as of July 1984) in 10^6 t.

	Manganese reserves in 10^6 t	% of World reserves
Republic of South Africa	2630	73.1
USSR	508	14.1
Gabon	172	4.8
Australia	152	4.2
Brazil	63	1.8
Various	72	2.0
World	3597	100

Statistical calculations predict that the individual reserves will be exhausted within 100 to 800 years.

The USA does not possess economically exploitable manganese ore reserves at current manganese prices. The USA imported:

USA has no economically extractable manganese ore deposits

- 178×10^3 t of manganese ore (as Mn) in 1985
- 258×10^3 t of ferromanganese (as Mn) in 1985
- 83×10^3 t of silicomanganese (as Mn) in 1984.

Only ca. 5% of the manganese ores extracted is utilized for nonmetallurgical purposes.

Manganese ore utilization:

- 95% for metallurgical purposes
- 5% for nonmetallurgical purposes

Industrially, there are three different qualities of ore:

- Metallurgical quality: manganese content $\geq 40\%$, $P < 0.1\%$, as lumps.
- Battery quality: > 48 to 58% manganese, > 68% MnO_2.
- Chemical quality: > 80% MnO_2, $\leq 3\%$ iron.

3.5.1.3 Manufacture of Manganese Compounds

3.5.1.3.1 Manganese(II) Compounds

Manganese(II) oxide

It is manufactured by the reduction of naturally occurring manganese(IV) oxide-containing ores (e.g. pyrolusite) with carbon or methane:

MnO:

- By reduction of MnO_2 with carbon or methane

$$MnO_2 + C \longrightarrow MnO + CO$$

at 400 to 1000 °C. By appropriate control of the reaction, it is possible to remove some of the oxygen from the manganese(IV) oxide thermally, thereby saving reducing agent. The process is carried out in conventional units such as rotary or shaft kilns. The manganese(II) oxide produced has to be cooled in an inert gas atmosphere to avoid reoxidation.

Manganese(II) sulfate

It is manufactured by reacting manganese(II) oxide or carbonate with sulfuric acid:

$MnSO_4$:

- From MnO and sulfuric acid with solution purification by precipitation of interfering cations

$$MnO + H_2SO_4 \longrightarrow MnSO_4 \cdot H_2O$$

Interfering cations such as iron, aluminum and molybdenum, and silica can be removed from the manganese(II) sulfate solution (after prior oxidation of iron(II) to iron(III) with manganese(IV) oxide, if necessary) by precipitation with calcium oxide, sodium carbonate or manganese(II) carbonate after adjustment to the appropriate pH. Transition metal ions such as cobalt, nickel or copper and traces of arsenic are precipitated as the sulfides.

Manganese(II) sulfate is formed as a byproduct in the oxidation of organic compounds with manganese(IV) oxide in the presence of sulfuric acid, e.g. the oxidation of aniline to hydroquinone.

Manganese(II) chloride

$MnCl_2$:

- From MnO_2 and hydrogen chloride
- From MnO and hydrochloric acid
- From Mn and chlorine

There are a number of processes for manufacturing manganese(II) chloride:

- Reaction of manganese(IV) oxide with hydrogen chloride (Weldon process):

$$MnO_2 + 4HCl \longrightarrow MnCl_2 + Cl_2 + 2H_2O$$

  The chlorine produced is absorbed in calcium hydroxide solution forming calcium hypochlorite.

- Reaction of manganese(II) oxide or carbonate with hydrochloric acid:

$$MnO + 2HCl \longrightarrow MnCl_2 + H_2O$$

  The heavy metal impurities in the solution obtained by the reaction of manganese(II) carbonate with hydrochloric acid are precipitated out by the addition of further manganese(II) carbonate.

- Reaction of chlorine with manganese or ferromanganese:

$$Mn + Cl_2 \xrightarrow{>700\,°C} MnCl_2$$

  The molten iron(III) chloride formed during ferromanganese chlorination can be removed by sublimation leaving manganese(II) chloride.

Manganese(II) carbonate

It occurs naturally as the mineral rhodochrosite. It is manufactured from manganese(II) sulfate by precipitation with alkali carbonates, or alkali or ammonium hydrogen carbonates:

$MnCO_3$:

- From $MnSO_4$ and alkali (ammonium) carbonate and hydrogen carbonate

$$MnSO_4 + 2NH_4HCO_3 \longrightarrow MnCO_3 + (NH_4)_2SO_4 + CO_2 + H_2O$$

After separating off and washing, the manganese(II) carbonate obtained has to be dried at 120 °C under inert gas to avoid oxidation.

3.5.1.3.2 Manganese(II,III) Oxide (Mn_3O_4) and Manganese(III) Oxide (Mn_2O_3)

Manganese(II,III) oxide (the mineral hausmannite) is formed by the heating of manganese oxides with other valency states in air at 890 °C; e.g.:

Mn_3O_4:

- By heating MnO_2 at 890 °C

$$3MnO_2 \xrightarrow{890\,°C} Mn_3O_4 + O_2$$

Manganese(III) oxide is formed when manganese(IV) oxide is heated at 500 to 600 °C:

Mn_2O_3:

- By heating MnO_2 at 500 to 600 °C

$$2MnO_2 \xrightarrow{500-600\,°C} Mn_2O_3 + 0.5\,O_2$$

3.5.1.3.3 Manganese(IV) Oxide

Manganese(IV) oxide exists in many modifications. The only modification which approaches a stoichiometric composition is β-MnO_2 (e.g. the mineral pyrolusite) and it is the least reactive. All the other modifications contain additional cations, such as K^+, Na^+, Ca^{2+} and Ba^{2+}, anions, such as OH^-, and water. The average valency state of the manganese is also less than 4 (down to 3.4). The almost amorphous γ-MnO_2 (the mineral nsutite) is particularly reactive.

Manganese(IV) oxide for batteries and raw material for the manufacture of ferrites can be produced by a number of processes:

Manufacture of manganese(IV) oxide by:

- Aftertreatment of MnO_2 minerals
- Reaction of MnO_2 minerals with nitrous gases
- Oxidation of $MnCO_3$
- Electrochemical oxidation of Mn(II) salts

- Aftertreatment of manganese(IV) oxide minerals (pyrolusite)
- Reaction of manganese(IV) oxide minerals with nitrous gases
- Oxidation of manganese carbonate
- Electrochemical processes starting from manganese(II) salt solutions

The products from the second and third processes are known as chemical manganese dioxide (CMD), that from the fourth process as electrolytic manganese dioxide (EMD). Furthermore, manganese(IV) oxide is also formed as a byproduct in the oxidation of organic compounds with potassium permanganate: e.g. oxidation of o-toluenesulfonamide to o-sulfobenzoic acid imide (saccharin).

Aftertreatment of manganese(IV) oxide minerals

Aftertreatment of MnO_2 minerals:

1st step: Thermal decomposition of MnO_2 to Mn_2O_3

2nd step: Dissolution of Mn_2O_3 in sulfuric acid forming $MnSO_4$ and activated MnO_2, which is suitable for battery applications

Manganese(IV) oxide minerals are thermally decomposed (Sec. 3.5.1.3.2) and then reacted with sulfuric acid to hydrated manganese(IV) oxide and manganese(II) sulfate:

$$Mn_2O_3 + H_2SO_4 \longrightarrow MnO_2 + MnSO_4 + H_2O$$

The manganese(IV) oxide obtained contains some of the impurities of the manganese(IV) oxide minerals. In France, 2×10^3 t/a of manganese(IV) oxide for batteries are manufactured using this process.

Reaction of MnO_2 minerals with nitrous gases:

1st step: Reaction of MnO_2 minerals with nitrous gases to a $Mn(NO_3)_2$ solution

2nd step: Decomposition of $Mn(NO_3)_2$ to nitrous gases and MnO_2, γ-MnO_2 being formed at $T < 350\,^\circ C$ and β-MnO_2 at $T > 350\,^\circ C$

Manganese(IV) oxide minerals react with nitrous gases as follows:

$$MnO_2 + N_2O_4 \longrightarrow Mn(NO_3)_2$$
$$Mn(NO_3)_2 \longrightarrow MnO_2 + N_2O_4$$

A slurry of finely ground manganese(IV) oxide minerals in water or dilute manganese(II) nitrate solution is treated with nitrous gases in absorption towers. In addition to manganese, Cu, Zn, Ni and Cd are also dissolved. These accompanying ions are precipitated as carbonates or hydroxides by adding carbonates (sodium carbonate, manganese(II) carbonate) or calcium oxide. The thus purified solution is concentrated to manganese(II) nitrate hexahydrate. Decomposition at temperatures below 350 °C yields γ-MnO_2 and above 350 °C β-MnO_2. The nitrous gases produced during the decomposition are utilized for the dissolution of manganese(IV) oxide minerals.

The thermal decomposition is technically difficult. In a modern process, a stirred manganese(II) nitrate melt is seeded with manganese(IV) oxide. 4×10^3 t/a of high purity β-MnO_2 are manufactured in the USA using this process.

Oxidation of manganese(II) carbonate

Manganese(II) carbonate is obtained using the process described in Sec. 3.5.1.3.1. The impurities are mainly precipitated and removed during the neutralization of the manganese(II) sulfate solution.

Oxidation of $MnCO_3$:

1st step: Oxidation with oxygen to a Mn(III,IV) oxide

2nd step: Further oxidation with sodium chlorate to ε, γ-MnO_2

The oxidation of manganese(II) carbonate proceeds initially with air to a manganese(III,IV) oxide:

$$MnCO_3 \xrightarrow{O_2} MnO_{1,85} + CO_2 \,.$$

This is suspended in sulfuric acid and oxidized with sodium chlorate to ε,γ-manganese(IV) oxide:

$$MnO_{1,85} \xrightarrow{H_2SO_4/NaClO_3} MnO_2 \,.$$

Electrolytic manganese(IV) oxide (EMD)

Industrially, the electrochemical processes for the manufacture of manganese(IV) oxide are currently the most important. A purified manganese(II) sulfate solution is electrolyzed at 90 to 95 °C on lead, titanium, or graphite anodes at current densities of 0.5 to 1.2 A/dm^2, whereupon manganese(II) is oxidized to manganese(III):

Electrochemical oxidation of $MnSO_4$:

1st step: Electrochemical oxidation of dissolved $MnSO_4$ to Mn(III)

2nd step: Disproportionation of Mn(III) to Mn(IV) and Mn(II), the MnO_2 deposited at the anode being removed discontinuously

$$Mn^{2+} \longrightarrow Mn^{3+} + e^-$$

Hydrogen is produced at the cathode. The Mn(III) formed disproportionates in a nonelectrochemical reaction into Mn(IV) and Mn(II):

$$2Mn^{3+} \longrightarrow Mn^{4+} + Mn^{2+}$$

The Mn(IV) formed is deposited on the anode as 1 to 3 cm layers of very disordered ε-MnO_2, which still contains Mn(III).

The MnO_2-coated anodes are periodically taken out of the electrolyzers and the manganese(IV) oxide mechanically removed, ground, washed and dried. The titanium and graphite anodes can then be reused. In order to maintain the Mn(II) concentration in the electrolyte constant during the electrolysis, fresh electrolyte is added continuously or intermittently and the spent electrolyte taken off and used for manganese(II) oxide dissolution.

3.5.1.3.4 Potassium Permanganate

$KMnO_4$ from MnO_2 minerals:

- Oxidation of MnO_2 with air by way of Mn(V) to Mn(VI) (in one or two steps), then electrochemical oxidation to Mn(VII).

There are several processes for the manufacture of potassium permanganate. Manganese(IV) oxide minerals are primarily used as raw materials for its manufacture, the processes passing through the following stages:

$$Mn(IV) \longrightarrow Mn(V) \longrightarrow Mn(VI) \longrightarrow Mn(VII)$$

The first two stages are accomplished with atmospheric oxidation, the third electrochemically.

Oxidation of manganese(IV) oxide minerals to manganate(VI) can be carried out in single- or two-step processes:

- *Two-step roasting (melt) processes*

In the first step, a slurry of 50% potassium hydroxide solution with finely ground manganese dioxide ore, in a molar ratio MnO_2 : KOH of 1 : 2.3 to 2.7, is oxidized with air at 390 to 420 °C in a fast reaction to manganate(V):

$$2MnO_2 + 6KOH + 0.5O_2 \longrightarrow 2K_3MnO_4 + 3H_2O$$

Rotary kilns or spray towers are used as reactors. In the second step, the manganate(V) formed is oxidized at 180 to 220 °C to manganate(VI) after optional grinding:

$$2K_3MnO_4 + H_2O + 0.5O_2 \longrightarrow K_2MnO_4 + 2KOH$$

This reaction proceeds much more slowly than the first step. 3 to 4 h are necessary if intermediate grinding is employed. In the second step a particular partial pressure of water has to be maintained. The types of reactor used are similar to those used for the first step. The yield of manganate(VI) is 85 to 90%.

- *Single-step liquid phase processes*

In this process, a mixture of finely ground MnO_2 ore and a 70 to 90% potassium hydroxide solution, in a MnO_2 : KOH molar ratio of $\geq 1:5$, is oxidized with air at 200 to 260 °C under slight positive pressure. The reaction time is 4 to 6 h. After crystallization the yield of manganate(VI) is 87 to 94%. The potassium manganate(VI) produced is then electrochemically oxidized, continuously or discontinuously, to potassium permanganate:

The electrochemical oxidation of manganese(VI) to potassium permanganate proceeds continuously or discontinuously on nickel or Monel anodes

$$K_2MnO_4 + H_2O \xrightarrow{e^-} KMnO_4 + KOH + 0{,}5H_2$$

Different types of cell are used: mono- or bipolar, with or without diaphragms. Nickel or Monel anodes and steel cathodes are used. The electricity yield is 60 to 90%. Depending upon the cell type, the permanganate formed crystallizes directly in the cell or in separate crystallizers. Recrystallization may be necessary. Potassium permanganate must be dried below 150 °C, because it decomposes exothermically above 200 °C.

During electrolysis, one mole of potassium hydroxide is produced per mole potassium permanganate. This has to be recovered.

3.5.1.4 Applications for Manganese Compounds

Manganese(II) compounds

Applications for Mn(II) compounds:

- Nutritional supplements in animal feedstuffs
- Fertilizer additives
- In the ceramic industry
- In metallurgy
- As raw materials for the manufacture of other Mn compounds
- Manufacture of the fungicide "Maneb"

Manganese(II) oxide:
Utilized as a starting material for manganese(II) salts, as additives for fertilizers (in the USA 20×10^3 t/a), as a nutritional supplement in animal feedstuffs and in the manufacture of oxide ceramic components.

Manganese(II) sulfate:
Utilized for the electrochemical manufacture of metallic manganese and EMD-MnO_2; as a raw material for the manufacture of the manganese(II) salts of organic acids ("siccative") and the fungicide "Maneb" (manganese-ethene-1,2-bis-dithiocarbamate); and as additives for fertilizers (80% in the USA, particularly for the cultivation of soybeans) and animal feedstuffs.

Manganese(II) chloride:
Utilized in the manufacture of corrosion resistant magnesium alloys, for the coloring of bricks and in dry batteries.

Manganese(II) carbonate:
Utilized in the manufacture of other manganese(II) salts, ferrites and welding rods.

Mn_3O_4 and Mn_2O_3:

- In the manufacture of magnetic materials and semiconductors

Mn_3O_4 and Mn_2O_3:
Both oxides were formerly used as starting materials for the aluminothermic production of manganese. The pure oxides are utilized in the manufacture of magnetic materials and semiconductors.

MnO_2:

- Mainly in batteries and in the manufacture of ferrites

Manganese(IV) oxide:
Utilized in dry batteries, in the manufacture of ferrites, as an oxidizing agent in organic synthesis, as a crosslinking agent for polysulfide rubbers and as a component of oxidation catalysts.

$KMnO_4$

- Oxidizing agent in organic synthesis, for oxidation of impurities in a multitude of application fields

Potassium permanganate:
It is utilized as an oxidizing agent in organic synthesis, for the removal of trace impurities (from e.g. the lower aliphatic alcohols), for effluent and flue-gas purification, in the production of potable water, for the bleaching of inorganic and organic materials and for the purification of zinc sulfate solutions for zinc electrolysis.

3.5.2 Manganese – Electrochemical Manufacture, Importance and Applications

There are little data available about the electrochemical manufacture of manganese metal. Delta Manganese (Pty) Ltd. in South Africa is the largest manufacturer of electrochemical manganese. It is mainly used for the manufacture of alloys with nonferrous metals such as copper and aluminum.

Alternative production processes for manganese are the reduction of low iron content manganese ores with silicon, which leads to 97% manganese and is not dealt with here, and the no longer industrially operated aluminothermic reduction of manganese(II,III) oxides. In the production of pure manganese metal, electrochemical manufacture predominates:

Electrochemical manufacture of manganese:

- By electrolysis of highly purified $MnSO_4$ solution. Mn is deposited on the cathode and is discontinuously removed. In addition to oxygen, MnO_2 is formed at the anode in a side reaction

$$MnSO_4 + H_2O \xrightarrow{e^-} Mn + H_2SO_4 + 0.5\,O_2$$

A highly purified manganese sulfate solution (see Sec. 3.5.1.3.1) serves as the electrolyte. The cathodic electrolyte contains 30 to 40 g/L manganese sulfate and 125 to 150 g/L ammonium sulfate and the manganese depleted anodic electrolyte 10 to 20 g/L manganese sulfate, 25 to 40 g/L sulfuric acid and 125 to 150 g/L ammonium sulfate. The anodes consist of lead alloyed with 1% silver, the cathodes of stainless steel or ®Hastelloy, type 316. The cells are operated at 35 to 40 °C, a cathodic current density of 2 to 5.5 A/dm^2 and a potential of 5 V. The electricity yield is 50 to 70%.

Cathodically, hydrogen ions are discharged as well as manganese ions. Anodically, oxygen is formed together with considerable quantities of manganese(IV) oxide which precipitate and have to be removed.

When the manganese layers on the cathodes reach a thickness of 1 to 4 mm, the cathodes are removed and the manganese metal knocked off and dried. It has a purity of over 99.6% with oxygen and sulfur as its main impurities. Most of the other impurities are present in concentrations below 20 ppm.

References for Chapter 3.5: Manganese Compounds and Manganese

Commercial Information:

General:

"Manganese Products", *Chemical Economics Handbook*, Stanford Research Institute, Menlo Park, California, USA (July 1980).

"Manganese, Salient Statistics", in *Chemical Economics Handbook Manual of Current Indicators*, Stanford Research Institute, Menlo Park, California, USA, pp 340–341 (Oct. 1986).

Directory of Chemical Producers, Stanford Research Institute, Menlo Park, California, USA, p. 1300 (1985).

Manganese(II) sulfate:

Chemical Marketing Reporter, p. 3 (Jan. 7, 1985).

"Manganese sulfate", *Chemical Profiles*, Schnell Publishing Company, Inc. (Oct. 1, 1984).

Manganese(IV) oxide:

Chemische Industrie, **XXXV**, p. 445 (1983), **XXXVII**, p. 835 (1985).

Japan Chemical Week **27**, p. 8 (1986).

Chemical Week, p. 42 (Feb. 19, 1986).

Potassium permanganate:

Eur. Chemical News, p. 10 (Oct. 6, 1986).

Manganese metal:

Chemische Industrie, **XXXII**, p. 847 (1980).

General Information:

Kirk-Othmer: *Encyclopedia of Chemical Technology*, 3rd Ed., Vol. 14, John Wiley & Sons, New York-Chichester-Brisbane-Toronto, pp 824–895 (1981).

Ullmanns Encyklopädie der technischen Chemie, 4th Ed., Vol. 16, pp 425–480, Verlag Chemie, Weinheim-Deerfield Beach-Basel (1978).

S. A. Weiss: *Manganese – the other uses*. Metal Bulletin Books Ltd., The Garden City Press Ltd., Letchworth, Hertfordshire, UK (1977).

Raw Materials:

Minerals Yearbook 1983, Vol. I Manganese, pp 593–606, US Department of the Interior, Washington (1983).

G. Roethe and W. Sames: "Strategische Rohstoffe der Republik Südafrika – Ein Diskussionsbeitrag – Teil I: Chrom und Mangan" (The strategic raw materials of South Africa – a discussion contribution – Part I: chromium and manganese), *Metall.* **41**, pp 186–190 (1987).

Manufacture and Applications of Manganese(IV) Oxide:

E. Preisler, *Chemie in unserer Zeit* **14**, pp 137–148 (1980).

K. J. Euler and R. Kirchhof, "Quelques Propriètes d'Oxides des Manganèse obtenus par Décomposition thermique" (Several properties of manganese oxides obtained by thermal decomposition), *Electrochimica Acta* **26**, pp 1383–1387 (1981).

4 Silicones

4.1 Structure and Properties

Silicones, poly(organosiloxanes), are the only industrially important polymers which possess an inorganic "backbone". The groups linked by carbon-silicon bonds to the backbone are mostly methyl or phenyl groups. In addition to polysiloxanes with linear structures the simplest of which is α,ω-bis(trimethylsiloxy)poly(dimethylsiloxane), there are also branched and crosslinked siloxanes.

Silicones = poly(organosiloxanes); e.g.

$$CH_3-\overset{CH_3}{\underset{CH_3}{Si}}-O-\left[\overset{CH_3}{\underset{CH_3}{Si}}-O\right]_n-\overset{CH_3}{\underset{CH_3}{Si}}-CH_3$$

n: up to 14 000 known

The term silicone is also used in a broader sense to include formulations of these polymers with other materials.

Poly(organosiloxanes) are built up of $R_3SiO_{1/2}$ (monofunctional, M), $R_2SiO_{2/2}$ (difunctional, D), $RSiO_{3/2}$ (trifunctional, T) and $SiO_{4/2}$ (tetrafunctional, Q) units. These can be chemically combined in almost any conceivable way.

Silicone functional units:

- $R_3SiO_{1/2}$, M
- $R_2SiO_{2/2}$, D
- $RSiO_{3/2}$, T
- $SiO_{4/2}$, Q

R is normally methyl or phenyl

Linear poly(organosiloxanes) are built up of a combination of monofunctional M units and difunctional D units, the former being the chain ends. Poly(organosiloxanes) solely consisting of difunctional units are either cyclic or linear with, for example, hydroxy or alkoxy endgroups. The incorporation of T and Q units leads to branched poly(siloxanes).

These different poly(organosiloxane) types form the basis of industrial silicone products. The product spectrum is vast, manufacturers with a complete product range having in excess of 1000 separate products in their programme.

Combinations of these functional units lead to a multiplicity of products

As a result of their exceptional properties silicone products are widely employed as raw materials and additives. They are stable at high temperatures, resistant to oxidation and weathering, hydrophobic and depending upon their structure can exhibit either anti-

Silicone properties:

- High temperature stability
- Resistant to
  - oxidation
  - weathering
- Hydrophobicity

- Depending upon structure
  - defoaming or
  - foam stabilizing
- Abhesive
- Electrically nonconducting
- Gas and vapor permeable
- Little change in physical properties with temperature
- Physiologically harmless

foaming or foam-stabilizing properties. They are abhesive, electrically nonconducting and exhibit high gas and vapor permeability. Their physical properties also vary little with temperature.

4.2 Economic Importance

Silicones have been manufactured on an industrial scale since 1950. The main producing countries in the West are:

- The USA with: Dow Corning, General Electric, Stauffer-Wacker, Union Carbide
- Great Britain with: Dow Corning, ICI
- France with: Rhône-Poulenc
- West Germany with: Bayer, Goldschmidt, Wacker
- Japan with: Shin-etsu, Toray (with Dow Corning), Toshiba (with General Electric)

Among the Communist Countries East Germany, Czechoslovakia, the P.R. of China and the USSR are silicone producers.

1985 worldwide capacity for (chloro)methylsilanes: $> 600 \times 10^3$ t/a

The worldwide (chloro)methylsilane capacity in 1985 was estimated to be in excess of 600×10^3 t/a of which 50% was in the USA, 30% in Western Europe and 10 to 15% in Japan. In Western Europe and Japan silicone elastomer production predominates, whereas silicone oils predominate in the USA.

Silicones are relatively independent of crude oil

Since silicone production is only partly based on petrochemical products, silicones are not affected by oil shortages and price increases in the same way as many purely organic products. However silicon manufacture is extremely energy intensive (Sec. 3.4.1).

4.3 Silicone Manufacture

4.3.1 Synthesis of Starting Materials

The most important starting materials are:

- (Chloro)methylsilanes
- (Chloro)phenylsilanes

with (chloro)methylphenylsilanes, (chloro)methyl(trifluoropropyl)silanes, (chloro)methylvinylsilanes and organo-functional silanes being of secondary importance.

4.3.1.1 (Chloro)methylsilanes

(Chloro)methylsilanes are manufactured using the "direct synthesis" process discovered by Rochow and Müller. In this process silicon powder (ca. 99% pure) is reacted with methylchloride in the presence of 2 to 6% "copper catalyst" (mostly copper oxide) and 0.05 to 0.5% zinc (zinc oxide) as activator (percentages

Manufacture of (chloro)methylsilanes:

- From Si + CH_3Cl

  in the presence of a Cu catalyst and Zn or ZnO activator in a fluidized bed reactor

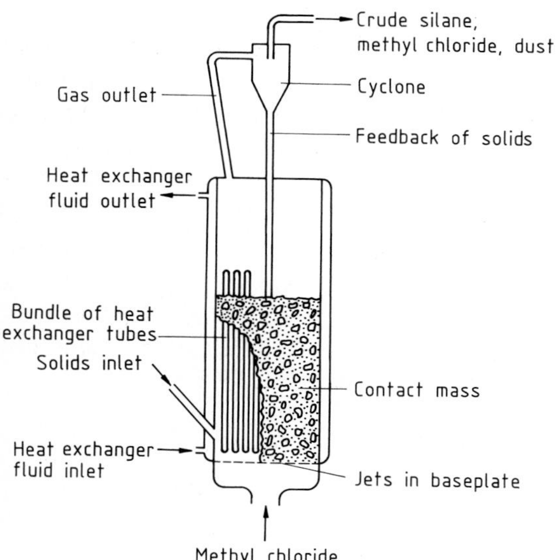

Fig. 4-1. Schema of a reactor for direct synthesis of (chloro)methylsilanes.

based on silicon used) in a fluidized bed reactor under pressure at ca. 300 °C. The reaction is exothermic.

$$Si + 2CH_3Cl \xrightarrow[\text{ca. } 300\,°C]{Cu} (CH_3)_2SiCl_2$$

Figure 4-1 shows a schema of a plant for the synthesis of (chloro)methylsilanes.

Direct synthesis managed so that $(CH_3)_2SiCl_2$ is the main product

The yield of (dichloro)dimethylsilane, the main product, has to be maximized for economic reasons. Important byproducts are (trichloro)methylsilane, (chloro)trimethylsilane, (dichloro)methylsilane, di- and polysilanes, which are formed together with other products including some siliconless products e.g. hydrocarbons.

The reaction mixture, whose components have in some cases very similar boiling points, is separated by fractional distillation.

4.3.1.2 (Chloro)phenyl- and (Chloro)methylphenylsilanes

Manufacture of (chloro)phenylsilanes:

- From Si and chlorobenzene

(Chloro)phenylsilanes can as with (chloro)methylsilanes be produced by "direct synthesis" from silicon and chlorobenzene, but require higher temperatures (ca. 500 °C). (Dichloro)diphenylsilane and (trichloro)phenylsilane are the main products:

$$Si + 2\,C_6H_5Cl \xrightarrow[\text{ca. } 500\,°C]{Cu\ (HCl)} (C_6H_5)_2SiCl_2$$

$$Si + 2\,C_6H_5Cl + HCl \xrightarrow{Cu} C_6H_5SiCl_3 + C_6H_6$$

The composition of the reaction mixture can be adjusted so that $C_6H_5SiCl_3$:$(C_6H_5)_2SiCl_2$ ratios between 1:1 and 1:3 are obtained.

- From Si-H-containing chlorosilanes and chlorobenzene

(Trichloro)phenylsilane can also be produced from trichlorosilane:

$$HSiCl_3 + C_6H_5Cl \xrightarrow{550-650\,°C} C_6H_5SiCl_3 + HCl$$

(Dichloro)methylphenylsilane can be manufactured analogously:

$$CH_3HSiCl_2 + C_6H_5Cl \xrightarrow{550-650\,°C} C_6H_5(CH_3)SiCl_2 + HCl$$

These syntheses proceed without catalysts or applied pressure.

4.3.1.3 Other Industrially Important Silanes

These include:

- (Chloro)vinyl- and (chloro)methylvinylsilanes, $CH_2{=}CH(CH_3)_nSiCl_{3-n}$, used for crosslinking silicone rubbers (Sec. 4.4.3.3 and 4.4.3.4)
- (Dichloro)methyl(3,3,3-trifluoropropyl)silanes, $CF_3CH_2CH_2(CH_3)SiCl_2$, used in the manufacture of modified poly(organosiloxanes)
- So-called organo-functional siloxanes such as trialkoxy(3-aminopropyl)silanes, $H_2N(CH_2)_3Si(OR)_3$, trialkoxy(3-mercaptopropyl)silanes, trialkoxy(3-methacryloxypropyl)silanes etc., used mainly as adhesion promoters for glass fiber-reinforced plastics (Sec. 5.2.3.1) and for the surface treatment of fillers (Sec. 5.8.4.1.3).

These silanes can be manufactured by the addition of the appropriate alkene or alkyne to hydrogen-containing silanes in the presence of platinum compounds (hydrosilation):

Manufacture of organo-functional silanes by hydrosilation:

organo-functional silanes contain such groups as

- $CH_2{=}CH-$
- $CH_2{=}C(CH_3)-COO(CH_2)_3-$
- $H_2N(CH_2)_3-$
- $H_2C-CH-CH_2-O(CH_2)_3-$ (with O bridging the first two carbons, epoxide)
- $HS(CH_2)_3-$
- $CF_3CH_2CH_2-$

$$-\overset{|}{\underset{|}{Si}}-H + H_2C{=}CH-R \xrightarrow{Pt} -\overset{|}{\underset{|}{Si}}-CH_2-CH_2-R$$

$$-\overset{|}{\underset{|}{Si}}-H + HC{\equiv}C-R \xrightarrow{Pt} -\overset{|}{\underset{|}{Si}}-CH{=}CH-R$$

In this way organosilicon compounds are easily prepared with organic groups other than methyl and phenyl e.g. (chloro)methylvinylsilanes, which are important for silicone rubber materials, and organo-functional silanes such as (trichloro)methacryloxypropylsilane; amino-functional silanes e.g. trialkoxy(3-aminopropyl)- or trialkoxy[3-(2-aminoethylamino)propyl]silanes; epoxysilanes and (3-mercaptopropyl)silanes.

4.3.2 Linear Poly(organosiloxanes)

4.3.2.1 Synthesis of Oligomeric Precursors

Manufacture of linear and cyclic dimethylsiloxane oligomers from $(CH_3)_2SiCl_2$ by:

- Hydrolysis
  - in the liquid phase
  - in the vapor phase
- Methanolysis

The most important silicone products are those based on (chloro)methylsilanes with the derivatives of (dichloro)dimethylsilanes predominant i.e. silicone elastomers and silicone oils. High molecular weight poly(dimethylsiloxanes) are synthesized from linear and cyclic dimethylsiloxanes which are produced either by hydrolysis or methanolysis of (dichloro)dimethylsilane.

Hydrolysis of (Dichloro)dimethylsilane

The complete hydrolysis of (dichloro)dimethylsilane leads to a mixture of cyclic and linear dimethylsiloxane oligomers, the latter with hydroxy endgroups [α,ω-dihydroxypoly(dimethylsiloxanes)]:

$$n\ Cl{-}Si(CH_3)_2{-}Cl + n\ H_2O \longrightarrow \left[{-}Si(CH_3)_2{-}O{-}\right]_n\ (\text{cyclic}) + 2\ n\ HCl$$

$$n\ Cl{-}Si(CH_3)_2{-}Cl + (n+1)H_2O \longrightarrow HO{-}\left[Si(CH_3)_2{-}O{-}\right]_n H + 2\ n\ HCl$$

Liquid phase hydrolysis leads to:

- Linear and cyclic siloxanes in a ratio 1:1 to 1:2
- Ca. 32% hydrochloric acid

Hydrolysis is carried out continuously either in the liquid phase with ca. 25% hydrochloric acid or in the gas phase at temperatures in excess of 100 °C. In liquid phase hydrolysis linear and cyclic siloxanes are formed in the ratio 1:1 to 1:2 together with ca. 28 to 35% aqueous hydrochloric acid. The latter can be converted into methyl chloride with methanol in a subsequent step.

Vapor phase hydrolysis:

- Disadvantages: more expensive and technically more complicated
- Advantage: gaseous hydrogen chloride formed

Vapor phase hydrolysis is technically more complicated due to greater corrosion problems but has the advantage of producing gaseous hydrogen chloride, which can be more easily utilized in methyl chloride synthesis.

Methanolysis of (Dichloro)dimethylsilane

Depending upon the reaction conditions, (dichloro)dimethylsilane methanolysis produces methyl chloride and either α,ω-dihydroxydimethylsiloxanes in yields > 98% or a mixture of cyclic poly(dimethylsiloxanes), according to:

Methanolysis:

products dependent on reaction conditions: CH_3Cl with either α,ω-dihydroxydimethylsiloxanes or a cyclic poly(dimethylsiloxane) mixture

$$n(CH_3)_2SiCl_2 + 2n\,CH_3OH \longrightarrow [(CH_3)_2SiO]_n + 2n\,CH_3Cl + n\,H_2O$$

The methyl chloride obtained is processed (e.g. to remove dimethylether) and then utilized in the direct synthesis process.

Purification of the Hydrolysis and Methanolysis Products

Linear and cyclic dimethylsiloxane oligomers are the starting materials for the manufacture of high molecular weight linear poly(dimethylsiloxanes). They can only be directly utilized if the (dichloro)dimethylsilane used in their synthesis is very (> 99.99%) pure, traces of (trichloro)methylsilane being particularly troublesome. If this proviso is not met, the oligomer mixture has to be heated above ca. 160 °C in the presence of potassium hydroxide, whereupon an equilibration reaction takes place in which Si–O–Si bonds are broken and freshly made. Cyclic dimethylsiloxanes such as hexamethyltrisiloxane "D_3", octamethyltetrasiloxane "D_4" and larger cyclic dimethylsiloxanes are formed and distilled off. The entire siloxane mixture is thus converted into cyclic materials, leaving the trifunctional impurities behind. D_4 is removed from the mixture of cyclic dimethylsiloxanes by distillation and the other cyclic dimethylsiloxanes again equilibrated.

If the $(CH_3)_2SiCl_2$ used in hydrolysis or methanolysis is insufficiently free from tri- or tetra-functional silanes, purification is necessary by way of cyclic oligomers, particularly octamethylcyclotetrasiloxane:

$$\begin{array}{ccc} (CH_3)_2Si-O-Si(CH_3)_2 & & \\ | \quad\quad\quad | & & \\ O \quad\quad\quad O & & \text{"}D_4\text{"} \\ | \quad\quad\quad | & & \\ (CH_3)_2Si-O-Si(CH_3)_2 & & \end{array}$$

4.3.2.2 Synthesis of High Molecular Weight Linear Poly(dimethylsiloxanes)

Linear poly(dimethylsiloxanes) with long chain lengths such as silicone oils and silicone rubbers are manufactured either from cyclic methylsiloxanes by alkaline or acidic ring opening polymerization or from linear α,ω-dihydroxymethylsiloxane oligomers by po-

Manufacture of long chain length linear poly(dimethylsiloxanes) by:

- Alkaline or acidic ring opening polymerization of cyclic siloxanes
- Polycondensation of linear α,ω-dihydroxysiloxane oligomers

lycondensation. The average chain length and hence the viscosity of the polymers can be regulated by the addition of chain termination agents such as trimethylsiloxy-containing siloxanes (in the simplest case hexamethyldisiloxane) or water.

Alkaline Ring Opening Polymerization

Alkaline polymerization:

with KOH as equilibrating catalyst at 160 to 180 °C producing a broad molecular weight distribution

Cyclic poly(dimethylsiloxanes), with octamethylcyclotetrasiloxane being preferred, equilibrate in the presence of very small quantities of potassium hydroxide (ca. 10 ppm) forming linear poly(dimethylsiloxane) with a broad molecular weight distribution and cyclic siloxanes of different ring sizes.

Average chain length regulated by addition of chain terminating agents such as $(CH_3)_3SiOSi(CH_3)_3$ or H_2O

The average chain length and hence the viscosity can be regulated within wide limits (5 to 10^7 mPas) by using hexamethyldisiloxane or water as chain terminating agents, whereupon trimethylsiloxy- or hydroxy-terminated linear polymers are formed. After neutralizing the reaction mixture with e.g. phosphoric acid derivatives, the cyclic siloxanes (ca. 12% of the reaction product) are distilled off.

Modification of the linear polymers during polymerization:

e.g. incorporation of vinyl groups along the chain or at chain ends

If the polymerization is carried out in the presence of cyclic siloxanes containing vinyl groups e.g. tetramethyltetravinylcyclotetrasiloxane, poly(dimethylsiloxanes) are produced with vinyl groups distributed along the chains. Polymers with vinyl endgroups are also manufactured.

Acidic Ring Opening Polymerization

Acidic polymerization:

- With strong mineral acids e.g.
  - H_2SO_4
  - Perfluoroalkylsulfonic acids
- Acid-activated clay particles

Strong mineral acids such as sulfuric or perfluoroalkylsulfonic acids or acid-activated clay particles can also be utilized as equilibration catalysts for the ring opening polymerization of octamethylcyclotetrasiloxane. Acidic equilibration, which also produces a mixture of linear polymers with a broad molecular weight distribution and cyclic siloxanes, is particularly favored for the manufacture of low viscosity and Si–H-containing poly(dimethylsiloxanes). The cyclic siloxanes can be removed by distillation after neutralizing the reaction product.

Nonequilibrating Polycondensation

The linear α,ω-dihydroxymethylsiloxane oligomers produced by hydrolysis or methanolysis of (dichloro)-dimethylsilane are converted into high molecular weight linear polymers by condensation at room temperature. Phosphonitrile chlorides are particularly effective catalysts. The reaction is terminated by neutralizing with amines.

Polycondensation of siloxane oligomers with OH end-groups:

preferred catalyst $(PNCl_2)_x$ with e.g. $x = 3$

The hydroxy endgroups of the siloxane polymers formed can be reacted with trimethylsilyl-containing substances such as (chloro)trimethylsilane or hexamethyldisilazane thereby producing polymeric siloxanes with trimethylsilyl endgroups.

Linear polysiloxanes with Si–H- or organo-functional groups on the silicon can also be manufactured by polycondensation, in some cases with subsequent equilibration e.g. with acid-activated clay catalysts.

4.3.3 Manufacture of Branched Poly(organosiloxanes)

Branched poly(organosiloxanes) such as those in e.g. silicone resins are manufactured using processes basically similar to those used for linear poly(organosiloxanes). The first reaction step is the hydrolysis of a (chloro)organosilane mixture containing (trichloro)-organosilanes as the branching component. In the manufacture of silicone resins the hydrolysis of chlorosilanes containing methyl and phenyl groups is mainly carried out in the presence of solvents (e.g. xylene) and higher alcohols. The resulting hydrolysate contains silanol and alkoxysilyl groups and is either converted directly to a resin by thermal polymerization or is used for compound resin manufacture (Sec. 4.4.4).

Manufacture of branched poly(organosiloxanes):

1st step: Hydrolysis of a (chloro)organosilane mixture, which contain (trichloro)organosilanes

2nd step: Thermal polycondensation

The hydrolysis reaction conditions can be controlled so that the siloxanes formed still contain chlorosiloxy groups. Such hydrolysis products are important starting materials for copolymers e.g. with poly(ethers) (Sec. 4.4.5).

The hydrolysis can be controlled to produce organosiloxanes containing Cl–Si- and RO–Si-groups as well as HO–Si-groups, which can not only react with each other but also e.g. with HO-functional poly(ethers)

4.4 Industrial Silicone Products

A wide range of silicone products are manufactured using the above-mentioned processes. The most important product groups are described below on the basis of their chemical constitution, their properties and their applications.

4.4.1 Silicone Oils

Silicone oils:

$$(CH_3)_3Si-O-\left[\begin{array}{c}R\\ |\\ Si-O-\\ |\\ R\end{array}\right]_n Si(CH_3)_3$$

R/R: CH_3/CH_3
CH_3/C_6H_5
C_6H_5/C_6H_5
CH_3/H

The industrially most important silicone oils are the α,ω-trimethylsilylpoly(dimethylsiloxanes). Hydrogenmethylsiloxy, methylphenylsiloxy or diphenylsiloxy groups are also commonly utilized as building blocks in these chains.

Properties of silicone oils:

- Pour points of −60 to −35 °C
- Viscosity weakly dependent upon temperature (up to 10_6 mPas)
- Thermal stability
- High specific resistance
- Low surface tension
- Odorless, tasteless, physiologically inert

Silicone oils are manufactured with viscositics in the range 5 to 10^6 mPas. They exhibit pour points between −60 and −35 °C and a weak dependence of viscosity upon temperature. In comparison with hydrocarbon oils they are much more stable to high temperatures stress and oxidative degradation. Silicone oils containing phenyl groups exhibit the highest thermal stability. These can be used for months at 250 °C in air and at temperatures up to 300 °C in closed systems. Silicone oils exhibit good electrical insulating properties e.g. a specific resistance of 10^{14} Ωcm and breakdown resistance of 14 kV/mm for poly(dimethylsiloxane) oils. Furthermore they have low surface tensions: ca. 21 mN/m and are odorless, tasteless and virtually physiologically inert.

Silicone oils used as:

- Heat transfer media
- Lubricants
- Hydraulic oils
- Transformer oils
- Brake fluids
- Paint flow improvers
- Gloss improvers
- Defoaming agents
- Mold releasing agents
- Components of skin creams and protective polishes

Silicone oils are utilized in many sectors with different requirements e.g. as heat transfer media, lubricants, hydraulic oils, brake fluids and transformer oils. In addition they are used as flow and gloss improvers in the paint industry, as foam stabilizers in poly(urethane) foams and as defoaming agents in the crude oil industry. Their water repelling (hydrophobic) properties are the basis of their utilization in cosmetics and car and furniture polishes.

4.4.2 Products Manufactured from Silicone Oils

Silicone Oil Emulsions

Aqueous silicone oil emulsions are produced from poly(dimethylsiloxane) or poly(hydrogenmethylsiloxane) oils with trimethylsiloxy groups and viscosities of ca. 1000 mPas and (preferably nonionic) emulsifiers in emulsification equipment (e.g. high pressure homogenizers). These emulsions contain between 3 and 50% by weight of oil. They are utilized in large quantities as mold releasing and deaeration agents in the manufacture of tires and for the hydrophobizing and bulking of fabrics in the textile industry.

Applications of silicone oil emulsions:

- Mold releasing agent
- Deaeration agent
- Hydrophobizing and bulking of fabrics

Silicone emulsions can also contain fine particulate solids, usually silicas. These so-called silicone compounds are extremely active antifoaming agents for aqueous media and are suitable for utilization in textile coloring and brightening systems.

- Antifoaming agent (silica-containing emulsions)

Silicone Pastes and Greases

Silicone oils can be converted into pastes by incorporating large quantities of highly dispersed silicas or calcium or lithium soaps. These products are utilized as sealants and speciality greases.

Silicone pastes and greases are produced by the incorporation of highly dispersed silica or Ca- or Li-soaps in silicone oils

4.4.3 Silicone Rubbers

Among the silicone rubbers there are different product groups which differ in their crosslinking mechanism and application areas.

4.4.3.1 Room Temperature Vulcanizable Single Component Silicone Rubbers

Crosslinking of single component silicone rubbers occurs in several stages. First a hydroxy group-terminated poly(dimethylsiloxane) reacts with a hydrolyzable (acetoxy, amido, oximo, alkoxy) group-containing silane or siloxane crosslinking agent in the presence of fillers and a condensation catalyst:

Room temperature vulcanized silicone rubber produced by mixing a hydroxy-terminated poly(dimethylsiloxane) with a crosslinking agent in the presence of fillers and a Sn-organic compound as condensation catalyst

$$HO\left[\begin{matrix}CH_3\\|\\Si-O\\|\\CH_3\end{matrix}\right]_n H + 2\,CH_3-SiX_3$$

$$\rightarrow CH_3-\underset{X}{\overset{X}{Si}}\left[O-\underset{CH_3}{\overset{CH_3}{Si}}\right]_n O-\underset{X}{\overset{X}{Si}}-CH_3 + 2\,HX$$

where X = $-O-\overset{O}{\overset{\|}{C}}-CH_3$ acetoxy

$-NH-\overset{O}{\overset{\|}{C}}-R$ amido

$-O-N=C\langle^{R}_{R}$ oximo

$-O-R$ alkoxy

Prevulcanized polymeric mixture is stable in the absence of moisture e.g. after filling in cartridges

Moisture in the air induces hardening

Such mixtures are stable in the absence of moisture e.g. when stored in closed cartridges. In the presence of moisture the reactive groups are hydrolyzed and the hydroxysilyl groups thus formed react with the formation of water (condensation), producing a crosslinked elastomeric material. The crosslinking starts on the surface and extends into the bulk with the diffusion of moisture into the polymer.

Poly(dimethylsiloxanes) terminated with hydroxy groups at both chain ends and viscosities between 5000 and 10^5 mPas are utilized in industrial single component silicone rubbers. Reinforcing fillers, preferably highly dispersed silicas, are incorporated in concentrations of 1 to 30% by weight to improve the mechanical properties together with noninteractive (extender) fillers. Up to 30% of trimethoxy-terminated poly(dimethylsiloxane), i.e. a silicone oil, is usually incorporated to obtain a sufficiently pliant silicone rubber. Organotin compounds such as dibutyltindilaurate are usually added to accelerate the condensation (hardening) process.

Most of the room temperature vulcanizable single component silicone rubbers are utilized as sealing compounds in the construction industry e.g. in so-called expanding joints, for plumbing applications and for the sealing of windows.

They are also used as adhesives for heat-resistant bonds and in the manufacture of "formed in place gaskets" particularly in the USA motor industry.

Room temperature vulcanizable single component silicone rubbers used as:

- Sealing compounds in the construction industry, sanitory sector, glass sector, automobile industry
- Adhesives for heat-resistant bonds and seals

4.4.3.2 Room Temperature Vulcanizable Two Component Silicone Rubbers

In the case of room temperature vulcanizable two-component silicone rubbers, the polymer and the crosslinking component are mixed immediately before application. Two different crosslinking systems are utilized, one based on polycondensation and the other on polyaddition.

Room temperature hardening two component silicone rubbers are cross-linked by:

- Polycondensation
- Polyaddition

In polycondensation systems a poly(dimethylsiloxane) with hydroxy end-groups and a viscosity of 10^3 to 10^5 mPas is crosslinked with a crosslinking component containing tetraalkoxysilanes, e.g. tetraethoxysilane, and a condensation catalyst, e.g. dibutyltindilaurate. With such rubber systems thick layers can be crosslinked quickly and homogeneously.

Polycondensation crosslinking systems consist of:

- A poly(dimethylsiloxane) with hydroxy endgroups
- Tetraalkoxysilanes or their condensation products as crosslinker
- Filler
- An organotin compound as condensation catalyst

In polyaddition systems the polymer component is a methylvinylsiloxy-containing poly(dimethylsiloxane) with trimethylsiloxy or dimethylvinylsiloxy endgroups and a similar viscosity to that used in condensation systems. These polymers are crosslinked by a hydrosilation reaction with a hydrogenmethylsiloxy-containing poly(dimethylsiloxane) with either trimethylsiloxy or hydrogendimethylsiloxy endgroups. Platinum compounds such as hexachloroplatinic acid are used as the catalyst, the reactions being carried out in some cases at temperatures just above room temperature.

Polyaddition crosslinking systems consist of:

- Vinyl-containing poly(methylsiloxanes)
- H–Si-containing poly(methylsiloxanes)
- Fillers
- Pt-compounds as catalysts

Room temperature vulcanizable two-component silicone rubbers are usually formulated as flowable materials. They contain reinforcing and inactive fillers and are utilized as molding compounds e.g. in the furniture or construction industries, as sealing compounds for cable connectors or electrical components and as molding compounds for dental impressions.

Flowable cold hardening two component systems used as:

- Molding compounds
- Sealants for electrical components

Filler-free materials with or without added solvent used for coating paper and plastic foils

Filler-free formulations of two-component silicone rubbers crosslinkable by condensation or addition are utilized, with or without added solvent, for the manufacture of adhesive-repellant laminating paper or foil.

4.4.3.3 Hot Vulcanizable Peroxide Crosslinked Silicone Rubbers

Hot vulcanizable silicone rubbers:

- Consist of long chain mainly vinyl group-containing poly(methylsiloxanes) and fillers
- Crosslinked with organic peroxides
- Processed with kneaders, rollers, extruders

Hot vulcanizable silicone rubbers consist mainly of very long chain, usually vinyl group-containing, poly(dimethylsiloxanes) with dimethylvinylsiloxy end-groups and viscosities of 10^6 to 10^7 mPas and highly dispersed fumed silica (10 to 35% by weight). The vinyl group content, which has a considerable influence on the crosslink density, is very low (<1 mole%). Crosslinking proceeds by a free radical mechanism with organic peroxides, e.g. dicumylperoxide or dichlorobenzoylperoxide, at temperatures between 130 and 250 °C. Peroxide crosslinking is currently the most important processing technique for silicone elastomers in the rubber and cable industries. The rubber mixtures are produced with kneaders, rollers and extruders.

4.4.3.4 Hot Vulcanizable Addition Crosslinkable Silicone Rubbers

Hot vulcanizable silicone rubber also crosslinkable by a hydrosilation reaction. Advantage: absence of organic peroxide scission products

The crosslinking of vinyl-containing silicone rubber can also be achieved by a platium-catalyzed hydrosilation reaction (addition crosslinking) with hydrogenmethylsiloxy-containing poly(siloxanes) as reaction partners. These are generally two component systems.

Addition crosslinking systems have the advantage over peroxide systems of having no decomposition products. They are therefore gaining in importance in the manufacture of computer components (interference eliminating components and seals).

Hot vulcanizable silicone rubbers can be processed by:

- Molding into molded articles in presses
- Extrusion into tubes and cables in extruders

Hot vulcanizable silicone rubbers can be extruded as tubes and cables.

Silicone rubber tubes are mainly utilized in medical technology e.g. as transfusion tubes and catheters. Silicone cables are widely utilized in applications for which thermal stability and weather and chemical resistance are required e.g. in household appliances, lighting applications and high voltage resistant connections to electrical motors and transformers.

Silicone rubber articles can also be manufactured in presses e.g. for crankshaft seals in internal combustion engines or for headlight and gearbox seals.

As a result of its physiological inertness, silicone rubber can be utilized in the manufacture of contact lenses and silicone rubber components can be inplanted into the human body e.g. as artificial heart valves.

Applications for hot vulcanized silicone rubber products:

- Silicone rubber tubing in medical technology
- Cables in the electrical industry
- Press-molded seals for e.g. the motor industry
- Inplants into the human body
- Contact lenses

4.4.3.5 Hot Vulcanizable Liquid Rubbers

A liquid rubber technology has become established since 1980 in which low viscosity two component silicone rubbers are vulcanized by addition crosslinking. Their low viscosity means that they can be processed in automatic injection molding machines, which enable very short vulcanization times to be achieved for small components (8 to 15 s). Molded articles such as the teats for babies' bottles, printed circuit boards, keyboards and protective caps for anodes are manufactured using liquid silicone rubbers.

Processing of liquid silicone rubber (LSR) with automatic injection molding machines for articles such as:

- Teats for babies' bottles
- Printed circuit boards and keyboards
- Protective caps for anodes

4.4.3.6 Silicone Rubber Properties

Silicone rubbers exhibit excellent thermal stability. The elastomeric properties hardy change after several thousand hours in hot air at 180 °C and the mechanical properties at 180 °C are almost as good as those at room temperature. Silicone rubbers thus exhibits markedly better tensile strength at 180 to 200 °C than organic rubbers, which exhibit 2 to 3 times the tensile strength of silicone rubbers at room temperature. The best mechanical values are achieved with hot vulcanized silicone rubbers with Shore A hardness values between 30 and 70, tensile strengths up to 12 N/mm^2 and tear strengths (according to ASTM 624 B) of up to 45 N/mm.

The hardness of silicone rubbers remain unchanged down to ca. – 50 °C and therefore they are usable in the unusually wide temperature range of – 50 to 180 °C (for short periods up to 300 °C). Silicone rubbers exhibit good stability to chemicals except for

Silicone rubber properties:

- Continuously stressable at temperatures between – 50 °C to 180 °C
- Mechanical properties hardly change between room temperature and 180 °C:
  - Shore hardnesses between 30 and 70 Shore A
  - tensile strengths up to 12 N/mm^2
  - tear strengths up to 45 N/mm
- Good insulation characteristics
- Electrical conduction in mixtures with carbon black
- Difficultly flammable and self-extinguishing upon the addition of platinum compounds

strong acids, strong bases and chlorine. Under normal environmental conditions they are stable for decades.

Silicone rubbers are good insulators both at room temperature and at higher temperatures and can be made electrically conducting by the incorporation of carbon black. They can be made flame resistant and self-extinguishing by adding platinum compounds. In the case of fire hardly any toxic products are formed except for carbon oxides, nonconducting layers of silicon dioxide being formed with good mechanical strength and insulation characteristics.

4.4.4 Silicone Resins

Silicone resins are branched polysiloxanes. Hardening proceeds by polycondensation at higher temperatures

Pure silicone resins are poly(organosiloxanes) with a high proportion of branched tri- or tetra-functional siloxy groups (Sec. 4.3.3).

Silicone resin coatings are

- Thermally stable
- Weather resistant
- Hydrophobic

They are hardened (polycondensed) to a highly crosslinked state by subjecting liquid silicone resins or silicone resin solutions to several hours heating (curing) at 180 to 250 °C. The thermally stable coatings formed, particularly when phenyl groups are present, do not lose their transparency, gloss or elasticity even at 200 to 250 °C. They are also hydrophobic and extremely weather resistant.

Silicone resins are utilized as paint raw materials and binders for e.g.

- Electrically insulating lacquers
- Corrosion protection lacquers, pigmented with zinc dust
- Thermally stable stoving enamels for decorative purposes
- Coil coating of metallic plates for facades
- Rendering plastics scratch resistant

Pure methyl- and methylphenylresins are utilized as raw materials for paints, binders and in building preservation. In the electrical industry they are utilized as electrically insulating lacquers (wire enamel) and for the bonding of glass filaments or mica insulating materials.

Corrosion protection-stoving enamels are produced upon pigmentation with zinc dust. These are utilized for the enameling of components which operate at high temperatures e.g. metal chimneys.

Silicone-polyester compound resins are utilized for:

- Heat resistant stoving enamels for:
  - cooking and roasting utensils
  - heating apparatus and cookers

- Coil coating of metallic plates for facades
- Rendering plastics scratch resistant

Solutions or emulsions of silicone resins are utilized in the preservation of buildings providing moisture protection and in combination with silicic acid esters as sandstone solidifiers.

Silicone resins also used for the protection of buildings for

- Hydrophobizing building materials and house facades
- Sandstone hardening in combination with silicic acid esters

4.4.5 Silicone Copolymers, Block Copolymers and Graft Copolymers

The properties of silicones can be modified by compounding with organic polymers. Industrially important are block copolymers of poly(organosiloxanes) and poly(ethers), impact resistant graft copolymers and silicone compound resins.

Silicone properties can modified by copolymerization with organic polymers

Poly(ethersiloxanes) contain a poly(methylsiloxane) polymer which may be branched and onto which poly(ether) blocks have been bonded. Its structure may be linear or comblike. The blocks are connected by Si – O – C- or Si – C-bridges. The Si – O – C-links can, for example, be produced by reacting branched (chloro)dimethylsiloxy-terminated poly(methylsiloxanes) (Sec. 4.3.3) with monohydroxy-functional poly(ethers).

Poly(ethersiloxanes) result from the reaction of e.g.:

- Branched (chloro)dimethylsiloxy-terminated poly(methylsiloxanes) with monohydroxy-functional poly(ethers)
- Poly(ethers) with H – Si-containing poly(methylsiloxanes) with an unsaturated endgroup

Si – C-linked poly(ethersiloxanes) can be obtained by reacting poly(ethers) with unsaturated end-groups with hydrogenmethylsiloxy-containing poly(methylsiloxanes).

Poly(ethersiloxanes) have surfactant properties. They are utilized as foam stabilizers in the manufacture of poly(urethane)-foams. Poly(ethersiloxanes) are also used in antifoaming formulations, as a paint additive and as auxiliary agents in the textile industry.

Poly(ethersiloxanes) are utilized as:

- Poly(urethane)-foam stabilizers
- Antifoaming agents
- Aids in the textile industry

Impact resistant grafted copolymers are obtained by grafting partly crosslinked particulate silicone rubbers onto e.g. styrene-acrylonitrile copolymers. High grade polymer alloys with improved cold toughness together with superior aging and light resistance can be produced by mixing these grafted silicone copolymers with

High grade polymer alloys manufactured by "blending" impact resistant silicone graft copolymers with thermoplastics thermoplastic resins such as poly(carbonates). These so-called "blends" are used in the manufacture of molded articles for household appliances and cars.

References for Chapter 4: Silicones

Kirk-Othmer: *Encyclopedia of Chemical Technology*, 3rd Ed., Vol. 20, pp 887–973, John Wiley & Sons Inc., New York-Chichester-Brisbane-Toronto-Singapore (1982).

Ullmanns Encyklopädie der technischen Chemie, 4th Ed., Vol. 21, pp 485–543, Verlag Chemie, Weinheim-Deerfield Beach-Basel (1982).

W. Noll: *Chemistry and Technology of Silicones*, Academic Press, New York (1968).

W. Büchner, "Novel aspects of silicone chemistry", *J. Organomet. Chem. Rev.* **9**, 409–431 (1980).

5 Inorganic Solids

5.1 Silicate Products

5.1.1 Glass

The art of glassmaking is about 5000 years old. The first marked technological improvement was the discovery of the glassmaker's blowing iron, probably in the 1st century BC. It is still one of the tools used in the manufacture of special glass products. The mass production of glass dates from the introduction of mechanical glass production and processing at the end of the last century.

5.1.1.1 Economic Importance

The worldwide glass production for 1977 was estimated to be ca. 63×10^6 t of which 26% was in the USA, 9% in the USSR and 8% in Japan. Together with the West European countries they account for 70% of the total glass production. This can be split into flat glass (e.g. window glass, mirror glass) and container glass (e.g. for bottles, containers, light bulbs, glass utensils), the ratio of ca. 4:10 varying strongly from country to country. The quantity of special glasses is insignificant compared with flat and container glass, but in value it accounts for > 10% of total glass turnover.

World glass production in 1977:

ca. 63×10^6 t comprising

- Flat glass: 30%
- Container glass: 70%

Special glasses:

- In quantity insignificant
- In value ca. 10% of total glass turnover

The 1981 USA production statistics were 3.2×10^6 t of flat glass and 12.1×10^6 t of container glass. In West Germany the 1985 production statistics were 1.1×10^6 t and 3.4×10^6 t respectively.

1981 USA glass production:

- Flat glass: 3.2×10^6 t
- Container glass: 12.1×10^6 t

5.1.1.2 General Information

Glass is an inorganic noncrystalline solidified (amorphous) molten product which in contrast with crystalline substances exhibits a diffuse X-ray diffraction dia-

Glasses are inorganic molten products, which solidify without crystallization

gram. It is notable for its local regular crystal-like arrangement of atoms and macroscopic nonperiodic liquid-like arrangement of atoms and is therefore also described as a supercooled liquid.

The industrially important silicate glasses are based on three dimensionally linked SiO_4-tetrahedra. Cations with different sizes and charges can be accommodated in the interstitial spaces.

Silicate glasses are by far the most widely used industrial glasses. They consist of a three dimensional network of oxygen-bridged SiO_4-tetrahedra and can be almost infinitely modified by incorporating the oxides of other elements (e.g. B_2O_3, P_2O_5, TiO_2) or cations (e.g. Na^+, K^+, Ca^{2+}, La^{3+}) into the network or the remaining interstitial spaces. This is the basis of the wide application spectrum and multiplicity of silicate glasses.

5.1.1.3 Glass Compositions

Quartz glass consists of SiO_2 and is notable for its:

- Excellent dielectric properties
- High chemical resistance
- Low thermal expansion coefficient
- High thermal stability
- Exceptionally high transparency to UV-light

Quartz ware contains included air bubbles

Quartz glass is by far the most important industrially utilized single component glass and is the only single component silicate glass. It is notable for its very good dielectric and chemical properties, low thermal expansion coefficient, high thermal stability and exceptionally high transparency to UV-light. Its high softening point and high working temperature (> 2000 °C) are disadvantageous. Thus quartz products obtained by sintering, which contain included air bubbles and are therefore opaque, are frequently used.

All other (silicate) glasses are multicomponent glasses, whose compositions can vary over a wide range depending upon the particular application (see Table 5.1.1-1).

Table 5.1.1-1. Typical Compositions of Industrial Glasses (average values in % by weight)*.

	Container glass		Flat glass	Laboratory glass	Consumer glass	
	white	green	(float glass)	(Pyrex®)	lead crystal glass	pressed glass
SiO_2	73.0	72.0	72.8	81.0	60.0	75.5
Al_2O_3	1.5	2.6	0.7	2.0	0.02	0.3
CaO	11.3	11.4	8.7	0.3	–	6.5
PbO	–	–	–	–	24.0	–
MgO	0.1	0.1	3.6	0.2	–	–
Na_2O	13.0	12.5	13.7	4.5	1.0	14.8
K_2O	1.0	0.5	0.2	0.1	14.9	2.0
B_2O_3	–	–	–	11.4	–	–
Fe_2O_3	0.04	0.5	0.1	0.15	0.02	0.01
Cr_2O_3	–	0.2	–	–	–	–

* The glasses frequently contain small quantities of SO_3, As_2O_3, TiO_2 etc.

The industrially most important glasses are the so-called soda-lime glasses, which account for ca. 90% of all the glasses produced and are utilized both for flat glass and container glass products. They contain large quantities of alkali and alkaline earth oxides, Na_2O, K_2O and CaO (fluxes) and are chemically more stable than the alkali silicates alone. The addition of small quantities of aluminum oxide improves their thermal stability. Magnesium oxide reduces their crystallization tendency and iron and chromium oxides are added to color the glasses.

Multicomponent glasses contain network modifiers, which enable them to melt and to be worked at lower temperatures than quartz glass

Aluminum oxide improves the thermal stability, boron oxide the durability

Glass for chemical laboratory apparatus differs in its high boron oxide and low alkali oxide content, which endow it with considerably better chemical stability and a reduced thermal expansion coefficient. Both boron oxide and the alkali(ne earth) oxides reduce the melting temperature of the glass.

Lead glasses are easily melted and are very easy to work. This together with their high refractive indices make them suitable for handmade fine glass articles and optical glasses. Their strong absorption of high energy radiation makes them suitable for utilization in cathode ray tubes and lead windows (for nuclear plants).

Lead glasses are suitable for fine glass articles (crystal glass), optical glasses and cathode ray tubes

Optical glasses are very high quality homogeneous glasses with well-defined refractive indices, dispersion and absorption properties. There are two main classes: flint glass (lead alkali silicates) and crown glass (alkali silicates). Particular optical properties can be achieved by the addition of boric acid, phosphoric acid and fluorides. Large quantities of rare earth oxides (e.g. La_2O_3, Y_2O_3, Nd_2O_3), thorium(IV) oxide, niobium(V) oxide and tantalum(V) oxide are also often used.

Glass-ceramics are a special case. They consist predominantly of crystalline phases with homogeneous fine particulate grains. In their manufacture preformed glasses are subjected to heat treatment to form crystal nuclei from which crystals are grown mainly by increasing the temperature further. The number and size of the crystals formed are mainly influenced by the nucleus-forming phase and by the glass composition. Precious metals, TiO_2, ZrO_2, sulfides and phosphates in quantities up to 3% are utilized as heterogeneous nuclei formers. Transparent glass-ceramics have very

"Glass-ceramics" in which fine crystalline phases predominate are formed by heating to the nucleation temperature and then further heating to the nucleus growth temperature

Nucleus formation with heterogeneous nuclei:

precious metals, TiO_2, ZrO_2, sulfides, phosphates

fine crystallites (with crystal size well below the wavelength of visible light e.g. 50 nm) and crystal and amorphous phases with virtually identical refractive indices.

Advantages of glass-ceramics:

- Higher thermal stability
- Very good thermal shock resistance

Glass-ceramic materials have a higher thermal stability than glasses with identical compositions and are in principle, unlike ceramic materials, nonporous. The industrially most important (e.g. for utensils and surfaces of modern domestic cooking stoves) are the lithium and magnesium aluminum silicate glass-ceramics. They are notable for their extreme thermal shock resistance, due to the very low or even negative thermal expansion coefficients of the crystalline phase (e.g. cordierite $2MgO \cdot 2Al_2O_3 \cdot 5SiO_2$, betaspodumene $Li_2O \cdot Al_2O_3 \cdot 4SiO_2$). Other glass-ceramic articles are ferroelectric, mechanically workable and photosensitive, selectively etchable glass-ceramics.

Glass-ceramic applications:

- Utensils
- Surfaces of domestic cooking stoves
- Special articles (ferroelectric, mechanically workable, selectively etchable)

5.1.1.4 Glass Manufacture

5.1.1.4.1 Glass Raw Materials

Raw materials:

- Sand
- Lime, dolomite
- Sodium carbonate
- Feldspars and other aluminum silicates
- Boric acid and boron minerals
- Waste glass from own production

The main component of all silica glasses is fine particulate sand. Another acidic glass oxide is boron oxide which is preferably incorporated into the mixture as boric acid, borax or colemanite (Sec. 3.1.1.3.7). The most important alkaline glass oxide raw materials are sodium carbonate, potash, lime and dolomite ($CaCO_3 \cdot MgCO_3$). Sodium sulfate is used as a clarifying agent. Aluminum oxide-containing glasses are manufactured from potash, lime and sodium feldspar and optionally other naturally occurring or synthetically manufactured aluminum silicates. Other cations required for glass manufacture are incorporated in the form of their oxides, carbonates, silicates or phosphates (e.g. ZnO, TiO_2, PbO_2, Li_2CO_3, $BaCO_3$, $ZrSiO_4$, $AlPO_4$).

Waste glass represents another important raw material which is particularly cost effective, saving both melting energy and raw materials. The amount of waste glass produced in the glass industry varies widely. For container glass it amounts to 10 to 15%, for flat glass 20 to 30% and for light bulb glass 50 to 70%. Recycled glass (i.e. not arising from the plant itself) is gaining in importance, but can only be utilized without problems in the manufacture of flat glass. The contribution of recycled glass to West European flat glass manufacture had already reached 10 to 15% (nationally up to 50%) in 1983, corresponding to ca. 2.5×10^6 t of glass.

Glass recycling gaining in importance

Glass raw materials together with clarifying and coloring components are weighed out in so-called mixing houses and homogenized in ribbon blenders. The quality of the resulting glass is strongly influenced by the precision of the weighing and the effectiveness of the mixing.

Glass quality depends upon the care taken with the raw material mixing

3 to 4% of water are often added to the mixture to reduce the amount of dust formed and to improve the homogenization. This can also be achieved by briquetting or pelletting the mixture.

The presence of small quantities of the oxides of nickel, vanadium, copper, manganese, chromium and particularly iron in the raw materials leads to undesirable discoloration of the glass, depending upon the redox conditions in the glass melt. In the case of particularly stringent specifications, natural raw materials have to be replaced by purified starting materials. The maximum permissible Fe_2O_3-content of UV-transparent glasses is, for example, 0.004%, that for spectacle glasses 0.020% and that for plate glass 0.10%.

Undesirable discoloration due to high iron contents in the raw materials

Maximum iron oxide content permitted for plate glass: 0.10% Fe_2O_3

Chemical decoloration is utilized for Fe_2O_3 contents below 0.1%. In this process oxidizing agents which yield oxygen at high temperatures [e.g. KNO_3, CeO_2, MnO_2 (glassmakers soap)] are added to the glass mixture. These oxidize iron(II) to its trivalent, less color intensive, state. At higher iron oxide contents the yellow-green color of the glass can be *physically* compensated by adding raw materials producing complementary colors e.g. CoO and Nb_2O_3.

Chemical decoloration (Fe^{2+} oxidized to Fe^{3+}) for Fe_2O_3 contents $< 0.1\%$

Physical decoloration (compensation with colored additives) for Fe_2O_3 contents $> 0.1\%$.

5.1.1.4.2 Melting Process

The melting process can be subdivided into the following phases: melting, clarification, homogenization and settling (temperature reduction before the subsequent molding). According to current ideas the melting process is initiated by silicate formation at temperatures of 800 to 900 °C. Eutectic intermediates such as $Na_2Ca(CO_3)_2$ and $(Na_2O)_{0.5-2} \cdot (CaO)_{1-2} \cdot (SiO_2)_3$ are formed in solid state and liquid phase reactions in equilibrium with unreacted silica, carbon dioxide and

Melting process can be divided into:

- Melting
- Clarification
- Homogenization
- Temperature adjustment before molding

water vapor. Other carbonates and other high temperature melting ingredients e.g. feldspars react in a similar way.

In this reaction the mixture sinters together and is then converted into the molten state. The melt obtained at temperatures of 1200 to 1650 °C (depending upon the composition of the mixture) is not yet homogeneous, furthermore it contains many gas bubbles due to moisture and air included in the raw materials mixture, as well as gases formed during the decomposition of sulfates, nitrates, carbonates and hydrates (SO_2, CO_2, H_2O).

Clarifying agents rapidly liberate gases at high temperatures. Sodium sulfate mainly used

These bubbles are removed by clarifying the glass melt by adding clarifying agents. These increase the decomposition pressure of the gases in the glass, the liberated gases rising up as bubbles in the melt. This effect can be aided by reducing the melt viscosity by increasing the melt temperature.

Sodium sulfate is very widely used for clarifying mass-produced glass, SO_3 being liberated. Sodium or potassium nitrate in combination with arsenic(V) oxide (higher melting glasses, 1450 to 1500 °C) or antimony(III) oxide (lower melting glasses, 1300 to 1400 °C) are also used.

Glass melt homogenized by blowing air through the melt or mechanical mixing

Inhomogeneities in the glass mixture, melt separation, evaporation and furnace wall corrosion give rise to local fluctuations in the glass composition in the melt. Therefore the melt has to be homogenized. This is achieved, for example, by blowing air or steam into the bottom of the melt vessel or by mechanical mixing of the glass melt.

The glass melt is then cooled to the temperature at which further processing to the final product takes place.

5.1.1.4.3 Melting Furnaces

Various melting furnaces are utilized in the glass industry. Large tanks are mostly utilized for the manufacture of flat glass or container glass. Smaller furnaces are used if the composition of the glasses in production frequently vary (e.g. optical glasses) e.g. multiple port furnaces or day tank furnaces. Unit melters are intermediate in size.

Tank furnaces. Mass produced glass is manufactured in continuously operating tank furnaces with capacities up to 900 t/d for flat glass and 300 t/d for container glass. An elongated usually rectangular tank (dimensions e.g. 10 m × 40 m, depth 0.5 to 1.5 m) is utilized as the melting vessel. The mixture is uniformly introduced at one end of the tank and the glass run off and fed into molds at the other. The temperatures required for melting and clarifying the glass differ from that of the final glass melt. This results in the tank being divided into melting and conditioning zones. The melting, clarification and homogenization take place in the melting zone. Conditioning of the glass is carried out at a lower temperature in the conditioning zone. For container glass manufacture the two parts of the tank are separated by a cooled wall with an opening, the so-called throat. Flat glass tanks are built without a dividing wall and the lower temperature of the processing tank achieved by opening the furnace cover. The melting zone can also be divided off by a float made of firebrick.

Melting furnaces:

- Tank furnace, divided into
  - a melting zone
  - a conditioning zone

The furnaces are directly heated by the combustion of oil-air or gas-air mixtures with the flames burning laterally over the tanks. The hot combustion gases are drawn off from the opposite side and led through prewarming chambers (regenerators) fitted with a grid of refractory bricks. In these either the combustion air (in oil heated furnaces) or the combustion air and the gas are preheated. The flame direction is reversed at 15 to 30 minutes intervals, so that the combustion gases heat up the regenerators on either side of the furnace alternately. Supplementary electrical heating (electrical boosting) by molybdenum electrodes dipped in the electrically conducting melt can markedly reduce heat and evaporation losses, the latter particularly in the case of lead- and boron-containing glasses.

Regenerators are used to utilize the combustion gas heat to preheat the gases before combustion

Supplementary electrical heating (electrical boosting) in the melting zone improves the furnace heating and reduces the evaporation losses

Important parameters for tank furnaces are the specific melting power (ca. 2 t/m^{2}d) and the specific heat consumption (ca. 4500 to 7000 kJ/kg glass). The latter is strongly dependent upon the glass composition and the melting temperature, since ca. 2/3 of the energy is required for maintaining the (high) temperatures

Tank furnace parameters:

- Specific melting power
- Specific heat consumption

required. The life span of a tank furnace depends upon the quality of the refractory materials, particularly the tank bricks, and is ca. 6 to 8 years.

Other smelting furnaces:

- Multiple port furnaces

Multiple port furnaces consist of ca. 6 to 14 ports constructed of burnt clay in which ca. 150 to 500 kg of glass mixtures are melted at temperatures below 1480 °C. As a result of their operating discontinuously, they require very much more energy than tank furnaces (ca. 14 500 to 30 000 kJ/kg glass) and are therefore only utilized for the smelting of high quality special glasses such as crystal glass, colored glass or optical glasses.

- Day tank furnaces

Day tank furnaces with melting tanks for 1 to 5t glass are fired with gas or oil and are named after the work cycle length which for loading, melting, clarification and discharge is 24 hours.

- Unit melters

Unit melters. Short tanks (2 to 3 m wide, 10 to 12 m long) are utilized for the continuous manufacture of glass in quantities up to a maximum of 100 t, when a rapid change from one glass sort to another is required.

- Electric tank furnaces

Electric tank furnaces with melting powers of 4 t/m^2d are, like multiple port furnaces, uneconomic for the mass production of glass. They are useful for the manufacture of glasses with high quantities of volatile components, since the evaporation losses on the relatively cold surfaces are comparatively small. The electricity consumption of furnaces with molybdenum electrodes operating at 80 to 120 V AC and current densities of 2 to 3 A/cm^2 is 0.8 to 1.2 kWh/kg glass.

Large quantities of refractory materials are required for the construction of melting furnaces. The surfaces of the tank furnaces in contact with the glass melt are lined with melt cast or isostatically pressed bricks (e.g. made of zirconia) and the cover over the glass tanks manufactured from mullite or silica materials.

5.1.1.5 Molding

Manual molding only carried out for complicated special articles, otherwise large automated plants used

Molding is carried out in fully automated plants with high throughputs. Manual molding is no longer of any importance, except for complicated special articles and esthetic products. It is carried out at temperatures between 800 and 1400 °C depending upon the composition of the glass. Typical processing procedures require viscosities between 10^2 and 10^5 Pas, the viscosity at the softening point being ca. $10^{7.6}$ Pas. If the temperature difference between these viscosity limits is large, then one refers to a "long" glass and if narrow to a "short" glass. Long glasses are easier to process than short glasses.

Flat glass manufacture. This is done by rolling the molten glass between rollers to the desired width and thickness. The glass strip is cooled to room temperature in an adjoining belt furnace. In the Fourcault process, which is preferred for thin glass (thickness < 1.5 mm), a glass strip is drawn through a burnt clay slit with pairs of rollers perpendicularly upwards in a shaft. In the Pittsburgh process the glass strip is also pulled perpendicularly upwards, but a block of refractory material is utilized at the pulling point under the glass smelt surface to stabilize the streaming conditions. This enables higher pulling speeds to be achieved than with the slit, but places particularly high demands on the homogeneity and constancy of the melt viscosity.

Manufacture of flat glass:

rolling of the glass smelt between rollers and subsequently cooling in a belt furnace

- Fourcault process
- Pittsburgh process

The best flat glass quality (mirror quality without grinding and polishing) is achieved at particularly high throughputs using Pilkington's float-glass process. The glass flows under a protective gas over a tank filled with molten tin (dimensions: several meters wide, up to 60 m long) and assumes the surface of the tin bath. After the float-glass chamber, the glass strip is cooled conventionally. Flat glass manufacture is currently dominated by the float-glass process.

- Float-glass process:

  glass melt solidifies on a surface of molten tin

Container glass manufacture. Pressure and blowing processes are utilized in the automated manufacture of container glass. Pressure processes are mainly used for the manufacture of wide-necked hollow glass objects and consist of placing precise quantities of smelt in hot steel compression molds and molding under pressure. Blowing processes also consist of placing precise quantities of smelt in a mold, but then deforming it by suction or blowing with compressed air.

Manufacture of container glass:

- Pressure process
- Blowing processes

Aftertreatment may be necessary depending upon the particular glass product required e.g. grinding, polishing, etching, frosting, application of a reflective coating. Glass objects can be made stronger by sub-

Strength of glass objects can be increased by prestressing the glass surface (quenching, ion exchange)

jecting the glass surface to compression strain. This is done by heating the glass almost to its softening temperature and then rapidly quenching it with a blast of air or by submersion in a liquid (thermal hardening). In chemical hardening the compression strain is produced by exchanging sodium ions for the larger potassium ions e.g. by dipping the glass objects in potassium nitrate salt melts.

5.1.1.6 Glass Properties and Applications

Important glass properties influenced by the glass composition are:

The glass properties most important for the melting process, molding and its applications are largely determined by its composition (Sec. 5.1.1.3).

- Thermal expansion coefficient

The thermal expansion coefficient for quartz glass is 5×10^{-7}/K, that for (thermally stable) borosilicate glass ca. 30×10^{-7}/K and that for high lead content glasses and window glass 80 to 90×10^{-7}/K. In composite systems, such as used in the electronics industry for fusing (soldering) glasses to metallic conductors or as casing components, the thermal expansion coefficients of the glass and metal have to be precisely matched.

- Density

The density of industrial glasses is dependent upon the atomic weights of the elements in the glass components. It varies from 2.2 (quartz glass) to 2.5 g/cm^3 (mirror glass) and in individual cases (e.g. lead crystal glass) is much higher.

- Electrical resistance
- Dielectric constant

Most glasses are excellent insulators and dielectrics. Their electrical resistance decreases strongly in damp atmospheres and at high temperatures, particularly in the case of alkali-rich glasses, its being between 10^{11} and 10^{19} Ωcm at room temperature. The dielectric constants vary between 5 and 7 for multicomponent glasses and 3.5 and 4 for silica glasses.

- Chemical resistance

The stability of glass, particularly bottle and container glass, to almost all chemicals and solvents is of paramount importance. However, glass is attacked by aqueous acids and alkalis their degrading it to (porous) silica glass by breaking up the network to form silicate anions:

acidic attack:

$$-\overset{|}{\underset{|}{Si}}-O-Me + HX \rightarrow -\overset{|}{\underset{|}{Si}}-O-H + MeX$$

alkaline attack:

$$-\overset{|}{\underset{|}{Si}}-O-\overset{|}{\underset{|}{Si}}- + MeOH \rightarrow -\overset{|}{\underset{|}{Si}}-O-H + Me-O-\overset{|}{\underset{|}{Si}}-$$

Me: alkali metal cations X: monovalent anion

This is the cause of the weathering of glass. Glasses containing up to 15 % by weight of CaO, MgO, Al_2O_3, B_2O_3 and particularly ZrO_2 are particularly stable to alkalis. Borosilicate and aluminosilicate glasses exhibit the highest stability to acids. Hydrofluoric acid dissolves silicate glasses very rapidly with the formation of hexafluorosilicic acid (H_2SiF_6).

- Transparency to visible light

An important property of most industrial glasses is their transparency to visible light (particularly important for window glass, mirror glass, light fittings, solar collectors and optical glasses for spectacles, lenses, prisms etc.). Colored glasses contain transition metal cations, lanthanide cations and actinide cations (e.g. Cr^{3+}, Co^{3+}, Ni^{2+}, Nd^{3+}, Pr^{3+}, U^{6+}) or colloidal deposits of precious metals (e.g. Cu, Ag, Au, Pt).

The flexural and tensile strengths of glass influenced by the surface finish

In contrast with other properties, the flexural and tensile strength of glass are less dependent upon glass composition and much more dependent on the surface finish (amount of indentations and damage). Its tensile strength values are typically between 50 and 140 N/mm^2, but can be increased for extremely light and thin-walled glass containers by special posttreatment steps (ca. 10^4 N/mm^2 is theoretically achievable). The compressive strength of glass is much higher being 600 to 1200 N/mm^2 with 2300 N/mm^2 for silica glass.

5.1.1.7 Sol-Gel Process

In the sol-gel process glass formation occurs in alcohol by hydrolysis and polycondensation of dissolved highly reactive starting materials (e.g. tetraethylorthosilicate, sodium methoxide, boric acid)

Whereas in the conventional manufacture of glass high melting point mainly oxidic starting materials are melted together at high temperatures to glass, in the sol-gel process glass formation occurs by the hydrolysis and polycondensation of reactive (organometallic)

starting materials in alcoholic solution. Thus tetraethylorthosilicate (TEOS) at 60 to 80 °C in aqueous ethanol forms a polysiloxane gel with increasing cross-linking as more and more alcohol and water are split off:

$$-\overset{|}{\underset{|}{Si}}-OR + HO-\overset{|}{\underset{|}{Si}}- \rightarrow -\overset{|}{\underset{|}{Si}}-O-\overset{|}{\underset{|}{Si}}- + ROH$$

$$-\overset{|}{\underset{|}{Si}}-OH + HO-\overset{|}{\underset{|}{Si}}- \rightarrow -\overset{|}{\underset{|}{Si}}-O-\overset{|}{\underset{|}{Si}}- + H_2O$$

(R: organic group)

Polysiloxane gel is condensed to a glass by heating to 1000 °C

which condenses to silica glass upon increasing the temperature to ca. 1000 °C, optionally under pressure. The reactions with other metallic alkoxides [e.g. NaOR, LiOR, $Ti(OR)_4$] and metallic hydroxides with partially hydrolyzed TEOS proceed similarly.

Using the sol-gel methods, glasses with higher purity and homogeneity can be manufactured as well as glasses inaccessible using conventional techniques

The properties of the very homogeneous and pure glasses obtained in this way are comparable with those of multiply melted glasses. The sol-gel process is in some cases the only way of manufacturing glasses with particular compositions without phase separation (e.g. binary silicate glasses with high ZrO_2 or La_2O_3 contents).

For technical and economic reasons the sol-gel process is, however, only suitable for the manufacture of high grade coatings on glass, ceramic or metallic surfaces and for the manufacture of monolithic glass blocks. Reflecting and antireflecting layers based on SiO_2-TiO_2 glass are manufactured in large quantities in this way by dipcoating and subsequent heat treatment.

References for Section 5.1.1: Glass

General Information:

Kirk-Othmer: *Encyclopedia of Chemical Technology,* 3rd Ed., Vol. 11, pp 807–880, Glass; pp 881–892, Glass-Ceramics, John Wiley & Sons Inc., New York-Chichester-Brisbane-Toronto (1980).

Winnacker-Küchler, *Chemische Technologie.* 4th Ed., Vol. 3, pp 98–158, Carl Hansen Verlag, München (1983).

H. Scholze: "Glas, Natur, Struktur und Eigenschaften". (Glass, nature, structure and properties.) 2nd Ed., Springer-Verlag, Berlin (1977).

A. Paul: "Chemistry of Glasses". Chapman & Hall, London (1982).

H. Rawson: "The Properties and Applications of Glass". Elsevier, Amsterdam (1980).

N. P. Bansal and R. H. Doremus: "Handbook of Glass Properties", Academic Press, New York (1986).

P. J. Doyle: "Glass making today", Portaillis Press, Redhill, England (1979).
E. B. Shand: "Glass Engineering Handbook", 2nd Ed., McGraw Hill, New York (1978).
J. I. Duffy, Ed., "Glass Technology: Developments since 1978", Noyes, Park Ridge, *Chem. Tech. Rev. Ser.* **No. 184** (1981).
P. W. McMillan: "Glass-Ceramics", 2nd Ed., Academic Press, New York (1979).
H. Dieslich: "New Routes to Multicomponent Oxide Glasses", *Angew. Chem. International Ed.* **10,** 373 (1971).
L. C. Klein: "Sol-Gel Glass Technology – A Review", *The Glass Industry* **14** (1981).

Specific Information:

Commercial Data:

The Glass Industry Directory. 1979 Issue, Glass Industry, New York (1979).
Chemical Economics Handbook, Stanford Research Institute, Menlo Park, California, USA: "Data Summary – Glass" (Aug. 1986).
Chemical Economics Handbook, Stanford Research Institute, Menlo Park, California, USA: "Glass – Value of Shipments" (July 1985).
Industrial Statistics Yearbook 1982, Vol. II, pp 511 – 516, United Nations, New York (1985).

5.1.2 Alkali Silicates

5.1.2.1 General Information and Economic Importance

The industrially important alkali silicates (sodium and potassium silicates) are characterized either by the SiO_2 : alkali oxide weight ratio or the SiO_2 : alkali oxide molar ratio. The latter is obtained by multiplying the weight ratio by 1.032 and 1.468 for sodium silicate and potassium silicate respectively.

Alkali silicates characterized by the weight or molar ratios of SiO_2 : alkali oxide

The alkali silicates can be divided into sodium and potassium silicates with SiO_2/alkali oxide ≥ 2 (products with 2 to 4 moles SiO_2/mole alkali oxide are industrially utilized) and sodium silicates with $SiO_2/Na_2O \leq 1$.

Classification of alkali silicates:

- SiO_2/alkali oxide ≥ 2
  - anhydrous solid or crushed silicate glasses
  - aqueous solutions (waterglasses, liquid glasses)
- $SiO_2/Na_2O \leq 1$
  - anhydrous products
  - products containing crystal water

Anhydrous products termed solid or crushed silicate glasses and the aqueous solutions made from them so-called water glasses or liquid glasses belong to the first group. The second group comprises anhydrous and crystal water-containing sodium silicates. Industrially important products contain 1 mole of SiO_2 to 1 to 1.5 moles of Na_2O.

In 1983, the USA, Western Europe and Japan each produced ca. 0.71×10^6 t of solid silicate glasses.

1983 production of solid silicate glasses in 10^6 t:

- USA: 0.71
- Western Europe: 0.71
- Japan: 0.72

5.1.2.2 Manufacture of Alkali Silicates

Anhydrous alkali silicates with SiO_2/alkali oxide ≥ 2 (solid silicate glasses) are manufactured by reacting pure quartz sand, which is clay-free and has a low iron

Manufacture of:

- Anhydrous alkali silicates with SiO_2/alkali oxide ≥ 2 from quartz sand and alkali carbonates or hydroxides

content, with alkali carbonates or hydroxides at 1300 to 1500 °C in tank furnaces lined with refractory bricks or in rotary furnaces:

$$\text{e.g. } Na_2CO_3 + 4\,SiO_2 \rightarrow Na_2O \cdot 4\,SiO_2 + CO_2\,.$$

The alkali silicate melt flows into casting molds in which it solidifies to transparent more or less colored lumps.

- Water glass solutions with SiO_2/alkali oxide $\geqslant 2$ by:
  - solution of solid silicate glass in water at ca. 5 bar and 150 °C

Water glass solutions with SiO_2/alkali oxide ≥ 2 are obtained by dissolving these materials in water under pressure (ca. 5 bar) at ca. 150 °C. The viscosity of these colorless alkaline transparent water glass solutions increases with alkali silicate concentration and at constant concentration with increasing SiO_2 : alkali oxide ratio. Water glasses with low SiO_2: alkali oxide ratios can be produced by adding alkali hydroxide during the dissolution process.

  - reaction of sand with aqueous NaOH under pressure at elevated temperatures

Aqueous solutions of alkali silicates can be produced with less expenditure of energy by reacting sand with sodium hydroxide e.g. in agitator autoclaves under pressure.

Alkali silicate powders containing ca. 20% water are produced by spray or drum drying waterglass solutions. They dissolve upon warming with a little water to water glass solutions.

- Alkali silicates with SiO_2/alkali oxide ≤ 1 i.e. Na_2SiO_3 by:
  - melting sand with sodium carbonate
  - reacting sand with solid sodium carbonate

The only industrially important anhydrous product in the group of *alkali silicates* with SiO_2/alkali oxide ≤ 1 is sodium metasilicate, Na_2SiO_3. It is produced by melting sand with sodium carbonate in a 1:1 molar ratio or reacting sand with solid sodium carbonate in rotary or drum furnaces.

K_2SiO_3, Na_4SiO_4 and K_4SiO_4 unimportant

Potassium metasilicate, anhydrous sodium orthosilicate and potassium orthosilicate are unimportant.

- Crystal water-containing sodium silicates i.e. $Na_2H_2SiO_4 \cdot xH_2O$ with x = 4, 5, 7, 8 by mixing:
  - water glasses with NaOH
  - Na_2SiO_3 with H_2O

Crystal water-containing sodium silicates i.e. hydrates of orthosilicic acid with the general formula $Na_2H_2SiO_4 \cdot xH_2O$ with x = 4, 5, 7, 8 are industrially important products in the SiO_2/Na_2O group with crystal water. $Na_3HSiO_4 \cdot 5H_2O$ is another industrial product.

These crystal water-containing sodium silicates are manufactured by either mixing water glasses with sodium hydroxide or sodium metasilicate with water to the required water content and bringing about crystallization with seed crystals.

5.1.2.3 Applications

Alkali silicates are utilized in large quantities in the manufacture of detergents and cleaners and as starting materials in the manufacture of silica fillers for rubbers and plastics and cracking catalysts for the petrochemical industry. They are also used in the synthesis of zeolites, silica gels and silica sols; in adhesives; as binders in water-based paints; in the flotation of ores; in the ceramic, cement and foundry industries; as flocculating agents in water purification and for the chemical consolidation of ground.

Alkali silicates utilized in:

- Detergent and cleaner manufacture
- Silicate filler manufacture
- Catalyst manufacture
- Zeolite manufacture
- Silica sols
- Silica gels
- Adhesives
- Ore flotation
- Water treatment
- Water-based paints
- Enamel, ceramic, cement and foundry industries
- Chemical consolidation of ground

Potassium water glass solutions are mainly utilized in the production of coatings for welding electrodes, as binders for luminescent pigments used in cathode ray tubes, for impregnating walls and as a binder for plasters.

Potassium water glass used:

- For welding electrode-coatings
- As a binder for cathode ray tube luminescent pigments
- For wall impregnation
- As a binder for plasters

References for Chapter 5.1.2: Alkali Silicates

Kirk-Othmer: *Encyclopedia of Chemical Technology,* 3rd Ed., Vol. 20, pp 855–880, John Wiley & Sons, Inc., New York-Chichester-Brisbane-Toronto (1982).

G. Lagaly, D. Klose, W. Tufar and E. Heinerth: *Ullmanns Encyklopädie der technischen Chemie*, 4th Ed., Vol. 21, pp 409–416 (1982).

M. A. Syndicus-Pessier, R. A. Davenport, O. Kamatari and G. Hemy, "Silicates and Silicas", *Chemical Economics Handbook,* Stanford Research Institute, Menlo Park, California, USA (1981).

L. S. Dent Glasser, "Sodium silicates", *Chem. Brit.* **18,** pp 33–39 (1982).

G. C. Schweiker, "Sodium Silicates and Sodium Aluminosilicates", *J. Amer. Oil Chem. Soc.* **55,** pp 36–40 (1978).

5.1.3 Zeolites

5.1.3.1 Economic Importance

The consumption of zeolites in three important sectors:

- Adsorption applications (Table 5.1.3-1)
- Detergents (Table 5.1.3-2)
- Catalysts (Table 5.1.3-3)

Main application sectors:

- Adsorption
- Detergents
- Catalysts

determine their production capacity and production. The statistics in the following three tables are estimates, the figures for zeolites used in detergents being fairly uncertain, since it has not been possible to ascertain whether the expansion plans announced by various firms have been realized or not.

Table 5.1.3-1. Estimated 1987 Zeolite Capacities for Adsorption Applications in 10^3 t/a.

World	USA	W. Europe	W. Germany	Japan	Various
49	22.4	20.6	11.6	3.8	2

The world consumption of zeolites in the three application sectors: industrial plant, double glazing and as a water-binding agent in poly(urethane) systems in 1986 was ca. 35×10^3 t of which ca. 17×10^3 t was consumed in North America.

Table 5.1.3-2. Estimated 1987 Zeolite Capacities for Detergent Applications in 10^3 t/a.

USA	W. Europe	W. Germany	Japan
58.5	180	150	260

For catalyst applications, zeolites are mainly utilized in inert matrices. The statistics in Table 5.1.3-3 relate to 100% zeolite.

Table 5.1.3-3. Estimated 1987 Zeolite Capacities for Catalyst Applications in 10^3 t/a.

USA	Europe	Japan	Other countries
23.3	5.5	1.4	9.3

5.1.3.2 Zeolite Types

Zeolites are aluminosilicates with Si/Al ratios between 1 and ∞. 150 types are known, of which 40 occur naturally

Zeolites are in principle aluminosilicates, although in recent years aluminophosphates and silicoaluminophosphates (SAPO) have been prepared with a zeolite structure. The latter have not yet achieved industrial importance. Aluminosilicate zeolites can be represented by the following general formula:

$$[(M^+, M^{2+}_{0,5})AlO_2]_x \cdot [SiO_2]_y \cdot [H_2O]_z$$

M^+ : e.g. alkali metal cations
M^{2+}: e.g. alkaline earth cations

The zeolites (currently 150 types of which 40 are naturally occurring) are chemically differentiated by the Si/Al ratio in their anionic framework. This varies between 1 (in zeolite A) and ∞ in silicalite (an aluminum-free crystalline silica-modification). The industrially most important synthetic zeolites apart from zeolite A are the synthetic zeolites X and Y with Si/Al ratios of 1.5 to 3. It is ca. 5 in synthetic mordenites and > 10 in the ZSM 5 type zeolites. The acid resistance and thermal stability of zeolites increase as the Si/Al ratio increases.

Zeolites occur in many structures. The basic units are always SiO_4- and AlO_4-tetrahedra, linked to one another by shared oxygen atoms. Figures 5.1.3-2 and 5.1.3-3 show the structures of zeolites A and X(Y), as examples of zeolite structures. These zeolites are built up of cubo-octahedra ("β-cages") with SiO_4- and AlO_4-tetrahedra at their corners, as shown in Fig. 5.1.3-1. The zeolite A structure is produced by linking the quadratic surfaces of the β-cages by cubes. In

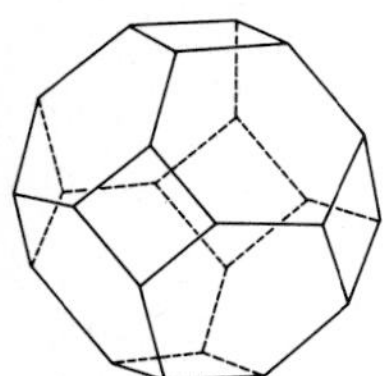

Fig. 5.1.3-1. Cubo-octahedra – "β-cages".

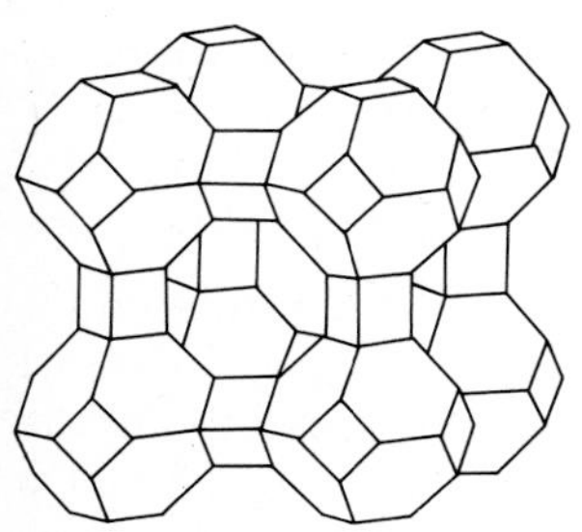

Fig. 5.1.3-2. Zeolite A structure.

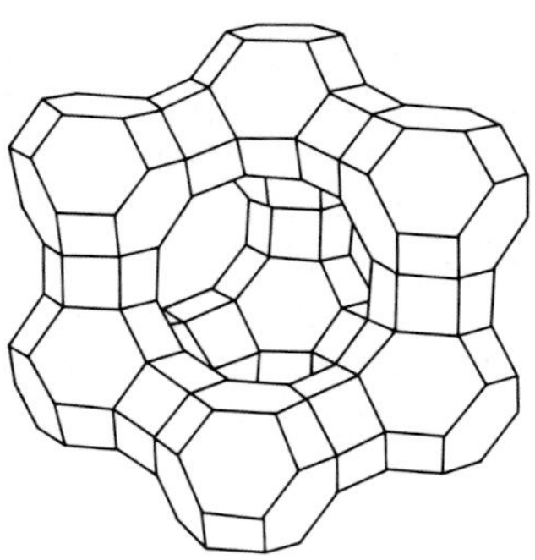

Fig. 5.1.3-3. Faujasite structure, zeolites X and Y.

faujasite (zeolite X and Y) the cubo-octahedra are linked to one another through hexagonal prisms. The cationic sites are not shown in these figures. The complicated structure of ZSM 5 zeolite systems is characterized by two crossing channel systems: one linear and the other zig-zag.

Zeolites contain cavities, whose volumes are type specific. These cavities are connected by pores with uniform diameters. The pore diameters can be modified by cation exchange

The essence of zeolites is the system of cavities in the frameworks, whose volumes are type specific. These cavities are connected to one another by pores, whose diameter is also type specific. The pore diameters (in nm) and pore volume (in %) for several types are given in Table 5.1.3-4.

Table 5.1.3-4. Pore Diameters (in nm) and Pore Volumes (in %) of Several Zeolite Types.

	A	X	Y	Synthetic mordenite	ZSM 5
Pore diameter	4.1	7.4	7.4	6.7 to 7*)	6
Pore volume	47	50	48	28	–

*) Pore opening not round
Inner surface area: 800 to 1200 m^2/g (BET)

The values in this table refer to zeolites with sodium cations. The cations are mobile in the framework and can be exchanged. Upon exchanging sodium cations in zeolite A with potassium ions the pore diameter decreases to 3 nm. Cation exchange also affects other properties.

5.1.3.3 Natural Zeolites

Ca. 40 zeolite types have been found in nature, of which several have attained industrial importance. The natural zeolites result indirectly from volcanic activity. They are formed by hydrothermal transformation of basalt, volcanic ash and pumice and are found e.g. in basalt cavities and in large sedimentary deposits. The most important industrially utilized natural zeolites are:

- Clinoptilolite
- Mordenite
- Chabazite
- Erionite

Deposits are mainly found in the USA, Japan, the USSR, Great Britain and Italy. Clinoptilolite is the most widely occurring natural zeolite.

Natural zeolites:

result indirectly from volcanic activity. Large deposits in the USA, Japan, USSR, Great Britain. Clinoptilolite is the most widely occurring natural zeolite

5.1.3.4 Manufacture of Synthetic Zeolites

5.1.3.4.1 From Natural Raw Materials

Zeolites, in particular zeolite A, can be manufactured from kaolinitic clays, which are particularly found in Central Europe, Great Britain, Japan and China. For transformation to zeolite, the kaolin has to be thermally converted, e.g. by shock-heating to > 550 °C, to metakaolin. The metakaolin formed is then suspended in sodium hydroxide solution and converted at 70 to 100 °C into zeolite A. Some of the impurities contained in the natural raw materials are retained in the final product.

Synthetic zeolites can be manufactured from:

- Natural raw materials such as kaolin
- Synthetic raw materials such as sodium aluminate and silica (utilized as water glass, silica fillers, silica sols)

5.1.3.4.2 From Synthetic Raw Materials

The sodium aluminate required for zeolite synthesis is obtained by dissolving aluminum oxide hydrate in sodium hydroxide. Silica is utilized in the form of water glass, fine particle silica (e.g. silica fillers) or silica sols. The cheaper water glass is preferred, but exhibits the lowest activity of the above-mentioned sources of silica. The reaction has therefore to be carried out in a special way when SiO_2-rich zeolites are manufactured from water glass. Potassium hydroxide and, especially in the synthesis of silica-rich ZSM range of zeolites, organic bases such as tetraalkylammonium hydroxide are utilized in addition to sodium hydroxide. The manufacture of the industrially important zeolite types A, X and Y is generally carried out by mixing sodium aluminate and sodium silicate solutions, whereupon a sodium aluminosilicate gel is formed. In this gel an as yet incompletely understood rearrangement reaction takes place: SiO_2- and Al_2O_3-containing species being formed in the gel and going into solution and there interacting to form the zeolite through crystallization. Nucleation takes place mainly on the inner surfaces of the gel. The crystallization centers are solvated sodium ions. The removal of components from solution upon zeolite growth is accompanied by further solubilization of the gel.

The required zeolite formed as the final product is generally metastable with respect to the associated byproducts, e.g. the byproduct hydroxysodalite is more stable than zeolite A, and the byproduct phillipsite is more stable than zeolites X and Y.

Technically important variables during zeolite synthesis:

- Stoichiometry of the reaction mixture
- Concentrations of the ingredients
- Temperature
- Shear energy used

As a result several variables have to be heeded during the zeolite synthesis, if usable material is to be obtained. Such variables are e.g.:

- The stoichiometry of the reaction mixture, which in all cases is different from that of the zeolite formed,
- The concentrations, which should be kept within well-defined concentration limits,
- The temperature or temperature sequence, which should be maintained at the prescribed values,
- The shearing energy during stirring.

The manufacture of SiO_2-rich zeolites may require a longer gel-aging at low temperatures before crystallization starts. In special cases, the addition of small quantities of seed crystals is used to control the synthesis.

In the synthesis of zeolite A for detergent purposes, for which small particles ($< 5 \times 10^{-3}$ mm) and a narrow particle size distribution are required, the manufacture has been much improved in recent years by adopting much shorter reaction times. This is achieved by adding seed crystals, increasing the shear forces and heating the reaction components to temperatures just under the crystallization temperature. Continuous manufacture of detergent zeolite has also been achieved.

At the end of the crystallization, the zeolite formed is filtered off (e.g. with the aid of filter presses) and washed repeatedly. The mother liquor and filtrates from the washings have to be recycled for ecological and economic reasons.

5.1.3.4.3 By Cation Exchange of Synthetic Zeolites

The capability of zeolites of exchanging the cation used in the synthesis (mainly sodium) with other cations is very important. There are different exchange equilibria with the different cations e.g. silver ions are bound extremely strongly. The exchange of sodium ions with potassium, ammonium, calcium, rare earth and transition metal ions such as nickel, cobalt, platinum, palladium etc. is industrially important. It can be carried out with the zeolite powder formed during the synthesis or with molded articles produced from the powder. Ammonium-exchanged zeolites with a Si/Al ratio > 1.7 can be converted into a stable H-form upon heating. The organic cations incorporated into the framework of e.g. ZSM 5 during synthesis can not, due to their size, be exchanged for other ions. They can only be removed pyrolytically.

Synthesized zeolites can be modified by ion exchange. Exchange of sodium by potassium, ammonium, calcium, barium, rare earth and transition metal elements is industrially important

Ammonium zeolites with a Si/Al ratio > 1.7 can be converted into a stable H^+-form upon heating

5.1.3.5 Pelletization of Zeolites

A range of applications requires zeolites in the form of pelletized articles. They are manufactured using many processes, such as granulation on dish granulators, extrusion or spray drying. Clays or silica-containing materials (silica sol, water glass) are used as binders in the pelletization process. When silica-containing binders are used, subsequent treatment with sodium aluminate-containing solutions followed by heating converts the binder into zeolite, whereupon pelletized articles solely consisting of zeolite are obtained.

Pelletization of zeolite powders e.g. possible by:

- Granulation
- Extrusion
- Spray drying

Clays or silica-containing binders are used

5.1.3.6 Dehydration of Zeolites

Prior to their use as adsorption agents or after technical utilization for the absorption of water, zeolites have to be dehydrated. This is carried out at > 400 °C by passing dry air over them e.g. in rotary kilns, on belt driers etc.

Removal of water from zeolites ("activation"):

by heating to > 400 °C in a stream of dry air

5.1.3.7 Applications for Zeolites

5.1.3.7.1 As Ion Exchangers

Zeolites as ion exchangers:

- Clinoptilolite used for the removal of ammonium ions from waste water
- Zeolite A used in detergents for the removal of calcium and magnesium ions from the washing liquid

In addition to the use of clinoptilolite for the removal of ammonium ions from waste water, the ability of zeolite A to exchange sodium ions for calcium and, to a lesser extent, magnesium ions from aqueous solutions has acquired considerable industrial importance. There are ecological reservations about the use of tripolyphosphate in detergents (danger of eutrophication in certain waters due to too high nutrient levels, see Sec. 1.5.1.2.2). In recent years detergents with less phosphate and even no phosphate at all have been introduced into the market in Europe, Japan and partly in North America. Zeolite A is often used in such detergents, due to its high (and with increasing temperatures higher) calcium-binding capacity. Magnesium is bound less efficiently. Since zeolites are, of course, unable to bind dirt, such detergents must either contain a certain quantity of polyphosphate or polyphosphate substitutes such as nitrilotriacetic acid, citrate or polycarboxylate. It is an open question whether such low phosphate or phosphate-free detergents have the same washing capacity as phosphate-containing detergents.

5.1.3.7.2 As an Adsorption Agent

Zeolites as adsorption agents for:

- Water removal from gases, air, liquid circuits and in double glazing
- Adsorption of carbon dioxide, hydrogen sulfide, mercaptans from gases

Zeolites are capable of strongly binding molecules which are small enough to penetrate their pore systems. This particularly holds for water and for other small polar and polarizable molecules and is the basis of their utilization as drying- or cleaning-agents for gases such as natural gas or air prior to liquification. This enables very low dew points to be achieved. Furthermore, they are utilized for water removal in closed liquid circuits, e.g. in refrigeration units for absorbing water and the adsorption of residual solvent in the space between the panes in double-glazed windows to prevent condensation. The utilization of zeolites for heat storage is currently under discussion: utilizing the heating up of moist air upon water absorption by zeolites followed by regeneration (dehydration) of the zeolites with the aid of solar energy. In addition to the removal of water, zeolites are industrially utilized for the removal of carbon dioxide, hydrogen sulfide, mercaptans etc. from gas mixtures.

5.1.3.7.3 For Separation Processes

High electrostatic fields occur in the cavities of zeolites. Nitrogen, which is polarizable, is therefore much more strongly bound in zeolites (calcium-exchanged zeolite A or X) than oxygen. By multistage adsorption-desorption-cycles under pressure ("pressure-swing process") it is possible to produce oxygen-enriched air, which can be utilized in sewage plants or steel works. Mixtures of n- and iso-alkanes can also be separated with the aid of calcium-exchanged zeolite A (molecular sieve effect), since only n-alkanes can penetrate into the zeolite cavities. Analogously in the separation of xylol isomer-mixtures, p-xylol is preferentially adsorbed by K-Ba-exchanged zeolite X.

Zeolites for the separation of materials (molecular sieves):

- Manufacture of oxygen-enriched air
- Separation of n- and iso-alkanes
- Separation of xylol isomers

5.1.3.7.4 As Catalysts

Important processes in which zeolites are utilized as catalysts are:

- In the isomerization of n- to iso-alkanes for fuel purposes. Pd- or Pt-containing H-zeolite Y is used as the catalyst.
- In the catalytic cracking of crude oil distillate for fuel manufacture. 95% of the "cat-crackers" in the USA use zeolite-containing catalysts, zeolite X or Y (at concentrations of ca. 15% in an inert matrix) with rare earth cations.
- In the manufacture of fuels, alkenes or arenes from methanol, the dewaxing of alkanes, the alkylation of benzene to ethylbenzene in which H-ZSM 5 is used. Pilot plants for methanol conversion have been operated for several years, a full-scale plant is currently being commissioned in New Zealand for the production of fuels (with methanol from natural gas). The manufacture of fuels from biomass is being investigated. The dewaxing of alkanes and the manufacture of ethylbenzene are industrial processes.
- In hydrocracking (conversion of crude oil fractions to petrol in the presence of hydrogen), H-zeolite Y

Zeolites as catalysts for:

- Isomerization
- Catalytic cracking
- Conversion of methanol into hydrocarbons
- Hydrocracking

optionally exchanged with calcium, manganese or rare earth ions is preferably utilized. The zeolite contains palladium or platinum as hydrogenation catalyst.

5.1.3.7.5 Miscellaneous Applications

Large quantities of zeolite-containing rock is worked in different countries (including West Germany in the Eifel region) and is utilized in the manufacture of cements, mortars and lightweight building blocks. In Japan, natural zeolites are used as a filler in paper.

References for Chapter 5.1.3: Zeolites

Commercial Information:

P. L. Layman, "Detergents shift focus of zeolites market", *Chem. Eng. News,* pp 10–13 (Sept. 27, 1982).
J. Haggin, "First methanol-to-gasoline plant nears startup in New Zealand", *Chem. Eng. News,* pp 39–41 (March 25, 1985).

Zeolite Catalysts:

S. M. Csicsery, "Shape selective catalysis in zeolites", *Chem. Br.,* pp 473–477 (May 1985).
W. O. Haag, R. M. Lago, P. B. Weisz, "The active site of acidic aluminosilicate catalysts", *Nature,* **309,** pp 589–591 (1984).
C. J. Plank, "The invention of zeolite cracking catalysts", *Chemtech.,* pp 243–249 (April 1984).
W. Hölderich und E. Gallei, "Industrielle Anwendung zeolithischer Katalysatoren bei petrochemischen Prozessen" (Industrial utilization of zeolite catalysts in petrochemical processes), *Chem.-Ing. Tech.,* **56,** pp 908–915 (1984).
J. Weitkamp et al., "Formselektive Katalyse in Zeolithen" (Shape-selective catalysis in zeolites), *Chem.-Ing. Tech.,* **58,** pp 623–632 (1986)

Applications:

A. Dyer, "Uses of natural zeolites", *Chem. Ind.,* pp 241–245 (April 2, 1984).
L. Puppe, "Zeolithe-Eigenschaften und technische Anwendungen" (Zeolite properties and industrial applications), *Chemie in unserer Zeit,* **20,** pp 117–127 (1986).

Reviews:

Kirk-Othmer: *Encyclopedia of Chemical Technology,* 3rd Ed., Vol. 15, pp 638–669, John Wiley & Sons, New York-Chichester-Brisbane-Toronto (1981).
Proceedings of the 6th International Zeolite Conf., Reno, USA, 1983, Ed.: D. Olson and A. Bisio, Butterworth, Guildford, G.B. (1984).
Proceedings of 7th International Zeolite Conf., Tokyo, 1986, Ed.: Y. Murakami, A. Jijima and J. W. Ward, Elsevier, Amsterdam (1987).

5.2 Inorganic Fibers

5.2.1 General Information

Inorganic fibers:

- Length to diameter: > 10 to 1
- Cross sectional area: < 0.05 mm^2

The term "fibers" includes both naturally occurring asbestos fibers and synthetic filament and short fibers such as glass fibers, carbon fibers and oxide fibers when they exceed a length to diameter ratio of 10:1 and have a cross sectional area below 0.05 mm^2.

Table 5.2-1. Survey of the Application Fields of Inorganic Fibers.

Application field	Foremost requirements	Fibers utilized
reinforcement of:		
plastics	high tensile strength, high elasticity modulus, low density	asbestos, glass fibers, carbon fibers, SiC fibers, boron fibers, oxide fibers (Al_2O_3)
seals, frictional linings	compression strength, elastic deformation behavior, thermal stability, abrasion resistance	asbestos, glass fibers, steel fibers
tire cord	stability to alternating mechanical stress	steel fibers and very short metal fibers
cement	chemical resistance to cement, mechanical stability	asbestos, cement-resistant glass fibers, steel fibers
thermal insulation and fire prevention:		
heat and cold insulation in buildings	low thermal conductivity	glass wool rock wool slag wool
fire prevention	high thermal resistance, noninflammability	aluminum silicate fibers
thermal insulation of high temperature plant	high upper utilization temperature	oxide fibers (Al_2O_3, ZrO_2)
miscellaneous:		
antistatic finish	electrical conductivity	metal fibers
incandescent filaments	high melting point, mechanical stability	tungsten fibers
light transmission	highest purity for virtually loss-free light transmission	special glass fibers
hot gas filtration	chemical stability	oxide fibers
filtration of liquids (wine and beer production)	high specific surface area, high filtration efficiency	asbestos

The reinforcement and insulation sectors are by far the most important application fields for inorganic fibers and pose very different demands on the fibers utilized, see Table 5.2-1.

Most important application fields:

- Reinforcement
- Thermal insulation

The fibers utilized for the manufacture of *insulation materials* should exhibit noninflammability, low thermal conductivity (high thermal resistance) and in the case of high temperature insulation applications should also exhibit thermal shock resistance, high upper utilization temperatures and long-term thermal stability.

Requirements for fibers for insulation materials:

- Noninflammability
- Low thermal conductivity

and additionally for high temperature applications:

- High thermal stability

Requirements for reinforcing fibers:

- High tensile strength
- High elasticity modulus
- Low density
- Low temperature dependence of the properties

Mechanical properties are crucial for fibers for the *reinforcement sector.* The tensile strengths and elasticity moduli of the utilized fibers should be particularly high and the densities as low as possible. The former is important for obtaining the highest possible firmness and stiffness in the composite materials formed by unidirectional or multidirectional incorporation of these fibers, preferably in plastics. Low density values are important to produce composites with the highest possible strength to weight ratios so that weight savings can be made in the construction of motor vehicles or missiles with resulting fuel savings. In Fig. 5.2-1 the mechanical properties of the most important reinforcing fibers are compared in a tensile strength/elasticity modulus diagram (in N/mm^2).

Preconditions for reinforcement:

- Wetting of the fibers by the matrix
- Good fiber/matrix adhesion

The transferability of the mechanical properties of the reinforcing inorganic fiber to the composite generally depends upon the wetting of the fibers by the surrounding matrix i.e. the strength of the bond between

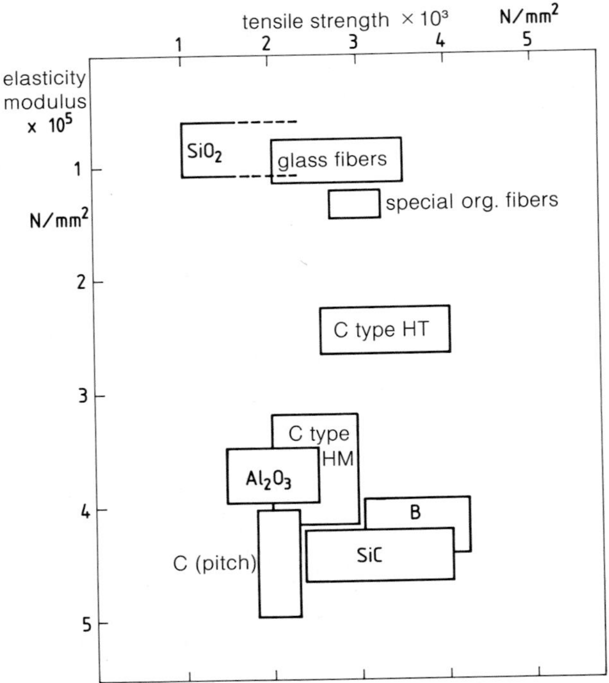

Fig. 5.2-1. Comparison of the most important synthetic inorganic reinforcing fibers in a tensile strength/elasticity modulus diagram.

the fiber and the matrix. Assuming a sufficient bonding strength, the mechanical properties improve as the volume fraction of the incorporated fibers increases. The degree of filling is maximized to incorporate into the composite material as much as possible of the tensile strength and elasticity of the inorganic fibers.

Mechanical properties proportional to the volume fraction of reinforcing fibers

Degrees of filling between 50 and 60 volume % can be achieved and in extreme cases even up to 80%, depending upon the location and mutual orientation of the embedded fibers. As the fraction of inorganic fibers increases the composite assumes more and more of the properties of the inorganic fibers. In many cases the desirable properties for a particular application can not be obtained by incorporating one type of fiber type but can be achieved by embedding a mixture of fibers e.g. mixtures of carbon and glass fibers.

Composite materials contain up to 60 volume % of reinforcing fibers (in extreme cases up to 80%)

Attainment of special properties by utilizing fiber mixtures

5.2.2 Naturally Occurring Inorganic Fibers – Asbestos Fibers

5.2.2.1 Deposits, Extraction and Economic Importance

Asbestos, the first inorganic fiber material used, is currently still exclusively produced from natural mineral deposits. It is formed by the hydrothermal conversion of basic and ultrabasic volcanic rock (olivine and pyroxene) to serpentine upon which the actual asbestos formation process takes place which leads to the two asbestos sorts with different structures: serpentine asbestos and amphibolite asbestos. These sorts have various sub-types (Table 5.2-2).

Ca. 95% of the worldwide deposits of asbestos are chrysotile asbestos with the largest deposits being in Canada and the USSR. The most important amphibole deposits are in South Africa (see Table 5.2-3). The advance of asbestos substitutes and the falling chrysolite asbestos production means that the chrysotile asbestos deposits will last for considerably longer than the 40 years first estimated.

Asbestos deposits:

- 95% chrysotile asbestos
- 5% amphibole asbestos

Deposits of chrysolite asbestos are sufficient for > 40 years

Table 5.2-2. Most Important Asbestos Types and their Compositions.

Asbestos type	Formula
I. Serpentine asbestos	
chrysotile asbestos	$Mg_3(OH)_4[Si_2O_5]$
(ortho-, klino-, parachrysotile)	
II. Amphibole asbestos	
1. alkali-free or low alkali content amphibole asbestos	
a) lime-free:	
anthophyllite asbestos	$(Mg,Fe^{2+})_7(OH)_2[Si_8O_{22}]$
amosite asbestos	$(Fe^{2+},Mg,Al)_7(OH)_2[(Si,Al)_8O_{22}]$
b) lime-containing:	
tremolite asbestos	$Ca_2(Mg,Fe)_5(OH,F)_2[Si_8O_{22}]$
actinolite asbestos	$(Ca,Na)_2(Fe,Mg,Al)_5(OH,F)_2[(Si,Al)_8O_{22}]$
2. alkali-containing amphibole asbestos	
crocidolite asbestos	$Na_2(Fe^{2+},Mg)_3Fe_2^{3+}(OH)_2[Si_8O_{22}]$

Table 5.2-3. Asbestos Production of the Most Important Producers in 10^3 t.

Country	1981	1982	1983	1984
Italy	137.086	116.410	139.054	147.272
USSR	2 105.000	2 180.000	2 250.000	2 300.000
South Africa				
amosite	56.834	43.457	40.656	33.237
crocidolite	102.337	87.263	87.439	88.884
chrysotile	76.772	81.140	93.016	75.414
Zimbabwe				
chrysotile	247.503	197.682	153.221	165.385
Canada				
chrysotile	1 121.845	834.249	857.504	837.982
USA	75.618	63.515	69.906	57.422
Brazil	138.417	145.998	158.855	136.083
P.R. China	106.000	110.000	160.000	160.000
Australia				
chrysotile	45.494	18.587	3.909	–
World	4 300.000	4 000.000	4 200.000	4 200.000

Table 5.2-4. Quantities of Asbestos Extracted according to Asbestos Type in 10^3 t.

	1980	
Asbestos types	10^3 t	% (World)
amphibole asbestoses		
crocidolite	141.4	2.9
amosite	51.6	1.1
anthophyllite	31.4	0.6
actinolite	2.4	< 0.1
tremolite	0.1	< 0.1
total	226.9	4.6
chrysotile asbestos		
Western world	2273.1	46.4
Eastern block	2400.7	49.0
total asbestos extracted	4900.7	100.0

The percentage of asbestos in the rock of extractable deposits is 4 to 10% (Quebec, Urals) with 3% being the borderline for economic extraction.

The industrial extraction of asbestos fibers by the dry process starts with the crushing and loosening of batches of asbestos-containing rock with jaw crushers and double-roll mills. The asbestos fibers are liberated from the dried raw product by repeated breaking up, crushing in pan mills and shreading in drums equipped with knifelike projections. The fibers are separated from the gangue by sieving and air classification and sorted into the different fiber lengths with grading drums.

Industrial extraction of asbestos:

1. using the dry process:

- Crushing of the rock
- Shreading
- Air classification
- Sorting

The wet process enables a narrower fiber length distribution to be obtained, which is important for amphibole asbestos. In this process the asbestos fibers are separated from the rock in the crushed mixture by using a hydrocyclone and the longer and shorter fibers separated with a rotary sieve.

2. using the wet process:

- Separation of asbestos fibers using hydrocyclones
- Sorting with a rotary sieve

5.2.2.2 Properties of Asbestos Fibers

The physical and chemical properties of the asbestos fibers are dependent not only on the asbestos type and deposit, but also on the fiber length and the degree

Asbestos properties are dependent on:

- Asbestos type
- Deposit

- Chemical composition
- Fiber length
- Degree of hydrolysis

of hydrolysis. In the two main groups namely hand-selected unshreaded crude asbestos with 3/8 and 3/4 inch or longer fibers on the one hand and ground fibers on the other, there are many subgroups on the basis of fiber length, degree of hydrolysis and fiber texture, which enable a classification to be drawn up which is important for industrial applications.

The different qualities are reflected in the prices of the asbestos fibers. Thus in 1979 for example crude asbestos No. 2 cost 2295 Canadian $/short ton, whereas the short fibers of group No. 7 only cost 113 to 218 Canadian $/short ton.

The different chemical composition and structure of chrysotile and amphibole asbestoses exert a considerable influence on their physical properties.

Chrysotile asbestos fibers consist of bundles of hollow microfibrils 15 to 40 nm in diameter

The macroscopic individual fibers of chrysotile asbestos consist of more or less parallelly oriented fibrils 15 to 40 nm in diameter, which are carpet roll-like structured hollow fibers according to electron microscopic investigations. The wall partitions of these "little rolls" consist of double layers with the composition $Mg_3(OH)_4Si_2O_5$ in which one brucite $Mg(OH)_2$- and one Si_2O_5-layer have been condensed with one another. This leads to curvature and hence a hollow fiber, due to the two crystal lattices not fitting precisely on one another. Chrysotile asbestos is as a result easily fragmented into very fine fibers, which are soft and flexible and hence of all the asbestos sorts lend themselves best to spinning. However, industrially chrysotile asbestos is always present as fiber bundles.

Due to this structure chrysotile asbestos can be easily fragmented into very fine spinnable fibers

In contrast amphibole asbestos consists of Si_4O_{11}-bands as the supporting entities. These are condensed with metallic hydroxide bands to double bands, which are held together by cations.

Amphibole asbestos fibers are solid crystalline needles

Amphibole asbestoses can not therefore be fragmented into such fine fibers and the fragments are more like hard crystalline needles.

Zeta-potential:

- Chrysotile asbestos positive
- Amphibole asbestos negative

The different structure types lead to very different physical and chemical properties. Unlike amphibole chrysotile fibers exhibit a positive zeta-potential, which is important for its utilization as a filtration and clarifying agent in the food and drink industries as well as for the separation of viruses and pyrogens from pharmaceutical solutions. Due to the accessibility of its hollow fibers, chrysotile asbestos is unstable to acids unlike amphibole. Other properties are summarized in Table 5.2-5.

Behavior with acids:

- Chrysotile asbestos unstable
- Amphibole asbestos fairly stable

Since the *toxic properties* of fibers (fibers longer than 5 nm and a diameter < 3 μm) in the lung and therefore also the fine dust which is formed during the processing of asbestos and asbestos-containing materials became generally known (occurrence of asbestosis, lung cancer and mesothelioma), the regulations governing safety provisions have been considerably tightened [maximum permissible concentration (MPC value) in a workplace in West Germany = 1 fiber/cm^3 from 1982]. This led to an intensification of the search for asbestos substitutes, which has led to the industrial introduction of several substitutes.

MPC-value for asbestos fibers = 1 fiber/cm^3 in West Germany from 1982

5.2.2.3 Applications of Asbestos Fibers

General Information

Asbestos due to its unique combination of many desirable properties such as e.g. thermal stability, noninflammability, resistance to rotting, stability to chemicals, spinnable fibers, good thermal insulation, mechanical strength and good wettability by organic and inorganic binders is utilizable in a broad range of applications in almost all industrial sectors. For example asbestos is used in asbestos cement plates and tubes, brake and clutch linings, asbestos textiles, floor coverings, plastic molding compositions, high pressure sealing plates, asbestos boards, asbestos papers, as filter materials, in molding compounds, as an insulation material, as a diaphragm material and in many other application fields.

Application-related properties of asbestos fibers:

- Noninflammability
- Thermal stability
- Resistance to rotting
- Stability to chemicals
- Spinnable
- Mechanical strength
- Thermal insulation properties
- Good wettability
- Very high specific surface area

There are more than 3000 uses for asbestos. Table 5.2-6 gives the most important of these together with the corresponding consumptions of asbestos in West Germany (1975), the USA (1980) und worldwide (1981). In the meantime a gradual change is in progress due to the increasing utilization of asbestos substitutes.

Table 5.2.5. Comparative Survey of the Properties of the Six Most Important Asbestos Sorts.

Properties	Chrysotile	Anthophyllite	Crocidolite	Actinolite	Tremolite	Amosite
ideal form	Mg_6 $[(OH)_4Si_2O_5]_2$	$(Mg, Fe)_7$ $[OH/Si_4O_{11}]_2$	Na_2Fe_5 $[OH/Si_4O_{11}]_2$	$Ca_2(Mg, Fe)_5$ $[OH/Si_4O_{11}]_2$	$Ca_2(Mg, Fe)_5$ $[OH/Si_4O_{11}]_2$	$MgFe_6$ $[OH/Si_4O_{11}]_2$
crystal structure	fibers	prismatic crystals & fibers	fibers	prismatic crystals & fibers	prismatic crystals & fibers	prismatic crystals & fibers
color	white, gray, green, yellowish	blue-green, white-gray, brown-green	blue to green	green	white, green, yellowish, brownish	gray, brown
flexibility	very high	low	good	low	low	good
Mohs hardness	2.5 to 4	5.5 to 6	4	6	5.5	5.5 to 6
fiber diameter (mm) (individual fiber)	2 to 4×10^{-3} (hollow fiber; wall thickness ca. 5×10^{-6})	1 to 2×10^{-4} (solid fiber)	1 to 2×10^{-4} (solid fiber)	1 to 2×10^{-4} (solid fiber)	1 to 2×10^{-4} (solid fiber)	1 to 2×10^{-4} (solid fiber)
fiber diameter (mm) (industrial fiber tufts)	0.75 to 1.5×10^{-3}	1.5 to 4×10^{-3}	1.5 to 4×10^{-3}	1.5 to 4×10^{-3}	1.5 to 4×10^{-3}	1.5 to 4×10^{-3}
density (g/cm^3)	2.3 to 2.7	2.8 to 3.2	2.8 to 3.6	3.0 to 3.2	2.9 to 3.2	2.9 to 3.3
specific surface area (m^2/g)	13 to 22 (max. 50)	7	2.7 to 4.6 (max. 10)	–	–	9
tensile strength (N/mm^2)	2000 to 6000 (max. 20 000)	30	7500 to 22 500	6	6 to 50	1000 to 3000 (max. 6000)
elasticity modulus (N/mm^2)	30 000 to 160 000	–	100 000 to 190 000	–	–	30 000 to 160 000
DTA peak for H_2O-scission °C	600 to 780	950	610	1040	–	780
structure breakdown temperature, °C	800 to 850	950	800	1040	–	600 to 900
pH value	9.5 to 10.3	9.4	9.1	9.5	9.5	9.1
acid stability	unstable	very good	good	satisfactory	good	satisfactory
alkali stability	good to satisfactory	good	good	good	good	satisfactory
electrical charge (zeta-potential)	positive	negative	negative	negative	negative	negative

Table 5.2-6. Consumption of Asbestos in Different Application Sectors in 10^3 t/a.

Application sector	W. Germany (1975)	USA (1980)	World (1981)
asbestos cement products	105.0	151.9	3126
floor covering	12.8	90.2	
roof covering		26.5	
sealing materials	7.0	12.3	
insulation materials		8.9	
construction materials excl. cement products			799
brake & clutch linings	11.0	43.7	
frictional linings			272
asbestos textiles	4.0	1.9	
asbestos board and paper	6.15	0.5	
filtration materials	3.0		
top layers of roads	1.0		
asbestos sprays	0.25		
molding materials	2.0		
plastics, coatings		12.4	
miscellaneous			543
total	162.20	358.7	4740

By far the largest quantities of asbestos are utilized in asbestos fiber composite materials for the reinforcement of inorganic (asbestos cement) or organic (PVC, rubber, hardenable plastics) bonding agents. The ca. 10% increase in the tensile strength of asbestos fibers up to 400 °C is significant in some of these application fields. This compares with glass fibers which at this temperature have only a fraction of their strength at room temperature (see Table 5.2-7).

Main application sectors for asbestos:

Asbestos fiber reinforcement of:

- Cement
- PVC
- Rubber
- Hardenable plastics

Table 5.2-7. Average Values for the Densities, Tensile Strengths and Elasticity Moduli of Asbestos Fibers.

Fiber type	Density (g/cm^3)	Elasticity modulus at 23 °C (N/mm^2)	Tensile strength at 23 °C (N/mm^2)	Tensile strength at 400 °C (N/mm^2)
chrysotile	2.3 to 2.7	up to 160 000	2000 to 6000 depending on fiber length (max. up to 20 000)	
crocidolite	2.8 to 3.6	up to 190 000	7500 to 22 500 depending on fiber length	5000 to 15 000
glass fibers	2.53 to 2.44	73 000 to 77 000	ca. 2000	600 to 900

Applications of Asbestos Fibers in Composite Materials

Asbestos has been utilized in the building industry since 1900

The most important fiber-reinforced inorganic-based product is asbestos cement, which has been used in the building industry since 1900. Asbestos fibers (normally chrysotile) are bonded by Portland cement and the mixture sets to a fiber-reinforced material with 10 to 20% asbestos, which has excellent properties (Eternit®, Fulgurit® etc.).

Asbestos cement: Portland cement reinforced with 10 to 20% (chrysotile) asbestos

Manufacture of asbestos cement:

slurry of cement and asbestos fibers is filtered, compressed in molds (200 bar) and allowed to set for 24 to 48 h

In the classical asbestos cement processes a slurry of asbestos fibers and cement is produced and deposited as a fibrous fleece in continuously operating cylindrical sieves. This fleece is then compressed under 200 bar pressure and allowed to set in molds for 24 to 48 h. This process can be accelerated by steam hardening.

Articles made out of asbestos cement:

- Boards
- Shingles
- Tubes
- Molded articles

Boards and shingles for covering roofs and other industrially molded components, including high pressure tubes suitable for use at pressures up to 40 bar, can be manufactured using this or similar processes.

Asbestos cement products exhibit:

- Good mechanical properties
- Low thermal conductivity
- Weather resistance
- Rot resistance

Direct processing of asbestos cement slurry as *asbestos spray*

Asbestos cement materials are widely used due to their good mechanical properties, low thermal conductivity and stability to weather, frost and rotting. The slurry of asbestos and cement can also be processed directly using a spray process as so-called asbestos spray for fire-retarding layers as well as for thermal (hot and cold) and sound insulation.

Substitution of asbestos in asbestos cement by

- Glass fibers
- Steel fibers
- Mineral wool
- Organic fibers

has been striven for

Considerable effort has been expended over a long period in the search for a replacement for asbestos in asbestos cement e.g. glass fibers, steel fibers, mineral wool and organic fibers. However, due to the price of possible alternative fibers and/or their instability with respect to the alkaline substances in cement, no totally satisfactory substitute has been found.

Frictional linings (clutch and brake linings):

Composite material consisting of

- 20 to 60% asbestos
- Phenol resins
- Rubber
- Fillers
- Metal chips

The high frictional coefficient (0.4 to 0.5 compared with < 0.1 for glass fibers) of asbestos fibers is crucial to its utilization in the frictional lining sector. In the manufacture of brake and clutch linings 20 to 60% asbestos is incorporated together with fillers, metal chips and preferably phenol resins and rubber into a composite material, which has to satisfy many requirements. Currently there are asbestos-free so-called "semimetallic" brake linings, which consist of mixtures of metal fibers, metal powders, cellulose fibers, aluminum silicate fibers and mineral wool bonded with synthetic resins.

Asbestos-reinforced organic binders (thermoplastics, duroplastics and elastomers) are widely utilized e.g. hardenable molding materials on the basis of asbestos-reinforced phenol or melamine resins for the manufacture of insulating components for combustion engines, components for electrical installations, cogwheels etc. Possible fiber substitutes are glass fibers, carbon fibers and other synthetic fibers. Possible non fiber substitutes are calcium carbonate, clay or talc.

Hardenable molding materials on the basis of phenol or melamine resins with asbestos are used in the manufacture of:

- Insulating components in combustion engines
- Components for electrical installations
- Cogwheels

Asbestos is incorporated in thermoplastics in the manufacture of poly(vinylchloride) tiles and floor coverings, although this is being increasingly replaced by glass fibers. Asbestos-reinforced rubber is utilized as a sealant and is used in the manufacture of It-sheets ("It" is an abbreviation of "Gummi-Asbest"), which is known under different trade names e.g. Klingerit®. Amphibole asbestos is preferred over the usually used chrysotile asbestos for acid-resistant seals. Other fibers e.g. graphite or poly(tetrafluoroethene) fibers are also being used more and more in these applications.

Asbestos-reinforced PVC for:

- Tiles
- Floor coverings

Asbestos-reinforced rubber for:

- Sealing materials ("It"-sheets)
- Rotary seals

Woven, flocculated or twisted asbestos textiles are frequently the basis of sealing components such as packing for rotary seals and gaskets. Asbestos paper and asbestos board, produced by the filtration and compression of an asbestos fiber slurry in analogy to paper production, are utilized for cylinder-head gaskets for water-cooled combustion engines.

Manufacture of asbestos paper and board by fitration and compression of an asbestos fiber slurry

Applications of Asbestos Fibers in Asbestos Textiles and Filter Materials

Chrysotile is the asbestos-type most suitable for *asbestos textiles*. Only 0.8% of the raw material is spinnable. In the manufacture of asbestos textile products so-called carrier fibers (cotton or organic synthetic fibers) have to be added to achieve the required strength. These reduce the maximum operating temperature of the textiles from the ca. 480 °C of yarn consisting of 99 to 100% asbestos, the temperature reduction being a function of the carrier fiber-content. In the high temperature sector asbestos-free mineral fiber textiles are being utilized more and more.

Manufacture of asbestos textiles from chrysotile and varying amounts of "carrier fibers" (cotton or organic synthetic fibers)

Asbestos used as a filter material for:

- Clarification of food and beverages
- Separation of germs, viruses and pyrogens

The utilization of chrysotile asbestos as a *filtration material* ranges from the clarifying filtration of foods and beverages (ca. 1.3×10^{10} l of wine is filtered annually through asbestos in the EEC) to degermination to the separation of viruses and pyrogens from highly sensitive pharmacological solutions. The high specific surface area, positive zeta-potential and ion exchange properties of chrysotile asbestos are a particularly favorable combination and make it difficult to find an asbestos-free substitute. Filters based on cellulose mixed with diatomaceous earth (Zeta-Plus-Filter®) have already been investigated as a substitute.

References for Section 5.2.2: Asbestos Fibers

Asbestos:

W. Noll: *Ullmanns Encyklopädie der technischen Chemie*, 4th Ed., Vol. 8, pp 67–69, Verlag Chemie, Weinheim-Deerfield Beach-Basel, 1974.
P. Eyerer: *GAK 8* – 31st Issue, p. 574 (1978).
K. Höbius: *GAK 11* – 32nd Issue, p. 833 (1979).
Chemical Economics Handbook, Stanford Research Institute, Menlo Park, California, USA, Oct. 1977.
P. Eyerer: *GAK 9* – 31st Issue, p. 663 (1978).
P. Eyerer: *GAK 10* – 31st Issue, p. 776 (1978).
H. Thiemecke: *Die BG*, pp 307–314, June 1981.
G. Clarke: *Industrial Minerals*, pp 19–37, March 1982.
E. Poeschel, R. König and W. Lohrer, *Staub-Reinhalt. Luft* **39**, No. 8, pp 272–278 (1979).
World Mineral Statistics (1980–1984).

5.2.3 Synthetic Inorganic Fibers

5.2.3.1 Textile Glass Fibers

5.2.3.1.1 General Information

Since it became known in the USA in the 1940's that plastics could be reinforced by glass fibers, they have developed into an important product group. These are known as textile glass fibers to distinguish them from insulating glass fibers.

Despite the high brittleness of glass, glass fibers are flexible, pliable products due to their small diameter, being in general between 10 and 25 μm but well below 10 μm for some applications. It is important that glass fibers behave completely elastically under load. Their elongation at fracture is ca. 3%.

5.2.3.1.2 Economic Importance

1984 worldwide production of textile glass fibers: ca. 1.35×10^6 t

The worldwide production of textile glass fibers in 1984 was ca. 1.35×10^6 t, almost 90% of which was in the USA, Western Europe and Japan (see Table 5.2-8).

Table 5.2-8. 1984 World Production of Textile Glass Fibers in 10^6 t.

World	USA	W. Europe	Japan	USSR/P.R.China	Others
1.35	0.63	0.30	0.25	0.10	0.07

5.2.3.1.3 Classification of Textile Glass Fibers

Differently constituted glass fibers are utilized in the different application fields. The industrially important glass fiber types are summarized in Table 5.2-9.

Table 5.2-9. Typical Compositions of the Important Textile Glass Types (in % by weight).

	A-Glass	C-Glass	D-Glass	E-Glass	R-Glass	S-Glass	AR-Glass
SiO_2	72.5	65.0	74.0	54.5	60.0	65.0	71.0
Al_2O_3	1.5 (with Fe_2O_3)	4.0 (with Fe_2O_3)	–	14.5	25.0	25.0	1.0
Fe_2O_3			0.2	0.5	0.3	–	–
B_2O_3	–	5.0	22.5	7.5	–	–	–
CaO	9.0	14.0	0.5	17.0	9.0	–	–
MgO	3.5	3.0	0.2	4.5	6.0	10.0	–
Na_2O	13.0	0.5	1.3	0.8 (with K_2O)	0.4	–	11.0
K_2O	–	8.0	1.5		0.1	–	–
Various	–	BaO: 1.0	–	TiO_2: 0.1	TiO_2: 0.2	–	ZrO_2: 16.0 Li_2O: 1.0

The E-glass fiber is by far the most important textile glass fiber and is utilized for reinforcing plastics and for applications in the electrical sector. A-glass is used for less demanding applications. C-glass exhibits particularly high chemical resistance and D-glass exhibits excellent dielectric properties. R-glass and S-glass are special glasses for high temperature applications and uses requiring high mechanical strength. AR-glass (**A**lkali-**R**esistant) is a ZrO_2-containing special glass with improved stability to alkalis.

Textile glass fibers with diameters from less than 10 to ca. 25 μm exhibit high strength and stability. They are manufactured in different compositions. By far the most important of these is E-glass, an alkali-deficient Ca/Al/B-silicate, used for the reinforcement of plastics

5.2.3.1.4 Manufacture

All industrially important glass fibers are manufactured from silicate melts. They retain their glassy X-ray amorphous structure in the solid state.

The silicate melts suitable for fiber production have to fulfill many requirements: first a sufficient fiber drawing potential which is dependent upon the ratio of surface tension to viscosity. The fiber shape is deter-

Textile glass fibers are manufactured from silicate melts with sufficient fiber drawing potential. The final shape of the initially molten filament is determined by the viscosity increase upon cooling

mined by the increase in viscosity with decreasing temperature. Good fiber formation requires a viscosity-temperature dependence which is relatively flat. The glass also must not exhibit a tendency to crystallization.

The manufacture of textile glass fibers will be illustrated below by that for E-glass fibers.

Raw materials for the manufacture of E-glass fibers:

kaolin, quartz sand, colemanite (calcium borate), boric acid, limestone, dolomite and fluorite

The preferred raw materials for the manufacture of E-glass fibers are kaolin (aluminum silicate), quartz sand, colemanite (calcium borate, $Ca_2B_6O_{11} \cdot 5\ H_2O$), boric acid, limestone, dolomite and fluorite, which have undergone careful quality control regarding impurity elements and particle fineness. The well-mixed raw materials are melted in flat elongated tank furnaces, which are gas heated. Stringent requirements have to be met by the brick-lining of the furnace, due to the high melt temperatures of ca. 1350 °C and the corrosiveness of the melt. Chromium oxide bricks are utilized for the sidewalls and zirconia bricks for the bottoms of the tanks.

The raw material mixture is melted at ca. 1350 °C in tank furnaces lined with chromium oxide and zirconia bricks

Uninterrupted spinning of the melt requires that it be very homogeneous and free from nonmolten raw materials and gas bubbles

The melt remains in the furnace for several days. A satisfactory yield during the subsequent drawing process requires that the clarification of the glass melt be as good as possible i.e. very homogeneous and free from nonmolten raw materials and gas bubbles. This is achieved by long residence times in the furnace, by air injection through one or more rows of platinum tubes at the bottom of the tank into the glass melt and by adding clarification agents.

The hot combustion gases are utilized indirectly using heat exchangers, mainly steel reversible exchangers, for preheating the air for the oil or gas burners.

Conversion of glass melts into fibers is achieved using the direct melt, marble melt and rod drawing processes with the direct melt process predominant

The clarified glass melts can be converted into fibers by three processes: direct melting, marble melting and rod drawing.

- Direct melt process:

  the molten glass is led through spinning jets by way of brick-lined (cooling) channels

The industrially most important process is the direct melt process in which the glass melt passes through a cooling zone before being spun. This zone consists of a brick-lined channel in which the melt is cooled to the ca. 1250 °C necessary for satisfactory spinning with spinning jets mounted on its side arms.

In the older marble melt process glass marbles manufactured separately are melted in the spinning jets. The marble melt process is only used when very fine fiber diameters are required or when special glass compositions are being used.

- Marble melt process:

  separately manufactured glass marble melted in the spinning jets

The rod drawing process is less important. In this process glass rods are fed at constant speed into the melting zone and continuous fibers are drawn onto a spool.

- Rod drawing process:

  glass rods drawn to fibers in the melting zone

In the direct melt and the marble melt processes the glass melt or glass marbles are fed into spinning jets (bushings) made out of a platinum-rhodium alloy which are maintained at ca. 1250 °C by resistive heating. The individual bushings normally have 400, 800, 1000 or 2000 orifices with diameters of 1 to 2 mm. Upon emerging from a jet the glass melt forms filaments which combine into a spinning strand below the bushings, so that each spinning strand consists of as many individual filaments as there are orifices in the jet. Sometimes several strands are manufactured from the filaments from a single jet ("spin dividing").The spinning strands are wound at speeds of up to ca. 200 km/h on winding machines producing "spinning cakes". The diameter of the glass fiber at constant drawing speed is determined by the quantity of glass per unit time emerging from each spinning orifice. This for a given orifice geometry, constant glass composition and glass state is only dependent upon the temperature of the glass.

The spinning jets (bushings) each with 400, 800, 1000 or 2000 orifices 1 to 2 mm in diameter are made of a Pt-Rh alloy and are maintained at the spinning temperature by direct resistive heating.

The filaments are combined into one or more spinning strands ("spin dividing") below the spinning jet

The spinning strands are wound at speeds of up to 200 km/h on winding machines producing "spinning cakes", immediately quenched below the bushings with a water spray and finally coated with size

On their way to the winding machine the filaments are immediately quenched with a water spray after formation and led past a sizing roller, where they are coated with so-called size. This is an aqueous dispersion which contains film-forming and adhesion promoting substances together with lubricants, antistatic agents, plasticizers, emulsifiers and wetting agents.

Sizes are aqueous dispersions with the following ingredients:

- Film-formers
- Adhesives
- Lubricants
- Antistatics
- Various aids such as emulsifiers, plasticizers, wetting agents

Their purpose is to protect the vulnerable glass fiber surface, guarantee good fiber processibility and provide a good bond between the glass fibers and the plastic being reinforced. Almost every organic matrix and every manufacturing process for glass fiber reinforced plastic components requires a size of a different composition.

Functions of the sizes:

- Protection of the glass surface
- Guarantee processibility
- Bonding of the glass fiber to the organic matrix

Film formers:
organic polymers

The film formers used are organic polymers such as poly(vinyl acetate), poly(esters), epoxides, poly-(urethanes), poly(acrylates), poly(alkenes). These protect the fiber during processing and bond the individual filaments together into spinning strands. The adhesion promoters used are organofunctional trialkoxysilanes e.g. trimethoxyvinylsilanes or those with the general formula $(RO)_3Si\text{-}(CH_2)_3\text{-}X$ such as triethoxy-(3-aminopropyl)silane. X can also be chlorine, mercapto-, methacryloxy-, or glycidyl-groups and R an alkyl- or hydroxyalkyl-group. These adhesion promoters improve the mechanical properties of glass fiber-reinforced polymers, due to their forming a chemical bond both with the surface hydroxy groups of the glass fibers and the polymer matrix. For glass fibers not destined for the reinforcement of plastics e.g. yarns and twists, adhesion promoter-free sizes are used. Lubricants such as fatty acid esters, fatty acid amides, alkanes, poly(alkenes) reduce frictional damage of the fibers and make their processing easier. Inorganic and organic salts are added to prevent electrostatic charging. The size amounts to 0.1 to 2% by weight of the dry fibers.

Adhesion promoters:
organofunctional alkoxysilanes

Lubricants:
fatty acid derivatives, alkanes, poly-(alkenes)

Antistatics:
inorganic and organic salts

After drying, the spinning cake is converted into glass fiber end-products:

- So-called “rovings” are formed by combining multiple spinning strands into glass fiber hanks
- Cut glass: cut spinning strands
- Milled glass fibers
- Textile glass mats

The spinning cake obtained after coating with size is dried in forced air drying kilns and then converted into glass fiber end-products. In the manufacture of so-called “rovings” many spinning cakes are combined in parallel glass fiber hanks and are wound on precision winding machines. Cut glass is produced by cutting the spinning strands from the spinning cake to the desired length (e.g. 3 to 6 mm) with rapidly turning cutting rollers. Milled glass fibers are produced by grinding precrushed spinning cakes. In the manufacture of textile glass mats chopped or unchopped spinning strands are deposited completely randomly and solidified with a binder.

- Glass fiber yarns and twists

Glass fiber yarns and twists are obtained by twisting glass filaments or glass staple fibers, which are short glass fibers mainly manufactured by the so-called jet blowing process in which the glass filaments emerging from the jet orifices are tangled together by uniform very fast gas jets. Glass fiber textiles are woven from rovings or yarn in specialized mills.

- Glass fiber textiles

New developments have led to the elimination of the spinning cake stage for certain glass fiber products, the rovings being wound directly ("direct roving"). Cut glass can also be directly produced by the cutting of spinning strands ("direct chops").

New developments avoid the necessity of producing the spinning cake as an intermediate product

5.2.3.1.5 Applications

The main application of glass fiber products (ca. 80%) is in the reinforcement of plastics. By embedding glass fibers in plastics important properties such as tensile strength, compression strength, flexural strength, elasticity modulus, impact resistance, thermal expansion, maximum use temperature and creep tendency can be improved. In addition glass fiber reinforced plastics exhibit easier processibility, lower density and better corrosion resistance compared with traditional materials such as wood or metals. They are currently used in a wide range of applications (see Table 5.2-10).

Ca. 80% of glass fiber production utilized for reinforcing plastics with resulting improvement in the following properties:

- Tensile, compression and flexural strengths
- Elasticity modulus
- Impact resistance
- Thermal expansion
- Maximum use temperature
- Creep tendency

Table 5.2-10. Applications Spectrum of Glass Fiber Reinforced Plastics in 1983.

	Europe (%)	USA (%)
Transportation	17	25
Electrical industry	15	9
Construction industry	20	20
Industry and agriculture	25	21
Sport and leisure activities	12	15
Consumer goods	3	6
Miscellaneous	8	4

Most important applications sectors for glass fiber reinforced plastics:

- Transportation
- Electrical industry
- Construction industry
- Industry, agriculture
- Sport, leisure sector

In 1983 ca. 2×10^6 t of glass fiber reinforced plastics were manufactured worldwide, the USA and Western Europe accounting for 0.9×10^6 t and 0.7×10^6 t respectively.

Worldwide production in 1983: ca. 2×10^6 t

Originally mainly unsaturated polyesters and in smaller quantities epoxy resins were reinforced. Later elastomers were reinforced e.g. for car tires. Since the mid-1960s glass fiber-reinforced thermoplastics, particularly polyamide, have increased considerably in importance. In West Germany, for example, they accounted for 39% in 1985. Car tires with glass cord reinforcement are particularly important in the USA market.

Unsaturated polyesters are predominant among glass fiber reinforced plastics, but reinforced thermplastics, particularly polyamides, are strongly increasing in importance

Other application sectors:

- Glass fiber fleece and textiles for bituminous roofing felt and for carpet backing
- Glass fiber textiles for printed circuit boards and grinding disks
- Noncombustible textiles e.g. fireproof curtain fabrics

Glass fiber fleece and textiles are utilized in the manufacture of bituminous roofing felt and for carpet backing. Glass fiber textiles are also used in the manufacture of printed circuit boards and in the fabrication of grinding disks. Since they do not burn, glass fibers are incorporated into noncombustible textiles such as fireproof curtain fabrics.

- Glass fiber filters
- Cement reinforcement with for:
  - nonloadbearing applications: alkali resistant AR-glass fibers
  - loadbearing applications: Polystal® glass fibers

Glass fiber filters are utilized in dust removal technology. Alkali resistant AR-glass fibers are utilized in the reinforcement of cement for nonloadbearing applications. The recently available Polystal®, a strong unidirectionally aligned polyester-coated E-glass fiber from Bayer AG, with tensile strengths of 1400 to 1500 N/mm^2 can also be utilized for loadbearing applications (e.g. for strengthening optical fiber cables).

5.2.3.2 Optical Fibers

Optical fibers consist of a:

- Core glass with a high refractive index
- Cladding glass with a lower refractive index

Step-index and graded-index fibers differ in their (radial) refractive index profiles

Optical fibers consist of thin flexible glass strands with a core of optical quality high refractive index glass (GeO_2- or P_2O_5-doped quartz glass) and a cladding of lower refractive index boro- or fluorosilicate glass. There are two types: step-index fibers (with a constant refractive index in the fiber core) and graded-index fibers (with continuously decreasing refractive index from the fiber core to the outside of the fiber). The latter exhibit particularly low transmission losses and dispersion (variation in signal transmission time) of the light pulse fed into the ends of the fibers.

Manufactured by:

- Rod/tube process
- Double crucible process
- Modified CVD process (with $SiCl_4$, $GeCl_4$, BCl_3, $POCl_3$ as starting materials)

Step-index fibers are manufactured using the rod/tube or the double crucible process, in which the fiber core or cladding glass are melted and drawn either simultaneously or sequentially to a ca. 125 μm thick glass fiber.

Graded-index fibers are manufactured from up to 25 mm thick and 1.5 m long preforms in a drawing tower at temperatures between 2000 and 2300 °C. The preform consists of a quartz glass tube which has been coated on the inside with glass-forming elements from gaseous compounds (e.g. $SiCl_4$, $GeCl_4$, BCl_3, $POCl_3$), which upon reaction with oxygen at 1500 to 1600 °C are oxidized to clear high purity glass. By multiply repeating this process (up to 1000 times) this modified CVD-process (CVD = **C**hemical **V**apor **D**eposition) can be used to produce almost any refractive index profile desired.

Immediately after the drawing process the fiber is coated (e.g. with a silicone resin, PVC, urethane-acrylate resins) to reduce its brittleness and significantly increase its tensile strength, so that it can withstand the tension in the manufacture and installation of cables.

As a result of the high quality requirements of the optical fibers, contamination with absorbing ions, in particular with iron, copper, chromium, vanadium and hydroxyl ions, and changes in the fiber diameter must be minimalized. This is achieved by strict control of the inert gas flow rate, drawing temperature, pulling speed and cooling rate and by operating at constant relative humidity and under clean room conditions (contamination < 1 ppm).

Transmission rates of 8 Gbit/s at transmission losses of a minimum of 0.2 dB/km over distances of up to 250 km have already been achieved. Their main application is for optical communications systems. By 1986 5×10^6 km of optical fiber cables had been installed worldwide compared with 5000 km in 1980. The most important manufacturing country is the USA followed by Japan and Great Britain.

Most important properties:

- Low transmission losses and dispersion
- High transmission rate
- Wide transmission band width
- Low weight

Installed cable length by 1986: ca. 5×10^6 km

5.2.3.3 Mineral Fiber Insulating Materials

5.2.3.3.1 General Information and Economic Importance

Mineral fiber insulating materials and insulating glass fibers consist, as with textile glass fibers, of glassy mineral fibers. The name is derived from the mineral starting materials mostly used for their manufacture. Glass, rock, slag and ceramic fibers come under this heading. Other commonly used terms are mineral, glass, rock and slag wools, since, unlike textile glass fibers, they are produced as short randomly ordered fibers. The end products are therefore known as mineral insulating wools.

Mineral fiber insulating materials consist of glassy amorphous fibers

The economic importance of mineral fibers is considerable (see Table 5.2-11). In 1985 in the USA alone, mineral fibers were produced to the value of $1.96 + 10^9$ $.

1985 production of mineral fiber insulating materials:

- USA 1.5×10^6 t
- Western Europe 1.2×10^6 t
- Rest of world 1.7×10^6 t
- Worldwide 4.4×10^6 t

Table 5.2-11. 1985 Mineral Fiber Production in 10^6 t.

World	USA	W. Europe	W. Germany	Rest of world
4.4	1.5	1.2	0.33	1.7

Slag fibers are of little importance. Ceramic fiber consumption is increasing with ca. 70 000 t being consumed in 1983 in Western Industrialized Countries

The most important mineral fiber products are glass and rock wool, the relative production of these products differing considerably from region to region being for example 1:1 in West Germany. Slag fiber production is only of minor significance. Ceramic fibers on the other hand are steadily gaining in importance with 1983 consumption figures of 45 000, 18 000 and 9000 t for the USA, Western Europe and Japan respectively and an annual increase of 5 to 10%.

The contribution of mineral fibers to the total consumption of insulating materials was ca. 60% in West Germany in 1985. In other countries the proportion is considerably higher e.g. ca. 90% in the Scandinavian Countries.

5.2.3.3.2 Manufacture

General Information

Mineral fibers are manufactured from silicate melts:

- Fiber diameter: 1 to 30 μm
- Fiber length: 1 cm to several dm

Mineral fibers are manufactured from silicate melts of suitable composition. The conversion of this melt into fibers occurs with considerably more efficient use of time and space than with the manufacture of textile glass fibers, since the melts are spun at much lower melt viscosities. After solidification the fibers consist of amorphous glasses (according to X-ray diffraction measurements) with diameters between 1 and 30μm and lengths between one cm and several dm depending upon the manufacturing process used. Glass wools exhibit the lowest temperature stability with ceramic fibers, due to their low flux contents, at the high end of the temperature stability range and rock and slag fibers in between.

Raw Materials and Mineral Fiber Compositions

Raw materials for glass wool:

sand, limestone, dolomite, feldspar, kaolin, alumina-containing volcanic rocks, sodium carbonate, sodium sulfate, potassium carbonate, boron minerals

In the manufacture of *glass wool*, the raw materials are those usually used in the glass industry i.e. sand, limestone, dolomite, feldspar, kaolin, aluminum oxide-containing volcanic rocks, sodium carbonate, sodium sulfate, potassium carbonate and boron minerals such as kernite ($Na_2B_4O_7 \cdot 4H_2O$). The purity requirements particularly as regards iron content are not particularly high.

Rock wool is manufactured from sedimentary or magmatic rocks (e.g. clay, marl, basalt and in particular diabase) with small quantities of flux (e.g. lime, dolomite). *Slag fibers* are produced from slags e.g. blast furnace slags with added mineral raw materials. Pure ceramic raw materials are utilized in the manufacture of low flux *ceramic fibers*, particularly kaolin and cyanite (Al_2SiO_5) together with alumina, quartz and zircon but also including increasing amounts of synthetic starting materials.

Raw materials for rock wool:

clay, marl, basalt, diabase and fluxes of limestone or dolomite

Slag fibers are produced from slags from metallurgical processes with added flux

Raw materials for ceramic fibers:

kaolin, cyanite together with alumina, quartz, zircon and synthetic starting materials

The suitability of a raw material composition for fiber formation is determined by the ratio of the viscosity increasing components SiO_2 and Al_2O_3 to the viscosity decreasing ones such as alkali, alkaline earth, iron, manganese and titanium oxides. It can vary over wide limits. Table 5.2-12 gives typical compositions of the different mineral fibers.

Viscosity increasing compounds:
SiO_2, Al_2O_3

Viscosity decreasing compounds:

alkaline (earth), iron, manganese and titanium oxides

Table 5.2-12. Compositions of Mineral Fibers (% by weight).

	Glass fibers	Rock fibers	Slag fibers	Ceramic fibers
SiO_2	65.0	43.5	40.6	52.9
Al_2O_3	3.5	13.4	12.5	45.1
B_2O_3	5.0	–	–	0.08
Fe_2O_3/FeO	0.5	14.0	1.0	<0.1
CaO	14.0	10.8	37.5	–
MgO	3.0	10.0	5.0	–
TiO_2	–	2.0	0.4	1.7
CaS/S	–	–	1.5	–
K_2O	8.0	1.4	0.3	–
Na_2O	0.5	2.5	1.5	<0.2

Manufacture of Melts

The raw materials are melted at temperatures of 1200 to 1600 °C in tank, cupola, electric arc or resistance heated furnaces. Tank furnaces similar to those used for the manufacture of glass are used for the production of glass wool. The melt vessel is a large rectangular tank with the mixture added at one end and the melt taken off at the other.

The raw materials for the different mineral fibers are melted in different furnaces:

- Glass wool: tank furnaces with gas or oil heating

• Rock and slag wools: cupola furnaces

Water cooled cupola furnaces several meters high and with shafts up to 1 m wide are used for the manufacture of rock and slag fibers. The raw materials are added alternately with coke. Metallic iron is formed from iron-containing raw materials and has to be regularly drawn off.

• Ceramic fibers: electric arc and resistance-heated furnaces

Electric arc and resistance heated furnaces are mainly used for ceramic fibers due to the high melting temperatures of the raw materials.

Manufacture of Fibers

Fiber formation by

- Centrifugal processes
- Jet processes
- Rod-drawing processes
- Two-step centrifugal jet processes

The manufacturing processes can be divided into pulling, centrifugal and blowing processes. These one stage processes can be combined with a second blowing stage. The pulling process is currently less important than the centrifugal, blowing and two stage centrifugal blowing processes.

Centrifugal processes:

fibers formed from melts with the aid of centrifugal forces

Centrifugal processes: In the centrifugal process molten glass is thrown out of holes in the base of a metal spinner rotating at high speeds and is twisted into fibers. In the Hager-Rosengarth centrifugal process the melt flows onto a rotating gas-heated ceramic disc. The melt is flung off by way of axial guiding grooves and stretched to fibers 12 to 30μm in diameter and 100 to 600 mm in length.

The cascade centrifugal process is used for the manufacture of rock and slag fibers

Thin fibers 2 to 10μm in diameter and with an average length of ca. 15 mm are obtained by the cascade centrifugal process widely used for the production of rock and slag wool. In this process the melt flows onto one of three to four rotors with horizontal axes. The melt is then transferred at increasing speed from one rotor to the outer casing of the following rotor. The melt droplets flung from the rotors are twisted into fibers and separated in a stream of air from the solidified melt droplets.

Blowing process:

melt is drawn into fibers with high speed gas jets

Blowing process: Blowing processes also supply thin fibers 3 to 12μm in diameter and several cm in length.

In the atomization process the melt jet falls into the path of a horizontal high velocity jet of steam or air which draws it into fibers. In the jet blowing process fine staple fiber is produced from the fine melt streams that flow under gravity through holes in the base of platinum bushings. These are drawn into fibers by the action of high speed acute-angled gas jets from slit jets or a multitude of single jets and are broken into staples by the turbulence.

Two-step centrifugal jet process: The centrifugal jet process predominates in the manufacture of glass wool. The melt is thrown out of holes in the base of a metal drum rotating at high speed and the melt strands subjected at right angles to high velocity hot combustion gases from a concentric chamber. This jet draws the strands into fibers and the turbulence breaks them into fibers 5 to 10μm in diameter and 20 to 400 mm in length.

Drum centrifugal blowing process predominates in glass wool manufacture

Processing of Fibers into Insulating Materials

The fibers from the fiber-shreading units fall into a fleece shaft with a perforated conveyor belt at the bottom. The fibers are deposited as a thick fleece due to the partial vacuum behind the belt. A lubricating oil emulsion (e.g. of vegetable oil) is added to the fleece shaft to reduce the frictional forces and hence the chafing and breakage of fibers. In the absence of binders the fibers can be used as loose wool or stitched or sewn onto a carrier material, such as wire gauze, crepe paper or corrugated paper, into mats.

Production of insulating materials by deposition of fibers as a fleece

Lubricating oils hinder fiber breakage

Binder-free insulating materials:

- Loose wool
- Stitched on carrier materials into mats

In the manufacture of bonded insulating materials the fibers in the fleece shaft or on the conveyor belt are sprayed with an aqueous binder, mostly a phenol-formaldehyde resin. The binder content in the bonded insulating materials is 3 to 4%. Compaction to the desired density and hardening of the resin binder occurs continuously by conveyor belt transport through a tunnel kiln. The compaction is achieved with a second belt which exerts the required pressure on the upper surface of the continuous sheet. This is often followed by laminating the sheet with e.g. paper, aluminum or plastic foil. Finally the product is rolled up or cut into sheets.

Bonded insulating materials e.g. plates and rolls are manufactured by spraying a binder resin, mostly phenol/formaldehyde, onto fibers and hardening under compaction in a tunnel kiln

The important properties of the mineral fiber insulating materials are their low thermal conductivity (0.03 to 0.04 W/mK), low bulk densities (10 to 200 kg/m^3), porosity, elasticity, temperature stability and

Important properties of mineral fiber insulation materials are:

- Thermal conductivity
- Bulk density
- Porosity

- Elasticity
- Temperature stability
- Burning behavior

burning behavior (which is strongly influenced by the binder).

The thermal insulation properties depend upon the structure and the bulk density, which can be influenced during processing by varying the thickness, length and arrangement of the fibers in the fleece, the proportion of nonfibrous materials, the so-called bead content and the degree of compaction of the fibers. At constant bulk density the thermal conductivity increases with increasing fiber thickness, due to a reduction in the homogeneity of the distribution of included air. Thermal conductivity also increases with increasing temperature.

5.2.3.3.3 Applications

Main application sector for mineral fiber insulation materials:

thermal and sound insulation and fire protection in the construction industry

Mineral fiber insulation materials are mainly utilized in the construction industry for thermal and sound insulation and fire protection. Products on the basis of both glass wool and rock and slag wools are used. The thermal conductivity of glass wool is more favorable than that of rock wool, which results in glass fiber-based insulating materials having lower bulk densities than those based on rock fibers. Mineral wool products are used for the insulation of refrigeration plants, cold storage chambers etc., where hard foam insulating materials e.g. expanded poly(styrene) or poly(urethane) cannot be used due to more stringent fire protection requirements. For this application the mineral fibers have to be protected against moistening e.g. by condensed water.

Insulation of refrigerated plants requires mineral fiber insulation protected against condensed water

In the industrial sector loose wool and molded components e.g. cup-shaped pieces are used for the insulation of industrial plant as well as sheets and rolls. Preformed insulating tubes are utilized for insulating pipelines, often consisting of mineral wool mats stitched onto gauze.

Insulation of industrial plants with loose wool, sheets and rolls, as well as preformed molded components

The choice of mineral fiber for applications in the industrial sector (remote heating, plant construction, containers) must take their upper temperature limits into account and a choice made on the basis of the operating temperatures of the object concerned. These upper temperature limits are 350 to 400 °C for glass wool, ca. 700 to 750 °C for rock and slag wool and ca. 1250 °C for ceramic fibers. The upper temperature

The upper temperature limits for glass wool are 350 to 400 °C, for rock and slag wools ca. 700 to 750 °C and for ceramic fibers ca. 1250 °C limits for combinations with polycrystalline fibers (e.g. aluminum oxide fibers, Sec. 5.2.3.5) are considerably higher.

For thermal insulation at very high temperatures (e.g. in furnaces for the metal and ceramic industries) composite systems on the basis of different mineral fiber sorts are utilized. Ceramic or aluminum oxide fibers (e.g. Saffil®) are arranged on the hot side with rock wool on the cold side. This considerably reduces the consumption of expensive ceramic fibers. Recently, single layer composite systems consisting of ceramic fibers and aluminum oxide fibers have become more widely used.

For thermal insulation at very high temperatures ceramic fibers are combined with rock wool for economic reasons. Extremely thin high temperature resistant insulating layers can be manufactured with composite systems on the basis of ceramic fibers and aluminum oxide fibers

High temperature-resistant fibers can often be utilized in furnace construction instead of linings of refractory brick. This enables the construction of lighter and more rapidly heated and cooled furnaces resulting in substantial savings in fuel.

References for Sections 5.2.3.1 to 5.2.3.3: Mineral Fiber Insulation Materials

General Information:

Ullmann's Encyclopedia of Industrial Chemistry, 5th Ed., Vol. 10, Fibers, Synthetic Inorganic, Verlag Chemie, Weinheim-Deerfield Beach-Basel (1987).

Kirk-Othmer: *Encyclopedia of Chemical Technology,* 3rd Ed., Vol. 13, pp 591–605, John Wiley & Sons Inc., New York-Chichester-Brisbane-Toronto (1981).

J. Griffiths: "Synthetic mineral fibers", *Industrial Minerals,* pp 20–43 (Sept. 1986).

B. von Falkai (Ed.): *Synthesefasern, Grundlagen, Technologie, Verarbeitung und Anwendung* (Synthetic fibers, fundamentals, technology, processing and applications), pp 248–261, Verlag Chemie, Weinheim-Deerfield Beach-Basel (1981).

C. Z. Carroll-Porczynski: *Advanced Materials,* Astex Publishing Comp., Guildford (1962).

Specific Information:

Textile Glass Fibers:

J. G. Mohr, W. P. Rowe: *Fiber Glass,* Von Nostrand Reinhold Comp., New York (1978).

K. L. Loewenstein: *The Manufacturing Technology of Continuous Glass Fibres,* 2nd Ed., Elsevier Publishing Comp., Amsterdam (1983).

D. Hull: *An Introduction to composite materials", Cambridge University Press,* Cambridge (1981).

B. A. Proctor, D. R. Oakley, K. L. Litherland: "Developments in the assessment and performance of GRC over 10 years", *Composites* **13,** pp 173–179 (1982).

Chemical Economics Handbook, Stanford Research Institute, Menlo Park, California, USA: "Textile glass fibers – data summary" (March 1984).

Chemical Economics Handbook, Stanford Research Institute, Menlo Park, California, USA: "Man-made fibers – data sheet" (Aug. 1986).

International Symposium: *Man-made Mineral Fibers in the Working Environment,* Copenhagen, 28/29 Oct. 1986, World Health Organization.

Optical Fibers:

W. A. Gambling: "Glass, light and the information revolution", *Glass Technol.* **27,** pp 179–187 (1986).

"Various papers in Technical Digest of Conf. Optical Fiber Communications", Feb. 1986, Atlanta, USA, Optical Society of America.

Mineral Fiber Insulation Materials:

J. Fryatt: "Basic properties of ceramic fibers and their effect on insulation performance", *Appl. Energy* **2,** pp 117–126 (1976).

P. Schubert, P. H. I. Harper and D. Barham: "Criteria involved in the prediction of spinability of mineral wool melts", *J. Can. Ceram. Soc.* **48,** pp 17–22 (1979).

F. H. Fidler: "Ceramic fiber in ceramic kilns: past, present and future", *Ceram. Eng. Sci. Proc.* **6,** pp 1432–1436 (1985).

L. Zagar, W. Schaefer-Rolffs: "Untersuchung von magmatischen Gesteinen in Hinblick auf ihre Verwendbarkeit für die Herstellung von Gesteinsfasern" (Investigations into magmatic rocks in connection with their utilization for the manufacture of rock fibers), *Sprechsaal* **114,** pp 498–512 (1981).

Chemical Economics Handbook, Stanford Research Institute, Menlo Park, California, USA: "Speciality inorganic fibers – product review" (Jan. 1985).

H. G. Grimm: "Carcinogenic risks of man-made mineral fibers from mineral fiber insulation products", *Staub-Reinh. Luft* **46,** pp 105–111 (1986).

5.2.3.4 Carbon Fibers

5.2.3.4.1 Properties

Formation of carbon fibers:

by inert thermal degradation of nonmeltable polymer fibers

The industrial manufacture of carbon fibers is based on the thermal degradation of nonmelting organic polymers, or organic polymers which have been rendered nonmelting, to carbon in inert atmospheres.

The level of mechanical properties attainable with carbon fibers is dependent upon the crystallinity and structure of the carbon formed. These in turn are critically dependent both on the quality and composition of the fiber starting material and on the manufacturing technology used.

Industrially there are two types of carbon fiber, whose different mechanical properties can be explained in terms of their different fiber structures (see Table 5.2-13)

Isotropic carbon fibers:

amorphous glassy carbon, poor mechanical properties

Isotropic carbon fibers, whose degree of crystalline order is exceptionally low and whose structure is similar to glassy carbon, have the poorest mechanical properties. Their strength is, however, sufficient for their utilization as insulating fibers, filter media and catalyst carriers.

Anisotropic carbon fibers:

mechanical properties dependent upon the degree of orientation of the graphite strands in the direction of the fiber axis (elasticity modulus!)

Anisotropic fibers, on the other hand, consist of intertwined graphite strands. Their extreme anisotropic properties are known from measurements on graphite single crystals. Graphite exhibits high bond strength and a high elasticity in the layer plane, whereas these values are 1 to 2 orders of magnitude lower perpendicular to the plane. The level of the mechanical properties of anisotropic fibers (particularly the elasticity modulus) are therefore dependent upon the extent to which the graphite strands are ordered in the direction of the fiber axis.

Table 5.2-13. Manufacture and Properties of the Different Types of Carbon Fiber.

Starting materials	Structure	Fiber class	Fiber properties: Tensile strength in N/mm^2	Fiber properties: Elasticity modulus in N/mm^2	Fiber properties: Density in g/cm^3	Thermal conductivity parallel to the fiber in W/mK	Applications & application forms
cellulose fibers, cotton wool, poly(acrylonitrile) fibers, pitch fibers, animal fibers (sheep's wool)	isotropic (amorphous glassy carbon)	insulating fibers, textile carbon fibers	900 to 1100	40 000	1.5 to 1.6	15	insulation fibers in the form of carbon wool, carbon felts, catalyst carriers, woven carbon fibers; carbon packing cord for corrosion-resistant shaft seals with high temperature stability, filter material for corr-osive chemicals
poly(acrylonitrile)-fibers		standard fiber Type HT (high tensile strength)	3500 to 3700 max. 7200	240 000 max. 300 000	1.7 to 1.75	15 to 20	reinforcing fibers for plastics as fiber yarn rope, stable fibers, yarns, twist-ed yarns and woven textiles
	anisotropic	reinforcing fibers Type HM (high modulus)	2000 to 2500	350 000 to 500 000	1.8 to 2.0	60 to 100	
fibers from mesophase pitch		standard	600 to 1000 max. 3000 max. 5000	30 000 to 33 000 with 700 000 with 250 000			

In the high modulus(HM-type)-fibers, also known as graphite fibers, the layer planes of the graphite fibrils are oriented parallel to the fiber axis with high long range order. In the case of high strength carbon fibers (HT-type) the layer planes are oriented along the fiber axis, but with poor long range order.

Strength of the carbon fibers due to the strand-like graphite crystals which intertwine with one another

The ca. 6 nm wide and over 100 nm long strand-like graphite crystals formed by the degradation of the organic polymer material are arranged into microfibrils which intertwine with one another. This is the basis of the strength of this fiber type. They represent the building blocks of all the anisotropic carbon fibers (HT and HM types).

5.2.3.4.2 Manufacture and Applications

The manufacture of carbon fibers generally occurs by the thermal degradation of suitable organic polymers.

The polymers utilized as starting materials for the reinforcing fibers (HT and HM types) have to fulfill the following requirements:

Requirements for starting materials for reinforcing carbon fibers:

- Nonmelting
- High yield of carbon upon thermal degradation
- Anisotropic, graphitic structure after carbonization

- No melting during thermal degradation. Melting polymers can be utilized after a prior stabilizing treatment to render them nonmelting,
- High yield of carbon upon thermal degradation,
- The carbon structure left after pyrolysis should have a graphitic structure which is as anisotropic as possible and have as few defects as possible.

Isotropic carbon felt, carbon wool and woven carbon by the pyrolysis of the analogous organic starting materials:

- 1st stage: precoking at 300 °C
- 2nd stage: carbonization at ca. 1000 °C

Carbon felt, carbon wool and woven carbon, which due to their isotropic structure only exhibit low mechanical strengths and low elasticity moduli, are manufactured by the pyrolysis of organic textiles. Depending on the form of the starting material utilized e.g. as woven textiles or felt, carbonization produces the corresponding woven carbon or carbon felt.

The manufacturing process proceeds in two stages. The first stage comprises the decomposition of the organic material at ca. 300 °C (precoking). In the second stage the precoked material is degraded at ca. 1000 °C to elemental carbon in the absence of air.

Applications of (isotropic) carbon felts, carbon wool and woven carbon:

Carbon felts are mainly utilized for thermal insulation at high temperatures e.g. in resistive or induction furnaces. Carbon wool is manufactured by the decomposition of cotton wool or similar raw materials and is suitable as a packing material for high temperature thermal insulation. Its resistance to chemical corrosion makes it suitable as a filter material for corrosive media, as a carrier material for catalysts and for corrosion-resistant linings in the construction of chemical plant.

- High temperature thermal insulation
- Packing materials
- Filter material
- Catalyst carrier

Table 5.2-14. Carbon Yields from the Pyrolysis of Different Starting Materials.

Starting material	Stabilizing treatment	Calcining temperature (°C)	Carbon yield in %
poly(vinylalcohol)	atmospheric oxidation at 200 °C	up to 2500	ca. 20
phenolic resins	–	up to 900	ca. 65
rayon	HCl; O_2	up to 1300 up to 2900	30 25
PAN homopolymers or copolymers	oxidation at 220 to 250 °C	1100	ca. 60
mesophase pitch	atmospheric oxidation below the softening point	1500	80

Carbon fibers manufactured from pitch are also isotropic when neither the pitch nor the woven fiber has undergone a special treatment. In this treatment pitch or coal extract is melt spun at temperatures between 250 and 400 °C and the fibers crosslinked and thus made nonmelting by treatment with oxidizing agents such as atmospheric oxygen or Lewis acids. Subsequent heating in inert atmospheres at temperatures above 1000 °C carbonizes the fibers. Suitable raw materials are thermally degraded poly(vinylchloride), crude oil bitumen or hard coal pitch, which are dissolved in high boiling point highly aromatic oils. The manufacture of high modulus carbon fibers from pitch is described in the schema in Fig. 5.2-3.

Manufacture of isotropic carbon fibers from pitch:

- 1st step: melt spinning
- 2nd step: render nonmelting (oxidation)
- 3rd step: inert carbonization

Raw materials for carbon fibers from pitch:

- Thermally degraded PVC
- Coal extracts
- Tar pitch
- Bitumen

PAN-fibers are currently mainly utilized for the industrial manufacture of anisotropic reinforcing fibers of

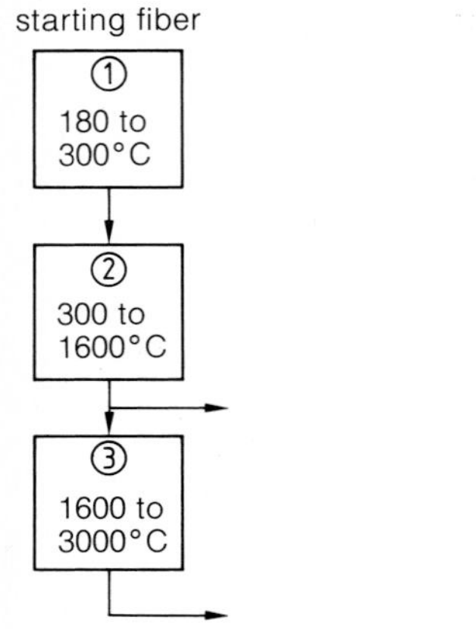

Fibers on the basis of PAN, cellulose:

① 1st Process step "stabilization":

Oxidative and/or other chemical treatment (with fiber stretching for PAN) as step prior to solid phase pyrolysis

② 2nd Process step "carbonization":

Thermal decomposition of stabilized, nonmeltable fiber intermediate product

Product: *carbon fiber* type HT (high tensile strength) from PAN

③ 3rd Process step "graphitization":

High temperature treatment (with fiber-stretching for cellulose - stretch graphitization)

Products: *carbon fiber* type HM (high modulus)

Fig. 5.2-2. Manufacture of reinforcing carbon fibers (schematic).

the HT- and HM-types. The production process is shown in the above flow chart (see Fig. 5.2-2).

Reinforcing carbon fibers from poly-(acrylonitrile) (PAN):

- 1st step: stabilizing oxidation with clamped fibers at temperatures up to 300 °C
- 2nd step: carbonizing in an inert atmosphere at temperatures up to 1600 °C
- 3rd step: graphitizing at temperatures up to 3000 °C

The PAN-fibers utilized as starting materials are first clamped to prevent shrinkage of the fibers during the subsequent oxidative pretreatment at temperatures below 300 °C. This clamping also produces preorientation in the direction of the fiber axis in the subsequently formed graphite crystals. In the second stage (carbonization stage) at ca. 1600 °C, high strength carbon fibers of the HT-type are formed, which are converted in a third stage by graphitization at temperatures up to 3000 °C into high modulus carbon fibers (HM-type). Significant improvements in the properties have been recently achieved by mutual fine tuning of the raw materials and the carbonization process. Thus the elongation at fracture has been increased from the 1.25 to 1.5% typical of the HT-type to the 2% of the so-called HST(high strain)-types with a tensile strength of 5000 N/mm^2, while retaining an elasticity modulus of 250 000 N/mm^2. Specific strength values similar to those of the organic aramide reinforcing fibers have thus been achieved.

HST-fibers (high strain) attained with 2% elongation at fracture

Industrially, carbon fibers are currently manufactured in the form of rope yarns (with up to 320 000 single filaments), as staple fibers of different lengths, as yarns, twisted yarns and also as woven textiles and felt. Carbon fibers are flexible and can be treated as textile fibers due to the small diameter of the individual fibers (down to ca. 8 μm).

In the manufacture of carbon fibers from regenerated cellulose (Rayon) stretching of the graphite strands at high temperatures is necessary to obtain the optimum properties. This takes place by so-called stretch graphitization in which the fibers are stretched by tensioning at temperatures above 2600 °C. This increases the degree of anisotropy and thereby the elasticity modulus. This process has decreased in importance due to considerable technical problems.

Reinforcing carbon fibers from regenerated cellulose (Rayon): stretch graphitization at temperatures above 2600 °C required for the manufacture of high modulus carbon fibers

A Union Carbide process for the manufacture of reinforcing carbon fibers from special pitch is expected to lead to considerable price reductions in the future. A schema of this manufacturing process is shown in Fig. 5.2-3.

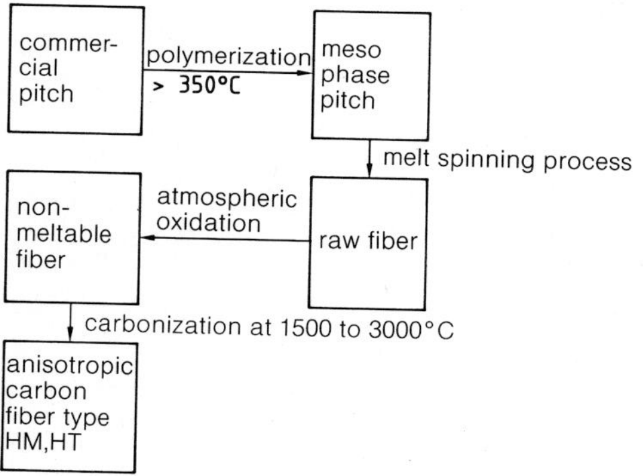

Fig. 5.2-3. Manufacture of reinforcing carbon fibers from pitch.

Bitumen or coal tar is polycondensed by thermal treatment at temperatures above 350 °C, whereupon the so-called mesophase is formed which already contains a significant proportion of preformed anisotropic graphite. This mesophase pitch is then melt spun and made nonmeltable by oxidizing the green fiber. It is subsequent carbonized at temperatures between 1500 and 3000 °C. The great advantage of this process is a carbon yield of ca. 80% compared with 60% or less when other starting materials are utilized. In addition neither tensioning nor stretch graphitization are required to produce carbon fibers with anisotropic struc-

Reinforcing carbon fibers (anisotropic) from bitumen or tar using a Union Carbide Corp. process:

- 1st stage: polymerization at temperatures > 350 °C to mesophase pitch

- 2nd stage: melt spinning
- 3rd stage: render nonmeltable (by oxidation)
- 4th stage: carbonization in an inert atmosphere at 1500 to 3000 °C

tures. Carbon fibers from mesophase pitch with mechanical properties close to those of carbon fibers obtained from poly(acrylonitrile) have been attained.

5.2.3.4.3 Economic Importance

Table 5.2.15. Consumption of Carbon Reinforcing Fibers in t.

Year	Japan	USA	Europe	Other regions	Worldwide (total)
1981	290	550	140	50	1050
1982	470	600	220	150	1440
1984	580	850	310	390	2130
1985					ca. 3000
1987*	800	2000	650	600	4050
Production capacity (on PAN basis)					
1986	2880	2000	1860	–	–

* estimated

1985 worldwide consumption of carbon reinforcing fibers: ca. 3000 t

Reinforcing carbon fibers are relatively recent industrial products which are currently being introduced. Since about 1970 they have been utilized in the fields of military aircraft, spacecraft and more recently in the sports article sector (golf clubs, fishing rods, tennis rackets, skis, boots, yachts, masts, sports cars etc.). Its breakthrough into civil aviation came in 1980 (1 t carbon fiber per Boeing 767!, side tail units of the Airbus representing the first fiber-reinforced primary structural component) and in the event of further price reduction this could be extended to industrial fields to achieve, for example, weight savings in motor vehicles, lighter engine components and light-weight components for machines and robots.

Its price evolution will determine the future development of this market. At an average price of ca. 75 US $ for fibers based on PAN, the worldwide turnover in 1985 was ca. 6×10^9 US $.

5.2.3.5 Aluminum Oxide Fibers

Applications of aluminum oxide fibers in:

- Reinforcing sectors
- Thermal insulation at high temperatures

Aluminum oxide fibers are utilized as reinforcing and thermally insulating fibers in a similar way to carbon fibers.

The most widely developed aluminum oxide fiber for the fiber-reinforcing sector is the α-Al_2O_3 fiber from Du Pont ("FP-fiber", see Fig. 5.2-4). Raw fibers are

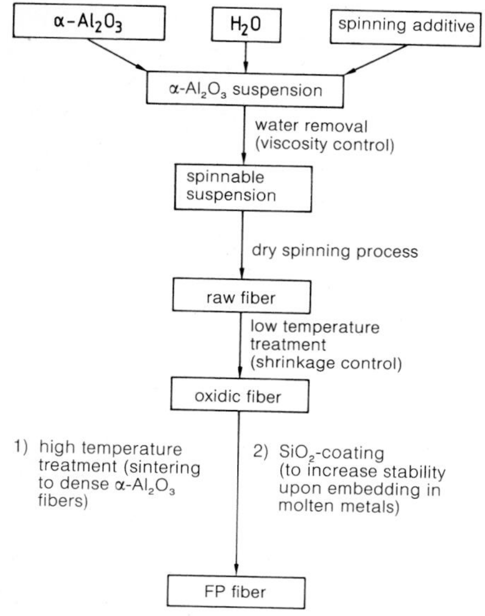

Fig. 5.2-4. FP-aluminum oxide fiber manufacture (Du Pont).

produced by spinning an extremely fine particulate aluminum oxide-dispersion with the help of additives. These are converted into densely sintered α-aluminum oxide fibers by high temperature treatment. According to Du Pont its manufacturing costs are comparable with those for the manufacture of high modulus carbon fibers. Aluminum oxide-fibers with a diameter of ca. 20 μm have the disadvantages of high density (3.9 g/cm^3) and brittleness.

Manufacture of reinforcing α-Al_2O_3 fibers:

- Spinning of an Al_2O_3-suspension to a raw fiber
- Sintering of the raw fiber

The fibers are manufactured in bundles of 210 individual filaments and exhibit an average tensile strength of 1380 N/mm^2 and an elasticity modulus of 379 000 N/mm^2. The average crystallite size is 0.5μm. It is expected that it will be particularly utilized for the reinforcing of aluminum and magnesium. For this application a variant coated with a layer of SiO_2 (to aid wetting) with a higher tensile strength (1900 N/mm^2) is preferred.

Properties of the Du Pont FP-fiber:

- Density: 3.9 g/cm^3
- Tensile strength: 1380 N/mm^2
- Elasticity modulus: 379 000 N/mm^2

with a SiO_2-coating:

- Tensile strength: 1900 N/mm^2

Manufacture of ICI's Saffil®-fibers:

- Blowing of a spinnable aqueous Al-salt solution
- Sintering the fibers

Another type of aluminum oxide-fiber is the Saffil®-fiber from ICI. The raw fibers are produced in the form of short fibers up to several cm in length by a blowing process from an aqueous medium. The spinnable or blowable solution is obtained from different aluminum salts by adding small quantities of SiO_2 and high molecular weight spinning aids [poly(ethyleneoxide)]. The raw fibers produced by blowing then have to be decomposed by heating at high temperatures, whereupon the individual fibers consolidate by sintering into a polycrystalline structure of a transitional aluminum oxide.

Applications of Saffil® fibers:

High temperature thermal insulation material up to 1650 °C

Forms in which it is utilized:

- Papers
- Wools
- Mats
- Fabrics
- Boards
- Molded articles
- Elastic cement in kilns

These Saffil® fibers are mainly utilized for thermal insulation applications at high temperatures, with a maximum of 1650 °C. They are marketed in the form of papers, wools, mats, felts, fabrics, boards and molded articles and are also used in the iron and steel industry. The very good insulation properties of the fibers enable considerable energy savings to be made, due to reduction in heat losses and increase in heating rates e.g. for ceramic firing kilns. The fibers are also used as a substitute for brick lining or as an elastic cement in furnaces and can considerably increase the lifetime of components exposed to high temperatures. A particular advantage of aluminum oxide fibers is that they can be utilized in reduction furnaces in which silicate fibers would be destroyed due to the formation of SiO.

Recently a special type of Saffil® for the reinforcing of metals (Al, Mg), Saffil-RF®, has come onto the market. In this it competes with glassy melt spun mullite fibers.

A number of Japanese companies have also begun to manufacture polycrystalline Al_2O_3-insulating fibers e.g. Denki Kagaku Kogyo. The total world capacity in 1986 was just under 1000 t/a and the price ca. 60 US $/kg.

5.2.3.6 Boron Fibers

Manufacture of boron fibers:

By vapor deposition of boron on carbon or tungsten fibers according to:

$$BCl_3 + 1.5H_2 \xrightarrow{1200 \text{ to } 1300\,^\circ C} B + 3HCl$$

Boron fibers are manufactured by chemical vapor deposition (CVD) of boron by the reduction of boron(III) chloride on substrate fibers heated to 1200 to 1300 °C:

$$BCl_3 + 3/2H_2 \longrightarrow B + 3HCl$$

Tungsten and carbon fibers are suitable substrates as they can be heated up to the reaction temperature by direct resistive heating using mercury contacts. The reaction tube, which is ca. 2 m high, allows the substrate fibers to pass individually through it.

The high density of the tungsten fibers, which have a diameter of ca. 12 to 15 μm, means that the fiber thickness has to be increased to 100 to 140 μm to achieve the required low density of the final fibers: ca. 2.6 g/cm^3. Such thick fibers are, however, very stiff and difficult to process and therefore deposition on carbon fibers is preferred.

Properties of boron fibers:

- Density: ca. 2.6 g/cm^3
- Tensile strength: 3000 to 4000 N/mm^2
- Elasticity modulus: ca. 400 000 N/mm^2

Commercially obtainable boron fibers exhibit elasticity moduli of 400 000 N/mm^2 and tensile strengths of 3000 to 4000 N/mm^2. The upper temperature limit for continuous operation is given as 370 °C. This increases to 600 °C for boron fibers coated with 2 to 6 μm thick barrier layers of silicon or boron carbide, which are marketed for incorporation into metallic matrices (particularly aluminum and titanium alloys).

Boron fibers have up to now been almost exclusively utilized in USA and Western Europe for the manufacture of light structures in the aerospace industry. They are marketed in the form of reinforcing strands in which the fibers are arranged parallel to one another in a partially polymerized epoxy resin so-called prepregs or embedded in an aluminum alloy foil.

Applications of boron fibers:

Reinforcement of aerospace components

The worldwide production capacity was 15 to 20 t/a in 1974 and ca. 40 t/a in 1980. In 1979 the price of boron fibers was 800 US $/kg. This may drop to as low as ca. 400 US $/kg in the future.

1980 world capacity: ca. 40 t/a

5.2.3.7 SiC-Fibers and SiC-Coated Carbon Fibers

Silicon carbide fibers can be manufactured in the same way as boron fibers by using trichloromethylsilane as the starting material. The deposition proceeds according to:

SiC-fibers manufactured by chemical vapor deposition (CVD) on W- or C-fibers:

$CH_3SiCl_3 \xrightarrow{1200\,°C} SiC + 3HCl$

$$CH_3SiCl_3 \xrightarrow{1200\,°C} SiC + 3HCl$$

and can also be performed on tungsten fiber or carbon fiber substrates. Compared with boron fibers, SiC-fibers exhibit the advantages of higher stability to chemicals and higher oxidation stability while retaining their strength and infusibility. Unlike the deposition of boron, no reaction takes place during the deposition of SiC on carbon fibers, which makes them especially suitable as substrates. Carbon fiber substrates are also to be preferred for economic reasons and the resulting reduction in overall fiber density. Commercial SiC-fibers exhibit monofilament tensile strengths between 2000 and 4000 N/mm^2 and elasticity moduli of ca. 450 000 N/mm^2. They are suitable for the reinforcement of plastics, metals and other matrices.

Properties of SiC-fibers manufactured using the CVD process:

- Tensile strength: 2000 to 4000 N/mm^2
- Elasticity modulus: ca. 430000 N/mm^2
- Density: ca. 3.4 g/cm^3 (W)
ca. 3.0 g/cm^3 (C)

Fibers manufactured by the deposition of SiC on carbon yarns, consisting of up to 10 000 individual fibers ca. 10 μm in diameter, should also be added to this category. In this case the SiC-skin produced by the thermal decomposition of trichloromethylsilane above 1000 °C only has a thickness of ca. 0.5 μm. All the monofilaments can be uniformly coated by a suitable choice of deposition conditions.

Fibers coated with SiC (as a reaction barrier) are suitable for reinforcing reactive matrices (Al, glass) up to 400 °C

Since the SiC-layers serve as a reaction and diffusion barrier, SiC-coated carbon fibers are suitable for the reinforcement of reactive matrices such as aluminum or glass. In the case of aluminum, the reaction barrier hinders the formation of aluminum carbide. As a result such composite materials are utilizable up to temperatures of 400 °C.

A pure SiC-fiber not based on substrate fibers is the β-SiC-fiber (Nicalon® fiber from Nippon Carbon Ltd.). In its manufacture (see Fig. 5.2-5) dichlorodimethylsilane is condensed with sodium to poly(dimethylsilane) and is converted either by treatment under pressure in autoclaves or reaction with the aid of a special catalyst to spinnable poly(carbosilanes). The spun fibers are rendered infusible and then subjected to a high temperature treatment whereupon β-silicon carbide fibers are produced with tensile strengths of 2500 to 3000 N/mm^2 and elastic moduli of ca. 200000 N/mm^2.

Properties of β-SiC-fibers from poly(carbosilanes) (Nicalon®):

- Tensile strength: 2500 to 3000 N/mm^2
- Elasticity modulus: ca. 200000 N/mm^2

SiC-fibers are also in the market introduction phase. Nippon Carbon Ltd. has a capacity of 12 t/a.

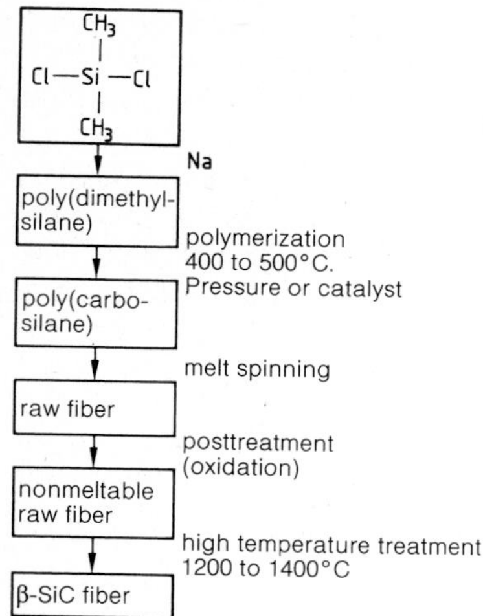

Fig. 5.2-5. Manufacture of the SiC-fiber Nicalon®.

5.2.3.8 Metal Fibers

All metal fibers are polycrystalline inorganic fibers, except for tungsten fibers which can include large single crystalline regions. They are mainly manufactured by physical deformation processes, chemistry playing a minor role. They are mentioned here only for the sake of completeness.

5.2.3.8.1 Manufacture

Metal fibers can be produced by machining processes; by foil cutting processes; powder metallurgically by sintering metal powders, which are extruded into fibers with the aid of organic binders; by metallizing non-metallic fibers and by controlled chemical dissolution of wires to the required fiber thickness. Thin metal wires and thick metal fibers can, in principle, be manufactured by the same methods.

Manufacture of metal fibers:

- Machining processes
- Foil cutting processes
- Extrusion processes
- Metallizing of fibers
- Dissolution of wires

and particularly for the manufacture of thin fibers:

- Mechanical drawing processes
- Melt spinning processes
- Taylor process

Special processes are also known particularly for the manufacture of thin metal fibers: mechanical drawing, melt spinning and the Taylor process.

In *mechanical drawing processes* rolled or predrawn wire are pulled through orifices, die stones or die rings, of decreasing cross-sectional areas on drawing banks in a multistage process. The increasing brittleness with

Mechanical drawing processes:

multistep drawing of wires through ever smaller orifices with annealing between steps to eliminate embrittlement. This process is utilized in the manufacture of steel fibers for tyre cords (ca. 150 μm in diameter)

decreasing fiber diameter can be eliminated by annealing between treatments. The manufacturing costs of metal fibers increase steeply with decreasing fiber diameter due to the increasing probability of fiber fracture. Fibers with diameters above 150 μm can be manufactured fairly inexpensively. Such methods are used for the production of the steel fibers important for the tire industry. The brass-coated fibers have diameters of ca. 150 μm.

Bundle drawing process:

mechanical drawing process with a wire bundle embedded in a matrix (Cu); fiber diameters down to 0.5 μm obtainable

Much thinner metal fibers are manufactured using the so-called *bundle drawing process* in which the wires are embedded in a ductile matrix (copper) and together subjected to a mechanical drawing process. The fibers remaining after removing the matrix have diameters down to 12 μm. Recently diameters down to 0.5 μm have been achieved with this process.

Melt spinning process:

sudden cooling of a jet of molten metal

In the *melt spinning process*, a metal melt is forced through dies in a thin jet into a liquid medium at such a speed that the cooling rate is faster than the rate of disintegration of the jet into droplets. Monsanto had planned to produce 75 μm thick tire cords using this process, but they are not yet in production.

Melt extraction process:

cooled rotating discs pull fiber from a metal melt

The *melt extraction process* seems to be more widely utilized industrially. In this process a metal fiber is produced by contact dipping of a cooled rotating disc in a metal melt or a melt droplet. Fibers with diameters down to 40 μm and lengths of up to several centimeters can thereby be produced.

Taylor process:

melt and draw a metal-filled glass tube. Fiber diameters down to 1 μm

The *Taylor process* is a variant of the melt spinning process in which a glass tube filled with a metal or metal alloy powder or wire is melted by passing through a furnace and then drawn to a metal-filled thread. Extremely thin threads down to 1 μm in diameter can be produced as single filaments (Taylor wires). Depending upon the application field the glass skin can either be removed or it can be retained as an insulating coating.

5.2.3.8.2 Properties, Applications and Economic Importance

Metal fibers exhibit a number of valuable properties such as electrical and thermal conductivity, high tensile strength, high elasticity modulus and high melting points (see Table 5.2-16).

Table 5.2-16. Properties of Metal Fibers.

Fiber	Melting point °C	Density g/cm³	Tensile strength N/mm²	Elasticity modulus N/mm²	Diameter μm
steel (tyre cord)	1400	7.8	2500	205 000	150
steel martensite hardened (Taylor wires)	1400	8	3800	210 000	7
tungsten	3400	19.3	3000 to 4000	350 000 to 420 000	12

Since the densities of the metal fibers are relatively high, they are only suitable as reinforcing materials when no extreme demands are made regarding weight saving, in particular the concrete and rubber sectors. Since 1959 steel cord has increasingly dominated the tire market. In 1986 700 000 t of steel fiber cord was used worldwide. In Western Europe $>90\%$ of cars are fitted with steel-braced radial tires. The 1978 prices for steel cord were 4 to 5 US $/kg and those for stainless steel fibers produced by the melt extraction process of ca. 6 to 9 US $/kg.

Stainless steel fibers produced by the bundle drawing process are utilized for seals, in silencers, for antistatic finishes and as filter materials.

Applications of steel fibers:

- Reinforcement of concrete and rubber (tire cord, seals)
- Silencers
- Rendering textiles antistatic
- Filter materials

An important application field for stainless steel fibers is the textile sector, in which carpets and protective clothing are rendered antistatic by incorporating 0.5 to 6% of these fibers. The current price of antistatic metal fibers is ca. 120 US $/kg. The tungsten fibers 12 μm thick used, for example, for the deposition of boron or SiC and as the filament in light bulbs cost ca. 600 US $/kg.

1986 consumption of steel fiber cord worldwide: 7×10^5 t

5.2.3.9 Whiskers and Miscellaneous Synthetic Short Fibers

Since the 1960's potassium titanate fibers ($K_2Ti_6O_{13}$) have been produced for the most diverse applications

Synthetic mainly single crystalline short fibers:

- $K_2Ti_6O_{13}$
- $CaSO_4 \cdot 1/2H_2O$
- $CaSO_4$

Applications as a filler for reinforcing and increasing dimensional stability

(pigments, reinforcing fibers for plastics) in the form of thin (0.2 to 0.5 μm) long (10 to 20 μm) single crystalline needles from salt melts (alkali halide, alkali borate) or reaction mixtures with the addition of fluxes (K_2MoO_4). Since Du Pont discontinued its production for ecological reasons, it has been exclusively produced in Japan. The largest manufacturer, Otsuka Chemical, had in 1986 a capacity of ca. 700 t/a. Its price is currently ca. 15 US $/kg.

$CaSO_4 \cdot \frac{1}{2} H_2O$ and $CaSO_4$ single crystal fibers ca. 2 μm in diameter and 60 μm in length (Franklin Fiber of US Gypsum) are utilized partly as a substitute for glass fibers as cheap reinforcement and fillers in thermoplastics and duroplasts to improve their dimensional stability and as reinforcement for plastic molded articles. The production capacity of US manufacturers is ca. 4000 t/a.

Whiskers:

extremely thin single crystal fibers, which, when defect free, exhibit extremely high tensile strengths (> 10 000 N/mm^2):

- β-SiC
- α-Si_3N_4

Defect-free single crystal fibers several μm in diameter and currently up to well over 1000 μm in length, so-called whiskers, possessing otherwise unachievable tensile strengths (up to more than 10 000 N/mm^2), have been marketed since about 1985. The most important are the β-SiC and α-Si_3N_4 whiskers, the Al_2O_3 whiskers being only of minor importance. The 1986 worldwide production capacity is certainly less than 100 t/a. The manufacturers are Arco Advanced Materials Co. (USA) and several Japanese companies (Tokai Carbon, Tateho Chemical etc.). Their current very high price (ca. 250 to 600 US $/kg) and the availability of polycrystalline filaments with very good properties hinder the wider utilization of whiskers. Application is therefore anticipated in fields in which polycrystalline fibers can not fulfill the extreme requirements demanded e.g. increase in fracture toughness of engineering ceramics at high temperatures. A first practical application was found in Al_2O_3 cutting ceramics with 15% SiC whiskers.

Application:

increasing the fracture toughness of Al_2O_3 cutting ceramics

References for Sections 5.2.3.4 to 5.2.3.9: Other Inorganic Fibers

ASTM Committee D-30 (1966).

H. Böder, D. Gölden, Ph. Rose and H. Weiruscher: *Z. Werkstofftech.* **11,** pp 275–281 (1980).

O. Vohler, P.-L. Reiser, R. Martina and D. Overhoff: *Angew. Chem.* **82,** pp 401–452 (1970).

Ullmanns Encyklopädie der technischen Chemie, 4th Ed., Vol. 11, pp 359–390, Verlag Chemie, Weinheim-Deerfield Beach-Basel, 1976.

E. Fitzer and J. Schlichting: *Z. Werkstofftech.* **11,** pp 330–341 (1980).

"New fibers and their composites", *Phil. Trans. R. Soc. Cond.* **A 294,** pp 407–597 (1980).
E. Fitzer: *Angew. Chem.* **92,** pp 375–386 (1980).
S. Writers: *Chemical Economy & Engineering Review* **12(9),** No. 141, pp 36–39, (Sept. 1980).
A. Budzinski: *Chem. Ind.* **XXXIII,** pp 269–271 (May 1981).
John A. Roberts: *Handbook of Fillers and Reinforcements for Plastics.* Ed. H. S. Katz, J. V. Milewski (1978).
Ind. Heat **47 n,** pp 28–29 (12 Dec. 1980).
J. Tanaka: *Chemical Economic & Engineering Review* **12(9),** No. 141, pp 40–45 (Sept. 1980).
S. Writers: *Technocrat* **18(2),** (Feb. 1985).

5.3 Construction Materials

5.3.1 General Introduction

Many inorganic substances are used in the construction industry:

Bonding agents are inorganic substances which can bond pieces of stone together. They are processed with water and harden through chemical or physical reactions to a stone-like mass. There are hydraulic bonding agents and air bonding agents. Hydraulic bonding agents harden both in air and water and when hard are water resistant. The different silicate cements (Sec. 5.3.3) and hydraulic hardening limes (Sec. 5.3.2.3) fall into this category. Air bonding agents only harden in air and are not water-resistant (Sec. 5.3.2.4).

Bonding agents: inorganic substances which harden in the presence of water or air which bond pieces of stone

Hydraulic bonding agents: harden in air *and* in water (e.g. cement)

Air bonding agents: harden only in air (e.g. calcined gypsum)

Calcined gypsum, anhydrite (Sec. 5.3.4), magnesia cement (Sec. 5.3.3.8) and silicate-free limes (Sec. 5.3.2) belong to this category. Mixtures of bonding agents with sand are known as *mortar,* mixtures of hydraulic bonding agents with coarse admixtures such as gravel, rubble or expanded materials (Sec. 5.3.6) are known as *concrete. Plasters* are mixtures of lime and sand or lime, sand and gypsum.

Mortar: mixture of bonding agents with sand

Concrete: mixture of gravel, rubble or expanded materials with hydraulic bonding agents

Natural stone or artificial *stones* such as bricks (Sec. 5.3.5), sand-lime bricks, aerated concrete bricks (Sec. 5.3.2.5) and expanded-clay blocks (Sec. 5.3.6.2.3) are used for building walls.

Building of walls with stones, bricks, sand-lime bricks, aerated concrete bricks

Tiles (Sec. 5.3.5), concrete roofing tiles, asbestos cement tiles (Sec. 5.3.3.7), porous gypsum boards (plasterboard) and metallic materials are used as roofing materials.

Roofing materials: tiles, concrete roofing tiles, asbestos cement tiles, metals

Glass, sound and thermal insulation materials

Other inorganic materials are also used, which are considered in other sections of this book: glass (Sec. 5.1.1), and sound and thermal insulation materials (Sec. 5.2.3).

5.3.2 Lime

5.3.2.1 Economic Importance

"Limestone": $CaCO_3$

"Quicklime":

"Slaked lime" or "lime hydrate": CaO $Ca(OH)_2$

The term lime normally covers both limestone ($CaCO_3$) and "quicklime" (CaO), which is produced from it by calcination. "Slaked lime" [$Ca(OH)_2$] is formed by reacting quicklime with water.

1984 lime consumption in:

- West Germany: 6.284×10^6 t

Table 5.3-1. Lime Consumption in West Germany 1982 to 1985 in 10^3 t (according to statistics from the Bundesverband der Deutschen Kalkindustrie).

Consumer	1982	1983	1984	1985 (est.)
iron and steel industry	2068	2094	2435	2435
chemical industry	388	403	411	400
construction materials industry	1328	1435	1279	950
building trade	696	733	603	500
agriculture	445	445	347	350
various incl. export	649	542	646	765
water treatment	439	509	563	600
Total	6013	6269	6284	6000

1981 lime production:

- World: 117.4×10^6 t
- EEC: 19.5×10^6 t
- USA: 17.1×10^6 t

Table 5.3-2. Lime Production in 10^6 t [according to the Mineral Yearbook Vol. I (1982)].

Year	1978	1979	1980	1981	1982 (est.)
USA	18.5	19	17.2	17.1	12.8
EEC	22.6	22.2	20.9	19.5	19.6
World	119.9	120.6	118.9	117.4	111.9

5.3.2.2 Raw Materials

Deposits of calcium carbonate:

- As calcite, aragonite and vaterite
- With $MgCO_3$ in the form of $CaMg(CO_3)_2$ as dolomite
- In mixtures with clays as marl

The starting material calcium carbonate is found in nature as calcite and more rarely as aragonite and vaterite. In natural deposits calcium carbonate is often found as mixtures with magnesium carbonate. The mineral $CaMg(CO_3)_2$ is known as dolomite.

Mixtures of limestone with clays are so-called marl. Other deposits of limestone are marble, travertine, shell limestone and chalk whose properties depend upon their origin.

- Further deposits as marble, travertine, shell limestone and chalk

5.3.2.3 Quicklime

"Lime burning" is the thermal decomposition of $CaCO_3$:

$$CaCO_3 \longrightarrow CaO + CO_2$$

"Lime burning":

$CaCO_3 \rightarrow CaO + CO_2$

The decomposition temperature of $CaCO_3$ is ca. 900 °C, that of $MgCO_3$ between 400 and 480 °C. Industrially the decomposition of $CaCO_3$ is carried out at 900 to 1100 °C.

Industrial decomposition of:

- $CaCO_3$ at 900 to 1100 °C

With dolomitic rock, which can contain up to 50% dolomite, the decomposition proceeds in two stages: first the magnesium carbonate is decomposed at 650 to 750 °C and then the calcium carbonate is decomposed above 900 °C. When the limestone is contaminated with clays (marl), compounds such as $2CaO \cdot SiO_2$, $3CaO \cdot Al_2O_3$, $2CaO \cdot Fe_2O_3$ and $4CaO \cdot Al_2O_3 \cdot Fe_2O_3$ are formed upon calcination. These compounds are also constituents of Portland cement clinkers (Sec. 5.3.3.3.2) and harden upon reaction with water (hydraulic limes or highly hydraulic limes), but not upon reaction with carbon dioxide.

- $MgCO_3$ at 650 to 750 °C

Hydraulic limes: by decomposition of marl

Lime burning formerly took place in holes in the ground and kilns fired with wood, coal or peat. These developed into the so-called "shaft kilns". Limestone is ground into coarse particles (∅ ca. 60 to 120 mm), mixed with coal and deposited into the vertical shaft kiln. The burning of the mixed-in coal produces a temperature of 900 to 1100 °C, whereupon carbon dioxide is liberated.

Industrial lime burning:

- In shaft kilns

The quality of the burnt lime is not uniform with this oven type as only coarse particles can be burnt. This led to the development of improved kiln types such as crossflow kilns, doubly inclined kilns (with additional gas or oil heating), parallel flow/countercurrent/regenerative kilns and ring shaft kilns. These new developments all achieved a more uniform temperature distribution and thus a more uniform product quality. In

- In modified shaft kilns

addition the processing of more finely ground limestone (∅ ca. 20 to 70 mm) became possible.

- In rotary kilns

Rotary kilns are also frequently used for lime burning. These enable the finest limestone particles (up to ca. 10 mm) to be uniformly calcined.

- In ring kilns

Ring kilns have a horizontal firing channel with floating flames. Loading and unloading are manual.

- On calcining grates

Continuous traveling grates are used as calcining hearths in which a mixture of limestone and fuel (mainly coal) is calcined by forced draught heating.

- In calcimatic furnaces (rotary grate kilns)

Calcimatic furnaces (rotary grate kilns) consist of a circular traveling grate on which the product to be burnt is fed as a thin layer through a stationary heating zone. Since the product is not moved during the calcining process, very fine products can be burnt in these kilns.

- In fluidized bed kilns (Dorr-Olivier process)

Dorr-Olivier have developed a unit in which the burning proceeds in a fluidized bed.

- In floating gas furnaces

In floating gas furnaces, the reaction is carried out in cyclones connected in series. This produces very reactive quicklime.

Table 5.3-3. Capacity in t/d and Heat Consumption in kJ/kg for the Burning of Lime in Different Kiln Types.

Kilns	Capacity	Heat consumption
shaft kilns	80 to 400	3500 to 4500
doubly inclined kilns	140	3900
parallel flow/regenerative kilns	100	3800
rotary kilns	up to 1000	5800 to 7800
calcimatic kilns	max. 300	5000 to 5700
fluidized bed kilns	100 to 200	4900

5.3.2.4 Slaked Lime

Curing reaction of slaked lime:

The suitability of lime as a bonding agent is based on the reaction of calcium hydroxide with atmospheric carbon dioxide:

$$Ca(OH)_2 + CO_2 \longrightarrow CaCO_3 + H_2O$$

Quicklime must therefore first be converted into slaked lime:

$$CaO + H_2O \longrightarrow Ca(OH)_2 + 65.5\,kJ/mol$$

This strongly exothermic process is technically known as "slaking" and is accomplished in different processes, which will be described in the following sections. The strength of the cured lime is based on the intergrowth and matting together of the calcium carbonate crystals formed.

Lime slaking:

$CaO + H_2O \rightarrow Ca(OH)_2$

Strength of cured lime based on intergrowth of $CaCO_3$ crystals

5.3.2.4.1 Wet Slaking of Quicklime

"Wet slaking" leads to a wet lime hydrate slurry (so-called "milk of lime"). This process was formerly carried out at the construction site by scattering lime into water, whereupon a lime hydrate slurry with high plasticity was obtained.

Wet slaking: results in a wet lime hydrate slurry ("milk of lime")

Wet slaking is currently only carried out industrially by large consumers such as mortar manufacturers.

5.3.2.4.2 Dry Slaking of Quicklime

In dry slaking calcium hydroxide is obtained as a dry powder. Use is made of the fact that so much heat is produced by the reaction of calcium oxide with water that the amount of water bound approximately equals the amount of water evaporated.

Dry slaking: yields dry $Ca(OH)_2$ powder

Industrially, a continuously operating overflow tank is used. About twice the stoichiometric quantity of water is added to the calcium oxide. The reaction mixture is heated up to 100 °C by the heat of reaction and the evaporating excess water entrains the very small lime hydrate particles and carries them upwards, where they are separated by an overflow.

Reaction operated:

- In overflow tanks

In the case of difficult to slake lime, i.e. particularly highly calcined, and magnesium oxide-containing lime, complete hydration can only be achieved under pressure. A 1:2 molar mixture of calcium oxide and water is loaded into a pressure vessel. During the reaction a pressure of 2 to 8 bar is attained under which even the poorly soluble components are hydrated. The steam is separated from the dry calcium hydroxide by rapid pressure release (Corson process).

- In autoclaves, in the case of difficult to slake and MgO-containing limes (Corson process)

5.3.2.4.3 Lime Hydrate from Calcium Carbide

"Acetylene lime": from calcium carbide

Very cheap "acetylene lime" was formerly produced as a byproduct of the manufacture of ethyne from calcium carbide:

$$CaC_2 + 2H_2O \longrightarrow C_2H_2 + Ca(OH)_2$$

Since most ethyne is now manufactured directly from crude oil or natural gas, this process is of decreasing importance.

5.3.2.5 Steam-Hardened Construction Materials

Sand-lime bricks and aerated concrete bricks:

from lime and silicate raw materials

Hardening by hydrothermal reaction and formation of calcium and magnesium silicates

Large quantities of quicklime are used in the manufacture of sand-lime bricks and aerated concrete bricks. These are construction materials which are manufactured from lime-containing and silicate-containing raw materials and whose strength is due to the hydrothermal reaction of the raw materials to calcium silicates.

Sand-lime bricks are manufactured by mixing lime, sand and water in the ratio ca. 1:13:0.7, shaping and then hardening in autoclaves at 180 °C and 7 to 9 bar.

Aerated concrete:

- By foaming of the reaction mixture of lime, sand and possibly cement
- By gas-forming reactions: e.g. formation of hydrogen from aluminium powder, water and $Ca(OH)_2$

Aerated concrete bricks are widely used as light building materials. They are obtained by mixing gas-forming additives into a moist mixture of lime, sand and possibly cement. Industrially, the following reaction is mostly used:

$$2Al + 3Ca(OH)_2 + 6H_2O \longrightarrow 3CaO \cdot Al_2O_3 \cdot 6H_2O + 3H_2$$

The well-known Ytong® building materials are manufactured using this process. Zinc or magnesium can be used instead of aluminum. Other gas-forming additives e.g. hydrochloric acid/limestone or hydrogen peroxide/chlorinated lime have been described, but have hardly been used industrially:

$$2HCl + CaCO_3 \longrightarrow CaCl_2 + H_2O + CO_2$$

$$H_2O_2 + CaCl(OCl) \longrightarrow CaCl_2 + O_2 + H_2O$$

5.3.2.6 Applications for Lime

Of limestone:

- Raw material for the manufacture of quicklime,
- As cut or hewn stone,
- As rubble for roads or railway track,
- As an additive for mortar or concrete,
- As a filler (Sec. 5.8),
- In the metallurgical industry as a sintering aid for fine ore pellets, as a slag-former, as a desulfurization agent,
- In the chemical industry in the manufacture of sodium carbonate and fertilizers [lime ammonium nitrate $NH_4NO_3 + CaCO_3$, calcium nitrate $Ca(NO_3)_2$] (Sec. 2.2 and 3.1),
- In the glass industry (Sec. 5.1.1),
- In cement manufacture,
- As a fertilizer (Chapter 2).

Of quicklime:

- In the iron and steel industry for removing sulfur and phosphorus,
- For the treatment of sewage,
- In the chemical industry for the manufacture of calcium compounds (calcium carbide, calcium cyanamide) and as a cheap base (Sec. 3.1),
- As a fertilizer (Chapter 2),
- In the sugar industry for the removal of oxalic and citric acids from cane juice,
- In the construction industry (with different specifications from DIN 1060).

Of lime hydrate:

- For the manufacture of whitewash paint,
- For the manufacture of caustic soda,
- For flue gas desulfurization (Sec. 5.3.4.6.2),
- For the manufacture of wollastonite (see Fillers: Sec. 5.8),
- In the construction industry (with different specifications from DIN 1060),
- As a fertilizer (Chapter 2).

5.3.3 Cement

5.3.3.1 Economic Importance

1984 cement production in 10^6 t:

- EEC: 114.9
- USA: 71.0
- World: 793.0

Table 5.3-4. Cement Production in the European Common Market in 10^6 t/a.

	1981	1982	1983	1984
Italy	43.0	41.5	40.1	38.8
West Germany	30.2	27.7	27.8	26.2
France	29.8	27.6	25.8	24.0
Great Britain	12.8	13.0	13.4	13.5
Belgium	6.6	6.3	5.7	5.7
Luxemburg	1.0	0.9	1.0	0.9
The Netherlands	3.3	3.1	3.1	3.1
Denmark	1.4	1.5	1.3	1.4
Eire	1.9	1.5	1.4	1.3
Total	130.0	123.1	119.6	114.9

Table 5.3-5. 1984 Cement Production in the World and the USA in 10^6 t/a (according to World Statistical Review No. 5 and No. 6).

	1981	1982	1983	1984
World	737.0	736.7	779.7	793.0
USA	75.2	65.5	71.7	71.0

5.3.3.2 Cement Compositions

The composition ranges of particular cement types and cement additives in the three phase system CaO-SiO_2-Al_2O_3 are given in Fig. 5.3-1.

5.3.3.3 Portland Cement

5.3.3.3.1 Raw Materials

Portland cement: consists of Portland cement clinkers and gypsum

The most used and most manufactured cement type is "Portland cement". It consists of so-called Portland cement clinkers and gypsum or anhydrite additives.

Raw materials for Portland cement clinkers: limestone or chalk, clay, marl

The raw materials for the manufacture of Portland cement clinkers are limestone marl i.e. naturally occurring mixtures of limestone and clay, or mixtures of limestone or chalk and clay prepared on site.

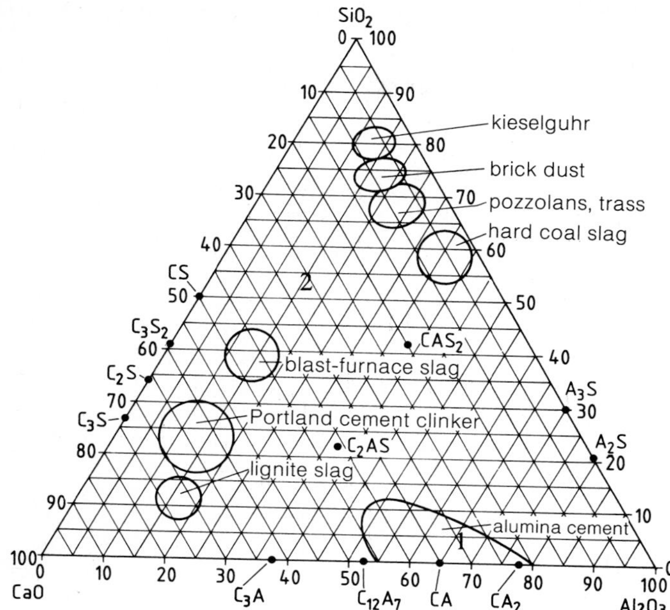

Fig. 5.3-1. Three phase diagram $CaO-SiO_2-Al_2O_3$ with composition ranges of the hydraulic bonding agents.
1) Strongly fire-resistant alumina cement
2) Acidic blast-furnace slag

For the composition of Portland cement clinkers see Fig. 5.3-1.

5.3.3.3.3 Manufacture of Portland Cement

Main component of Portland cement: tricalcium silicate

Since tricalcium silicate endows Portland cement with its desirable properties such as rapid hardening and high strength, it is important to ensure that as much tricalcium silicate as possible is obtained in the manufacture of Portland cement clinkers by using appropriate mixing ratios of the raw materials and sintering in the existence range of tricalcium silicate (between 1250 and 2070 °C).

Important in the manufacture of Portland cement clinkers:

- Accurate adjustment of the lime content (free lime in the clinker causes "lime bursting")

It is important to adjust the lime content as accurately as possible. According to H. Kühl the maximum lime content which can be bonded by the components SiO_2, Al_2O_3 and Fe_2O_3, can be calculated using the formula:

5.3.3.3.2 Composition of Portland Cement Clinkers

Table 5.3-6. Clinker Phases and their Properties.

Name	Formula	Abbreviation*)	Technical properties of cement
tricalcium silicate (alite)	$3CaO \cdot SiO_2$	C_3S	fast hardening, high heat of hydration, high strength
dicalcium silicate (belite)	$2CaO \cdot SiO_2$	β-C_2S	slow, steady hardening, low heat of hydration
tricalcium aluminate	$3CaO \cdot Al_2O_3$	C_3A	in larger quantities: fast solidification, high heat of hydration, contraction tendency, and sensitivity to sulfate water
calcium aluminum ferrite	$2CaO(Al_2O_3,Fe_2O_3)$	$C_2(A,F)$	slow hardening, resistant to sulfate water
free lime	CaO	C	unproblematical in small amounts, in larger quantities induces bursting and too rapid hardening
free magnesium oxide	MgO	M	in larger quantities induces magnesia bursting

*) The following abbreviations are usual in the cement chemistry:
C: CaO, S: SiO_2, A: Al_2O_3, M: MgO, N: Na_2O, K: K_2O,
C: CO_2, H: H_2O.

$$CaO_{max} = 2.8 \cdot SiO_2 + 1.1 \cdot Al_2O_3 + 0.7 \cdot Fe_2O_3$$

At higher lime contents free lime is present in the clinker. The $Ca(OH)_2$ or $Mg(OH)_2$ formed upon hydrating free lime (CaO or MgO) takes up more space than the original oxide. CaO (or MgO) in coarse crystalline lumps therefore gives rise to so-called "lime bursting" (or "magnesia bursting"), since the reaction with water is very slow and continues after the cement is hard. The quantity of lime added is, therefore, always kept below the calculated limit.

- Finest possible grinding of the raw materials, to ensure a quantitative reaction

In the preparation of the reaction mixture, the raw materials also have to be ground as finely as possible to ensure as complete a reaction as possible during burning.

There are several processes for the manufacture of the clinkers:

In the *wet process* the starting materials are ground wet to a "raw slurry", which is then dried in a rotary kiln and burnt. The wet process ensures a simple and accurate dosing of the components and requires little energy during grinding. The subsequent burning process is, however, very energy intensive, due to the quantity of water which has to be evaporated. This process is only economic when the starting materials have high water contents (> 20%) or when wet treatment is unavoidable e.g. with "slurries" of chalk.

Preparation processes:

- in the wet process the raw materials are ground wet to a raw slurry, which is then molded and burnt

In the *half-wet process* water is removed from the wet-prepared raw slurry in filter presses and then pressed in paste molds to pellets, which are burnt in rotary kilns.

- in the half-wet process the wet ground raw slurry is dewatered mechanically before sintering

The *half-dry process* differs from the half-wet process in that the raw mixture is produced by dry grinding and mixing of the starting materials and the homogenized raw meal is granulated to pellets with water on a granulating table.

- in the half-dry process the dry-ground raw mixture is moistened and granulated prior to sintering

In the *dry process* the starting materials are also dry ground and mixed to "raw meal". During grinding the product is dried with the hot flue gases of the firing kiln to a residual moisture content below 1%, in which state it is fed into the firing kiln.

- In the dry process the raw materials are processed dry

The burning of the raw materials in all the published processes is carried out at ca. 1450 °C, usually in rotary kilns. Table 5.3-7 summarizes the individual processes and the kiln types used.

Burning temperature for Portland cement clinker: ca. 1450 °C

In rotary kilns the raw meal is preheated by the hot flue gases. The use of special preheaters (grate preheaters, cyclone preheaters) can make heat recovery still more efficient.

Heat recovery in rotary kilns: with grate preheaters or cyclone preheaters

The largest rotary kilns with grate preheaters (Lepol kilns) have capacities up to 3000 t clinker per day, those with cyclone preheaters up to 5000 t per day. New developments have concentrated upon carrying out part of the calcination in a preliminary step to reduce the thermal load on the rotating tube. The desired clinker throughput can then be achieved with smaller rotary tubes (plants with up to 8000 t/day).

Shaft kilns are only economic for very small plants (< 300 t/day) and their number has declined sharply since 1974.

Table 5.3-7. Comparison of the Cement Manufacturing Processes and the Kiln Types Used.

		Wet process	Half wet processes and half dry processes		Dry processes		
kiln type used		long wet rotary kiln	short rotary kiln	shaft kiln	long dry rotary kiln	suspension preheater kiln	precalciner kiln
equipment for heat exchange		chains, crosses, preheating chamber	hearth preheater		chains, crosses, lifters built into chamber	cyclone or chamber preheater	cyclone or chamber preheater
feed material		raw slurry	blocks from filter-pressed raw slurry	pellets and fine particulate coal or coke	dry raw meal, granules	dry raw meal	dry raw meal
H_2O content in %		30 to 40	10 to 20	up to 10	0.5 to 1	0.5 to 1	0.5 to 1
kiln dimensions	diameter in m	3 to 7.6	2.5 to 6	2 to 3	3 to 7	3 to 6	6
	length in m	80 to 230	30 to 100	8 to 10	70 to 230	40 to 110	100
heat consumption in GJ/t clinker		5 to 6	3 to 3.8	3.1 to 4.2	3.6 to 4	3 to 3.8	ca. 3.3
throughput t/d		300 to 3800	300 to 3300	up to 300	300 to 3000	300 to 5400	planned up to 8500

The heat consumption of wet processes is significantly higher than with dry processes. Thus few wet kilns are still in operation.

Few wet kilns left due to higher energy consumption

The burnt clinker is then ground. Before grinding, up to 6.5% gypsum ($CaSO_4 \cdot 2H_2O$) or anhydrite ($CaSO_4$) is added to delay solidification together with other hydraulic materials such as blast furnace slag or pozzolans, as required.

Ready-to-use Portland cement: by grinding the clinkers with added gypsum

5.3.3.3.4 Applications for Portland Cement

Portland cement is widely used in the construction industry for the manufacture of concrete with admixtures such as gravel, sand and expanded materials, with steel reinforcement as reinforced concrete and as a binder between bricks and other building blocks.

Uses of Portland cement:

- As a binder in concrete
- To bond together bricks and other building blocks

5.3.3.4 Slag Cement

Lime is added during the manufacture of raw iron in blast furnaces to separate off the iron ore gangue, mainly consisting of clays. Easily melting mixtures of lime, silica and aluminium oxide similar in composition to Portland cement (see Fig. 5.3-1) are formed at the reaction temperatures of ca. 1600 °C. Upon addition of a suitable "activator" (such as lime hydrate) the rapidly cooled glassy solidified blast furnace slag possesses hydraulic properties. It is therefore termed "latent hydraulic". Slowly cooled (i.e. crystallized) blast furnace slag does not possess hydraulic properties. Blast furnace slags and melts are, therefore, rapidly cooled with water upon being vented from the blast furnace producing a glassy "foundry sand" as sand-like grains. Mixtures of Portland cement clinkers with less than 30% foundry sand are termed iron Portland cement, mixtures with higher slag contents Portland blast furnace slag cement. The cement clinker is here used as an activator for the blast furnace slag (clinker activation).

"Foundry sand" (blast furnace slag): formed upon addition of lime during iron ore gangue separation in blast furnaces

Iron Portland cement: mixtures of Portland cement clinkers with $< 30\%$ smelting sand

Portland blast furnace slag cement: mixtures of Portland cement clinkers with $> 30\%$ foundry sand

Activation of the latent hydraulic properties of foundary sand can also be achieved with gypsum (sulfate activation). Mixtures of foundary sands with a high aluminum oxide and lime content with ca. 15% calcium sulfate (as raw gypsum or natural anhydrite) are known as supersulfated slag cement.

Supersulfated slag cement: activation of the latent hydraulic properties of foundry sand with gypsum

Properties of slag cements:

- Lower heats of hydration
- Higher chemical resistance

Slag cements with high slag contents have lower heats of hydration than pure Portland cement, which is advantageous when massive concrete structures are being built (e.g. dams). The lower calcium hydroxide content results in their being somewhat more chemically stable than Portland cement. Slag cement is used in similar applications to Portland cement.

5.3.3.5 Pozzolan Cements

Pozzolans: natural or synthetic silica-containing materials, which react with the $Ca(OH)_2$ formed during cement hardening

Lime hydrate is formed during the hardening of Portland cement (Sec. 5.3.3.9), which hardly contributes to the strength of the cement stone. Pozzolans are silica-containing natural or synthetic materials, which can form calcium silicate hydrates with lime and water even at low temperatures. Their addition to Portland cement can therefore improve the strength of the cement, but their slower hardening is a drawback. The reactivity of the pozzolans depends upon particle size. However, too fine particles, such as e.g. highly dispersed silica (so-called "kieselguhr"), strongly increase the amount of water required during concrete mixing.

Natural pozzolans: tarras, santorin earth

Synthetic pozzolan: fly ash

Naturally occurring pozzolans are tarrass (a volcanic rock found along the Rhine, in Bavaria and in Austria), santorin earth (Greece) and the Italian pozzolan (Pozzuoli near Naples). A synthetic material with similar properties is fly ash, which is produced in coal-fired power stations.

Properties of pozzolan cements:

- Good chemical resistance

Pozzolan cements exhibit good chemical resistance.

The composition range of the pozzolans is shown in Fig. 5.3-1.

5.3.3.6 Alumina Cement

Alumina cements consist of calcium oxide and aluminum oxide compounds (see Fig. 5.3-1), mainly CA, CA_2 and $C_{12}A_7$ (for abbreviations see Table 5.3-6).

Alumina cement:

- Consists of CaO/Al_2O_3 compounds
- Manufactured from limestone and bauxite

It is manufactured from limestone ($CaCO_3$) and bauxite (mainly Al_2O_3). The melting point of mixtures of limestone and bauxite is in the range ca. 1500 to 1600 °C, the sintering temperature range being just below the melting point of the mixture.

Industrially, alumina cement is produced either in rotary or shaft kilns by fusing the raw material mixture (melt cement) or using electrofusion kilns (electro-melt cement). Alumina cement exhibits a very high initial strength, which, however, particularly at higher temperatures, strongly decreases with time. It is therefore forbidden in West Germany for load-bearing structures made of reinforced concrete. Alumina cement is less stable to alkali due to the possible formation of freely soluble alkali aluminates.

Melt cement: by fusing the raw material mixture in rotary kilns

Electro-melt cement: by fusing in electrofusion kilns

Properties of alumina cement:

- High initial strength, which decreases with time

Alumina cement is used in refractory materials at temperatures up to 1700 °C. "Rapid setting cement" consists of mixtures of alumina cement and Portland cement (used e.g. in the sealing of water leaks).

Uses:

in refractory materials and in rapid setting cement

5.3.3.7 Asbestos Cement

After mechanical dewatering the aqueous slurry of cement and asbestos sets to a very rugged material from which pipes, containers, roofing-tiles etc. can be manufactured (Eternit®, Fulgurit®). See also Sec. 5.2.2.3.

Asbestos cement: by setting a slurry of asbestos-reinforced cement

5.3.3.8 Miscellaneous Cement Types

"Expanding" or shrinkage-compensating cements do not shrink upon hardening. Addition of 10 to 15% calcium sulfate or aluminate even effects an expansion ("managed gypsum-bursting").

Expanding cement: mixture of Portland cement and calcium sulfate or aluminate

"White cement" is manufactured from particularly pure raw materials with very low iron oxide contents.

White cement: low iron oxide Portland cement

So-called "sorel cement" is not a real cement. Mixtures of caustic burned magnesite (reactive MgO, produced by heating magnesium carbonate just above its decomposition temperature) and magnesium salts harden to a stone-like mass. "Sorel cement" is obtained by reacting magnesium oxide with concentrated magnesium chloride solution:

Sorel cement: mixture of MgO and magnesium salts

$$3\,MgO + MgCl_2 \xrightarrow{H_2O} 3\,MgO \cdot MgCl_2 \cdot 11\,H_2O$$

$3\,MgO \cdot MgCl_2 \cdot 11\,H_2O$ is formed in fine needle-shaped crystals and develops similar strengths to Portland cement or gypsum by matting of the needles.

Woodstone or xylolith: sorel cement with added sawdust

The material is not weather resistant and is used for plasters and industrial flooring. With added sawdust it is known as woodstone or xylolith.

Magnesia cement exhibits, even when dry, a considerable electrical conductivity. As a result flooring produced from it does not exhibit static charging.

Heraklith®: stone-like mass from MgO and magnesium sulfate

Magnesium sulfate also forms a stone-like mass with magnesium oxide, which is used for the manufacture of lightweight building board (Heraklith®).

5.3.3.9 Processes in the Solidification of Cement

After mixing the cement:

- C_3A and gypsum react to ettringite

Upon mixing cement powder with water, part of the tricalcium aluminate C_3A and gypsum dissolve and react together forming ettringite:

$$3\,CaO \cdot Al_2O_3 + 3\,CaSO_4 + 32\,H_2O \rightarrow 3\,CaO \cdot Al_2O_3 \cdot 3\,CaSO_4 \cdot 32\,H_2O$$

ettringite

- Delayed setting of Portland cement by deposition of ettringite on cement particles

The ettringite crystals on the surface of the cement particles are so fine that they can not bridge the gap between the cement particles and therefore do not form a solid structure. This is the basis of the delay in solidification upon adding gypsum to cement. Without gypsum the tricalcium aluminate immediately reacts with water to calcium aluminate hydrate, which fills the space between the cement particles with its large crystals and leads to very rapid solidification of the cement slurry.

- Solidification of Portland cement: after 1 to 3 hours by recrystallization of small ettringite crystals to larger crystals

Portland cement (containing gypsum) solidifies after 1 to 3 hours. The needle-shaped ettringite crystals recrystallize forming larger needles which bond the cement particles together and strengthen the crust.

Solidification of Portland cement:

hydration of tricalcium silicate and dicalcium silicate to a tobermorite-like phase $3CaO \cdot 2SiO_2 \cdot 2H_2O$

The hydration of tricalcium silicate C_3S and dicalcium silicate C_2S (for abbreviations see Table 5.3-6) are responsible for the further strengthening of Portland cement. This reaction only begins in earnest after ca. 4 hours. A gelatinous hydrate coating forms rapidly round the cement grains from which fine long needles of calcium silicate hydrate grow developing a reticulate network between the cement grains, which bonds the

Table 5.3-8. Reaction Products of the Components of Portland Cement Clinkers with Water.

Clinker components		Reaction products	
Name	Formula	Formula	Name
tricalcium silicate (alite)	$3CaO \cdot SiO_2$	$3CaO \cdot 2SiO_2 \cdot nH_2O$	tobermorite-like phase
dicalcium silicate (β-belite)	$2CaO \cdot SiO_2$	$3CaO \cdot 2SiO_2 \cdot nH_2O$	
in the absence of sulfate			
tricalcium aluminate	$3CaO \cdot Al_2O_3$	$4CaO \cdot Al_2O_3 \cdot 19H_2O$ dry $4CaO \cdot Al_2O_3 \cdot 13H_2O$	tetracalcium aluminate hydrate
calcium aluminum ferrite	$CaO(Al_2O_3,Fe_2O_3)$	ditto, Al partially replaced by Fe	
in the presence of sulfate			
tricalcium aluminate	$3CaO \cdot Al_2O_3$	$3CaO \cdot Al_2O_3 \cdot 3CaSO_4 \cdot 32H_2O$ or $3CaO \cdot Al_2O_3 \cdot CaSO_4 \cdot 12H_2O$	ettringite 'monosulfate'
calcium aluminum ferrite	$CaO(Al_2O_3,Fe_2O_3)$	ditto, Al partially replaced by Fe	

cement particles together. Later, smaller needles of calcium silicate hydrate fill the gaps left. The more reactive tricalcium silicate hydrolyzes much faster than dicalcium silicate:

$$2(3CaO \cdot SiO_2) + 6H_2O \longrightarrow 3CaO \cdot 2SiO_2 \cdot H_2O + 3Ca(OH)_2$$

$$2(2CaO \cdot SiO_2) + 4H_2O \longrightarrow 3CaO \cdot 2SiO_2 \cdot H_2O + Ca(OH)_2$$

The tricalcium aluminate still present reacts with the calcium hydroxide formed to tetracalcium aluminate hydrate. Calcium aluminum ferrite forms the analogous hydration product, which reacts more slowly. Part of the aluminum is replaced by iron:

Subsequent reactions:

tricalcium aluminate reacts with $Ca(OH)_2$ to tetracalcium aluminate hydrate

$$3CaO \cdot Al_2O_3 + Ca(OH)_2 + 12H_2O \longrightarrow 4CaO \cdot Al_2O_3 \cdot 13H_2O$$

$$4CaO \cdot Al_2O_3 \cdot Fe_2O_3 + 4Ca(OH)_2 + 22H_2O \longrightarrow 4CaO \cdot Al_2O_3 \cdot 13H_2O + 4CaO \cdot Fe_2O_3 \cdot 13H_2O$$

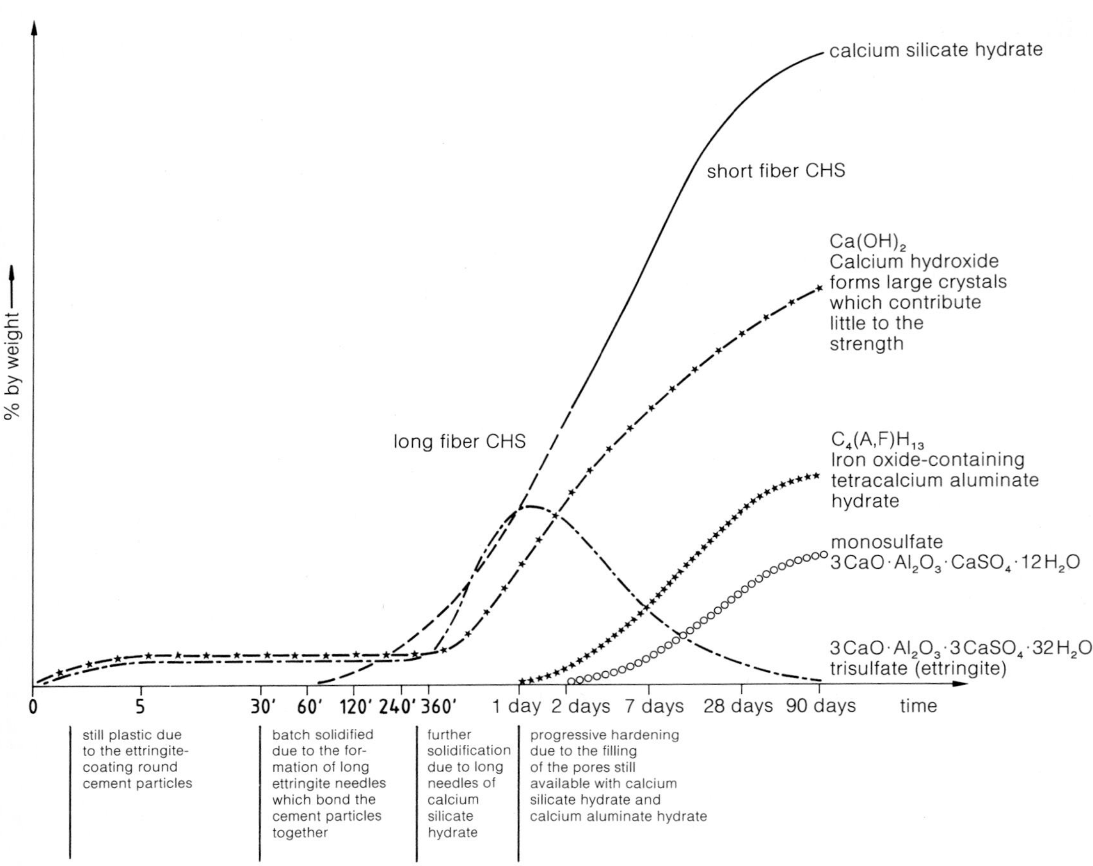

Fig. 5.3-2. Processes in the setting of cement [According to F. W. Locher in *Zementtaschenbuch (Cement Handbook)* 1979/1980, p. 19, Bauverlag, Wiesbaden (1979).]

Final phase in the consolidation:

ettringite reacts with tricalcium aluminate forming "monosulfate"

The ettringite initially formed reacts with the tricalcium aluminate still present and with the calcium hydroxide produced in the meantime to form "monosulfate":

$$2(3\,CaO \cdot Al_2O_3) + 3\,CaO \cdot Al_2O_3 \cdot 3\,CaSO_4 \cdot 32\,H_2O + 4\,H_2O \longrightarrow 3(3\,CaO \cdot Al_2O_3 \cdot CaSO_4 \cdot 12\,H_2O)$$

"monosulfate"

The temporal changes in the proportions of the phases occurring during the solidification of Portland cement are schematically shown above in Fig. 5.3-2.

5.3.4 Gypsum

5.3.4.1 Economic Importance

Table 5.3-9. Natural Gypsum Production in 10^6 t [from the Mineral Yearbook Vol. I (1982)].

	1980	1981	1982 (est.)
West Germany	2.25	2.25	2.27
EEC	13.8	13.5	13.3
USA	11.2	10.4	9.6
World	77.4	76.3	73.1

1981 natural gypsum production in 10^6 t:

- FRG 2.25
- EEC 13.5
- USA 10.4
- World 76.3

To the West German figures must be added a further ca. 1.5×10^6 t/a of synthetic gypsum, mainly from wet process phosphoric acid production, most of which is dumped. Only ca. 0.1×10^6 t/a is used in the construction sector after purification. Ca. 60×10^6 t/a of gypsum is estimated to be produced worldwide as a byproduct from the manufacture of wet phosphoric acid.

Synthetic gypsum production:

- FRG: 1.5×10^6 t/a, mainly from wet process (phosphoric) acid production
- World: 60×10^6 t/a from wet process (phosphoric) acid production

In West Germany about 0.5 to 1×10^6 t/a of natural anhydrite are used in the cement industry and 0.4 to 0.5×10^6 t/a in mining. 0.25×10^6 t of synthetic anhydrite is produced annually in West Germany in the manufacture of hydrofluoric acid. Of this 100000 t is used in the cement industry and the rest is used as a binder in the construction industry. The worldwide production of synthetic anhydrite is estimated to be ca. 2 to 3×10^6 t/a.

Natural anhydrite production:

- FRG: 0.9 to 1.5×10^6 t/a

Synthetic anhydrite production:

- FRG: 0.25×10^6 t/a from hydrofluoric acid production
- World: 2 to 3×10^6 t/a

Ca. 62×10^6 t/a of sulfur dioxide is produced annually as a component of the flue gases in the combustion of coal, crude oil and natural gas. Technology to convert this to gypsum has been recently developed in the USA and Japan. At the moment ca. 0.13×10^6 t/a of gypsum is produced from flue gas in West Germany. This is expected to rise to ca. 3.9×10^6 t/a in 1990 and eventually to supplant ca. 50% of the present natural gypsum production.

- FRG: 0.13×10^6 t/a from flue gas desulfurization

5.3.4.2 Modifications of Calcium Sulfate

The following modifications of calcium sulfate are known:

$CaSO_4 \cdot 2H_2O$:	natural gypsum, dihydrate
$CaSO_4 \cdot 1/2H_2O$:	bassanite, hemihydrate
$CaSO_4$:	anhydrite

Natural deposits:

- As natural gypsum $CaSO_4 \cdot 2H_2O$
- As anhydrite $CaSO_4$
- As bassanite $CaSO_4 \cdot 1/2H_2O$

In nature calcium sulfate is found as the dihydrate (natural gypsum), as the anhydrous anhydrite and rarely as the hemihydrate in the form of bassanite. Only the deposits of natural gypsum and anhydrite are of commercial interest. In the construction industry only the settable modifications, the hemihydrate and anhydrite, are important. These are manufactured by dehydrating the dihydrate. The properties and formation conditions of the different calcium sulfate modifications are given in Table 5.3-10.

α- and β-hemihydrates have different crystal habits and specific surface areas

The α- and β-forms of the hemihydrate do not differ in their crystal form. Their different solubilities and reaction behaviors are due to different crystal habits and specific surface areas. α-hemihydrate crystallizes in well-formed large crystals, whereas β-hemihydrate crystallites are extremely small. α- and β-hemihydrates are regarded as being two extreme forms of the same phase. They possess the ability to take up further water "zeolithically" forming the "monohydrate" described in the literature.

"Monohydrate" formed by the "zeolithic" take up of water by the hemihydrate

Anhydrite III: formed upon dehydration of the hemihydrate

The "soluble" anhydrite III occurs in three extreme forms, which are unimportant in the industrial manufacture of gypsum.

Anhydrite II: identical to naturally occurring anhydrite; rehydration behavior dependent upon manufacturing temperature

Anhydrite II is formed at higher temperatures from anhydrite III or hemihydrate and is identical to naturally occurring anhydrite. It occurs in three different intermediate forms. The rehydration rates of these intermediate forms depend upon the treatment temperature. The hard plaster obtained above 700 °C partly consists of CaO, since decomposition of anhydrite begins above 600 °C:

"Hard plaster" contains ca. 10% CaO

$$CaSO_4 \longrightarrow CaO + SO_3 .$$

The individual modifications are formed from the dihydrate at increasing temperatures in the following order:

dihydrate → hemihydrate → anhydrite III → anhydrite II (→ anhydrite I)

Table 5.3-16. Formation Conditions and Properties of the Calcium Sulfate Modifications.

Formula	Name	Occurring forms	Crystal form	Stability range in °C	Formation conditions laboratory in °C	Formation conditions industrial in °C	Properties	Applications
$CaSO_4 \cdot 2H_2O$	natural gypsum, dihydrate	–	monoclinic	< 40 only stable phase under normal conditions	< 40	< 40	"hardened plaster", high rigidity matted needle-shaped crystals. Setting process strongly accelerated by mixing with hemihydrate or anhydrite	
$CaSO_4 \cdot 1/2H_2O$	bassanite, hemihydrate	α-form	rhombohedral	metastable	> 45 in an aqueous medium	80 to 180 in an aqueous medium	very high rigidity, rapid and slow hardening	molding plaster, industrial and dental plaster
		β-form	rhombohedral	metastable	45 to 200 in dry air	120 to 180 in dry air	moderate rigidity, moderately fast hardening	plaster of Paris and stucco; wall plaster in mixtures with anhydrite
$CaSO_4$	anhydrite III, "soluble anhydrite"	α-anhydrite III β-anhydrite III β-anhydrite III	hexagonal	metastable	50 vacuum 100 air	290	reacts with water very rapidly to the hemihydrate, low rigidity	drying agent
$CaSO_4$	anhydrite II, "insoluble anhydrite" (≙ natural anhydrite)	AII poorly soluble AII insoluble AII estrich gypsum	rhombic	40 to 1180		< 500 500 to 700 > 700	slow to very slow reaction with water (hence use of activators), high rigidity	filler, imitation marble, hard plaster, wall plaster in mixtures with β-hemihydrate
$CaSO_4$	anhydrite I	–	cubic	> 1180	> 1180	–		industrially not of interest, since only exists at temperatures > 1180 °C

5.3.4.3 Natural Gypsum

Raw material treatment:

- Extraction
- Grinding
- Thermal dehydration

Natural gypsum is either extracted in open cast or underground mines. It is ground, homogenized and then thermally dehydrated, the process used depending upon the required properties of the final product.

Dry hydration yields β-hemihydrate and/or anhydrite

In the manufacture of β-hemihydrate, anhydritic and multiphased plasters, the dehydration is carried out dry. The β-hemihydrate is obtained at 120 to 180 °C and anhydritic or multiphase plasters at higher temperatures. Directly or indirectly heated rotary tubes or kettles, which are mostly indirectly heated, are used in their large-scale manufacture. Other processes are sporadically used e.g. the grinding-pyrolysis process in which the dehydration takes place in the pulverizer during the grinding. A carrier gas combustion process and a dehydration process in a heated auger (Holoflite combustion process) have also been described.

Industrial realization:

- Mainly in kettles and rotary furnaces
- Sporadically in grinding-pyrolysis, carrier gas combustion or Holoflite combustion processes
- In continuous grate furnaces

  especially for the manufacture of high temperature plaster

Continuous perforated grate kilns (Knauf) have been successfully used in the manufacture of high temperature plaster as well as the conventional rotary kilns. In this process the material to be pyrolyzed is placed on a continuously moving belt which is transported under a burner hood. The hot combustion gases are sucked through the layer of material being pyrolyzed with the help of ventilators. Since the material being pyrolyzed does not move, very little dust is produced in this process.

Capacities:

- 500 t/d for rotary kilns
- 1000 t/d for continuous grate kilns

Modern rotary tube plants have capacities of up to ca. 500 t/d, continuous grate furnaces up to 1000 t/d. The residence times are less than 1 hour. In kettles throughputs up to 150 t/d with residence times of several hours are achieved. Box shaft-kilns are no longer viable due to the lengthy (several day) calcination times.

α-hemihydrate produced by dehydration in autoclaves

α-hemihydrate can only be manufactured using wet processes in autoclaves at temperatures of 120 to 150 °C. Those processes described in the literature which operate at atmospheric pressure in acids or aqueous salt solutions, have not yet been industrially utilized.

Since the product properties depend upon the plant used, most modern plaster factories use several dehydration processes. Multiphase plasters are then produced by mixing. By observing precisely defined conditions these can also be obtained in a single process step (see Table 5.3-11).

Manufacture of multiphase plasters:

- By mixing of different plasters
- By observing well-defined reaction conditions during the dehydration

Table 5.3-11. Phase Compositions in % of the Different Plaster Types.

Plaster types	Production unit	Manufacturing temperature in °C	$CaSO_4 \cdot H_2O$	$CaSO_4 \cdot 1/2 H_2O$		$CaSO_4$		Residue
				α-form	β-form	anhydrite II	anhydrite III	
molding plaster	autoclave	130 to 135	0.1	94	2	–	1.9	2
plaster of Paris	kettle	180	2	2	73	10	1	10
	rotary tube	120 to 180	2	2	71	19	5	–
wall plaster	mixed from plaster of Paris and high temperature plaster	–	2		26	14	58	–
high temperature plaster	continuous grate	200 to 900	2	–	6	18	74	–
hard plaster	rotary tube	800 to 1000	–	–	–	–	90	10 (CaO)

5.3.4.4 Natural Anhydrite

Natural anhydrite is extracted by open cast or underground mining. A settable product is obtained by fine grinding to < 0.2 mm particles and the addition of "activators" (ca. 2% heavy metal or alkali sulfates, or calcium hydroxide).

Working up of natural anhydrite:

- Extraction
- Fine grinding
- Addition of activators

5.3.4.5 Synthetic Anhydrite from Hydrofluoric Acid Manufacture

The large scale manufacture of hydrofluoric acid is carried out by the reaction of fluorspar (CaF_2) with concentrated sulfuric acid in rotary furnaces at ca. 200 °C (see Sec. 1.7.1):

$$CaF_2 + H_2SO_4 \xrightarrow{200\,°C} CaSO_4 + 2\,HF$$

Anhydrite from hydrofluoric acid manufacture:

$$CaF_2 + H_2SO_4 \xrightarrow{200\,°C} CaSO_4 + 2\,HF$$

The calcium sulfate byproduct is in the form of anhydrite II. This contains as undesirable impurities

Purification of residual H_2SO_4 by reaction with CaO

unreacted fluorspar and small amounts of free sulfuric acid (0.5 to 10%). The latter is immediately removed from the anhydrite emerging from the rotary furnace by adding calcium oxide to neutralize the residual sulfuric acid in a mixing screw.

$$CaO + H_2SO_4 \xrightarrow{200\,°C} CaSO_4 + H_2O$$

Settable product by:

- Fine grinding
- Addition of activators

After a final grinding (fine grinding) to the required product fineness, the synthetic anhydrite is suitable for final processing. Should activators be required, these still have to be added (Sec. 5.3.4.4).

5.3.4.6 Synthetic Gypsum

5.3.4.6.1 From the Manufacture and Purification of Organic Acids

Synthetic gypsum from the manufacture or purification of organic acids by way of their Ca-salts

Small quantities of calcium sulfate dihydrate are formed as byproducts in the production or purification of organic acids such as citric, oxalic or tartaric acids by way of their calcium salts:

$$\text{e.g.: } CaC_2O_4 + 2\,H_2O + H_2SO_4 \rightarrow CaSO_4 \cdot 2\,H_2O + H_2C_2O_4$$

5.3.4.6.2 Synthetic Gypsum from Flue Gas Desulfurization

Synthetic gypsum from flue gas desulfurization by absorption of SO_2 with $Ca(OH)_2$ and subsequent oxidation to $CaSO_4 \cdot 2H_2O$

Processes for flue gas desulfurization are becoming increasingly important, due to increasingly stronger demands for environmental protection. The SO_2 present in flue gas is absorbed in water, precipitated with quick- or slaked lime or limestone and oxidized.

$$2SO_2 + 2Ca(OH)_2 \longrightarrow 2CaSO_3 \cdot 1/2H_2O + H_2O$$

$$3H_2O + 2CaSO_3 \cdot 1/2H_2O + O_2 \longrightarrow 2CaSO_4 \cdot 2H_2O$$

$$2SO_2 + 2Ca(OH)_2 + O_2 + 2H_2O \longrightarrow 2CaSO_4 \cdot 2H_2O$$

Industrial realization:

- In experimental plants using the Hölter process (FRG)

In the Hölter process sulfur dioxide is absorbed in an aqueous suspension of calcium hydroxide and calcium chloride. In West Germany only experimental plants are operating at present e.g. in the power station Weiher III in Quierscheid (Saarland, FRG).

Gypsum manufacture from flue gases is economically interesting for countries, such as Japan, with little naturally occurring gypsum (Mitsubishi and Chiyoda processes). Similar processes have also been developed in the USA (Lewis-Wäscher and Combustion-Engineer processes).

- In Japan with the Mitsubishi and Chiyoda processes

5.3.4.6.3 Synthetic Gypsum from Phosphoric Acid Production

The largest quantities of synthetic gypsum arise in the production of phosphoric acid by the reaction of natural phosphates with sulfuric acid (see also Sec. 1.5.1):

1.7 t gypsum/t raw phosphate is produced in the production of wet process phosphoric acid

$$Ca_5(PO_4)_3F + 5H_2SO_4 + 10H_2O \longrightarrow 3H_3PO_4 + 5CaSO_4 \cdot 2H_2O + HF$$

The 1.7 t gypsum/t raw phosphate produced is so impure that it is unsuitable for direct use in the construction industry. These impurities include admixtures of organic substances; soluble alkali and magnesium salts; poorly soluble salts (e.g. Na_2SiF_6); and phosphates and fluorides which are isomorphously incorporated in the gypsum crystal lattice.

Direct use in the construction industry hindered by impurities such as organic or inorganic admixtures or substances incorporated into the gypsum crystal lattice

The impurity content can be affected by the way in which the phosphoric acid manufacturing process is carried out: single stage processes, such as the classic Prayon dihydrate process, or newer processes which lead to the hemihydrate result in a high impurity level, making further purification steps necessary. Two stage processes, which either lead to the dihydrate by way of the hemihydrate (Central-Prayon process) or to the hemihydrate by way of the dihydrate (Nissan process), produce products which can be further processed without further purification, since the impurities incorporated into the lattice are liberated during conversion of dihydrate to hemihydrate or vice versa. Should further purification of the synthetic gypsum be necessary, there are two types of process:

Reduction in impurities with phosphoric acid manufacture using two stage processes:

- Central-Prayon process
- Nissan process

Type I: "Dry processes" which lead to β-hemihydrate plaster and multiphase plaster,

and

Type II: "Wet processes" which yield α-hemihydrate plaster.

Dry Processes

Dry purification processes lead to β-hemihydrate and multiphase plasters

Dry processes always consist of two stages: washing to remove the soluble impurities with, if necessary, organic impurity removal by flotation, and dry calcination. Washing and flotation are unnecessary in the case of synthetic gypsum produced by the Central-Prayon or the Nissan processes, as they are already fairly pure.

Industrial realization using the processes of:

- Rhône-Poulenc
- Knauf
- CdF-Chimie/Air Industries
- CERPHOS

The following processes differ only in the way they are technically carried out:

Rhône-Poulenc process:

Washing in a hydrocyclone, drying and calcination either separately or in a single step.

Knauf process in different variants:

After washing and flotation, drying is carried out with partial dehydration. The resulting mixture of dihydrate and hemihydrate is recrystallized with the addition of lime hydrate and stoichiometric quantities of water, then pelletized and calcined in a continuous grate or rotary kiln. This recrystallization also eliminates the needle-form present in synthetic gypsum, which can cause problems in further processing.

CdF-Chimie/Air Industries process:

The washing is carried out in a hydrocyclone. The subsequent thermal treatment is carried out in three steps: drying, dehydration to the β-hemihydrate and anhydrite III and rehydration of the anhydrite III to β-hemihydrate.

CERPHOS process (Centre d'Etudes et de Recherche des Phosphates Mineraux):

In the washing step the fine fraction, containing a particularly high impurity concentration, is separated in a hydrocyclone. Drying and calcining is carried out as in conventional plaster production.

Wet Processes

Wet purification processes lead to the α-hemihydrate

The wet processes also consist of a washing step (if necessary with flotation) and a dehydration process, which is carried out in the aqueous phase in autoclaves (preferably in continuously operating autoclaves). The only industrially operated process is the Guillini process. This process can be carried out with all types of gypsum byproduct from phosphoric acid manufacture, regardless of the types of impurity, without modifying the phosphoric acid manufacturing process. The wet treatment is particularly economic when further direct treatment of the plaster, e.g. to plasterboard, is possible, i.e. when no drying of the material is necessary.

Industrial realization in the Guillini process in autoclaves

Phosphate is enriched in the reaction medium up to 7 g/L (as P_2O_5). The yield of phosphate can thus be improved by recycling the water.

5.3.4.7 Processes in the Setting of Plaster

"Gypsum setting" is the reaction of hemihydrate, anhydrite II or anhydrite III with water i.e. the opposite reaction to that of the manufacture of plaster:

"Setting" is the rehydration of hemihydrate or anhydrite

Plaster manufacture by dehydration	Settable modifications	Setting by rehydration
$CaSO_4 \cdot 2H_2O \xrightarrow{-H_2O}$	$\left\{ \begin{array}{l} CaSO_4 \cdot 1/2H_2O \\ CaSO_4\ \mathbf{A\ II} \\ CaSO_4\ \mathbf{A\ III} \end{array} \right\}$	$\xrightarrow{+H_2O} CaSO_4 \cdot 2H_2O$

The plaster powder should be suspended as homogeneously as possible when making a slurry of plaster in water to avoid settling. This ensures thorough wetting of the individual plaster particles and the dissolution of the outermost layers of the hydrated hemihydrate (or anhydrite) particles.

Slurrying with water to a homogeneous suspension

After this so-called "induction period", dihydrate crystals from the saturated solution start to nucleate round the particles. Further growth of the needle-shaped dihydrate crystals follows which, for the correct ratio of water to plaster, mat to a close-grained structure in which the still unhydrated components and mixing water are embedded. After the solid structure is formed, the solid mass of plaster still has to dry. The transition of hemihydrate and anhydrite II to dihydrate occurs directly without intermediates, but anhydrite III is converted to the dihydrate by way of the hemihydrate. The hardening properties of plaster de-

After the "induction period" (initial solution of hemihydrate or anhydrite), nucleation of dihydrate occurs. Further crystal growth leads to a close-grained structure of dihydrate crystals which accounts for the rigidity of plaster

Excess water is removed by drying

pend upon many factors e.g. composition, type of manufacture, particle size, particle distribution and additives (accelerators or decelerators).

5.3.5 Coarse Ceramic Products for the Construction Industry

Coarse ceramic products are tiles and hard bricks

Manufacturing process similar in some ways to that for fine ceramic products

The clay products used in the construction industry are so-called coarse ceramic products. These include bricks, roofing tiles, clinkers (hard bricks), drainpipes and protection covers for cables. The manufacture of such products hardly differs from that described in Sec. 5.5.3 for the general manufacture of ceramic products.

Loam, marl, tileclay, claystone and slate clay are used as raw materials in the manufacture of tiles. The chemical composition of the raw materials is similar to that for the manufacture of fine ceramic products (Sec. 5.5). The quantities of impurities can, however, be higher.

Light tiles are obtained by:

- Incorporating light additives, such as perlite, vermiculite, kieselguhr
- Introducing pores and cavities by pyrolysis of organic additives
- Incorporating foaming additives

Light construction materials are of particular interest to the construction industry. The manufacture of light tiles can be achieved using different processes:

- Addition of light additives, such as vermiculite, perlite or diatomaceous earth (kieselguhr), enabling a reduction in the raw density of the ceramic from 1.9 kg/dm^3 to 0.8 kg/dm^3.
- Addition of suitable organic materials, such as wood (sawdust), carbon black, peat etc. These additives decompose upon firing leaving pores and cavities behind. Particularly suitable organic materials are expanded polystyrene beads or synthetic resin beads (Poroton® process).
- Addition of gas-forming materials. If, for example, aluminum powder is added to the moist batch the material foams. Hardening can also be carried out in pressure vessels, such as used in the manufacture of sand-lime bricks (Sec. 5.3.2.5).

Light bricks are manufactured almost exclusively by the incorporation of expanded polystyrene beads and/or sawdust.

5.3.6 Expanded Products

5.3.6.1 General Information

Inorganic expanded products are materials with closed pore structures and low densities, which exhibit very good insulation properties. Certain clays, shales, glasses and blast furnace slags are used in their manufacture.

Expanded products are materials with closed pore structures and low densities

Manufactured from clays, shales, glasses and blast furnace slags

It has long been known that clays can swell upon rapid heating. The first technical application of these swelling proerties was described by Stephen John Hayde in 1912 (USA Patent 1 255 878). These expanded clay granules were marketed under the name Haydite. In Europe, the first attempts to produce expanded clay were in Denmark. This product was marketed under the name LECA (*L*ight *E*xpanded *C*lay *A*ggregates). A similar process was developed in the USSR and as Keramzit was widely used. Expanded clay has been manufactured in West Germany using the LECA process since 1955. The expanded clay and shale capacities in West Germany are ca. 1.5 to 2×10^6 m^3/a.

Expansion of clays by rapid heating

Expanded product capacity in West Germany:

1.5 to 2×10^6 m^3/a

5.3.6.2 Expanded Products from Clays and Shales

5.3.6.2.1 Raw Materials

Materials suitable for the manufacture of expanded clays or expanded shales are those which form a glass-like ("pyroplastic") phase at high temperatures (ca. 1100 to 1200 °C), whose viscosity is so high that gas produced in the product can not escape and therefore expands the mass. The volume change of a clay granulate as a function of temperature is shown in Fig. 5.3-3.

Expandable raw materials form a glass-like "pyroplastic" phase at 1100 to 1200 °C and are expanded by the gases formed at these temperatures

The expandability of clays is favored by:

- High plasticity
- Large fine particle content
- Fe_2O_3 content of at least 3 to 6%
- Organic impurities (their content can be 0.6 to 2.5% or above). Particularly favorable is a ratio of organic C-content to Fe_2O_3 content of 0.04 to 0.20
- Low lime content (< 5%)

The expandability of clays depends upon

- Plasticity
- Fine particle content
- Fe_2O_3 content
- Organic C-content
- Lime content

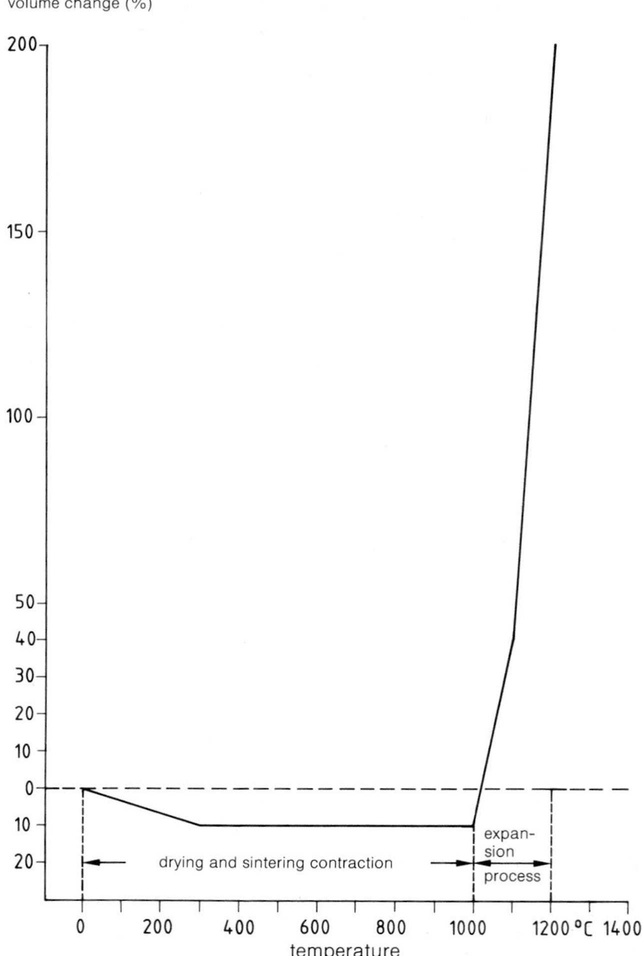

Fig. 5.3-3. Volume change in expanded-clay manufacture as a function of temperature (schematic).

• Mineral composition	• High proportion of clays of the illite and mica group; low kaolinite content (see Fig. 5.3-4)
• Sulfur content	• Sulfur-containing admixtures
• Sintering and melting points	• Small temperature difference between sintering and melting points; these should be at ca. 1200 °C
• Softening and expanding ranges	• Small difference between the softening temperature and the temperature at which the maximum amount of gas is produced
	• Expanding process temperature range of at least 50 to 100 °C

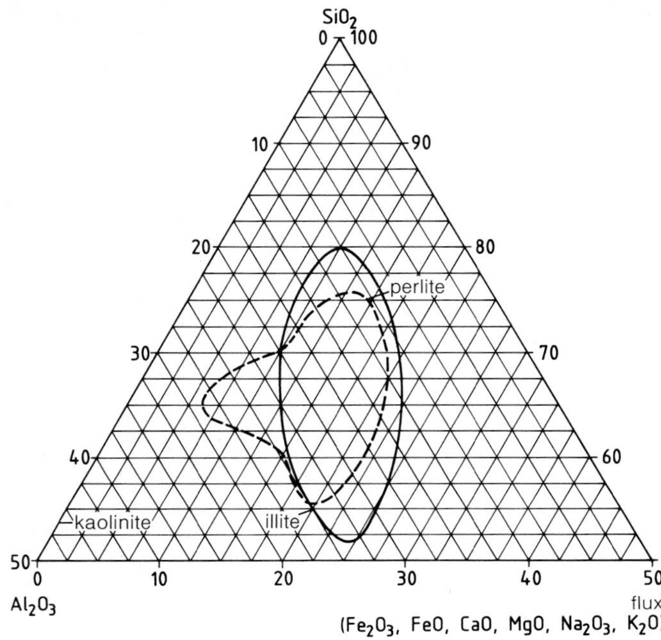

Fig. 5.3-4. Chemical composition of expandable clays.
——— according to C. M. Riley, ------ according to H. Wilson.

If a natural clay does not possess the required expandability, it can be made expandable by the addition of so-called expanding aids. Suitable additives are: organic materials such as bitumen, heavy oil or lignin sulfate; carbon; iron oxide, iron hydroxide or gypsum. The addition of SiO_2 or Al_2O_3 is advisable at too high flux contents.

Additives for increasing expandability: e.g. organic materials, iron oxide, gypsum

5.3.6.2.2 Gas-forming Reactions in the Manufacture of Expanded Products

The most common expanding-gases are CO_2, CO, O_2, SO_2 and H_2. They are formed in the:

- Decomposition of organic impurities at elevated temperatures:

Expanding gases are: CO_2, CO, O_2, SO_2, H_2

Gas-forming reactions are: decomposition of

- Organic materials

$$C_{org.} + O_2 \longrightarrow \underline{CO_2}$$

- Sulfur-containing compounds

- Decomposition of sulfur-containing compounds:

$$2\,CaSO_4 \rightarrow 2\,CaO + \underline{2\,SO_2} + \underline{O_2}$$
$$FeS_2 \rightarrow FeS + S_{gas}$$
$$S_{gas} + O_2 \rightarrow \underline{SO_2}$$

- Carbonates

- Decomposition of carbonates:

$$CaCO_3 \longrightarrow CaO + \underline{CO_2}$$

- Iron oxides

- Decomposition of iron oxide:

$$6\,Fe_2O_3 \xrightarrow{>1150\,^\circ C} 4\,Fe_3O_4 + O_2$$

According to C. M. Riley this is the most important gas-forming reaction.

At temperatures above 1100 °C, the following reaction scheme has been proposed:

$$\begin{array}{lcl} Fe_2O_3 + C & \rightarrow & 2\,FeO + CO \\ Fe_2O_3 + CO & \rightarrow & 2\,FeO + CO_2 \\ 4\,FeO + 4\,Al_2O_3 & \rightarrow & 4\,FeAl_2O_4 \\ \hline 2\,Fe_2O_3 + C + 4\,Al_2O_3 & \rightarrow & 4\,FeAl_2O_4 + \underline{CO_2} \\ & & \text{hercynite} \end{array}$$

The spinel formation can be confirmed by X-ray diffraction.

5.3.6.2.3 Manufacture of Expanded Products

Processing of the raw materials:

- Crushing
- Grinding
- Shaping

Processing is different for plastic clays and lumpy shale. The plastic clays are crushed, ground, mixed, partially dehydrated and then shaped into pellets of different sizes with a granulation table or similar apparatus before being thermally treated (half-wet process). Lumpy raw materials, such as shale, are only ground and, if necessary, classified (dry process) before thermal treatment. The adjustment of the bulk density of the final product to the required value is carried out during the subsequent thermal treatment.

A prerequisite for the manufacture of products with low bulk densities is a rapid temperature increase in the range from 600 °C to the expanding temperature (1100 to 1200 °C). The gas production reactions then occur at the same time as the glass phase is formed and the total amount of gas produced can be used for expanding the granulate. Products with higher densities arise when the heating process is carried out more slowly, because gas production has started before the glass phase is formed and hence some gas escapes before the particle surface is sintered.

Expanding process occurs at ca. 1100 to 1200 °C

The thermal treatment is carried out using different processes:

Technical realization:

a) LECA process

In this process the thermal treatment is carried out with (preferably directly heated) rotary kilns. Rotary kilns with grate prewarmers (MIAG) or shaft prewarmers (Krupp) are employed for predrying the materials. Upon leaving the furnaces the product still has a temperature of ca. 900 to 1000 °C. A cooling drum is coupled to the rotary kiln, which ensures uniform cooling of the product. The cooling down to about 400 °C should not be too fast, since structural changes may occur. Too high differences in temperature between shell and core will lead to structural stresses resulting in hairline cracks which considerably reduce the compressive strength of the granulate.

- LECA process with rotary kilns

b) Sintering process

In this process the granulated raw material is mixed with a solid fuel (mostly coal) and conveyed on a travelling grate under a firing hood (gas-fired). The combustion gases are sucked through the product.

- Sintering by mixing fuel with the raw material

The advantage of this process is that the material is very rapidly heated to the expansion temperature. As a result very little of the expansion gas escapes before the expansion process begins and poorly expandable products can also be economically processed (e.g. in Lurgi plants).

c) Circulation flow process

The dried material is fed into a swirl chamber in which hot combustion gases are fed in from below with such

- Circulation flow process

a high velocity that they entrain the falling granulate particles. The intensity of the gas jets decreases as they go upwards so that the granulate can fall again, whereupon the circulation starts again. The expansion only takes ca. 40 s.

d) Manufacture of large blocks from expanded clay granulate

• Manufacture of expanded-clay blocks by ceramic bonding

The manufacture of larger blocks from expanded clay granulate is carried out by sintering expanded clay granulate, e.g. from rotary kilns, while still hot, in heated molds (so-called "ceramic bonding").

5.3.6.3 Expanded Products from Glasses (Foam Glass)

Both naturally occurring and industrial glasses, preferably waste glass, are suitable for the manufacture of expanded products from glasses.

Perlite: expands at ca. 1000 °C upon evaporation of the entrapped water

Perlite is a glassy vulcanic rock with water trapped in its fine pores. The rock is pulverized and then heated to ca. 1000 °C, whereupon the material melts and the entrapped water evaporates expanding the molten glass to many times its initial volume.

"Foam glass": waste glass finely ground with powdered coal or $CaCO_3$

Synthetic glasses can also be processed to so-called "foam glass". Waste glass is preferably used and it can be processed by grinding with powdered coal or calcium carbonate and/or magnesium carbonate to a very fine powder. After optional addition of a binder, this is granulated and then fired at ca. 700 °C. The decomposition of calcium or magnesium carbonate begins at this temperature, the carbon dioxide formed foaming the material. At the same time the glass becomes more difficult to melt due to the incorporation of calcium or magnesium oxide.

Expansion due to $CaCO_3$ decomposition

No cooling is required due to the sintering of the material.

5.3.6.4 Applications for Expanded Products

Advantages of expanded products: low weight, good insulation properties

Expanded products have the dual advantages of low weight and good insulation properties. They are therefore utilized in the following application fields:

Applications in:

• Lightweight concrete

• As an additive in lightweight concrete for prefabricated units, cavity blocks and construction concrete (e.g. in skyscraper building) (Fresh concrete density 0.5 to 1.0 kg/dm^3 versus 2.3 to 2.4 kg/dm^3 for normal concrete with sand and gravel as additives).

- As an additive in insulation concrete with light cement bonding.
- Loosely packed for insulation purposes in floors and roofing constructions.
- In the manufacture of so-called UP-lightweight concrete by processing expanded clay or expanded shale granules with organic binders. Legupren® lightweight concrete consists of foamed unsaturated polyester resins with lightweight additives and is used for prefabricated units.
- As an insulation layer in road building.
- Keeping of plants in hydroculture. Here expanded clay granules are used as substrates, which are as free as possible from sulfur and which can take up large quantities of water.

- Insulation concrete
- Loosely packed state for insulation
- UP-lightweight concrete
- Road building
- Hydroculture

Literature for Chapter 5.3: Construction Materials

Lime:

Robert S. Boynton: *Chemistry and Technology of Lime and Limestone*. John Wiley & Sons, New York (1980).

Robert S. Boynton, "Lime and Limestone", in Kirk-Othmer: *Encyclopedia of Chemical Technology,* 3rd Ed., Vol. 14, p. 343, John Wiley & Sons, New York (1981).

H. Gundlach, "Kalk" (Lime) in: *Ullmanns Encyklopädie der technischen Chemie*. 4th Ed., Vol. 13, p. 497, Verlag Chemie, Weinheim-Deerfield Beach-Basel (1977).

Kalk-Taschenbuch (Lime Handbook) (61st Edition). Herausgegeben vom Bundesverband der Deutschen Kalkindustrie e.V.

H. Gundlach: *Dampfgehärtete Baustoffe* (Steam-hardened Construction Materials), Bauverlag, Wiesbaden (1973).

Cement:

R. A. Helmuth, F. M. Miller, T. R. O'Connor and N. R. Greening, "Cement", in Kirk-Othmer: *Encyclopedia of Chemical Technology,* 3rd Ed., Vol. 5, p. 163, John Wiley & Sons, New York (1979).

Cement Research Progress 1977, (1978).

Cement Research Progress 1978, (1979).

F. W. Locher und I. Dreizler, "Zement" (Cement), in: *Ullmanns Encyklopädie der technischen Chemie*. 3rd Ed., Vol. 19, p. 1, Verlag Urban u. Schwarzenberg, München-Berlin (1969).

Zementtaschenbuch. (Cement Handbook), Herausgegeben vom Verein Deutscher Zementwerke, 47th Edition, Bauverlag, Wiesbaden (1979).

H. Kühl: *Zementchemie*. (Cement Chemistry) (3 Bände) 3rd Ed., VEB Verlag Technik, Berlin (1956–1961).

W. Czernin: *Zementchemie für Bauingenieure*. (Cement Chemistry for Building Engineers). Bauverlag, Wiesbaden (1977).

F. Keil: *Zement*. (Cement). Springer Verlag, Berlin-Heidelberg-New York (1971).

Plasters:

R. J. Wenk and P. L. Henkels, "Calcium compounds (calcium sulfate)", in Kirk-Othmer: *Encyclopedia of Chemical Technology,* 3rd Ed., Vol. 4, p. 437, John Wiley & Sons, New York (1978).

R. A. Kuntze, "Plasters", *Cem. Res. Prog. 1977* pp 231–254 (1978).

R. A. Kuntze, "Gypsum and Plasters", *Cem. Res. Prog. 1978*, pp 215–237 (1979).

F. Wirsching, "Gips" (Plaster), *Ullmanns Encyklopädie der technischen Chemie*. 4th Ed., Vol. 12, p. 289, Verlag Chemie, Weinheim-Deerfield Beach-Basel (1976).

H. E. Schwiete und A. N. Knauf: *Gips – Alte und neue Erkenntnisse in der Herstellung und Anwendung der*

Gipse. (Plaster – Old and New Insights into the Manufacture and Use of Plasters). Merziger Druck- und Verlags-GmbH, Merzig (1969).

Clay Products:

G. Pilitz und H. Schmidt, "Ziegeleierzeugnisse" (Products of Brickworks) in: *Handbuch der Keramik* (Ceramics Handbook), Gruppe II G, Verlag Schmid GmbH, Freiburg (1973).

K. Schoner: "Tonindustrie, Ziegel und Klinker" (Clay Industry, Tiles and Bricks), in: *Ullmanns Encyklopädie der technischen Chemie*. 3rd Ed., Vol. 17, p. 482, Verlag Urban u. Schwarzenberg, München-Berlin (1966).

Expanded Products:

H. Kromer und W. Potschigmann, "Blähton" (Expanded Clay), in: *Handbuch der Keramik* (Ceramics Handbook), Gruppe II M, Verlag Schmid GmbH, Freiburg (1977).

C. M. Riley, "Relation of Chemical Properties to the Bloating of Clays", *J. Amer. Ceram. Soc.* **34**, 121 (1951).

R. D. Hill, "Some Causes of Bloating in Expanded Clay and Shale Aggregates", *Austr. J. Appl. Sc.* **11**, 374 (1960).

5.4 Enamel

5.4.1 General Information

Enamels are mainly applied to (alloyed) steel, cast iron and aluminum. Layers adhere very firmly to the substrate

The term "enamel" can be defined as "a solidified, preferably vitreous, mass of inorganic, mainly oxidic, composition formed upon melting frits. It is melted (or is to be melted) onto metallic objects and may consist of a single or multiple layers, with or without additives". Industrially, enameling is mainly carried out on steel (and steel alloys), cast iron and aluminum components. The applied layers adhere very firmly to the substrate as a result of chemical reactions which take place between the metal and the molten frits during the firing process. The sheet steel enamel bond can withstand a tensile stress of 1000 kp/cm^2.

Recent commercial statistics for enamel production are not available. However, it is estimated that several hundred thousand tons of enamel are produced annually worldwide.

5.4.2 Classification of Enamels

There are different, partially overlapping, classifications of enamel. These can be divided into:

- Layer arrangement classifications
- Color and opacifying system classifications.

Layer Arrangement

Enamels can be applied in single or multiple layers.

In multilayer application:

In principle, enamels can be applied as a single layer or in multiple (two) layers. The classic processes operate with ground coat and cover coat enamels.

Above all, the ground coat should adhere well to the metal and be capable of dissolving the iron oxide formed during firing. The fired ground coat should contain fine bubbles (10 to 40 μm) entrapped in the glass layer. This increases its elasticity. Typical compositions for a hard and soft ground coat enamel are given in Table 5.4-1.

- Ground coat enamel
- Cover coat enamel

Ground coat enamels, which should adhere strongly to the substrate, contain cobalt oxide and also nickel oxide as adhesion-promoting oxides and should possess a fine bubble structure to increase the elasticity of the layer. Cover coats supply the chemical inertness, surface smoothness and the surface glossiness or matness of the enamel

Table 5.4-1. Chemical Compositions of a Hard and a Soft Ground Coat Enamel (% by weight).

	SiO_2	B_2O_3	Me^I_2O	CaO	Al_2O_3	P_2O_5	CoO/NiO	F	BaO
"Hard"	45 to 55	8 to 14	15 to 18	5 to 10	3 to 5	0 to 2	1 to 2	1 to 2	–
"Soft"	35 to 45	14 to 20	16 to 22	5 to 10	0 to 5	0 to 2	1 to 3	1 to 5	1 to 5

The cobalt and nickel oxide (adhesion-promoting oxide) contents ensure good adhesion between the metal and enamel layer.

The cover coat is applied and fired after firing the ground coat. The purpose of the cover coat is to endow the enameled object with its typical properties:

- Chemical inertness
- Color
- Surface smoothness
- Surface glossiness or matness.

In two recently introduced two-layer processes one of the firing steps can be omitted:

With electrostatic powder coating (see Sec. 5.4.3.4.2), it is possible to apply two layers one after the other and to stove them together (two coat – one fire), provided that the ground coat melts at a somewhat lower temperature, has a smaller electrical resistance and a ca. 50 mN/m higher surface tension than the cover coat. In the "Combismalt" process, the application of two wet coats followed by a single firing is made possible by using a special ground coat slip formulation.

In electrostatic powder coating (dry) and the "Combismalt" process (wet), the ground coat and cover coat enamels are separately applied, but fired together (two coat – one fire process)

Particularly thin and therefore impact resistant layers are provided by the one-coat direct-on white and direct-on color enamels. These processes impose much higher demands on the steel quality and its pretreatment (low carbon steels, stronger pickling). The ab-

Single layer direct-on white and direct-on color enamels provide particularly thin and impact resistant layers. Prior nickel-coating of the steel is required

sence of a ground coat with its adhesion-promoting oxides makes nickel coating of the objects (2 g nickel per m^2 sheet steel) a prerequisite for good adhesion.

White enamels:

Opacification occurs during firing by crystallization of oxides (TiO_2, ZrO_2).

Colored enamels:

Coloration by dissolving Co-,Cu-,Cr-,Mn-oxides in the glass or by adding colorants (Cr-, and Fe-oxides, spinel-, rutile-, zirconium- and tin colorants, cadmium pigments) during grinding

Coloring and Opacifying Systems

White enamels are nowadays obtained by so-called "recrystallization opacity":

The opacifying agent (preferably titanium dioxide or alternatively zirconium dioxide) is dissolved in the glass during the smelting of the frit and due to the rapid quenching of the smelt is unable to crystallize, despite the glass being supersaturated with it at low temperatures. Nucleation and precipitation of the opacifying agent take place during enamel firing at ca. 820 °C, whereupon particles are formed with the optimum size for the pigments. The boron-titanium enamel, in particular, has a very high covering power and layer thicknesses of 100 μm are sufficient to cover the dark background. A typical ingredient mixture for a boron-titanium enamel is given in Table 5.4-2.

Opacifying agents can also be added (directly to the mill) while grinding the frit.

Table 5.4-2. Typical Composition of a Boron-Titanium Enamel in % by Weight.

Ingredient	% by weight
SiO_2	35 to 45
B_2O_3	12 to 20
Me^I_2O	15 to 20
TiO_2	15 to 22
Al_2O_3	0 to 5
P_2O_5	0 to 5
F	1 to 4

Two different processes are used for the coloration of transparent and semi-opaque enamels:

- Coloration in the melt, the glass itself being colored.
- Addition of colorants during grinding, which preferably should not dissolve in the glass during firing.

Both methods are to a limited extent (only for bright colors) also suitable for recrystallizing titanium enamels.

Cobalt, copper, chromium and manganese oxides are mainly used for melt coloration. The colorants used are, in the main, those used in the ceramic industry e.g. oxides such as chromium(III) or iron(III) oxides; spinels such as $CoAl_2O_4$; rutile, zirconium and tin colorants; cadmium yellow and cadmium red.

Formulas must be adjusted to take into account any interaction between the colorants and the glass during firing.

5.4.3 Frit Manufacture and Enameling Procedure

5.4.3.1 Survey

The enameling of objects is a complex process of which frit manufacture and firing are only individual steps.

Steel sheet

- Choice of enamelable steel
- Steel pretreatment
  (depends on enameling process)
  - degreasing
  - pickling (if needed)
  - nickel-coating (if needed)

Enamel frit

- Choice and mixing of raw materials
- Frit smelting in continuously or discontinuously operated furnaces
- Smelt-quenching and pulverization of the solidified frits

Enameling

- Conversion of frits into coatable forms:
  - production of slips by *wet-milling*, or
  - dry-grinding for *powder coating*
- Enamel application:
  - dipping
  - spraying
  - vacuum coating
  - electrophoretic dip-enameling
  - electrostatic spray-coating of
- slips (wet)
- powders (dry)
- Firing:
  - ground and cover coat
- With separate firings
- With simultaneous firing
  - direct on enameling

Schema 5.4-1. Flow sheet of steel enameling.

The enameling of steel is represented by the flow sheet in Schema 5.4-1 and covers the following processes in detail:

- Manufacture of enamel frit
- Conversion of frit into usable materials
- Application of enamel to steel
- Firing of the enamel layer.

5.4.3.2 Manufacture of Enamel Frits

5.4.3.2.1 Raw Materials

Raw materials for enamel manufacture:

Oxides, carbonates, nitrates and fluorides of

- Alkali and alkaline earth metals
- B and Al
- Si, Ti and Zr
- P and Sb
- Co and Ni
- Mn and Fe

Oxides, carbonates, fluorides and nitrates of the following elements are utilized:

- Lithium, sodium, potassium (potassium also as potassium feldspar),
- Magnesium, calcium, barium,
- Boron, aluminum (boron also as borax),
- Silicon, titanium, zirconium (zirconium as zirconium silicate),
- Phosphorus, antimony (phosphorus as alkali phosphate),
- Cobalt, nickel,
- Manganese, iron.

Fluorine can be incorporated as fluorspar, cryolite or alkali hexafluorosilicate.

5.4.3.2.2 Smelting of Frits

Smelting of enamel frits:

Discontinuously in rotary kilns or continuously in gas- or electrically-heated furnaces at ca. 1250 °C.

The smelts obtained are quenched in water or led over cooling rollers. Flue gas purification is necessary

The ingredients are mixed and smelted according to the secret formulae of each company for the particular application. Various furnace types are used:

- Rotary kilns with direct gas heating (discontinuous, for small lots)
- Gas- or electrically-heated furnaces (in the case of electrical heating, the high temperature electrical conductivity of the glasses is utilized) for continuous production.

Normal enamels melt at ca. 1250 °C. The raw materials are smelted, but not, as in glass manufacture, refined.

The frits smelted in rotary kilns are quenched in water, thus forming granules. The smelt from continuous furnaces passes over water-cooled rollers resulting in flake-shaped frits, which then have to be crushed.

Part of the raw materials mixture evaporates during smelting: fluoride (as hydrogen fluoride, silicon and boron fluorides) in addition to water and carbon dioxide. Effective scrubbing of the flue gases is therefore necessary, which is easier with an electrically heated furnace than with a gas-heated one.

5.4.3.3 Conversion of the Frits into Coatable Forms

In wet coating, which is still more widely used, aqueous suspensions, so-called slips, are used (see also Sec. 5.4.3.4.1). Slips are produced by grinding granules or flakes in lined ball mills, if necessary with cooling, together with:

Conversion of frits into "slips" (aqueous suspensions) by grinding granules or flakes in ball mills together with:

- Water
- Electrolyte
- Enamel clays
- Inert materials
- Coloring and opacifying agents, if necessary

- Water
- Enamel clay, 4 to 10%
- Electrolytes, 0.3 to 1%
- Inert materials (quartz, Al_2O_3, $ZrSiO_4$), 5 to 30%
- Coloring and opacifying agents, 0 to 8%

(all relative to the weight of the frits)

The milling time depends on the degree of filling of the mill. Grinding times of ca. 4 hours are desirable. The amount of water is adjusted so as to give a slip density between 1.6 and 1.8 g/ml, deionized water being generally used.

The amounts of clay and electrolyte (potassium, sodium, magnesium salts and borax) also depend upon the enamel type and coating process. Their addition should bring about a well-flowing, slightly thixotropic slip, which does not settle easily on standing.

The enamel powder for electrostatic powder coating (see Sec. 5.4.3.4.2) is produced by dry grinding the flakes in special mills and simultaneously or subsequently coating them with organic materials to increase the specific resistivity of the powder surface to 10^{13} to 10^{16} Ω-cm. Special silicones are especially suitable coating materials.

Electrostatic coating powder is coated with organic materials to obtain a surface resistivity of 10^{13} to 10^{16} Ω-cm either during or after milling

5.4.3.4 Enamel Application on Sheet Steel

A number of coating processes are used industrially:

- Wet applications processes:
  - immersion
  - flow coating
  - vacuum coating
  - electrophoretic dipping
  - electrostatic spraying
- Dry application processes:
  - dredging (for cast iron)
  - electrostatic powder coating

5.4.3.4.1 Wet Application Processes

Wet application processes:

slips are applied to metallic substrates by immersion, spraying, vacuum coating or electrophoretic processes

In the immersion process, the article to be enameled is dipped into the slip and taken out again, the excess slip running off. In the flow coating and casting processes, the slip runs down over the articles. Application on the inner walls of hollow objects, e.g. hot water tanks, can be achieved by sucking in the slip under vacuum and then allowing it to drain off. These processes can, at least in part, be automated.

In the electrophoretic dipping process, the sheet acts as the anode. The enamel particles, negatively charged in the aqueous slip, migrate to the article being enameled at current densities of 3 to 6 A/dm^2. Deposition of enamel takes place predominantly in field line-favored regions (addition of special additives), which become increasingly insulating as the coating proceeds. This retards deposition in these regions, whereupon deposition occurs in less favored regions (reversal). At the same time, the deposited layer becomes dehydrated by electroosmosis.

After deposition, the layers have to be dried, infrared radiation at 70 to 90 °C or air convection driers being preferred. The latter utilize waste heat from the furnaces by means of recuperators (in the case of gas-heated furnaces).

5.4.3.4.2 Dry Application Processes

Dry application processes:

- For cast iron: dredging of enamel powder over the glowing article, melting to a smooth surface in a furnace, multiply applied

Cover coats for cast iron articles, e.g. cast iron bath tubs, are dredged with a powder from a sieve onto the glowing article after the firing of the ground coat has been completed. The enamel powder melts and upon subsequent treatment in a furnace spreads, thus giving a smooth surface. Multiple application yields thicker layers. (The cover coat can also be applied to cast iron using a wet process.)

- For metal sheets: electrostatic coating of charged enamel powder on earthed substrates. Very high enamel frit utilization

Powder electrostatic coating has gained in acceptance over the last few years for the enameling of metal sheets. Ionization of the frit particles occurs by the adsorption of negatively charged ions formed by air ionization at the corona electodes placed in the outlet of the powder gun. The charged particles are deposited on the earthed article. Nondeposited enamel powder is recovered and reused, so that there are practically no losses of enamel powder.

5.4.3.5 Firing of Enamels

A large number of furnace types has been developed for firing enamels. The most recent are the indirectly heated so-called tunnel furnaces with electrically- or gas-heated steel tubes. In these furnaces, the articles being enameled are transported continuously through the furnace on racks or hanging on chains and on emerging from the furnace transfer their heat to articles entering the furnace. The articles thus pass through different zones: preheating, firing and cooling.

The firing temperature is ca. 820 °C. This furnace type optimally uses the heat energy. Modern furnaces with rapid heating up times, due to the use of less dense and hence low specific heat insulation materials, can be switched off overnight and during weekends.

5.4.4 Applications for Enamel

Enameled articles are used both in the home and in industry. Typical domestic applications are for kitchen utensils, cookers, bath tubs, sanitary equipment and hot water tanks. Enameled steel sheet is also used in the decoration of buildings.

Utilization of enamel:

- In households for utensils, bath tubs, sanitary equipment, cookers, hot water tanks
- For architectural enameling
- In industry e.g. for containers in the chemical and food industries, for traffic signs

In industry, enameled containers are used in the production of chemicals and food. Enameled sheet steel is used for signs and in the building industry. Modern applications for enamel are its use for acid resistant steel sheet in flue gas purification units in power stations and in the manufacture of enameled printed circuit boards.

References for Chapter 5.4: Enamel

Commercial Information:

H. Noth, "Die wirtschaftliche Situation der Emailindustrie" (The economic situation of the enamel industry), *Mitteilungen des Vereins Deutscher Emailfachleute E.V.*, **29**, pp 13–24 (1981).

General Information:

A. I. Andrews: *Porcelain Enamels*. The Garrand Press, Champaign, ILL., USA, 1961.

A. H. Dietzel: *Emaillierung*. (Enameling.) Springer-Verlag, Berlin-Heidelberg-New York, 1981.

H. H. Emons und Mitarbeiter: *Grundlagen der technischen anorganischen Chemie*. (Fundamentals of industrial inorganic chemistry). VEB Deutscher Verlag für Grundstoffindustrie, Leipzig, 1983, or Salle Verlag, Frankfurt/Main and Verlag Sauerländer, Aarau, pp 383–389.

A. Petzold und H. Pöschmann: *Email und Emailliertechnik*. (Enamel and enameling technology.) VEB Deutscher Verlag für Grundstoffindustrie, Leipzig, 1986.

New Developments:

H. Benninghoff, "Email – Emaillieren, Neueste Ergebnisse der internationalen Emailforschung" (Enamel – enameling, most recent results from international research), *Metall Oberfläche*, **8**, pp 362–371 (1982).

W. Warnecke, C. Straßburger und H. W. Hennicke, "Emaillierverhalten und Emailhaftung mikrolegierter Tiefziehstähle" (Enamel behavior and enamel adhesion of micro-alloyed drawn steel), *Thyssen Technische Berichte 1982*, Vol. 1, pp 79–82.

5.5 Ceramics

5.5.1 General Information

Ceramics are nonmetallic predominantly crystalline inorganic compounds

The term "ceramics" is a collective term for products consisting of nonmetallic inorganic compounds normally made usable by high temperature treatment. In the USA, Japan and several other countries this term includes glasses and other products. In this chapter they are specifically excluded being covered in Sections 5.1.1 and 5.2.3.1.

The borderlines between ceramics and metals and ceramics and glasses have become somewhat blurred with the development of cermets (ceramic-metallic composite materials) and glassy ceramics. Modern ceramics include a wide range of materials and products, which range from single crystals and dense polycrystalline materials by way of glass-bonded composites to insulating foams and mainly glassy substances.

The properties of ceramic products are determined by their chemical and mineralogical composition and by their microstructure

The properties of ceramic materials are influenced not only by their chemical and mineralogical compositions but also critically by their manufacturing process-dependent microstructure. Under the term microstructure is meant the spatial distribution of the individual phases as well as the shape, size and orientation of the particles and pores.

Continuous demands for ever higher quality ceramic products and the emergence of new fields of application (e.g. fuel rods for nuclear reactors, ceramic gas turbines and diesel motors, refractory furnace linings for coal gasification plants, packages and substrates for integrated circuits) increasingly require the development of new technologies, although the basic knowledge over the manufacture of ceramics originates in antiquity.

5.5.2 Classification of Ceramic Products

Ceramic materials can be classified into different groups on the basis of their chemical compositions or their characteristic properties. Neither of these criteria gives a consistent classification. A classification according to chemical composition makes it possible to take the processing characteristics of the different materials (e.g. ceramic clay products) into account.

Classification of ceramic products:

- Silicate ceramics
- Oxide ceramics
- Nonoxide ceramics

In this chapter the classification proposed by Schüller and Hennicke will be used in which ceramic products are divided into three groups (see Fig. 5.5-1):

- Silicate ceramic materials with SiO_2 as the main constituent
- Oxide ceramic materials
- Nonoxide ceramic materials

With respect to quantity silicate materials are the most important ceramics.

These three groups can be subclassified on the basis of their level of homogeneity into fine ceramic products with microstructural features under 0.2 mm in size and coarse ceramic products with pore or crystallite sizes above 0.2 mm.

Subclassification into:

- Fine ceramic products
- Coarse ceramic products

Further subclassification can be performed on the basis of product porosity (porous to nonporous).

Further subclassification on the basis of porosity:

- Porous
- Nonporous

The classification on the basis of chemical composition is not entirely consistent and thus must be modified

Ceramic materials:

1. Silicate ceramics						
Microstructure	coarse			fine		
	porous	nonporous		porous	nonporous	
Water absorption capacity (% by wt.)	(WAC > 6%)	(WAC < 6%)		(WAC > 2%)	(WAC < 2%)	
1.1 *Clay ceramics*			earthenware	fine earthenware	stoneware	porcelain
	bricks clay pipes terracotta fireclay bricks	vitrified bricks structural ceramics	pottery	whiteware calcareous-earthenware feldspathic-earthenware	fine stoneware sanitary ware split tiles Wedgwood porcelain	hard porcelain soft porcelain insulator porcelain bone china
1.2 *Other silicate ceramics*						
	silica refractories forsterite-refractories	fusion-cast bricks		cordierite high alumina insulators	cordierite	steatite Li-Al silicates
2. Oxide ceramics						
Microstructure		coarse		fine		
		alumina magnesia		alumina magnesia zirconia titania		
		chromite		aluminum titanate ferrites titanates spinels		
3. Nonoxide ceramics						
Microstructure		coarse		fine		
		carbon graphite		nitrides carbides silicides carbon		

and expanded by subdivision according to application (e.g. refractory ceramics).

5.5.3 General Process Steps in the Manufacture of Ceramics

The manufacture of ceramic products generally proceeds by the following process steps:

- Processing and preparation of raw materials (grinding, classification, mixing)
- Forming (molding, pressing, turning, casting)
- Drying or prefiring (removal of water or organic binders)
- Ceramic firing
- Posttreatment and finishing (mechanical processing, glazing)

Manufacture of ceramics:

- Processing of raw materials
- Forming
- Drying
- Firing
- Posttreatment

These process steps will be illustrated below in the context of the manufacturing processes of important clay ceramic products.

5.5.4 Silicate Ceramic Materials

5.4.4.1 Composition and Raw Materials of Clay Ceramic Products

The properties of the different clay ceramic products is critically influenced by the relative concentrations of the three main components in their "batches", these being clay/kaolin, quartz and feldspar. Fig. 5.5-2 shows a three component diagram of these ingredients in which the compositional variations of the batches used for the manufacture of a number of important industrial products are marked.

Clay ceramic batches basically mixtures of:

- Clay/kaolin
- Quartz
- Feldspar

Clay minerals impart moist clay ceramic raw material batches with the plasticity important for processing. Quartz is used as a "lean clay" and reduces the shrinkage upon firing. The feldspar constituent acts as a flux due to its alkali content.

Kaolin is formed as a weathering product of fluorspar or fluorspar-containing rocks (granite, porphyrite, gneiss etc. in primary deposits). Its main mineral is kaolinite, which crystallizes as fine platelets and has the composition $Al_2(OH)_4Si_2O_5$. The kaolins still con-

The main kaolin mineral is kaolinite. For porcelain manufacture the content of coloring impurities should be low

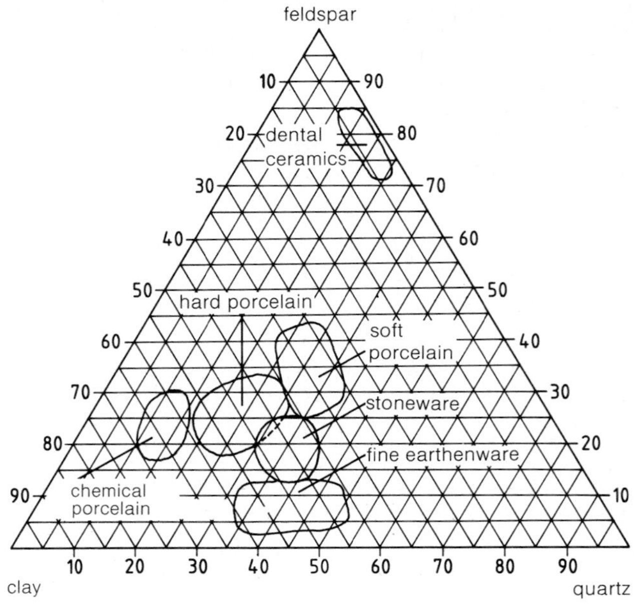

Fig. 5.5-2. Typical compositions of clay ceramic batches.

tain feldspars and quartz in varying quantities. The kaolins utilized in porcelain manufacture have a low content of coloring impurities such as iron oxides.

The main mineral of illitic clay is illite

Clays are found in secondary deposits. The main constituent of illitic clays is illite. This is formed as a hydrolytic degradation product of potash mica and has the following approximate composition:

$$(K,H_3O)_yAl_2(OH)_2(Si_{4-y}Al_yO_{10}) \text{ with } y = 0.7 \text{ to } 0.9.$$

It is widely used as a raw material in the manufacture of clay ceramic products.

Clay minerals are layer silicates

Clay minerals belong to the large group of layer silicates. This layer structure is responsible for their ability to form plastic pastes with water and imparts plasticity into the batches of clay ceramic products.

The most important European kaolin and clay deposits are found in West Germany (Oberpfalz, Westerwald), England (Cornwall, Devon), Czechoslovakia (Bohemia), East Germany (Saxony) and Spain (Galicia).

Finely divided clay-free quartz is used as a lean clay for porcelain, fine earthenware and stoneware batches. This is obtained by grinding pure quartz sand, such as that found in many sites in Mid and West Germany. Quartz powder with a still higher purity is obtained by the grinding of quartz gangue.

Fine particulate quartz is obtained by grinding quartz sand or quartz gangue

Feldspars are found in nature as constituents of coarse crystalline, late magmatic deposits (pegmatite) or in feldspar-rich sands (arcose). It is often found together with kaolins and is produced as a byproduct in kaolin processing (Sec. 5.5.4.2). Clay ceramic batches with high feldspar contents fire to products with glassy finishes (porcelain, dental ceramics). In porcelain batches potash feldspar $KAlSi_3O_8$ (orthoclase) is preferred to soda feldspar $NaAlSi_3O_8$ due to its more favorable melting behavior. The worldwide production of feldspars in 1982 amounted to 3.8×10^6 t with the USA, Italy, West Germany, the USSR and Japan being the main producing countries.

Feldspars are produced from pegmatic or arcose deposits or as a byproduct in the elutriation of kaolin

Smaller quantities of other raw materials such as aluminum oxide (calcined alumina), calcium-containing fluxes (marble, calcite, limestone, chalk) or lithium-containing fluxes (eucryptite, spodumene) are also added.

5.5.4.2 Extraction and Treatment of Raw Kaolin

Raw kaolin is mainly extracted by open cast mining. Secondary deposits can be exploited using excavators, whereas primary deposits of kaolin are often washed out of the rock with a high pressure water jet.

To obtain a uniform mineralogical composition with high alumina content (Al_2O_3 content > 37%), the raw kaolin is elutriated and simultaneously classified in hydrocyclones, whereupon kaolinite and the other layer silicates are separated from the accompanying quartz and feldspar on the basis of their different particle size distributions.

Separation of kaolin and other clay minerals from the accompanying quartz and feldspar by elutriation and particle size classification in hydrocyclones

After leaving the hydrocyclone, the water content of the fine product, so-called kaolin milk, is reduced to 25 to 30% in settling tanks and filter presses. The filter cake is often only dried to a water-content of 10 to 15%.

Kaolin suspensions (kaolin milk) concentrated by sedimentation and filtration

Different feldspar qualities and particle size-classified quartz sand are obtained as byproducts of the kaolin elutriation.

5.5.4.3 Manufacture of Clay Ceramic Batches

The particle size distributions of the raw materials influence the properties of the ceramic products

The properties of the clay ceramic products vary with the particle size distribution of the raw materials. This is varied depending upon the sort of product being manufactured.

Thus, whereas finely ground raw materials are required for fine ceramic products, mixtures of raw materials with different particle sizes are mainly used for coarse ceramic products.

Paste production proceeds by:

- Wet grinding of treated raw materials

Coarse grinding of the raw materials is carried out with crushers, pan grinders, hammer or cross beater mills and is often carried out by the suppliers. Thus only fine grinding is necessary during ceramic manufacture for which wet grinding in ball mills is satisfactory.

- Mixing of components in suspension
- Magnetic separation of iron impurities

The clay constituents of the ceramic pastes are produced during raw material processing in sufficient fineness $< 40\ \mu m$ so that they can be immediately made into a slip by adding water in stirring tubs. This is mixed with the suspension of feldspar and quartz and the mixture filtered over a sieve to remove any coarse particles still present. Magnetic separators are used to remove iron-containing impurities which have entered the paste during milling and processing and which would otherwise introduce undesirable coloration in the ceramic product.

- Adjustment of the shaping process-dependent water content

The processing of batches is determined by the intended forming process. Slips must be castable at water contents of up to 30 to 35% and are therefore stabilized with added deflocculants. Clay-containing slips typically contain 0.2 to 0.4% by weight of water glass and other sodium silicates or calcined sodium carbonate.

For the manufacture of turning pastes, the slip is dewatered with filter presses to water contents of 20 to 25% by weight, comminuted in a sieve kneader and formed in vacuum extruders into suitably sized slugs.

Complete drying of the paste is carried out mostly by spray drying. The resulting spray-dried granules can be utilized both in the dry pressing process and with added water as casting slips and turning pastes. Spray drying is therefore suitable for central batch processing plants.

Drying of pastes to flowing powders by spray drying

As a result of increasing energy costs, interest is increasing in dry processing technology, which avoids the expensive evaporation of water. Mixing of the appropriate finely preground raw materials is carried out in intensive mixers (edge-runner mixers, Eirich mixers) and then just sufficient water added to the homogenized mixture for the intended forming process. In these dry processes the removal of iron impurities is extremely difficult.

Dry processing of ceramic batches much less energy intensive than wet processing. Removal of iron impurities poses severe problems

One aspect of the processing of clay-rich batches for high quality products is so-called "draining" or "maturing" storage in damp atmospheres for several weeks or months (in China some of the clay for porcelain manufacture has been drained for several hundred years!) During storage, water probably seeps between the crystal platelets of the clay minerals with the aid of bacteria and a homogeneously moistened paste with improved plasticity is formed.

Storage in moist atmospheres produces more plastic clay ceramic batches

5.5.4.4 Forming Processes

The choice of a particular shaping process depends upon the geometric structure of the component being manufactured. The required end product properties such as density and strength and the batch size are also important.

The choice of shaping process depends on the:

- Shape of the product
- Required properties
- Batch size

The individual forming processes require batches with different water contents (Table 5.5-1)

Table 5.5-1. Water Contents of Porcelain Batches for the Different Forming Processes.

Batch types	Water content in % by weight
casting slips	33 to 36 (+ deflocculants)
turning and roller pastes	22 to 24
extruded pastes	22 to 24
leather hard pastes	15 to 18
moist molding pastes	8 to 12 (+ plastification)
dry molding batches	1 to 4

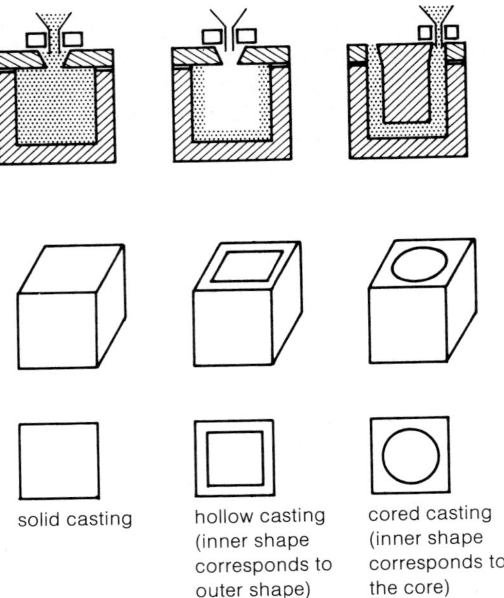

Fig. 5.5-3. Casting processes for shaped ceramic articles.

5.5.4.4.1 Casting Processes

Casting processes require high water contents. Can be divided into:

Casting processes require the highest water contents. These are particularly used for the manufacture of nonrotation symmetrical and complicated components. Casting processes can be divided into hollow casting, cored casting and solid casting (see Fig. 5.5-3).

The casting processes are based on the ability of plaster of Paris molds to extract water from the casting slip by capilliary forces. In time a thick solid layer is formed at the plaster surface. In hollow casting the residual casting slip is poured off once the solid layer adhering to the plaster surface has reached the required thickness. In core casting and solid casting the slip is topped up until the whole cavity is filled. On standing, water is abstracted and solidification and shrinkage occurs, which aids the release of the molded article from the mold.

- Hollow casting
- Cored casting, solid casting

Despite widespread development work, plaster of Paris molds are still utilized today due to their excellent absorptive capacity even taking into account the limited reuse of the molds: a maximum of 100 castings.

A particular prerequisite for the success of the casting process is precise control of particular rheological properties of the casting slip. This which is achieved by controlling the water content and adding dispersing agents and deflocculants (such as alkali silicates, alkali phosphates, humic acid or tannic acid).

Prerequisite for the success of the casting process: precise control of the rheological properties of the casting slip

5.5.4.4.2 Plastic Forming

Plastic forming is based on the plasticity of water-containing clay ceramic pastes and is in the form of the potter's wheel the oldest known sort of forming. This turning of ceramic products can be automated on a large scale using the roller process.

Turning on a potter's wheel is the oldest shaping process

In jiggering, the ceramic paste is mounted on a rotating plaster mold, which forms the inside of the molded article, and the outer surface is shaped by a profiling tool rotating in the same direction but at a higher speed. Jiggering is suitable for the manufacture of flat articles such as plates. Hollow articles (cups, bowls) are formed by roll turning in which the mold forms the outside of the article and the roller head forms the inner contours. Since the molds are made of plaster, the water is abstracted as with the casting process and the article shrinks away from the mold.

In the large scale roller process the two variants are:

- Jiggering of flat articles
- Roll turning of hollow articles

In extrusion molding so-called "slugs" are produced as preforms. These are subsequently shaped into their final form. In the manufacture of insulators the piston is dried to a "leathery hardness", which does not undergo any further shrinkage. These preforms are further processed with profiling tools.

Extrusion molding used to make differently shaped slugs and tubes so-called preforms

A new plastic forming process is currently under development, which is analogous to the injection molding process used for plastics. In "injection pressing" clay ceramic paste containing about 16% water is injected into the mold and further compressed by closing the slightly opened tool. This process can also be used for manufacturing nonrotationally symmetric articles.

Injection pressing is analogous to the injection molding of plastics and can also be used for nonrotationally symmetrical article manufacture

5.5.4.4.3 Forming by Powder Pressing

Dry pressing techniques can be used for simple geometric shapes and high numbers of articles. Dry or moderately moistened powder is compressed under

Dry pressing process utilizable for simple geometric shapes and long production runs. Stringent demands placed on the powder properties

pressure, during which the strength required for further processing is attained. Certain requirements have to be met by the powder regarding flow and compaction properties. Granulate from spray driers is a particularly suitable raw material.

The molded article should have as high density and strength as possible

In addition to the granule shape, the primary particle size distribution of the powder has an influence upon the attainable density of the molded article. The static friction between the individual particles and the compression mold can be reduced by lubricating additives, which largely suppress the development of structural stress.

The compression molds are mostly metallic, with carbidic materials being used for very hard powders. In presses in which pressure is applied at one side the pressure distribution in the powder mass is not very uniform, which leads to nonuniform density distribution in the molded article. Thus particularly for large components pressure is applied from both sides as long as the cost is justifiable.

Pressure application from both sides produces molded articles with less structural stress

Compression pressures of up to 2 kbar are used depending on the composition of the powder and the required "green density" of the unsintered molded article.

Highly mechanized plants can produce up to 300 unsintered articles/min/press. The achievable dimensional precision is very high: up to a few 10's of microns in thickness and ca. 1% in weight. Greater tolerances are allowable for large format refractory products.

Throughput of dry presses is high. Dimensional precision of the greenware (unfired article) is good

Isostatic pressing is a special type of dry pressing process. In this process the powder is poured into suitable preformed rubber matrices, which are placed in a perforated metal holder. The powder compression takes place in a pressure chamber with oil or hydrostatic pressures of 2 to 4 kbar. Pressure application on all sides produces a very uniformly compressed molded article, but one usually requiring further mechanical working due to the resilience of the rubber matrices not being completely geometrically constant. Insulators for sparking plugs (Sec. 5.5.5.1.2) are among the ceramic products manufactured on a large scale using this process.

Isostatic pressing processes produce molded articles with very uniform density distributions, but which mostly have to be worked further

5.5.4.5 Drying Processes

The greenware has to be dried to remove the residual moisture. This is accompanied by shrinkage of the article, which in the event of uneven drying is uneven and can result in cracking. Clay-rich paste with high shrinkage must therefore be very carefully dried.

Drying of unsintered molded articles is accompanied by shrinkage due to water loss

Moist air processes in which the article is thoroughly heated at high relative humidity and the temperature then slowly increased with decreasing humidity are particularly suitable. Shrinkage ends at water contents of ca. 8 to 10% at which point the temperature can be more strongly increased.

Drying takes place under precisely defined temperature, humidity and time conditions

Drying is carried out in rocking, box and tunnel driers often utilizing the combustion heat from kilns. Rapid drying (ca. 10 min) with IR irradiation has recently been introduced for thin-walled articles (crockery). This compares with drying times for large format articles of up to 24 h and more (e.g. sanitary ware).

5.5.4.6 Firing of Ceramics

5.5.4.6.1 Physical-Chemical Processes

An important step in the manufacture of ceramic products is their thermal treatment at relatively high temperatures: $> 800\,^\circ C$. This step termed firing leads to the strengthening of the unsintered molded articles into durable final products with the desired shape. The processes which take place are gathered together in the technological term "sintering".

Ceramic products are rendered durable by thermal treatment: firing

During the firing of clay ceramic pastes a number of physical-chemical processes take place, which are important for the mechanical and thermal properties of the final article:

During the firing of clay ceramic products the following physical-chemical processes take place:

- The kaolinite OH-groups split off at 400 to $600\,^\circ C$ forming metakaolinite which is stable up to $950\,^\circ C$. At temperatures $\geq 1150\,^\circ C$ cristobalite and mullite ($3\,Al_2O_3 \cdot 2\,SiO_2$) are formed in the form of flakes (so-called primary mullite) by way of a spinel-type intermediate phase:

- Formation of mullite flakes and cristobalite

$$Al_2(OH)_4(Si_2O_5) \rightarrow Al_2O_3 \cdot 2\,SiO_2 + 2\,H_2O \quad \text{(metakaolinite)}$$

$$Al_2O_3 \cdot 2\,SiO_2 \rightarrow \gamma\text{-}Al_2O_3 + 2\,SiO_2$$

$$3\,Al_2O_3 + 6\,SiO_2 \rightarrow 3\,Al_2O_3 \cdot 2\,SiO_2 + 4\,SiO_2 \quad \text{(mullite)} \quad \text{(cristobalite)}$$

- Melting of feldspar
- Dissolution of SiO_2

- The feldspar components melt at or below 985 °C, 985 °C being the melting point of the potash feldspar-silica eutectic. This melting point can be reduced to 920 to 950 °C in the presence of impurities. The feldspar melt dissolves the SiO_2 present in the starting mixture and also that formed by the thermal decomposition of metakaolinite. This results in an increase in melt viscosity which compensates the normally observed decrease in viscosity with increasing temperature. For this reason clay ceramics exhibit good dimensional stability in a wide firing temperature range.

- Precipitation of needle-shaped mullite

- Needle-shaped secondary mullite is then formed in a further chemical process during the thermal treatment. This makes a major contribution to the strength of the final article. The precipitation of secondary mullite from the feldspar melt occurs on sites with slight alkali deficiency.

Diffusion processes determine the rate constants of the physical-chemical processes

The rates of these reactions, which represent only part of the very complex chemical processes involved, are diffusion limited. By varying the firing temperature and firing time articles can be obtained either with very high transparency due to almost complete dissolution of the crystalline phases or high mechanical strength with high quartz contents.

5.5.4.6.2 Firing Conditions

Firing of mass-produced articles carried out in continuously operating tunnel kilns

For the manufacture of the mass-produced articles of the brick, refractory or crockery porcelain industries continuously operating tunnel kilns with gas, oil or electrical heating are currently almost exclusively used in which the articles travel through a well-defined temperature-time profile. Such kilns are 20 to 150 m long with transit times between 1 and > 72 h. Muffle kilns operating discontinuously are still important for small production runs and speciality products. Most of the kilns used incorporate low heat capacity fiber insulation to reduce energy costs.

The firing atmosphere is particularly important with white articles. In oxidizing atmospheres traces of iron in the starting materials oxidize to Fe_2O_3 and Fe_3O_4 depending on the firing temperature, coloring the articles brown to blackish blue. Even quantities of Fe_2O_3 below 0.35% by weight are troublesome. Upon firing in a reducing atmosphere, achieved with an air deficient combustion mixture, FeO is formed which dissolves in the melt phase giving a gray, greenish or bluish coloration. Under such conditions a white color is obtainable with iron(II) oxide contents of up to 0.6% by weight.

Discoloring iron impurities are Fe_2O_3 and Fe_3O_4 in oxidizing and FeO in reducing firing atmospheres

Glazed porcelain and fine earthenware articles are mostly fired twice. In the first firing (biscuit firing) at 900 to 1000 °C the article attains the strength required for the application of a glaze. No sintering of the article takes place and the articles can be stacked one on top of another. After biscuit firing the articles are coated with glazing slip and the glost firing carried out at product dependent temperatures between 1100 and 1500 °C. During this firing the articles must not touch one another. This is achieved by using so-called kiln furniture mounted on the tunnel kiln trolley consisting of racks made of clay- or directly-bonded silicon carbide or cordierite ($2MgO \cdot 2Al_2O_3 \cdot 5SiO_2$). Hollow articles may require supports made of shaped porcelain (so-called cases) to prevent deformation under their own weight.

For glazed articles the firing takes place in two steps:

- Biscuit firing

- Glost firing:

during the glost firing the unsintered molded articles are supported by kiln furniture

5.5.4.6.3 Glazes

The rough surfaces of fine ceramic products are smoothed with flux-rich glazes consisting of the same raw materials used for the product together with added low melting oxides (alkali oxides, alkaline earth oxides, lead oxide and boron oxide). The glazing mixture is applied by dip-coating or spraying onto the raw or precalcined articles. So-called glazing slips i.e. aqueous suspensions of the glazing mixture are required for both processes. Water soluble substances have to be converted into insoluble compounds by short fusion prior to the production of the slip (frit glazes).

Glazing mixtures mostly produce a homogeneous melt upon firing, which solidifies on cooling to a glassy finish

Glazing mixtures can be rendered insoluble by fritting

Salt glazes (thin layer glazes) are obtained by blowing rocksalt into the firing chamber. The sodium chloride reacts with the water vapor in the furnace atmosphere forming HCl and Na_2O, which form low melting sodium aluminum silicates on the surface of the article being fired.

5.5.4.7 Properties and Applications of Clay Ceramic Products

Only a survey of the fine ceramic clay-containing products will be given, the coarse ceramic products having been for the most part dealt with in the sections concerned with construction materials (Sec. 5.3.5) and refractory products (Sec. 5.5.3.5).

5.5.4.7.1 Fine Earthenware

Fine earthenware articles are porous and white or ivory in color

Porous clay ceramic products can absorb more than 2% by weight of water. This group of products includes colored earthenware (pottery) and the mostly white fine earthenware (Table 5.5-2). Due to its permeability to liquids and gases the latter can be used in the manufacture of ceramic filters and diaphragms.

Table 5.5-2. Fine Earthenware Products and their Properties.

Products	Composition of the batch in % by weight: clay minerals	quartz	feldspar	calcium carbonate	Firing conditions		Advantages	Disadvantages
calcareous earthenware (soft earthenware)	40 to 45	up to 40	–	15 to 25	biscuit glost	1150 °C 1000 to 1060 °C	cheap, low density	brittle, not stable to temperature variation
mixed earthenware	45 to 50	48 to 42	1 to 3	5 to 7.5	biscuit glost	1180 °C 1000 to 1100 °C	harder, less brittle than calcareous earthenware	
feldspathic earthenware (hard earthenware)	40 to 55	55 to 42	5 to 3	–	biscuit glost	1240 °C 1000 to 1100 °C	hard and denser	more expensive and heavier
							than calcareous & mixed earthenware	

Glazing renders it water impermeable and suitable for the manufacture of household utensils, sanitary ware, wall panels and tiles (majolica, fayence).

Feldspathic earthenware (hard stoneware) or calcareous whiteware (soft stoneware) are produced depending upon the flux used.

5.5.4.7.2 Stoneware

The term stoneware includes a range of different ceramic products, whose batches mainly consist of presintered mainly illitic clay and are fired to a dense product.

Stoneware articles are impervious to water, lightly glazed and fracture conchoidally

Fine stoneware for household utensils (real stoneware) is manufactured by casting and is grey, yellow or red. A well-known example of such fine stoneware is "Wedgwood porcelain".

Industrial fine stoneware (for laboratory apparatus, insulators) is manufactured from low quartz clays which fire to a light color, producing a porcelain-like article.

Acid-resistant stoneware receives a salt glaze and is resistant to all acids except hydrofluoric acid. The acid resistance can be increased by adding zirconium oxide or phosphates, the alkali resistance by adding barium oxide, magnesium oxide or chromium oxide.

Acid and alkali resistance of stoneware can be improved with additives

5.5.4.7.3 Porcelain

Due to the high flux content in the raw material mixture and the resulting high proportion of melt phase upon ceramic firing, porcelain is dense, white and glassy, thin walls being translucent and transparent.

Porcelain articles when thin-walled are translucent and fracture conchoidally

Porcelains are classified according to their chemical composition as well as on the basis of their firing temperatures. The hard porcelain favored in Mid-Europe for the manufacture of utensils and technical articles is produced from classical mixtures consisting of 50% kaolin, 25% feldspar and 25% quartz and are fired at 1380 to 1460 °C. The stability to temperature change particularly required for chemical apparatus porcelains and cooking utensils can be increased by reducing the quartz content, but at the expense of transparency.

Hard porcelain consists of 50% kaolin, 25% quartz, 25% feldspar. Firing temperature: 1380 to 1460 °C

By partial or complete replacement of quartz by alumina the higher strength alumina porcelain is obtained. The processing of batches is made more difficult by the abrasive action of alumina.

Soft porcelains have higher feldspar and quartz contents. The firing temperatures are 100 to 150 °C lower

If the feldspar and quartz content is increased over that in hard porcelain, soft porcelain is obtained which is fired at 100 to 150 °C lower temperatures. Highly plastic clay (ball clay) is added to obtain plastic and easy to work batches despite the lower kaolin content. Sanitary porcelains for the manufacture of bathroom articles are soft porcelains.

A special soft porcelain is vitreous china, which is used for sanitary articles

A particular type of soft porcelain widely manufactured in the USA and Great Britain is vitreous china, which produces dense nontransparent articles.

Bone china is a high quality product, whose ingredient mixture contains phosphate (40 to 50% bone ash). Applications for crockery and ornamental objects

Bone china is used particularly in Great Britain for crockery and ornamental objects and exhibits particularly high transparency and high whiteness. Its firing mixture consists of 40 to 50% bone ash, 20 to 30% feldspar and 20 to 30% kaolin. The phosphates introduced by the bone ash or apatite possess strong glass-forming properties. The firing temperature has to be reduced to 1200 to 1280 °C to prevent too much softening during firing.

Dental ceramics consist almost entirely of feldspars

So-called dental porcelains produced with particularly feldspar-rich mixtures (70 to 90% feldspar, 15 to 25% quartz, 2 to 8% calcite or kaolin) do not really belong to the clay ceramic products. Due to the high flux content a glossy surface is formed upon firing and no glazing is necessary.

5.5.4.7.4 Rapidly Fired Porcelain

Newly developed rapidly fired porcelains only require firing times of ca. 3 h

Conventional porcelain types require tunnel kiln transit times of 24 to 72 h. Firing times of 1 to 3 h are attainable with the recently developed rapidly fired porcelains. With such short firing times the furnace throughput can even be kept high when the articles are mounted as a single layer on the kiln trolley. Thus the kiln furniture can be largely dispensed with and the inert mass of the firing kiln reduced. The object of this development was a reduction in energy consumption.

The batches for rapidly fired porcelains must be adapted to the short firing times

The batches for rapidly fired porcelains have to be adapted to the short firing times to obtain product properties comparable with those of conventionally fired porcelains, despite the diffusion-controlled slow chemical reactions. Suitable mixtures contain lower concentrations of quartz and fluxes and contain pre-fired raw materials such as calcined kaolin (fireclay). Satisfactory transparency can be obtained by using very fine particulate quartz sorts, which dissolve sufficiently rapidly in the glass phase.

5.5.4.8 Economic Importance of Clay Ceramic Products

There are no reliable statistics regarding the worldwide manufacture of clay ceramic products, since much of the production is carried out by a multitude of small companies and there is no international classification for the individual product groups.

The total West German production of clay ceramic products was ca. 310×10^3 t in 1980.

The quantities for the different product groups were as follows:

- Crockery, household utensils, ornamental objects of
  - fine earthenware 28 500 t
  - stoneware 33 500 t
  - porcelain 87 000 t
- Earthenware and clayware for garden products 65 500 t
- Porcelain sanitary articles 67 500 t
- Chemical technical products 21 500 t
- Insulators of porcelain, steatite 2 500 t

5.5.5 Refractory Ceramics

5.5.5.1 Definition and Classification

The refractoriness of ceramic materials is determined by the Seger cone test

Ceramic materials are termed refractory, when they exhibit a cone-fusion point of at least SK 17 (average cone-fusion point 1500 °C) in the Seger cone test according to DIN 51 063. Highly refractory materials must have a cone-fusion point of at least SK 37 (1830 °C).

The so-called Seger cone values correspond to different ceramic bodies, which soften and deform at particular temperatures. The cone-fusion point is that temperature at which the point of the cone reaches the level of the cone feet. The different Seger cones are sequentially numbered with increasing cone-fusion point. The softening of a ceramic mass is characterized by the number of the Seger cone which has the same cone-fusion point. For many of the current refractory materials the cone-fusion point is too high to be measured. The refractory characteristics are then determined according to DIN 51 053 which measures the softening at high temperatures under pressure (pressure deformation or pressure flow).

The softening behavior of ceramic materials is compared using the Seger cone test

A characteristic property of highly refractory materials is their softening behavior under pressure

Oxide products have achieved industrial prominence as refractory materials mainly because they can be cheaply manufactured from natural raw materials.

Choice of suitable refractory materials dependent upon:

- Thermal stability
- Thermal expansion
- Thermal-shock resistance
- Chemical resistance
- Thermal conductivity
- Abrasion resistance

Apart from their thermal stability a number of other factors are important in the choice of refractory materials e.g. thermal expansion coefficient, thermal-shock resistance, chemical resistance, thermal conductivity and abrasion resistance. These properties depend on the microstructure of the ceramic material in addition to its chemical composition.

The melting and decomposition temperatures of the important constituents of refractory materials are given in Table 5.5-3:

Table 5.5-3. Melting and Decomposition Temperatures of Important Constituents of the Refractory Materials.

Material	Formula	Melting point in °C	Material	Formula	Melting point in °C
cristobalite	SiO_2	1723	mullite	$3Al_2O_3 \cdot 2SiO_2$	1850
corundum	Al_2O_3	2050	cyanite	$Al_2O_3 \cdot SiO_2$	1325 [2]
chromium(III) oxide (eskolait)	Cr_2O_3	2435	andalusite	$Al_2O_3 \cdot SiO_2$	1350 [2]
periclase	MgO	2825	sillimanite	$Al_2O_3 \cdot SiO_2$	1530 [2]
calcium oxide	CaO	2572	wollastonite	$CaO \cdot SiO_2$	1540
zirconium oxide (stabilized)	ZrO_2	2690	forsterite	$2MgO \cdot SiO_2$	1890
rutile	TiO_2	1830	chromite	$FeO \cdot Cr_2O_3$	2180
silicon carbide	SiC	2760 [1]	picrochromite	$MgO \cdot Cr_2O_3$	2330
silicon nitride	Si_3N_4	1900 [1]	spinel	$MgO \cdot Al_2O_3$	2135

[1] decomposition temperature

[2] decomposition to mullite and crystobalite

The refractory products are classified according to their chemical compositions in Table 5.5-4.

In addition they are categorized according to their temperature treatment:

- Unfired products in which bonding with inorganic or organic binders takes place at temperatures below 150 °C
- Heat-treated products in which bonding with inorganic or organic binders takes place at temperatures between 150 and 800 °C
- Fired or melted products in which mainly ceramic bonding takes place at temperatures above 800 °C

Furthermore they are classified according to their shape. Molded products include all sizes of brick and fusion-cast products. Unmolded products include mixtures for monolithic constructions (pressing compounds, plastic compounds) in which the ceramic bonding occurs during the application of the material.

In addition refractory materials are categorized on the basis of their temperature treatment:

- Unfired
- Heat treated
- Fired or melted

and their form:

- Molded
- Unmolded

Table 5.5-4. Classification of Refractory Products.

Category	Main ingredient % by weight	formula	Other major ingredient
silica products	> 93	SiO_2	–
fireclay products			
• acidic	67 to 76	SiO_2	Al_2O_3
• half acidic	22 to 30	Al_2O_3	SiO_2
• basic	30 to 45	Al_2O_3	SiO_2
alumina-rich products			
• alumina-enriched products	50 to 55	Al_2O_3	SiO_2
• cyanite bricks	55 to 65	Al_2O_3	SiO_2
• sillimanite bricks	60 to 70	Al_2O_3	SiO_2
• mullite bricks	72 to 75	Al_2O_3	SiO_2
• bauxite bricks	75 to 85	Al_2O_3	SiO_2
• corundum bricks	60 to 99.5	Al_2O_3	SiO_2
neutral products			
• zirconium silicate bricks	< 67	ZrO_2	SiO_2
• chromium ore bricks	< 50	Cr_2O_3	MgO,FeO,Al_2O_3
basic products			
• magnesia bricks	> 80	MgO	–
• magnesia/chromite bricks	55 to 80	MgO	Cr_2O_3
• chromite/magnesia bricks	25 to 55	MgO	Cr_2O_3
• forsterite bricks	< 55	MgO	SiO_2,FeO

5.5.5.2 Silica Products

Refractory silica products are manufactured from quartzites

The raw materials for the manufacture of silica-rich refractory construction materials are quartzites from different sources and of different compositions fulfilling the following specification:

$SiO_2 > 95\%$, $Al_2O_3 + TiO_2 < 2.5\%$

$Fe_2O_3 < 1.5\%$, $Na_2O + K_2O < 0.1\%$.

Phase transition of quartz makes firing of silica products more difficult

In the manufacture of silica bricks quartzites are crushed, ground, classified and mixed according to their grading curves. 1.5 to 2% lime is added to improve its sintering behavior. The firing of silica bricks must be very carefully carried out to ensure the conversion of quartz into its high temperature modification. The required highest possible conversion of quartz into cristobalite and tridymite needs firing times of up to 14 days at 1450 °C.

Applications:

- Tank furnaces for glass manufacture
- Coking ovens

Silica bricks have almost completely lost their former importance in the steel manufacturing sector and are mainly utilized in tank furnaces for glass manufacture and in coking ovens.

5.5.5.3 Fireclay Products

On the basis of their chemical composition fireclay products belong to the clay ceramic materials. However, their preferred utilization for refractory furnace linings makes it more reasonable to discuss them with the other refractory products.

Fireclay products are the cheapest refractory construction materials

Although the range of refractory products is continuously expanding, fireclay products are still the most important quantitywise. This is due to their possessing in a single product: a relatively high melting point, chemical resistance and a low coefficient of thermal expansion as well as a batch with the plastic properties so important for molding.

Fireclay produced by firing clay minerals

The clay minerals mentioned in Sec. 5.5.4.2 are used as raw materials in the manufacture of fireclay products. Direct firing of the molded article is disadvantageous due to the large shrinkage during firing. The so-called "chamotte" is therefore first produced in a pre-calcining process. This is then crushed, classified and mixed with plastic ball clay. The molding can be carried out by any of the processes used for clay ceramic products. The main firing takes place at 1250 to 1500 °C. The resulting fire brick consists of 25 to 50% mullite, 25 to 60% glass phase and up to 30% of cristobalite and quartz.

In fireclay products, mullite and SiO_2 are bonded together by a glass phase

The bricks are used in blast furnaces, for the lining of casting ladles and the refractory linings of tank furnaces for the glass industry.

Applications:

- Blast furnace linings
- Casting ladle linings
- Tank furnace linings

5.5.5.4 Alumina-rich Products

In contrast with pure aluminum oxide ceramics, the raw materials used in the manufacture of alumina-rich refractory products are, for economic reasons, natural products. The choice of aluminum silicates: cyanite, andalusite or sillimanite (chemical composition $Al_2O_3 \cdot SiO_2$) or low iron bauxite with an Al_2O_3-content > 85% and a SiO_2-content < 10% depends on the aluminum oxide content required. Natural mixtures of alumina hydrates (bauxite) and kaolin with Al_2O_3-contents of 48 to 70% are fired to so-called mullite chamottes. To obtain still higher Al_2O_3-contents, industrially produced corundum has to be added.

Raw materials for alumina-rich refractory products:

- Aluminum silicates
- Bauxite
- Bauxite + kaolin
- Corundum

During prefiring at temperatures above 1500 °C the starting materials react as follows:

bauxite + kaolin → mullite chamotte

aluminum silicate → mullite + cristobalite (together:)

calcined alumina + kaolin → sintered mullite

bauxite → mullite + corundum + glass (together:)

sintered bauxite

The manufacture of refractory bricks is carried out by hand ramming or mechanical pressing of mixtures of granular raw materials and binders (sulfite waste liquor, ball clay) and then firing at temperatures between ca. 1200 and 1800 °C.

Manufactured by:

- Addition of binders
- Molding
- Firing at 1200 to 1800 °C

Alumina-containing refractory linings are widely utilized in the steel industry. Bricks with high alumina contents are, due to their resistance to alkali slags, also used in the cement and glass industries.

Applications:

- Steel industry
- Cement industry
- Glass industry

5.5.5.5 Basic Products

Essential raw material for basic refractory products is magnesite

Basic refractory products contain various quantities of magnesium. Naturally occurring magnesite is used as a raw material and upon firing is converted into sintered magnesia:

$$MgCO_3 \xrightarrow{800 \text{ to } 900\,^\circ C} MgO\ (\text{caustic}) + CO_2$$

$$MgO\ (\text{caustic}) \xrightarrow[1900\,^\circ C]{1600 \text{ to}} MgO\ (\text{sintered magnesia})$$

The known magnesite deposits are no longer sufficient to supply completely the increasing demand for magnesium oxide. This has lead to the increasing utilization of so-called seawater magnesia. The oceans contain on average 5% of dissolved magnesium salts ($MgSO_4$, $MgCl_2$), which by metathesis with lime or calcined dolomite (CaO, MgO) can be converted into insoluble magnesium hydroxide. The hydroxide is separated, washed, dried and subsequently fired, as is magnesite, to caustic magnesia or sintered magnesia.

"Seawater magnesia" produced by precipitation of magnesium hydroxide from seawater

Manufacture of magnesia bricks:

- Dry compression process
- Sintered at 1550 to 1800 °C

Magnesia bricks are usually molded with the addition of sulfite waste liquor and magnesium sulfate using the dry pressing process and are sintered in box or tunnel kilns at 1550 to 1800 °C under a slightly reducing atmosphere. Low iron magnesia bricks are high quality refractory products. They are mainly used to line steel-producing furnaces and in the crowns and covers of tank furnaces for the glass industry.

Magnesium oxide seldom used in pure form. Addition of chromium ore improves its properties. Applications:

- Open hearth furnaces
- Rotary kilns for the cement industry
- Furnaces for the nonferrous metals industry

The stability of magnesia bricks to temperature change can be improved by the addition of chromium ore. Sintering at 1750 °C leads to so-called direct bonding by the formation of the spinels $MgO \cdot Cr_2O_3$ and $MgO \cdot Fe_2O_3$. Such bricks are utilized in open hearth furnaces, in the rotary kilns of the cement industry and in the furnaces of the nonferrous metals industry.

Other industrially important basic products are forsterite and dolomite bricks. Forsterite bricks are manufactured by firing naturally occurring low iron olivine, a mixed crystal of $2MgO \cdot SiO_2$ (forsterite) and $2FeO \cdot SiO_2$ (fayalite).

The starting material for refractory dolomite bricks is calcined dolomite, which is produced from dolomite by deacidifying at temperatures of ca. 1000 °C and then firing at 1500 to 1800 °C. They are very much cheaper than magnesite products, but are also less slag-resistant. To prevent hydration dolomite bricks have to be impregnated with tar.

Dolomite bricks are a cheap refractory construction material with limited application possibilities

5.5.5.6 Speciality Refractory Products

The industrially most important refractory products are supplemented by speciality products for particular applications sectors. The most important of these are zirconia bricks. They are produced from natural zircon ($ZrSiO_4$) with optional addition of quartz or alumina using standard ceramic processes. They are used in tank furnaces in the glass industry, aluminum melting furnaces and high temperature furnaces.

Zirconia bricks are the most important speciality refractory product

Chromium oxide-corundum bricks (mixed crystals of Al_2O_3 with 5 to 10% Cr_2O_3) exhibit very good resistance to alkaline slags. They are utilized in parts of blast furnaces (Fig. 5.5-4) and demand for them is expected to increase in the future.

Importance of chromium oxide-corundum bricks is steadily increasing

Pure chromium oxide bricks with 95% Cr_2O_3 are manufactured from synthetic Cr_2O_3 and are utilized in the refractory linings of tank furnaces for the production of E-glass fibers.

Pure chromium(III) oxide bricks due to their high price only utilized for special applications

5.5.5.7 Refractory Silicon Carbide Products

Despite its comparatively high price, silicon carbide is a significant refractory product due to its outstanding properties e.g. high thermal conductivity, high hardness and mechanical strength. It is used in zinc distillation kilns and in the manufacture of muffles, capsules and kiln furniture for the clay ceramic industry (Sec. 5.5.4.7.2). Recently silicon carbide has been used as refractory linings in blast furnaces and other sectors of the steel producing industry.

Properties of SiC ceramics:

- High thermal conductivity
- High hardness
- Mechanical strength

A temperature resistant binder is necessary to bond SiC-bricks (carborundum bricks). Clays or other silicates are usually used, the particles being bonded by a glass phase. The bricks are fired at ca. 1500 °C in an oxidizing atmosphere. Silicon carbide oxidation is limited by the formation of a passivation layer of SiO_2 on the SiC particles. This participates in the formation of the binder phase.

Coarse SiC products are bonded with clay minerals or silicon nitride

Clay-bonded bricks have a SiC-content of 40 to 90%. Bricks without a glassy binder are obtained by bonding with silicon nitride. Since the SiC-rich bricks exhibit better stability to temperature change Si_3N_4-bonding is increasing in importance.

5.5.5.8 Economic Importance

The refractories industry produces 25 to 30×10^6 t/a worldwide and is economically very important

Refractory products are of considerable economic importance. The worldwide production for the period 1975 to 1978 has been estimated to have been 25 to 30×10^6 t/a with 4.6×10^6 t, 2.4×10^6 t and 1.5×10^6 t having been produced in the USA, Japan and West Germany respectively in 1977. Table 5.5-5 gives the consumption spectrum of refractory construction materials in the USA, Japan and West Germany:

Table 5.5-5. Consumption Spectrum for Refractory Construction Materials in %.

Industry sector	USA	Japan	West Germany
iron and steel	50.7	69.7	57.7
nonferrous metals	6.5	1.9	2.7
glass	4.9	1.9	4.7
cement and lime	3.8	4.2	4.5
chemical	2.7	1.4	1.5
energy	0.8	0.1	1.1
ceramic	9.3	4.2	6.8
various	21.5	16.6	21.0

The development of the steel industry determines the consumption of refractory products

The iron and steel industry is by far the greatest consumer of refractory products. The consumption of refractory construction materials in the steel industry therefore largely determines the quality standards, the raw material requirements, the quantities produced and the technology of the refractories industry.

Market for unmolded refractory products increasing

In recent years there has been a clear drop in demand for molded products and an increase in unmolded products capable of manufacture with low energy input. In addition the proportion of high quality products has increased considerably at the expense of silica bricks and fire bricks.

The recent fundamental changes in the field of refractory linings in the steel industry, which are not yet at an end, are illustrated in Fig. 5.5-4 in which classic and modern blast furnaces are compared with one another.

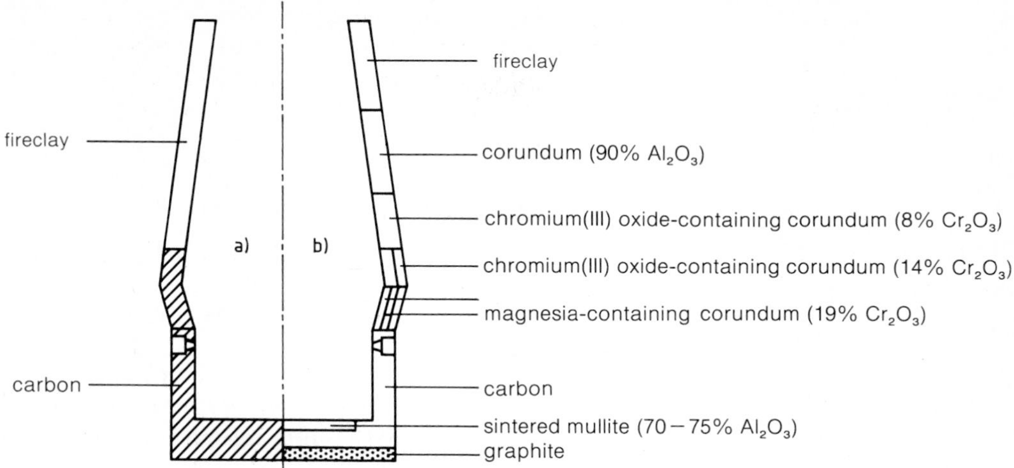

Fig. 5.5-4. A comparison of classic and modern blast furnace linings.
a) Classic blast furnace lining from the beginning of the 1960's.
b) Example of a modern lining for a 14 m diameter blast furnace.

5.5.6 Oxide Ceramic Products

5.5.6.1 General Information

In contrast with clay ceramic products, oxide ceramics consist mainly of a single pure crystalline phase of a high melting point metal oxide. Oxide ceramics are mainly nonporous fired fine ceramic products. They contain little or no low melting silicate phase.

Oxide ceramics are nonporous fine ceramics consisting mainly of single phase metal oxides

The manufacture of such materials follows the same procedure as that described in Sec. 5.5.3. In contrast with clay ceramic products, however, a few characteristic chemically determined process modifications have to be taken into consideration:

- Oxide ceramic raw materials have normally to be very pure. Such materials are not found in nature and therefore have to be manufactured by chemical processes.

Raw materials for oxide ceramics have to be very pure

- Oxide ceramic batches do not have the plasticity of clay ceramic batches. The rheological properties of the batch necessary for the particular molding process must therefore be obtained by the addition of organic compounds and water. In practice binders with partial plastifying properties such as starches,

Oxide ceramic batches are not plastic

The oxide ceramic batches are plasticized with binders and plastifying additives celluloses, dextrin, molasses, gelatines, lignins, bitumen, natural or synthetic resins or waxes are satisfactory.

- In the injection molding process the oxide powder is mixed with thermoplastics or waxes and hot pressed in the mold. The molded article can be removed from the mold, when it is cold. In subsequent heat treatment the organic additives are completely pyrolyzed.

Solidification obtained by sintering (200 to 300 °C below the melting point). Sintering temperature influenced by the particle size of the oxide particles

- The sintering of the pure oxide ceramic products is normally carried out without a melt phase at temperatures 200 to 300 °C below the melting point of the oxides. In the case of very finely divided oxide powders mixed with sintering aids to increase the diffusion of cations or anions in the oxide lattice (by a defect mechanism), dense sintering can be achieved at lower firing temperatures.

Shrinkage during firing must be taken into account in the manufacture of greenware

- Sintering is accompanied by volume contraction due to decreasing porosity. This shrinkage results in a certain dimensional tolerance in the finished ceramic component, which is too large for many applications e.g. for machine tool manufacture. Posttreatment by grinding, lapping or polishing with diamond or boron carbide tools is therefore necessary.

5.5.6.2 Aluminum Oxide

The industrially most important oxidic ceramic material is sintered aluminum oxide. The raw materials used are so-called calcined alumina and melted corundum.

Raw materials for aluminum oxide ceramics are:

- Calcined alumina (fine particulate)
- Melted corundum (coarse crystals)

Calcined alumina (α-Al_2O_3) is obtained by heating the aluminum hydroxide $Al(OH)_3$, which is produced in the bauxite digestion using the Bayer process, at temperatures above 800 °C. Depending upon the process parameters the aluminum oxide consists of 0.3 to 0.5 μm primary particles, which are mostly agglomerated into 10 to 100 μm particles. Coarser crystals are obtained by the fusion of bauxite or precalcined raw materials with carbon in electric furnaces. The impurities are reduced to metal and can be separated from the oxide melt. Purer products are obtained by the fusion of Bayer alumina. Upon quenching the melt, mm sized aluminum oxide particles of so-called melted or electrocorundum are formed. The alumina can also be sintered at temperatures of ca. 2000 °C into 0.01 to 1 mm sized platelet-shaped crystals.

• Sintered corundum (coarse crystals)

The manufacture of molded articles is usually carried out with mixtures of aluminum oxides with different particle size distributions. This is particularly important when pore-free end products are required, because this enables a higher volume concentration of aluminum oxide to be obtained than the 74% of ideally cubic close packed spheres by filling the gaps with smaller particles. The particle size distributions used in practice are usually determined using empirically determined approximate formulae (Andreasen or Fuller distribution curves) which take into account the morphology of the individual particles.

Manufacture of denser molded articles requires raw materials with precisely adjusted particle size distributions

Sintering of aluminum oxide ceramics is carried out at temperatures between 1500 and 1800 °C. The sintering rate is increased and undesirable particle growth prevented by adding MgO, but Y_2O_3, TiO_2, Cu_2O or MnO can also be used in quantities of 0.1 to 5% by weight.

The important properties of aluminum oxide ceramics are their high temperature stability (melting point of Al_2O_3: 2050 °C), good thermal conductivity, high electical resistivity and chemical resistance. Its mediocre thermal shock resistance is a disadvantage. All these properties are dependent upon the chemical pu-

Important properties of aluminum oxide ceramics:

- High temperature resistance
- Good thermal conductivity
- High electrical resistivity
- Chemical resistance

Table 5.5-5. Application Sectors for Aluminum Oxide Ceramics.

Required property	Application sectors
wear resistance	drawing dies for wire, thread guide
hardness, toughness	abrasive particles, cutting ceramics
corrosion resistance	components for acid or alkali pumps, crucibles,grinding balls, lamp tubes for sodium vapor lamps
refractoriness	melting crucibles, protective tube for thermocouples
high electrical resistivity	insulators (sparking plugs), thin disc substrates for minaturized integrated circuits
compatibility with living tissue	material for artificial limbs (e.g. for hips and teeth)

drawback:

- Mediocre thermal shock resistance

rity and particle size distribution of the oxide powder and the density, structure and pore size distribution of the ceramic. The choice of manufacturing process is closely linked with the intended application of the specific article.

The various possible applications of aluminum oxide ceramics corresponding to specific properties are listed in Table 5.5-5.

5.5.6.3 Zirconium(IV) Oxide

The natural raw material for zirconium oxide ceramics is the mineral zircon ($ZrSiO_4$)

The mineral baddeleyite (ZrO_2) is found in nature in small quantities. The main raw material for zirconium oxide ceramics is thus zircon ($ZrSiO_4$) from which impurity-free ZrO_2 is produced by fusing with lime and coke to reduce the SiO_2.

Zirconium(IV) oxide exists in monoclinic, tetragonal, orthorhombic and cubic modifications

Zirconium(IV) oxide exists in three solid phases: monoclinic, tetragonal and a cubic fluorite structure. The conversion of the monoclinic form into the tetragonal form occurs with volume reduction upon heating from 1100 to 1200 °C. The corresponding increase in volume upon cooling results in the fracture of pure sintered molded ZrO_2 articles. Addition of 10 to 15% (by weight) of CaO or MgO stabilizes the cubic high temperature modification.

Sintering of such stabilized ceramics is carried out at temperatures of ca. 1800 °C in air.

Stabilized zirconium oxide can be used up to 2600 °C. Important application sectors are:

- Furnace construction
- Resistive heating elements
- Solid electrolyte
- Neutron reflectors in nuclear reactors

Due to its high melting point (melting point of pure ZrO_2: 2700 °C) stabilized zirconium(IV) oxide ceramics are utilized in furnace construction, for the manufacture of melting crucibles and in the steel industry for continuous casting nozzles. Other important application sectors are resistive heating elements and solid electrolytes in fuel cells. The latter utilizes the unusually high conductivity of zirconium oxide ceramic materials (oxygen ion conduction) particularly that of Y_2O_3-stabilized ZrO_2 (Nernst mass).

In nuclear reactor technology hafnium-free zirconium(IV) oxide is used as a neutron reflector, due to its low neutron capture cross section.

PSZ-ceramics exhibit high mechanical strength and higher thermal shock resistance

In Partly Stabilized Zirconia (so-called PSZ-ceramics) the quantities of additives incorporated are insufficient for complete stabilization. These ceramics contain monoclinic and tetragonal components as well as the cubic phase and exhibit improved mechanical properties and higher thermal shock resistance. TZP-ceramics (**t**etragonal **z**irconia **p**olycrystals) consist of extremely fine-grained tetragonal crystals and exhibit very high mechanical strength.

5.5.6.4 Beryllium Oxide

Beryllium oxide ceramics exhibit the highest thermal conductivity of all the ceramic products and are the best high temperature electrical insulators. Despite these outstanding properties beryllium oxide ceramics have only found limited application due to their high cost and poisonousness. They are manufactured by sintering dry or plastically pressed fine particulate beryllium oxide at 1400 to 1450 °C in a hydrogen atmosphere.

Beryllium oxide ceramics have technically valuable properties. Their application is limited due to their high price and poisonousness

Typical application sectors are: casting molds for molded vanadium components, crucibles for high frequency induction furnaces and substrates for integrated circuits. In nuclear reactor technology beryllium oxide is mixed with nuclear fuel as a moderator for fast neutrons.

5.5.6.5 Uranium Oxide and Thorium Oxide

The recent increased use of nuclear energy has led to the manufacture of uranium(IV) oxide ceramics becoming an important branch of the ceramics industry. The fuel rods for the light water reactor, the most usual reactor type, preferably consist of sintered uranium(IV) oxide pellets ca. 6 mm in diameter with a length to diameter ratio of 0.9 to 1.2. The total fuel load for a light water reactor is 15×10^6 of these pellets. In 1982 it was estimated that within the next 5 years ca. 40×10^3 t of uranium(IV) dioxide would be utilized just in light water reactors.

Large quantities of UO_2-ceramics are needed in nuclear reactors

The fuel rods in light water reactors preferably contain UO_2-pellets

Pellets of U-235 enriched uranium(IV) oxide powder are manufactured using the usual dry pressing process at pressures of 2800 to 4200 bar. The subsequent sintering at 1650 °C in a nitrogen/hydrogen atmosphere produces sintered bodies with 96% of the theoretical density and a grain size of 10 μm.

UO_2-pellets are manufactured using the dry pressing process

Interest in the so-called "sol-gel process" for the remote-controlled manufacture of plutonium-containing fuel rods is increasing due to its high safety. In this

"Sol-gel" process enables the remote-controlled manufacture of spherical particles of $(U,Pu)O_2$

process a filter cake of freshly precipitated uranium-(IV) oxide is converted ultrasonically into a UO_2-gel, which after drying is fired at 1150 °C. The resulting microspheres, 40 to 60 μm in diameter, are then poured into casing tubes using vibratory techniques.

Coated ThO_2 or UO_2 fuel balls are utilized in high temperature reactors

Coated spherical ThO_2- or UO_2- particles are utilized in the fuel of gas-cooled high temperature reactors. Their 50 to 1500 μm core of uranium(IV) oxide is manufactured using conventional sintering techniques. This is then pyrolytically coated with many layers of carbon and silicon carbide (Sec. 5.7.5.1).

The uranium-thorium mixed oxide for fast breeder reactors is processed in a similar way.

Thorium oxide ceramics are only slightly radioactive

Thorium oxide ceramic materials are also important in conventional applications. Thorium oxide crucibles, for example, can, due to their extremely high melting point, be used up to ca. 3300 °C and also exhibit good corrosion resistance.

5.5.6.6 Other Oxide Ceramics

Industrially less important or unusual oxide ceramics consist of:

- Magnesium oxide

Pure sintered magnesium oxide ceramics have little importance compared with magnesium oxide-containing refractory products. They are used as high temperature materials.

- Magnesium aluminum oxide (spinel)
- Yttrium oxide

Nonporous spinel oxide ceramics of $MgAl_2O_4$ are notable for their very good high temperature corrosion resistance to alkaline media. Sintered yttrium oxide exhibits still better corrosion resistance at high temperatures (Y_2O_3 melting point: ca. 2400 °C), but its utilization possibilities are limited by its high price.

- Titanium dioxide

In recent years sintered titanium(IV) oxide has been utilized in addition to aluminum oxide in the textile industry in which it is utilized as an abrasion-resistant material for thread guides.

- Aluminum titanate

Ceramics on the basis of aluminum titanate Al_2TiO_5 are increasing in importance. These porous ceramics exhibit extremely low coefficients of linear expansion and low thermal conductivities. Aluminum titanate ceramics stabilized with MgO-, SiO_2- or ZrO_2-containing compounds are utilized for thermal insulation purposes in the exhaust systems of piston engines.

Al_2O_3 with 5 to 10% Na_2O, so-called β-aluminum oxide, is a new type of oxide ceramic solid electrolyte. Materials of this type are currently being evaluated as solid electrolytes in sodium-sulfur batteries.

• So-called β-aluminum oxide

5.5.6.7 Electro- and Magneto-Ceramics

5.5.6.7.1 Titanates

$BaTiO_3$, which has a perovskite crystal lattice, is a ferroelectric material i.e. a material in which the change in polarization P with varying applied electric field E traces a dielectric hysteresis loop analogous to the hysteresis loop exhibited by ferromagnetic materials.

Most important electroceramic material: sintered $BaTiO_3$

Technically useful properties of such perovskite ceramics are their high permittivities (relative dielectric constants), the semiconductor properties of certain chemical compositions and their piezoelectric properties.

Barium titanate is manufactured by a solid-state reaction at 1000 °C:

Large scale manufacture of $BaTiO_3$ from barium carbonate and titanium dioxide

$$BaCO_3 + TiO_2 \longrightarrow BaTiO_3 + CO_2 .$$

Very pure barium titanate is obtained by the coprecipitation of barium and titanium compounds (e.g. as oxalates or citrates) followed by calcination. In the manufacture of electroceramics this is followed by the usual ceramic processes: grinding, molding and a firing at 1300 °C.

The permittivity of these ceramic dielectrics varies between 1000 and 12 000 depending upon the particle size of the $BaTiO_3$-powder and its chemical composition. These also affect the temperature dependence of its permittivity.

Ceramic dielectrics exhibit extremely high permittivities

Barium can be partly substituted by strontium or lead and the titanium by zirconium or tin.

The physical properties of barium titanate are influenced by partial substitution with other ions

For its application as a cold conductor small amounts of barium or titanium are substituted by higher valency state ions such as antimony, niobium or the rare earths together with traces of copper, iron or manganese. Such cold conductors exhibit a strong in-

Application of $BaTiO_3$-ceramics:

- Dielectrics for capacitors
- Cold conductors
- Piezoelectrics

crease in specific resistance (10 to 100 Ωcm) with temperature e.g. an increase of a factor of 10^3 to 10^6 between 50 and 100 °C.

Piezoceramics are currently no longer manufactured from $BaTiO_3$, but from lead titanate zirconate which also crystallizes in a perovskite lattice.

In 1986 ceramic capacitors were produced worldwide with a value of approximately 2×20^9 US$.

5.5.6.7.2 Ferrites

General Information

Ferrites are magnetoceramic materials. There are two types:

- Cubic ferrites (soft ferrites) $M^{2+}Fe_2^{3+}O_4$
- Hexagonal ferrites (hard ferrites) $M^{2+}Fe_{12}^{3+}O_{19}$

The term "ferrites" has become established as a general term describing all magnetoceramic materials.

Ferrites are classified according to their crystal structure into:

1. "Cubic" ferrites crystallizing in the spinel ($MgAl_2O_4$) structure with the general chemical composition $M^{2+}Fe_2^{3+}O_4$.
2. "Hexagonal" ferrites crystallizing in the magnetoplumbite ($PbFe_{12}^{3+}O_{19}$) structure with the general chemical composition $M^{2+}Fe_{12}^{3+}O_{19}$.

The magnetic properties of these compounds and hence their applications are closely coupled to their crystal structure. This is the basis of their classification into ferrites with "soft" magnetic properties (cubic) and "hard" magnetic properties (hexagonal).

Soft Ferrites

Soft (cubic) ferrites with high permeabilities are used as coil- and transformer cores

The magnetic properties required for coil- and transformer cores are: the highest possible permeability and the lowest possible coercive force so that a high induction is produced with low magnetic fields and demagnetization in an AC field occurs easily and with low losses (soft ferrites). These requirements are met by ferrimagnetic compounds with high saturation magnetization at low crystal anisotropy.

Ferrites with the spinel structure form mixed crystals over the whole concentration range

As Table 5.5-6 shows, several ferrites with the spinel structure possess the required properties. Mixed crystal formation of these compounds with each other or

Table 5.5-6. Magnetic Properties of Simple Ferrites with the Spinel Structure.

Compound	Saturation magnetization M_s at 20°C in G*	Crystal anisotropy constant at 20°C in J/m^3	Curie temperature in K
$MnFe_2O_4$	400	-2.8×10^3	573
$FeFe_2O_4$	480	-1.1×10^4	858
$CoFe_2O_4$	425	$+2.0 \times 10^5$	793
$NiFe_2O_4$	270	-6.2×10^3	858
$CuFe_2O_4$	135	-6.0×10^3	728
$MgFe_2O_4$	120	-2.5×10^3	713

* Conversion of technical units into SI-units:
1 G (Gauss) = 10^{-4} T (Tesla)
1 Oe (Oersted) = $10^3/(4\pi)$ A/m (Ampère/meter)
1 G × Oe = $10^{-1}/(4\pi)$ J/m^3 (Joule/meter^3)

with other spinels produces a range of materials with different magnetic properties, giving the possibility of finding materials for special applications. Thus for example MnZn- and NiZn-ferrites are preferred for high frequency applications. Compounds in the system MnO-MgO-Fe_2O_3 with an iron deficiency exhibit an almost rectangular hysteresis loop (rectangular ferrites).

Rectangular ferrites are utilized in core memory systems in computers

The manufacture of high quality ferrites mainly requires chemically prepared raw materials, particularly metal oxides. These are first ground and mixed in a wet mill. The actual starting materials are formed during subsequent annealing.

Manufacture of ferrite powder by:

- Annealing oxide mixtures

In the case of ferrites with very complex compositions annealing of the oxide raw materials often produces insufficiently homogeneous compounds. In such cases a homogeneous mixture of thermally decomposable compounds (metal hydroxides, carbonates or oxalates) is produced wet chemically by coprecipitation. This preliminary product is then spray dried and presintered.

- Annealing of coprecipitated hydroxides, carbonates, oxalates

Finally the usual process steps in oxide ceramic technology are carried out: wet grinding, molding and sintering to the ceramic article.

Since a low porosity is desirable for soft ferrites to optimize their permeability, the hot pressing process is used for 1 molding high quality products. This process combines dry pressing and sintering in a single step. The pressure used is ca. 150 to 2000 bar and the temperature is about half the melting point temperature of the lowest melting eutectic. The compression molds are made of graphite, Al_2O_3- or SiC-ceramics or of refractory metals such as molybdenum, tungsten, niobium or tantalum.

Soft ferrites with low porosity are obtained by hot pressing

Other applications for cubic ferrites:

- Recording and erasure heads for tape recorders
- Ferrite antennae
- High frequency components

The most important application of cubic ferrites as regards quantity is in the manufacture of coil- and transformer cores. In terms of numbers, recording and erasure heads for tape recorders, ferrite antennae for the frequency range 0.5 to 1.5 MHz and other components for high frequency technology are all important

Hard Ferrites

Hard ferrites are permanent magnetic materials

For permanent magnets, materials with the highest possible remanence and coercive force are required, so that they cannot easily be demagnetized by external magnetic fields. These requirements are fulfilled by the hexagonal ferrites (hexaferrites) $BaFe_{12}O_{19}$ and $SrFe_{12}O_{19}$.

The magnetic properties of permanent magnets are strongly influenced by the particle size distribution

The theoretical maximum coercive force with such materials is 17 000 Oe. However in practice, with the optimum particle size of 1 μm only values of 3000 to 4000 have been achieved. This is due to lattice defects and the formation of coarse particles during sintering.

Hexaferrites are magnetically anisotropic

A special feature of hexaferrites is their pronounced magnetic anisotropy. The preferred direction for magnetization is perpendicular to the plane of the platelet-shaped hexagonal crystals. By parallelly orienting the platelets in the finished ceramic (anisotropic ferrite) the required remanence of > 4000 G and virtually rectangular hysteresis loops can be obtained.

A quality guide for permanent magnets is the maximum energy product $(B \times H)_{max}$

A guide to the quality of hard ferrites can be obtained from the maximum energy product $(B \times H)_{max}$, which can be determined from the hysteresis loop. This is proportional to the available energy per unit volume permanent magnet. For anisotropic ferrites the technically attainable value for $(B \times H)_{max}$ is 4 MG Oe. Isotropic ferrites without crystal orientation with $(B \times H)_{max}$ values of 1 MG Oe are used for cheap mass-produced articles.

Isotropic hexaferrites are manufactured in a similar way to soft ferrites using natural or waste iron oxide as a raw material.

The process for manufacturing anisotropic hexaferrites differs only in the molding process. The hexaferrite powder obtained by firing the oxide mixture at 1100 to 1300 °C is wetted with ca. 40% water and wet ground. The suspension is pumped into the compression mold, whose molding plug contains a filter and run off pipes for the water. The crystal platelets are oriented by applying a magnetic field during the compression step. Subsequent sintering at 1150 to 1250 °C requires the exact maintenance of a well-defined temperature profile due to the anisotropic shrinkage.

Molding of anisotropic ferrite ceramics takes place in the presence of an external magnetic field

Hexaferrites are mainly utilized in the manufacture of DC-motors, alternators, magneto-igniters, magnetic couplings, dynamic loudspeakers and magnetic adhesion systems. In 1986 the estimated turnover for hard ferrites in Japan and Western Europe was 710×10^6 US$.

Hexaferrite permanent magnets have many applications:

- DC-motors
- Alternators
- Magneto-igniters
- Magnetic couplings
- Dynamic loudspeakers
- Magnetic adhesion systems

1986 hard ferrite turnover in Japan and Western Europe

710×10^6 US$

5.5.7 Nonoxide Ceramics

Included in the term “nonoxide ceramics” are all non-electrically conducting materials in the boron-carbon-nitrogen-silicon-aluminum system. The industrially most important representatives apart from carbon (Sec. 5.7.4) are silicon carbide (SiC), silicon nitride (Si_3N_4), boron carbide (B_4C) and boron nitride (BN).

SiC, Si_3N_4, B_4C and BN are apart from the most important nonoxide ceramics

In contrast with the oxide ceramics the nonoxide ceramics are not thermodynamically stable in oxidizing environments. Their stability in other chemical media is excellent. Materials of silicon-containing compounds can nevertheless be used under oxidizing conditions up to ca. 1600 °C, because a passivating layer of SiO_2 is formed at the surface which strongly hinders further oxidation.

Nonoxide ceramics are very stable at high temperatures to all chemical media except oxygen

Nonoxide ceramics are utilized in applications requiring good mechanical properties at high temperatures. Their mechanical properties (high strength and

Utilized in applications requiring good mechanical properties at high temperatures

hardness) above 1000 °C are superior to those of oxide ceramics and metals. However, their abrasion resistance and chemical resistance coupled with their low densities make them also interesting materials at lower temperatures e.g. in engines.

5.5.7.1 Economic Importance

Silicon carbide with a current world production of 7×10^5 t is an industrially important product

Silicon carbide is apart from carbon the only nonoxide ceramic product of industrial importance. The worldwide production of unworked SiC was 456×10^3 t in 1975, corresponding to 83% capacity utilization. The USA and Canada accounted for 121×10^3 t of this, Western Europe 156×10^3 t and Japan 72×10^3 t. The world production for 1979/80 was estimated to be ca. 700×10^3 t/a. The largest European production plants are in Norway and the Netherlands.

Ca. 45% of the total SiC-production is utilized for metallurgical purposes, 35% are used as grinding and polishing agents and 20% for other applications such as ceramics, mainly refractory SiC-ceramics (Sec. 5.5.5.7).

5.5.7.2 Manufacturing Processes for Silicon Carbide

SiC is manufactured on a large scale using the Acheson process by reacting SiO_2 with carbon in an electric furnace

Silicon carbide is manufactured industrially by the electrochemical reaction of high purity quartz sand with carbon, preferably petroleum coke (pitch coke) or anthracite, in an electric resistance furnace (Acheson process):

$$SiO_2 + 3C \longrightarrow SiC + 2CO$$

Addition of sodium chloride ensures the removal of troublesome impurities as volatile chlorides. The presence of sawdust increases the porosity of the reaction mixture and eases outgassing.

Acheson process requires temperatures above 2000 °C

The electric furnaces in which the reaction is carried out are ca. $15 \times 3 \times 3$ m^3 in size and are lined with refractory material. Electrodes at opposite ends are connected to a graphite core. The furnace is filled round this core with the reaction mixture and electrically heated to 2200 to 2400 °C. The heating up time is ca. 18 h and the reaction time a further ca. 18 h. After cooling, the sides of the furnace are removed and the unreacted material on the edges removed. The silicon carbide, which has formed round the graphite core, is broken up and separated into different qualities.

The purest SiC is bright green (99.8% SiC), the color changing with decreasing SiC-content from dark green (99.5% SiC), to black (99% SiC), to gray (90% SiC). 70 t of raw mixture yields 8 to 14 t of high grade SiC and 6 to 12 kWh of energy are required to produce 1 kg of raw silicon carbide.

The different qualities of SiC can be distinguished by their color

SiC-manufacture is energy intensive (6 to 12 kWh/kg SiC)

The raw silicon carbide is processed by crushing in jaw crushers or hammer mills and subsequent fine grinding in ball mills. Very pure SiC qualities are obtained by chemical treatment with sulfuric acid, sodium hydroxide or hydrofluoric acid.

Silicon carbide is remarkable for its unusually large variety of different structures, which differ in their stacking sequences of hexagonal and cubic layers. All hexagonal and rhombohedral forms are often simply described as α-SiC. The commercially available SiC produced by the Acheson process is α-SiC.

Silicon carbide occurs in a large number of different crystal structures

The Acheson process produces α-SiC

The manufacture of cubic β-SiC, which is favored at temperatures below 2000 °C, or mixtures of α- and β-SiC is carried out by reacting metallic silicon with carbon. Extremely pure products can be obtained by deposition from the gas phase (**C**hemical **V**apor **D**eposition). β-SiC powder with good sintering properties and small crystallite size is e.g. obtained by the thermal decomposition of alkyl silanes or alkyl dichlorosilanes in plasmas or flow reactors at temperatures above 1000 °C.

Cubic β-SiC is produced by reaction of the elements or by gas phase processes (CVD process)

In another process metallic silicon is evaporated in an electrical arc and reacts with methane to silicon carbide. A disadvantage of all gas phase processes is their present high cost compared with the Acheson process.

Gas phase process produces high quality, but expensive, products

5.5.7.3 Fine Ceramic Silicon Carbide Products

Future applications for silicon carbide ceramics are predicted in high temperature-resistant components for machines and apparatuses. Its utilization is planned in e.g. gas turbines and diesel engines with turbocharging and in high temperature heat exchangers. Silicon carbide ceramic components can

The development of fine ceramic SiC-products opens up new application possibilities:

- Gas turbines
- Turbocharged diesel engines
- High temperature heat exchangers

be operated at temperatures up to 1400 °C without cooling and are about 60% lighter than traditional high temperature alloys, with estimated possible energy savings of 10 to 27% in gas turbines and diesel engines and of 42% for industrial heat exchangers.

Dense SiC-ceramics produced by hot pressing at:

350 bar and 1900 to 2000 °C

Only highly dense ceramics with high thermal shock resistance can be utilized for such strongly mechanically and thermally strained components. These demands are met by hot-pressed silicon carbide produced at pressures of 350 bar and sintering temperatures of 1900 to 2000 °C. The necessary but very expensive mechanical finishing is disadvantageous for large scale manufacture of hot-pressed silicon carbide.

Pressureless sintering of SiC is only possible with extremely fine powders

It is technically easier to manufacture a molded component from SiC-powder by slip casting or dry pressing, working it mechanically and then sintering pressureless at 1950 to 2100 °C. Due to the low sintering activity of silicon carbides, such processes have only been recently successfully carried out with the advent of fine particulate SiC-powders (specific surface area $> 5\ m^2/g$) with low oxygen-contents ($< 0.2\%$). Pressureless sintering is possible with added boron or aluminum and free carbon or boron carbide.

Reaction sintering by reacting liquid silicon sucked into the pores with graphite

Reaction sintered SiC contains excess silicon

So-called "reaction sintered" silicon carbide is manufactured by another process. The molding of a mixture of SiC-powder, graphite and a plastification agent is carried out by conventional techniques. Then the plastification agent is burnt off by presintering leaving a still porous component into which liquid silicon is sucked. This reacts with the graphite powder to SiC. An excess of silicon (typically 10 to 15%) fills the remaining pores. The maximum utilization temperature of reaction sintered SiC depends upon the silicon content but is limited to 1420 °C, the melting point of silicon.

5.5.7.4 Fine Silicon Nitride Ceramic Products

Si_3N_4 ceramics have similar properties to SiC ceramics

Silicon nitride ceramics possess similar properties to silicon carbide ceramics and can be utilized in the same application sectors.

Silicon can be directly reacted with nitrogen to the nitride. A mixture of α- and β-Si_3N_4 is produced

Si_3N_4 is manufactured by nitriding silicon at 1200 to 1400 °C:

$$3Si + 2N_2 \longrightarrow Si_3N_4$$

or SiO_2-C mixtures at ca. 1500 °C:

$$3SiO_2 + 6C + 2N_2 \longrightarrow Si_3N_4 + 6CO .$$

These processes produce mixtures of α- and β-Si_3N_4. By suitable choice of process parameters the proportion of the desired α-phase can be kept to 90 to 95%. Particularly pure Si_3N_4-powder can be manufactured by the tempering of silicon imide, which is produced by ammonolysis of silicon halides.

The compression and sintering of the powder to finished ceramic can be carried out by hot pressing at 140 bar and 1650 to 1750 °C. Addition of MgO, Y_2O_3, ZrO_2 or $SiBeN_2$ brings about an improvement in densification. The production of Si_3N_4-ceramics with 99% of the theoretical density has been recently achieved by pure sintering processes with dry pressed powders. It has been found that Si_3N_4 decomposition can be counteracted by sintering in nitrogen pressures above atmospheric pressure.

Ceramic components of Si_3N_4 are manufactured by:

- Hot pressing
- Pressureless sintering
- Reaction sintering
- Gas pressure sintering

The manufacture of reaction-bonded Si_3N_4 begins with a preform of silicon. This preform is nitrided in an atmosphere of pure nitrogen or of nitrogen/hydrogen. The reaction proceeds without volume change, but the products have a minimum porosity of 10%. 98% of the theoretical density can be achieved by postsintering in the presence of sintering aids.

The incorporation of metal oxides in the α- and β-Si_3N_4 lattice produces new ceramic materials. The development of this class of materials began with the incorporation of Al_2O_3 (SiAlON), but oxides such as MgO, BeO, Y_2O_3, ZrO_2 etc. can also be incorporated.

Incorporation of Al_2O_3 or BeO in Si_3N_4 produces new ceramic materials such as SiAlON and SiBeON

The practical utilization of silicon carbide and silicon nitride ceramics or of SiAlON in the above-mentioned application sectors currently mainly extends to purely experimental components. Commercialization is expected in the near future.

5.5.7.5 Manufacture and Properties of Boron Carbide

In the industrial manufacture of boron carbides, boron(III) oxide (1) or boric acid (2) are reacted with

Coarse particulate boron carbide is produced by reacting boron/oxygen compounds with carbon. The worldwide annual production amounts to several hundred tons

carbon in resistively heated furnaces at 2400 °C (cf. SiC-manufacture):

$$(1)\ 2B_2O_3 + 7C \longrightarrow B_4C + 6CO$$

$$(2)\ 4H_3BO_3 + 7C \longrightarrow B_4C + 6CO + 6H_2O$$

The tetraboron carbide produced by these processes consists of coarse hard particles suitable for grinding applications or as starting materials for the manufacture of metallic borides (Sec. 5.6.8).

Fine particulate tetraboron carbide is produced by the reduction of boron(III) oxide with magnesium or aluminum in the presence of carbon:

$$2B_2O_3 + 6Mg\ (4Al) + C \longrightarrow B_4C + 6MgO\ (2Al_2O_3)$$

Finely divided boron carbide can be processed to ceramics. Applications:

- Armored plate
- Protective material in nuclear reactors

Fine particulate powder with particle sizes below 5 μm can be compressed by hot pressing at 2100 to 2200 °C to dense ceramic products. These are used in the manufacture of armored plating, abrasion resistant components such as sand blasting nozzles or in nuclear reactors as a protective material against neutrons (high neutron capture cross section of boron).

5.5.7.6 Manufacture and Properties of Boron Nitride

Boron(III) nitride crystallizes in a hexagonal and a cubic modification

Boron nitride BN exists in a hexagonal modification with a graphite-like crystal structure and a cubic high temperature/high pressure modification with a diamond-like structure.

Hexagonal boron(III) nitride is produced by reacting boron(III) oxide with ammonia (1) or nitrogen and carbon (2)

The most widely used industrial process for the manufacture of hexagonal boron(III) nitride is the reaction of boron(III) oxide with ammonia at 800 to 1200 °C in a calcium phosphate matrix:

$$(1)\ B_2O_3 + 2NH_3 \longrightarrow 2BN + 3H_2O$$

After the reaction the matrix is dissolved in cold hydrochloric acid and unreacted boron(III) oxide washed out with hot alcohol. A raw product is obtained with 80 to 90% purity. Postreaction in nitrogen at 1800 °C or in ammonia at above 1200 °C increases the BN-content.

A very pure crystalline boron(III) nitride is obtained by reacting boron(III) oxide with carbon and nitrogen at 1800 to 1900 °C:

$$(2)\ B_2O_3 + 3C + N_2 \longrightarrow 2BN + 3CO$$

Hexagonal boron(III) nitride is not yet used in large quantities. Interest in this material is increasing, due to its wide application possibilities.

BN-powder is due to its graphite-like properties predestined for application as a high temperature lubricant or as a mold lubricant in metal casting. Molded articles of boron(III) nitride are used as melting crucibles for high purity metals or semiconductor alloys and as refractory linings for plasma burners, rocket jets and combustion chambers.

Hexagonal boron(III) nitride has graphite-like properties. Applications:

- High temperature lubricant
- Mold lubricant
- Melting crucible for metals
- Refractory lining for combustion chambers

Cubic boron(III) nitride is manufactured from hexagonal BN in a high pressure synthesis at 50 to 90 kbar and 1500 to 2200 °C in the presence of alkali or alkaline earth metals as catalysts. Cubic boron(III) nitride is after diamond the hardest material known. It is utilized in the grinding agent sector instead of diamond, due to its better chemical resistance at high temperatures.

Cubic boron(III) nitride (borazon) is produced from hexagonal BN in a high temperature/high pressure synthesis. Applications:

- Diamond substitute in the grinding agent sector

5.5.7.7 Aluminum Nitride

Aluminum nitride (AlN) is manufactured by similar processes to those used for silicon nitride. The powder is sintered with a sintering aid such as Y_2O_3 mainly using the hot pressing process. AlN-ceramics are industrially important due to their high thermal conductivity, which is superior to that of almost all other ceramic materials. They are thus utilized as substrates and housings for integrated circuits in applications in which the thermal conductivity of the usual materials (e.g. Al_2O_3) is insufficient.

References for Chapter 5.5: Ceramics

General Information:

W. D. Kingery, H. K. Bowen and D. R. Uhlmann, *Introduction to ceramics*. 2nd Ed., John Wiley & Sons Inc., New York-Chichester-Brisbane-Toronto, 1976.

H. Salmang and H. Scholze: *Die physikalische und chemischen Grundlagen der Keramik*. (The physical and chemical fundamentals of ceramics.) 5th Ed., Springer-Verlag, Berlin-Heidelberg-New York, 1968.

S. Kienow, "Keramik" (Ceramics), in K. Winnacker and L. Küchler Eds.: *Chemische Technologie*, 3rd Ed., Vol. 2, pp 282–342, Carl Hanser Verlag, München, 1970.

H. W. Hennicke, U. Hofmann, W. Lehnhäuser and K.-J. Leers: "Keramik" (Ceramics), in: *Ullmanns Ency-*

klopädie der technischen Chemie, 4th Ed., Vol. 13, pp 711 – 735, Verlag Chemie, Weinheim-Deerfield Beach-Basel, 1977.

Commercial Information:

Statistisches Bundesamt: *Produktion im Produzierenden Gewerbe nach Waren und Warengruppen.* (Production in producing industries according to product and product group) 3rd Quarter 1980, Verlag Kohlhammer GmbH, Stuttgart, Mainz, 1981.

Classification:

K. H. Schüller and H. W. Hennicke, *Ber. dt. keram. Ges.* **62,** pp 259 – 263 (1985).

Raw Materials and Refining Processes:

J. V. Hamme, "Ceramics (Raw Materials)" in Kirk-Othmer: *Encyclopedia of Chemical Technology,* 3rd Ed., Vol. 5, pp 237 – 253, John Wiley & Sons Inc., New York-Chichester-Brisbane-Toronto, 1979.

H. Trawinski: "Die Aufarbeitung von Kaolin" (The refining of kaolin), in: *Handbuch der Keramik.* (Ceramics handbook). Group I B 1, Verlag Schmid GmbH, Freiburg 1967 and 1973.

H. Trawinski, *Ber. dt. keram. Ges.* **53,** pp 117 – 121 (1976).

Shaping Processes:

Ceramics Fabrication Processes, in: *Treatise Mater. Sci. Technol.* **9,** (1976).

H. Mostetzky: "Formgebung in der Keramik" (Shaping processes for ceramics), in: *Handbuch der Keramik.* (Ceramics handbook.) Group I D 1, Verlag Schmid GmbH, Freiburg, 1978.

R. F. Stoops: "Ceramics (forming processes)", Kirk-Othmer: *Encyclopedia of Chemical Technology,* 3rd Ed., Vol. 5, pp 253 – 259, John Wiley & Sons Inc., New York-Chichester-Brisbane-Toronto, 1979.

S. Kienow and H. W. Hennicke, *Ber. dt. keram. Ges.* **57,** pp 267 – 272 (1980).

H. Seilstorfer and G. Moser, *Metall* **34,** pp 925 – 929 (1980).

Firing of Ceramics:

Th. M. Hare: "Ceramics (thermal treatment)", in Kirk-Othmer: *Encyclopedia of Chemical Technology,* 3rd Ed., Vol. 5, pp 260 – 266, John Wiley & Sons Inc., New York-Chichester-Brisbane-Toronto, 1979.

H. O. Gardeik and R. Scholz: "Brennen in der Keramik" (Firing of ceramics) in: *Handbuch der Keramik.* (Ceramics handbook.) Group I E 3, Verlag Schmid GmbH, Freiburg, 1979.

D. L. Johnson, *Mater. Sci. Res.* **11,** pp 137 – 149 (1973).

Clay Ceramic Products:

K. H. Schüller: "Porzellan" (Porcelain), in: *Ullmanns Encyklopädie der technischen Chemie*, 4th Ed., Vol. 19, pp 391 – 410, Verlag Chemie, Weinheim-Deerfield Beach-Basel, 1980.

K. H. Schüller: "Porzellan" (Porcelain), in: *Handbuch der Keramik.* (Ceramics handbook.) Group II a, Verlag Schmid GmbH, Freiburg.

O. Hufnagel and H. Stern: "Steingut und Fayence" (Fine earthenware and faience), in: *Handbuch der Keramik.* (Ceramics handbook.) Group II B, Verlag Schmid GmbH, Freiburg.

P. Fischer: "Kanalisations-Steinzeug" (Sewer stoneware), in: *Handbuch der Keramik.* (Ceramics handbook.) Group II D 1, Verlag Schmid GmbH, Freiburg, 1972.

H. Vogel: "Chemisch-technisch Steinzeug" (Stoneware for chemical purposes), in: *Handbuch der Keramik.* (Ceramics handbook.) Group II D 2, Verlag Schmid GmbH, Freiburg, 1977.

Ch. Hahn and D. Rothemund: "Dentalkeramik" (Dental ceramics), in: *Handbuch der Keramik.* (Ceramics handbook.) Group II L, Verlag Schmid GmbH, Freiburg, 1977.

H. Mörtel, *Ber. dt. keram. Ges.* **55,** pp 59 – 61 (1978).

Speciality Ceramics:

K. Heldt and A. Reckziegel: "Oxidkeramik" (Oxide ceramics), in: *Ullmanns Encyklopädie der technischen Chemie*, 4th Ed., Vol. 17, pp 515 – 529, Verlag Chemie, Weinheim-Deerfield Beach-Basel, 1979.

W. Heywang, H. Thomann and S. Ferckel: "Ferroelektrika" (Ferroelectrics), in: *Ullmanns Encyklopädie der technischen Chemie*, Vol. 11, pp 413 – 423, Verlag Chemie, Weinheim-Deerfield Beach-Basel, 1976.

K. Schüler: "Magnetische Werkstoffe" (Magnetic materials), in: *Ullmanns Encyklopädie der technischen Chemie*, Vol. 16, pp 369 – 406, Verlag Chemie, Weinheim-Deerfield Beach-Basel, 1978.

F. W. Ainger: "Ferroelectric ceramic oxides", in: B. Cockayne, D. W. Jones Eds., *Modern oxide materials.* Academic Press, London-New York, 1972.

W. Kampcyk and E. Röß: *Ferritkerne – Grundlagen, Dimensionierung, Anwendung in der Nachrichtentechnik.* (Ferrite cores – fundamentals, dimensioning, applications in telecommunications technology.) 1st Ed., Siemens Aktiengesellschaft, Berlin-München, 1978.

F. G. Brockmann, "Magnetic ceramics – a review", *Amer. Ceram. Soc. Bull* **47,** pp 186 – 194 (1968).

C. A. M. van den Brock and A. L. Stuijts, "Ferroxdure", *Philips Technische Rundschau* **37,** Nr. 7 (1977/78).

J. Smit and H. P. J. Wijn: *Ferrite.* Philips Technische Bibliothek, 1962.

J. T. A. Roberts, "Ceramic utilization in the nuclear industry", Part I and Part II, *Powder Metall. Int.* **11,** pp 24 – 29 and pp 72 – 80 (1979).

Refractory Ceramics:

J. H. Chesters: *Refractories, production and properties.* The Iron and Steel Institute, London, 1973.

K.-E. Granitzki, H. Ambs, W. Krönert, E. Müller and H. Schweinsberg: *Feuerfeste Stoffe im Gießereibetrieb.* (Refractory materials in foundries.) Gießerei Verlag GmbH, Düsseldorf, 1979.

F. Harders and S. Kienow: *Feuerfestkunde*. (Refractory technology.) Springer-Verlag, Berlin-Heidelberg-New York, 1960.

H. Ambs, F. Bens, G. Breuer and H. Schäfer, "Feuerfeste Erzeugnisse" (Refractory products), in: *Ullmanns Encyklopädie der technischen Chemie*, 4th Ed., Vol. 11, pp 549–565, Verlag Chemie, Weinheim-Deerfield Beach-Basel, 1976.

M. Koltermann, *Radex Rundschau* Heft **4,** pp 1121–1127 (1979).

K.-E. Granitzki, *Gießerei* **67,** pp 122–127 (1980).

Nonoxide Ceramics:

R. H. Smoak, T. M. Korzekwa, S. M. Kunz and E. D. Howell, "Carbides (silicon carbide)", in: Kirk-Othmer: *Encyclopedia of Chemical Technology,* 3rd Ed., Vol. 4, pp 520–535, John Wiley & Sons Inc., New York-Chichester-Brisbane-Toronto, 1978.

E. Gugel, "Nichtoxidkeramik" (Nonoxide ceramics), in: *Handbuch der Keramik*. (Ceramics handbook.) Group II K 2, Verlag Schmid GmbH, Freiburg, 1975 and 1977.

E. Gugel, "Net-shape processing of nonoxide ceramics", Annawerk Keramische Betriebe GmbH, *Mitteilung Nr. 311,* Rödental.

H.-G. Nitzsche and A. Fickel, "Über Technologie und Eigenschaften von CERANOX-Werkstoffen" (Technology and properties of CERANOX-materials), Annawerk Keramische Betriebe GmbH, *Mitteilung Nr. 163,* Rödental.

W. Bunk and M. Böhmer Eds.: *Keramische Komponenten für Fahrzeug-Gasturbinen*. (Ceramic components for gas turbines for vehicles.) Statusseminar im Auftrag des BMFT, Springer-Verlag, Berlin-Heidelberg-New York, 1978.

K. R. Nathan, *Science* **208,** pp 841–847 (1980).

T. Dickson, *Industrial Minerals,* pp 65–71 (Dec. 1980).

E. Fitzer and D. Hegen, *Angew. Chem.* **91,** pp 316–325 (1979).

5.6 Metallic Hard Materials

5.6.1 General Information

The industrially important hard materials can be divided into two groups, metallic and nonmetallic:

Hard materials are divided into:
- Metallic hard materials
- Nonmetallic hard materials

- The metallic hard materials include the carbides, borides and silicides of the elements in the subgroups of groups IV to VI of the periodic table together with thorium and uranium carbide.
- The nonmetallic hard materials include diamond (Sec. 5.7.2), corundum and other hard minerals together with silicon carbide (Sec. 5.5.7.2 and 5.5.7.3) and boron carbide (Sec. 5.5.7.5).

As the metallic hard materials, particularly the carbides, alloy easily with iron group metals, they are important in the production of cemented carbides (Sec. 5.6.5.4). Apart from their great hardness, which is in the range 9 to 10 on the Mohs scale, they exhibit comparable electrical and thermal conductivities to metals, very high melting points and high chemical resistance.

The hardnesses of metallic hard materials are almost without exception between those of corundum and diamond

In crystal structure the metallic carbides and nitrides are almost without exception interstitial compounds in which the carbon or nitrogen atoms are found in

The metallic carbides and nitrides are for the most part interstitial compounds and form mixed crystals with one another

the octahedral interstices of the metal lattice. The borides and silicides as well as a few carbides (e.g. Cr_3C_2) possess more complex structures.

Only tungsten and titanium carbide are produced in large quantities

The most important metallic hard materials as regards quantity are tungsten carbide WC and titanium carbide TiC with 15000 to 18000 t and 1200 to 1500 t respectively being produced worldwide in 1979. The other metallic hard materials are all produced in smaller quantities than titanium carbide. In individual cases they can be very important as components of alloys in the realization of particular material properties.

5.6.2 General Manufacturing Processes and Properties of Metal Carbides

Carbides mostly manufactured by reacting metals or metal oxides with carbon (lamp black) at high temperatures (1200 to 2300 °C). Processes utilized are:

Carbides are produced by reacting elemental carbon, or gases which yield carbon, with metals or metal compounds at sufficiently high temperatures (1200 to 2300 °C). This carburization is preferably carried out in a protective gas atmosphere or in a vacuum. The commonly used processes are:

- Carburization in the solid phase
- Carburization of loose or compressed powders of metal or metal oxide (occasionally metal hydrides) with reactive carbon in a protective gas.
- Carburization in metal melts (menstruum process)
- Manufacture in carbon-containing metal baths, preferably in molten iron alloys, molten nickel or molten aluminum with subsequent separation from the regulus (menstruum process).
- Carburization with gases
- Carbonization of metals or metal oxides with carbon-yielding gases (e.g. methane).
- Gas phase deposition (Chemical Vapor Deposition)
- Deposition from gas phase reactions of metal halide-hydrogen-hydrocarbon mixtures (CVD process).

The first process is by far the most important

By far the most widely used manufacturing process for metal carbides is the first process. The finest possible metal or metal oxide powder and carbon are either dry-mixed or ground together in a suitable organic liquid. Low ash and low alkali carbon blacks are used, since the purity of the starting materials strongly influences the purity of the carbide formed. Contamination with oxygen or nitrogen compounds is avoided by carburizing under a protective gas atmosphere.

The mixture or grinding product is optionally compressed and depending upon the required reaction temperature and reaction atmosphere reacted in one of the following furnace types:

The following furnace types are utilized for carburization reactions:

- Gas-heated, lined furnaces into which the reaction mixture is placed in a graphite or fire brick crucible. The combustion gas also serves as the protective gas atmosphere.

- Gas-heated furnace

- Electrically-heated resistive furnaces mainly so-called continuous carbon-tube furnaces in which the graphite tube serves as the electrical resistance and in which the reaction mixture is continuously transported in boats through the furnace. Hydrogen is used as the protective gas atmosphere.

- Continuous carbon-tube furnaces

- High frequency induction furnaces in which the graphite reaction crucible in high vacuum sealed containers is inductively heated with an induction coil coupled to a high frequency source. Temperatures up to 2800 °C can be achieved.

- High frequency induction furnaces

The proportions of raw materials in the mixtures are empirically adjusted to give carbon contents in the carbides as close as possible to the stoichiometric

Technical carbide grades do not in the main have stoichiometric compositions and often contain free carbon

Table 5.6-1. Properties of the Metal Carbides.

Compound	Theoretical carbon content % by weight	Actual carbon content % by weight	Density in g/cm^3	Melting point [1] in °C	Vickers microhardness HV 0.05
TiC	20.05	19 to 19.5	4.93	2940 to 3070	3000
ZrC	11.64	11.4	6.46	3420	2700
HfC	6.30	6.3	12.3	3820 to 3930	2600
VC	19.08	18.5 to 18.9	5.36	2650 to 2684	2900
NbC	11.45	up to 11.3 [2]	7.78	3613	2000
TaC	6.23	6.1 to 6.3	14.48	3825 to 3985	1800
Cr_3C_2	13.33	13 to 13.3	6.68	1810	1350
β-Mo_2C	5.89	up to 5.8	9.18	2485 to 2520	1500
WC	6.13	up to 6.1	15.7	2720 to 2775	1200 to 2500 [3]
ThC	4.92		10.64	2652	850
ThC_2	9.37		8.65	2655	600
UC	4.80		13.65	2560	920
UC_2	9.16		11.86	2500	620

[1] melting range given when literature data inconsistent
[2] after two carburizations
[3] different crystal surfaces have different hardnesses (hardness anisotopy)

quantities while keeping the quantity of free carbon as low as possible.

The most important properties of the metal carbides are summarized in Table 5.6-1.

5.6.3 Carbides of the Subgroup of the IVth Group

5.6.3.1 Titanium Carbide

Manufacture of TiC in induction furnaces:

$$TiO_2 + 3C \xrightarrow{2000 \text{ to } 2200\,^\circ C} TiC + 2CO$$

Titanium carbide is the most important metallic hard material after tungsten carbide. It is manufactured from pure TiO_2 and carbon black in induction furnaces at 2000 to 2200 °C:

$$TiO_2 + 3C \longrightarrow TiC + 2CO\,.$$

In small quantities TiC is also produced from titanium scrap in nickel or iron baths using the menstruum process (Sec. 5.6.2).

Titanium carbide has the highest hardness of all metal monocarbides

Titanium carbide has the highest hardness of all the metal carbides (see Table 5.6-1), but is rarely used on its own in cemented carbide technology since the high oxygen content of the titanium carbide makes the hard metal too brittle. It is thus mainly utilized in the manufacture of mixed crystals such as (Ti,W)C, (Ti,Ta,W)C and (Ti,Ta,Nb,W)C for hard metals.

Thin layers of TiC on cemented carbides increase the abrasion resistance (application as a cutting material)

One possibility of exploiting the hardness of pure TiC is by coating cemented carbides. Thin layers (5 to 8 μm) of TiC or Ti(C,N) can be deposited using the CVD process by thermal decomposition of titanium tetrachloride in a methane/hydrogen atmosphere.

5.6.3.2 Zirconium Carbide and Hafnium Carbide

Zirconium and hafnium carbide are only of minor importance

Zirconium carbide ZrC and hafnium carbide HfC can be manufactured in a similar way to titanium carbide. They are not very important in cemented carbide technology. The recent strong reduction in the price of HfC has enabled its application as a substitute for the dearer tantalum carbide.

5.6.4 Carbides of the Subgroup of the Vth Group

5.6.4.1 Vanadium Carbide

Vanadium carbide is manufactured by the carburization of pure vanadium pentoxide (mostly produced from ammonium vanadate) in vacuum at 1700 °C. Vanadium carbide exhibits a high hardness, but its brittleness prevents its wide utilization in cemented carbides. In WC-Co alloys it is added in quantities of $< 1\%$ by weight as a particle growth inhibitor.

Vanadium carbide only used in small quantities as a particle growth inhibitor in WC-Co alloys

5.6.4.2 Niobium Carbide and Tantalum Carbide

Niobium carbide NbC and tantalum carbide TaC are obtained by the reaction of their pentoxides with low ash carbon black at 1600 to 1800 °C in a hydrogen atmosphere. A higher carbon content can be obtained by postcarburization at 1600 °C.

Cheap niobium-containing tantalum carbide can be obtained from tantalum scrap, tantalum slag or ferrotantalum in an iron melt using the menstruum process. The carbide is separated chemically from the melt cake by dissolving the iron in warm hydrochloric acid.

Niobium-containing tantalum carbide manufactured with the menstruum process is industrially important

Pure TaC is utilized in cemented carbides in which it acts in a similar way to VC as a particle growth inhibitor. NbC-TaC mixed crystals are preferred for economic reasons and replace TiC in cutting materials of the WC-TiC-Co-type (Sec. 5.6.5.4) and improve scaling resistance.

5.6.5 Carbides of the Subgroup of the VIth Group

5.6.5.1 Chromium Carbide

The chromium-carbon phase diagram shows the existence of three carbide phases: $Cr_{23}C_6$, Cr_7C_3 and Cr_3C_2. Only Cr_3C_2 has industrial importance and it is manufactured by the carburization of chromium(III) oxide under a hydrogen atmosphere.

Chromium carbides are not interstitial compounds

Chromium carbides are mainly utilized in cemented carbides with a nickel bonding phase. These alloys are notable for their good corrosion and scaling resistance

Cr_3C_2 is a common component of corrosion and scaling resistant cemented carbides with a nickel bonding phase as well as abrasion resistance and are therefore utilized in high temperature applications.

5.6.5.2 Molybdenum Carbide

Application of Mo_2C in TiC-Mo_2C-Ni cutting materials

Of the different molybdenum-carbon phases only dimolybdenum carbide β-Mo_2C has any industrial importance. It is manufactured by reacting molybdenum(VI) oxide or metallic molybdenum with carbon black at 1350 to 1800 °C in carbon tube short circuit furnaces under hydrogen atmospheres. Due to its low hardness the use of Mo_2C is virtually limited to TiC-Mo_2C-Ni cutting materials.

5.6.5.3 Tungsten Carbide

WC is the industrially most important metal carbide

Two tungsten-carbon compounds exist in the tungsten-carbon system: W_2C and WC as well as low melting point eutectica in the systems W/W_2C and W_2C/WC. Monotungsten carbide is by far the most important metal carbide in cemented carbide metallurgy (Sec. 5.6.5.4).

Tungsten carbide is produced by the carburization of high purity metallic tungsten

WC is manufactured by first reducing β-tungstic acid H_2WO_4, tungsten trioxide WO_3 or ammonium paratungstate $5NH_3 \cdot 12WO_3 \cdot 5H_2O$ with hydrogen at 700 to 900 °C to high purity (> 99.9% W) tungsten powder. Carburization is performed at 1500 °C either under hydrogen in electrical resistance furnaces or under vacuum in induction furnaces. The total carbon content of the WC produced is between 6.05 to 6.20% by weight compared with the stoichiometric value of 6.13%, with the over-stoichiometric carbon content being present in a free state.

Properties of WC influenced by:

- Raw material particle size
- Raw material reduction conditions
- Tungsten carburization conditions

The powder properties of WC are influenced by the particle form and particle size of the raw material, the reduction conditions and the carburization conditions.

Ultrafine tungsten carbide is manufactured in a plasma:

$$WO_3 \xrightarrow[3000\,^\circ C]{CO \text{ or } CH_4} WC$$

There is increasing interest in ultrafine tungsten carbide particles < 100 nm in diameter (Sec. 5.6.5.4), which are manufactured by reacting very finely divided tungsten(VI) oxide with CO or CH_4 at temperatures of ca. 3000 °C in a plasma. However, these ultrafine carbide particles contain a relatively high concentration of oxygen, which limits their range of applications.

5.6.5.4 Cemented Carbides Based on Tungsten Carbide

Cemented carbides are sintered materials consisting of very hard and high melting point metal carbides in a matrix of low melting point bonding metals of the iron group, particularly cobalt.

Cemented carbides are sintered alloys with one or more hard material phases

The most important property of tungsten carbide in its application in cemented carbides is its ability to dissolve partially in compressed powder mixtures of WC and iron metals, particularly cobalt, at 1300 to 1500 °C. In the case of sintering with a liquid phase, WC partly crystallizes out of the binder phase of the WC-Co-alloy upon cooling. It becomes embedded in the tough but hard (not brittle) binder phase.

Cobalt is the preferred binder for the hard material phase in which WC or other hard materials are embedded

Cemented carbides are manufactured by powder metallurgical processes, which in many respects are similar to ceramic processes. A mixture of the carbides with the metallic bonding phase is first subjected to wet grinding, producing a finely dispersed distribution of the components. Removal of the grinding liquid is followed by vacuum drying and, in the case of high quality products, reductive annealing at moderate temperatures to remove traces of oxygen. Subsequent molding is by dry pressing, extrusion molding or hot pressing. After presintering at 900 to 1150 °C and finishing, sintering is carried out at 1400 to 1700 °C in a vacuum or under hydrogen. In the manufacture of pore-free materials postcompression with hot isostatic presses may follow.

Powder metallurgical processes similar to ceramic processes

In the manufacture of cemented carbide the oxygen content should be minimized. Manufactured by:

- Intimate mixing by wet grinding
- Molding
- Presintering at 900 to 1150 °C
- Finishing
- Sintering at 1400 to 1700 °C
- Optional postcompression with hot isostatic presses

Table 5.6-2. Chemical Composition and Applications of Tungsten Carbide-Containing Cemented Carbides.

Cemented carbide quality	Chemical composition	WC content in % by weight	Application
K-quality	WC-Co	88 to 96	cutting tools for short-chip materials, drilling plates for impact stress applications
P-quality	(W, Ti, Ta, Nb)C-Co	65 to 85	cutting tools for long-chip materials
M-quality	(W, Ta, Nb)C-Co	70 to 80	wide range of uses
G-quality	WC-Co	75 to 96	chip-less shaping

The chemical composition and applications of the different qualities of tungsten carbide-containing cemented carbides are summarized in Table 5.6-2.

WC-Co materials from ultrafine tungsten carbide are speciality products with increased abrasion resistance and therefore higher cutting efficiency.

5.6.6 Thorium Carbide and Uranium Carbide

Uranium and thorium carbides are not typical hard materials

The thorium carbides (ThC, ThC_2) and uranium carbides (UC, UC_2) exhibit metallic properties, but in other properties differ distinctly from the carbides of the subgroups of the IVth, Vth and VIth groups. Their hardnesses are substantially lower (see Table 5.6-1), they are pyrophoric and are easily hydrolyzed in water or weak acids. Their only economic importance is as carbidic fuels in nuclear reactor technology particularly for high temperature and breeder reactors, which are currently in the evaluation phase.

The importance of uranium carbide will increase with the introduction of high temperature nuclear reactors

Carbide powder is manufactured by carburization of the metal, metal oxide or metal hydride with nuclear-pure graphite at ca. 2000 °C. Carbide powder pellets (Sec. 5.5.5.1.5) or beads up to 0.6 mm in diameter are molded and sintered using ceramic process technology. Reaction sintering is commonly used for the manufacture of uranium carbide fuel beads, in which uranium oxide and carbon are first mixed then annealed to form the carbide and then are sintered to 90 to 95% of the theoretical density. An additional coating of the beads with multiple layers of pyrolytically deposited graphite and silicon carbide retains the fission products in nuclear fuel.

Carbidic fuels utilized in the form of pellets or graphite- or SiC-coated beads

5.6.7 Metal Nitrides

Metallic nitrides have similar properties to the corresponding carbide. They are of minor industrial importance

The metal nitrides of the transition metals of the subgroups of the IVth, Vth and VIth groups are very similar to the corresponding carbides in their structures (interstitial compounds of the MN type) and their properties such as hardness, melting point and electrical conductivity (see Table 5.6-3). Tungsten and molybdenum nitrides are exceptions with different structures and decomposition with nitrogen loss above 800 °C.

The following processes are used for the manufacture of metal nitrides:

- Nitriding of metal oxides under nitrogen or ammonia in the presence of carbon.
- Nitriding of metals or metal hydrides under nitrogen or ammonia.
- Reaction of metal chlorides or metal oxychlorides with ammonia.
- Reaction of metal chlorides with nitrogen or ammonia in a hydrogen plasma with gas phase deposition (CVD process).

Table 5.6-3. Properties of Metal Nitrides.

Compound	Melting point in °C	Vickers hardness HV 0.05
TiN	2950	2100
ZrN	2985	1600
HfN	3390	1700
VN	2180	1560
NbN	2205	1400
TaN	3095	1150
CrN	1085, 1500?	1100
Cr_2N	1590	1580
Mo_2N	decomposition	not determined
W_2N	decomposition	not determined

In comparison with the metal carbides the industrial demand for the nitrides is very small. This is amongst other things due to the poor wettability of the nitrides by iron group metals, which makes the manufacture of cemented alloys more difficult.

TiN-coatings are industrially important, increasing the abrasion resistance of cemented carbides and producing a decorative golden yellow coloration on articles of jewelry.

Titanium nitride coatings utilized as abrasion resistant layers on cemented carbides and as decorative layers (golden yellow) on articles of jewelry

Crucibles of titanium nitride and zirconium nitride are utilized for the melting of lanthanum alloys. ZrN, HfN and TaN are also used as electrodes in electronic valves.

5.6.8 Metal Borides

Metal borides possess excellent hard material properties, but are difficult to work

Metal borides are notable for their high melting points, high hardness, good electrical conductivity and chemical resistance (see Table 5.6-4). They are manufactured by the following processes:

- Melting of metal-boron mixtures in arc furnaces
- Reaction sintering of compressed metal-boron mixtures
- Reduction of mixtures of metal oxides and boric oxide with aluminum, silicon or magnesium with simultaneous boride formation
- Reaction of metal oxides with boron carbide to mainly carbon-containing borides
- Deposition from the gas phase reaction of volatile boron and metal compounds

Table 5.6-4. Properties of Metal Borides.

Compound	Melting point in °C	Vickers hardness HV 0.05
TiB_2	2850	3400
ZrB_2	3040	2250
HfB_2	3200	2900
VB_2	2450	2100
NbB_2	3000	2600
TaB_2	3150	2500
CrB	2050	2140
CrB_2	2150	2100
MoB	2350	2500
MoB_2	2100	2350
WB	2400	3750
W_2B_5	2300	2600

Borides cannot be processed with bonding metals to cemented alloys

Despite their interesting properties wide application of borides is not possible, because they cannot be processed with bonding metals to cemented alloys and the more expensive hot press process must be used. Furthermore borides form low melting point eutectics with iron-group metals.

The high resistance of titanium diboride to many metal melts, particularly aluminum, has led to its use as an electrode and crucible material for electrometallurgical processes. The chromium borides, CrB and CrB_2, are utilized in abrasion resistant layers on the basis of Ni-Cr-B-Si and in scaling-resistant composite materials.

Titanium diboride and the chromium borides are virtually the only borides utilized industrially:

- TiB_2 as an electrode and crucible material
- CrB, CrB_2 for abrasion resistant layers

5.6.9 Metal Silicides

The silicides of the metals of the subgroups of the IVth, Vth and VIth groups are in principle manufactured using the same processes as those utilized for the corresponding borides.

The silicides exhibit the lowest melting points and hardness values of the metallic hard materials (see Table 5.6-5). Their brittleness makes them unsuitable for utilization in cemented alloys. Silicides have only been utilized industrially in metallurgical sectors in which their scaling resistance and chemical resistance are important. They are also deposited using the CVD process e.g. as protective layers on high melting metallic surfaces.

Silicides are unsuitable for use in cemented alloys

Molybdenum disilicide is utilized in electrical heating elements, which can be operated at temperatures of up to 1600 °C in air.

Molybdenum disilicide is utilized for the manufacture of electrical heating elements, which can be operated in air at temperatures up to 1600 °C

Table 5.6-5. Properties of Metal Silicides.

Compound	Melting point in °C	Vickers hardness HV 0.10
$TiSi_2$	1545	950
$ZrSi_2$	1550	1025
$HfSi_2$	1545	975
VSi_2	1680	1090
$NbSi_2$	1950	1050
$TaSi_2$	2300	1250
$CrSi_2$	1520	1150
$MoSi_2$	2030	1290
WSi_2	2165	1200

References for Chapter 5.6: Metallic Hard Materials

R. Kiefer and F. Benesovsky: *Hartstoffe*. (Hard materials.) Springer Verlag, Vienna, 1963.

R. Kiefer and F. Benesovsky: "Carbides (Industrial heavy-metal)", in: Kirk-Othmer: *Encyclopedia of Chemical Technology,* 3rd Ed., Vol. 4, pp 490–505, John Wiley & Sons Inc., New York-Chichester-Brisbane-Toronto, 1978.

R. Kiefer and F. Benesovsky, "Carbides (cemented)", in: Kirk-Othmer: *Encyclopedia of Chemical Technology,* 3rd Ed., Vol. 4, pp 483–489, John Wiley & Sons Inc., New York-Chichester-Brisbane-Toronto, 1978.

H. Haag and H. Tulhoff: "Carbide und andere pulvermetallurgische Vorstoffe" (Carbides and other powder metallurgical raw materials), in: Winnacker-Küchler, *Chemische Technologie,* 3rd Ed., Vol. 6, pp 498–518, Carl Hanser Verlag, München, 1973.

F. Benesovsky and W. Schedler: "Hartmetalle" (Cemented carbides), in: *Ullmanns Encyklopädie der technischen Chemie*, 4th Ed., Vol. 12, pp 515–521, Verlag Chemie, Weinheim-Deerfield Beach-Basel, 1976.

R. Kiefer and F. Benesovsky: "Pulvermetallurgie" (Powder metallurgy), in: *Ullmanns Encyklopädie der technischen Chemie*, 4th Ed., Vol. 19, pp 563–576, Verlag Chemie, Weinheim-Deerfield Beach-Basel, 1980.

F. Benesovsky: *Pulvermetallurgie und Sinterwerkstoffe* (Powder metallurgy and sintering materials), Druckschrift der Metallwerk Plansee AG & Co. KG, Reutte, 1980.

H. Grewe: "Hartstoffe" (Cemented carbides), in *Handbuch der Keramik*. (Ceramics handbook). Group II K 3, Verlag Schmidt GmbH, Freiburg, 1976.

H. Grewe and J. Kolaska, *Metall* **33,** pp 23–27 (1979).

5.7 Carbon Modifications

5.7.1 Introduction

Allotropic forms of carbon:

- Graphite
- Diamond

Modifications of graphite:

- Hexagonal graphite
- Rhombohedral graphite

Modifications of diamond:

- Cubic diamond
- Hexagonal diamond

Carbon exists in two allotropic forms: graphite and diamond. Under normal conditions graphite is the thermodynamically more stable form. Two modifications of graphite are known: a hexagonal and a rhombohedral. Diamond also exists in two modifications: a cubic and a hexagonal.

These modifications obviously represent ideal modifications and a number of deviations in the arrangement of the carbon layers are observed. In this chapter industrially manufactured forms of carbon e.g. carbon black or activated carbon will be considered together with carbon modifications as their structures are in principle based on those of diamond and graphite.

5.7.2 Diamond

Classification of diamonds according to origin:

- Natural diamonds
- Synthetic diamonds

The systematic search for natural diamond deposits and their commercial mining started at the end of the last century with the discovery of the seams in South Africa. Synthetic diamonds were first reproducibly produced in 1953 by a team of scientists in Stockholm.

Only very pure almost colorless perfectly crystalline stones with certain minimum dimensions can be worked to cut diamonds. Diamonds that do not fulfil these requirements are known as industrial diamonds and are exclusively used as industrial materials.

Classification of diamonds according to application:

- Gem diamonds
- Industrial diamonds

5.7.2.1 Economic Importance

The worldwide production of natural diamonds has increased strongly in recent years and is still dominated by African countries. Australia has recently become an important new source of diamonds.

Table 5.7-1. Production of Natural Diamonds in 10^6 ct (1 carat = 0.2 g).

	1980*	1981*	1982*	1983*	1984*	1985**
Africa	29.8	26.3	27.2	36.8	44.6	46.2
USSR	10.9	10.6	10.7	10.8	11.0	12.0
Australia	–	–	0.5	6.2	5.7	7.1
S. America	1.3	1.6	1.0	0.6	1.1	0.8
Asia (excl. USSR)	0.9	1.0	1.0	1.0	1.0	0.5
Worldwide	42.9	39.5	40.3	55.4	63.4	66.5

* source: World Mineral Statistics
** source: Industrial Minerals, December 1986

The mining of gem diamonds is currently particularly high accounting for ca. 40% of the production compared with ca. 25% in 1980. The demand for industrial diamonds can nowhere near be satisfied from natural deposits. Ca. 85% of the 150×10^6 ct of industrial diamonds consumed in the Western World in 1985 were of synthetic origin.

Estimates from the US Bureau of Mines predict an annual increase in the world demand for diamonds, mainly industrial diamonds, of 2 to 9% up to the year 2000. The known world reserves of natural diamonds are well over 1×10^9 ct.

More than half of the world trade in diamonds is controlled by De Beers through the Central Selling Organisation (CSO).

Total turnover of diamonds through the CSO in 1985: 1.823×10^9 \$

5.7.2.2 Mining of Natural Diamonds

Diamond deposits:

- Primary: kimberlite rock
- Secondary: deposits (alluvial) formed by erosion, weathering and washing out from kimberlite veins

The primary source of diamonds is rock of volcanic origin, which is called kimberlite after the first important deposit discovered in South Africa. These deposits are strongly eroded by weathering and erosion and therefore most of the diamonds are found in surrounding areas in old water courses, alluvial deposits and sometimes on seabeds. Off the coast of Namibia, for example, diamond-containing clays are sucked up from the seabed and immediately concentrated on site.

Extraction processes for the extraction of diamonds from rock utilize the following properties of diamonds:

- High density
- Hydrophobicity
- Luminescence under X-ray excitation
- High refractive index

Standard underground and open cast mining techniques are used for extraction on land with even hewing by hand being widely practiced. The gangue is separated by mechanical methods and the diamonds concentrated by processes in which they can be separated on the basis of their high density of 3.51 g/cm^3. The high density minerals separated together with the diamonds are removed by utilizing the nonwettability of diamonds by water. Diamonds are retained by fatty surfaces and in the skin flotation process they float on the surface. Ferromagnetic minerals can also be removed by magnetic separation. A widely used extraction process uses X-ray excited luminescence to detect the diamonds, whereupon sluicing is started e.g. the diamond is deflected with a synchronized air jet directed perpendicularly to the stream of individual particles from the other minerals. Electrostatic charging processes have been found to be satisfactory for the separation of very small diamonds. Hand sorting and hand classification are indispensible, particularly for gem diamonds.

The choice of separation process depends strongly upon the type of deposit.

5.7.2.3 Manufacture of Synthetic Diamonds

Diamond are synthesised by:

- Diffusion controlled phase change
- Diffusionless phase change

The synthesis of diamonds requires high pressures and high temperatures. The processes used can be divided into two groups: those operating with catalysts (diffusion controlled phase change) and those operating without catalysts (diffusionless phase change).

More than 90% of all synthetic diamonds are manufactured catalytically from graphite using such catalysts as

- Iron
- Cobalt
- Nickel
- Manganese
- Chromium
- Alloys and carbides of these and other transition metals.

Catalysts for the diffusion-controlled synthesis of diamonds:

transition metals, their alloys and their carbides

Graphite, the starting material, and the catalysts are introduced into the usually metallic reaction vessel encapsulated in a series of holders designed to maintain the temperature and pressure and electrically heated. The pressure is produced hydraulically and exerted on the reaction vessel by a plunger made of a very hard material (e.g. tungsten carbide), whose geometry varies with the apparatus used. Upon heating under pressure, the catalyst metal melts dissolving the graphite and carbon precipitates as diamond from the supersaturated solution. At the end of this "recrystallization", which only lasts a few minutes, the diamonds are present as a mass of agglomerated single crystals. The catalyst metal is then removed mechanically and chemically and the diamonds sorted on the basis of size and shape. This synthesis is reproducible making it possible to produce large quantities of diamonds of uniform quality. General Electric and De Beers together are estimated to profitably produce more than 80% of the synthetic diamonds produced in the Western World using this method [up to 1 mm in diameter (ca. 0.2 ct)].

Reaction conditions for the diffusion-controlled phase change:

- Temperature: ca. 1500 to 2500 °C
- Pressure: 50 to 100 kbar

By comparison the process without catalysts has the following important disadvantages: lower yield, contamination with graphite resulting in stones with distorted crystal lattices, maximum diameter of ca. 60 μm, much higher process pressures and possibly much higher process temperatures. In the dynamic process operated by DuPont the pressure and temperature are produced for a few microseconds in a shock wave apparatus. The starting material is also graphite, which should be as crystalline as possible. Static high pressure synthesis processes without catalysts are industrially unimportant.

Conditions for diffusionless phase change:

- Temperature: up to ca. 3500 °C
- Pressure: > 120 kbar

5.7.2.4 Properties and Applications

Outstanding properties of diamonds:

- Hardness of 10 on the Mohs scale
- Refractive index of 2.42

Diamonds are exceptionally hard, but are easily cleaved in certain directions in the three main lattice planes (100), (110) and (111). Accurate knowledge of these directions is of fundamental importance in the cutting of raw diamonds to strongly refracting cut diamonds. Gem diamonds are still exclusively natural.

Applications of natural diamonds:

- Gem stones
- Industrial diamonds:
  - carbonados
  - ballas
  - boort (boart)

The types of naturally occurring industrial diamonds are carbonado, balla or boort (or boart). Carbonados are black glossy aggregates of intergrown diamond crystallites, which contain graphite and other impurities. This intergrowth makes carbonados far less brittle than diamond single crystals. They are preferably used for drill tips. The almost spherical ballas have a similar structure and are often used for coating coarse particulate grinding tools. The rest of the natural industrial diamonds are known as boort and comprise finely crystalline material with little intergrowth. It fractures easily and lends itself to the manufacture of more or less fine particulate diamond powder.

Diamonds utilized as a hard material in: drilling, cutting, honing, grinding, polishing etc.

Natural and synthetic diamonds are like other hard materials utilized for drilling, cutting, honing, grinding, polishing etc. (Sec. 5.6) by embedding in a resin or metallic matrix. In liquid suspensions they are used for lapping. They are also used as wire drawing stones. The materials worked with the aid of diamonds include diamond itself, glass, ceramics, tungsten carbide, rock, metals, metal alloys and plastics. Special application sectors are medicine (precision cutting tools) and the electronics industry (semiconducting diamonds by doping with boron, beryllium or aluminum).

Special application sectors:

- Precision cutting tools in medicine
- Semiconducting diamonds in electronics

5.7.3 Natural Graphite

5.7.3.1 Economic Importance

The World reserves of natural graphite are estimated to be several 100×10^6 t. Its high crystallinity makes natural graphite interesting for special applications. Demand is increasing, particularly for high quality types. Synthetic graphite is predominantly used in the processing industries.

Table 5.7-2. Production of Natural Graphite in 10^3 t.

	1980	1981	1982	1983	1984
P.R. China	160.0	180.0	185.0	185.0	185.0
USSR	80.0	70.0	75.0	80.0	80.0
South Korea	60.6	34.9	27.0	33.3	58.6
P.R. Korea	25.0	25.0	25.0	25.0	25.0
India	55.0	72.8	61.8	41.6	39.1
Mexico	44.9	42.3	36.2	44.3	41.5
Brasil [1]	27.3	33.8	21.5	27.6	32.5
Austria [2]	36.7	23.8	24.5	40.4	43.8
Czechoslovakia	15.7	20.3	22.0	26.7	27.0
Madagascar	9.9	16.0	15.2	13.5	14.0
Zimbabwe	7.4	11.2	8.2	19.9	12.3
Rumania	12.5	12.5	12.5	12.6	12.4
West Germany [3]	11.3	10.4	10.6	12.0	12.4
Norway	10.4	6.9	7.5	8.1	9.5
Sri Lanka	7.8	7.5	8.8	5.8	6.0
World	572	576	548	583	604

[1] includes processed and directly shipped products
[2] raw graphite
[3] includes production from imported raw graphite

5.7.3.2 Deposits and Mining

Natural graphite occurs in the form of flakes (scales) and pieces (coarse lumps) as well as in a very finely divided form. The latter is called amorphous graphite, which is a misnomer because even in its finest form it consists of crystals, however small. It is predominantly found in its hexagonal modification, although the rhombohedral form is also found. The degree of crystallinity and the carbon content vary from deposit to deposit.

The crystallite size of natural graphite is between 0.1 μm and several mm

Carbon content of raw graphite ores ca. 20 to 100% by weight depending upon the deposit

Flake graphite is found in beds in metamorphic rock, such as gneiss or marble and is formed from carbon-rich sediments (e.g. at Passau from algae colonies). Lump graphite is found where magma has solidified in stratified rock (limestone, slate). It is possibly formed by the transformation of coal into graphite at high temperatures in the veins and cavities at the edge of stratified rock formations (Sri Lanka). Finely crystalline (and microcrystalline) graphite is also formed from carbon-containing sediments in metamorphic stratified rock. Seams of graphite originating from coal seams are also found (Mexico).

Graphite deposits are found in or at the edges of stratified rock

Graphite ore is extracted by underground and open cast mining. Beneficiation takes place by sorting according to carbon content, crushing and enrichment.

Concentration of the graphite is often unnecessary

The lump graphite found in Sri Lankan deposits is the purest natural graphite known with a carbon content of up to 100%. Such high purity is not required for many applications. Thus enrichment, if needed, only has to be carried out to the particular purity level required. This assumes that the impurities do not have undesirable properties as regards its intended application e.g. abrasive impurities in graphite destined for solid lubricant applications.

Concentration normally carried out using flotation techniques

Highly purified graphite types (> 99.5% C) only obtainable using additional chemical treatment:

- Sodium carbonate digestion
- Hydrofluoric acid leaching

A typical enrichment process is flotation which is used, for example, in the Graphitwerk Kropfmühl AG (near Passau, West Germany) to produce flake graphite types with carbon contents between 70 and ca. 97% as follows: the raw ore is crushed, ground with water in a rod mill, freed from most of the gangue in a rake classifier and then subjected to multistep flotation separation interspersed with wet grinding steps. In the flotation cells graphite flakes surrounded by air bubbles float to the surface forming a graphite scum which is sucked off, centrifuged and dried. Purer graphite types with carbon contents > 99.5% require additional chemical refining, e.g. sodium carbonate digestion or hydrofluoric acid leaching. The combined use of sieving, grinding and classification enables marketable types to be obtained with well-defined particle sizes down to < 5 μm while retaining the flake shape.

Microcrystalline graphite is also beneficiated with the aid of flotation, but the processes are complicated by its small particle size and the resulting more homogeneous distribution of the accompanying minerals. Oils are preferred as flotation media.

5.7.3.3 Properties and Applications

The important properties of graphite are:

- Softness
- Low density
- Easy delamination
- High temperature stability, particularly thermal-shock resistance

- Chemical resistance (corrosion resistance)
- Very high thermal conductivity
- Very high electrical conductivity

It exhibits anisotropic physical properties as a result of its crystallographic structure of planar well-separated parallel sheets of carbon atoms. Thus graphite single crystals exhibit far higher electrical and thermal conductivities parallel to the layers of carbon atoms than perpendicular to them. Macroscopically, however, this anisotropy is seldom observable due to the random orientation of the individual particles.

The properties of graphite are strongly anisotropic, the electrical conductivity varying by a factor of 10^4!

Ca. 80% of natural graphite production is utilized for high temperature applications. The largest consumer is the steel industry in which graphite is utilized e.g. as an "abhesive agent" (dressing): the surfaces of channels, chill molds and molds being impregnated with graphite to prevent the penetration of molten steel and reactions with it. Furthermore it is utilized for covering melts, for cementation and as a deoxidant. Molds, channels and crucibles are not only surface-treated with graphite, but manufactured using up to 50% by weight of graphite. Ceramic binders, pitch or tar are utilized as bonding agents. These high quality types are preferably used in metal casting in which graphite-containing tamping compounds are also used.

Main application sectors for natural graphite:

- Iron and steel industry
- Foundry technology
- Metallurgy
- Refractory industry, crucible and retort manufacture

The easy delamination of graphite makes it a very suitable lubricant: in a dry form as a paste with greases or as a suspension in liquids. For the same reason graphite is used as the "lead" in lead pencils. Its electrical conductivity is utilized in dry batteries, brushes for motors and generators, contact rods, conductive paints and antistatic applications. Its high chemical resistance is utilized in corrosion resistant paints. Particularly pure graphite (carbon content $\geq$ 99.99%) is used as a matrix material for fuel rods in nuclear reactors. Flake graphite is used in the manufacture of graphite intercalated compounds (Sec. 5.7.5.1 and 5.7.5.3).

Other application sectors for natural graphite:

- Lubricant industry
- Pencil lead manufacture
- Electrical industry
- Conducting and anticorrosion paints
- Plastics industry
- Nuclear technology

5.7.4 Mass-Produced Synthetic Carbon and Synthetic Graphite

5.7.4.1 Economic Importance

Exact and comprehensive statistics regarding the production of and demand for carbon and graphite products are unavailable due to the large number of producers and the multiplicity of their industrial applications.

The 1985 consumption for the most important application sectors have been estimated from the production figures of the most important carbon and graphite consumers and the product-related consumption levels (Table 5.7-3).

Table 5.7-3. Estimated Consumption of Mass-Produced Carbon and Graphite Products in 1985 in 10^3 t (source: Vohler).

Application	Quantity in 10^3 t
carbon electrodes for aluminum production	7500
graphite electrodes for electrosteel manufacture	1000
carbon electrodes for the manufacture of alloys, carbides, silicon, phosphorus etc.	550
carbon and graphite linings	225

Although as regards quantity graphite products account for at most 20%, the value of the graphite products exceeds that of the nongraphite products.

5.7.4.2 General Information about Manufacture

Industrial manufacture of carbon by purely thermal decomposition of carbon-rich raw materials in inert atmospheres

Industrial production of graphite by heating graphitizable carbon in an inert atmosphere normally in the absence of catalysts

Carbon and graphite can be manufactured by many processes. The industrially operated processes are mainly purely thermal and utilize solid, liquid or gaseous carbonizable or graphitizable materials. Synthetic carbon is produced in the first process step and this is converted into graphite in a second step.

In mass production only solid and liquid preferably meltable starting materials are important. The raw materials used are almost all more or less graphitizable. Nongraphitizable carbon is as special carbon industrially less important and is also produced by gas phase pyrolysis of special carbon and graphite types and is considered in Sec. 5.7.5. Carbon fibers have already been discussed in Sec. 5.2.3.4.

Since under economically acceptable conditions carbon neither melts nor is sintered, production has to be carried out in such a way that a molded article is formed suitable for its intended application, which only needs further mechanical working. The raw materials, solids and binders, are therefore mixed, molded into the required form, fired to the carbon article and if required finally graphitized.

Basic processes involved:

- Mixing of raw materials
- Molding
- Carbonization
- Graphitization

5.7.4.3 Manufacture of Synthetic Carbon

5.7.4.3.1 Raw Materials

The raw materials utilized in the manufacture of carbon and graphite products can be divided into solids and binders.

Raw materials:

- Solid materials:
  - petroleum coke
  - pitch coke
  - furnace coke
  - anthracite
  - carbon black
  - natural graphite
- Binders: different sorts of pitch

The following solids are generally used:

- Petroleum coke from residues or intermediates from crude oil refining
- Pitch coke from coal tar pitch
- Furnace coke from coal
- Anthracite
- Carbon black
- Natural graphite

Cokes are formed from the appropriate materials by liquid phase pyrolysis. More or less structured cokes are formed depending upon the pyrolysis conditions and the composition of the starting materials. The ease of graphitization increases with increasing crystallinity and increasing crystallite size of the coke. Petroleum cokes are the preferred raw materials with well-ordered strongly anisotropic needle coke, a type of petroleum coke, being one of the most valuable raw materials. For some applications, however, isotropic cokes are used. Apart from crystallinity other important quality criteria are:

Most important solid is petroleum coke

- Low sulfur content
- Low content of volatile components
- Low ash content (metallic impurities, silicon)

- Low coefficient of thermal expansion
- High density

The addition of carbon black, anthracite and natural graphite improves certain properties. For economic reasons anthracite is used in the manufacture of carbon electrodes.

Most important binder is coal tar pitch

Various pitch sorts are used as binders for the solids. They must endow the mixture with favorable molding characteristics, have a high carbon content and fire to an, if possible, graphitizable coke which binds the solids permanently together. The most important binding agent is coal tar pitch.

5.7.4.3.2 Processing

The volatile components in green petroleum coke can be reduced to less than 0.5% and the density increased by calcining at 1300 to 1400 °C

The so-called "green" petroleum coke produced by delayed coking still contains volatile components, which escape at high temperatures leading to excessive porosity and cracking. In addition shape stability cannot be assured. To avoid this it is first subjected to calcination in gas fired rotary kilns or rotary hearth kilns at 1300 to 1400 °C. This is also necessary with anthracite, which can be calcined at still higher temperatures.

The ground solids are mixed with ground or molten binder at high temperatures

Possible introduction of additives

The cokes (or other solids) are then ground and classification carried out according to particle size. To obtain a molded article with the lowest possible unfilled volume, an accurately determined mixture of a number of fractions with different particle sizes is used rather than a uniform particle size. These are intimately mixed with the binder mainly using double arm kneaders, vortex mixers and mixing screws. Additives such as iron oxide can then be introduced to retard sulfur outbursts or oils to aid molding during graphitization.

The proportion of binder should be kept as low as possible while retaining sufficient plasticity after the molding process

The ratio of binder to solids is dependent upon the particle size distribution of the solids, since all the particles have to be coated as homogeneously as possible to ensure a certain plasticity in the mass after the molding process. On the other hand the probability of firing faults increases with increasing binder content and thus a compromise has to be sought.

5.7.4.3.3 Densification and Molding

The so-called green mixture is then heated to a temperature just above the softening point of the binder and introduced into the mold in which is it later to be fired. In almost all the molding processes it is densified at the same time to obtain dimensional accuracy, to avoid cavities and prevent cracking during firing. During densification orientation of the anisometric coke particles occurs in the direction of the applied force, so that the unsintered molded article normally exhibits anisotropic properties. The degree of anisotropy can be varied within certain limits by the type of molding process used.

The molding is carried out with the following machines:

Unsintered molded articles are produced under pressure at temperatures just above the softening temperature of the binder

- Ram extruders
- Screw extruders
- Cavity presses
- Isostatic presses
- Vibrating compressers
- Ram molds

Extruders are mainly used, although the number of vibrating compressers has increased in recent years not least because unsintered molded articles with lower binder contents are easier to fire. Isostatic pressing is suitable for quasi-isotropic unsintered molded articles.

5.7.4.3.4 Carbonization

Firing of the unsintered molded article to carbon articles is carried out at temperatures between ca. 800 and 1300 °C in gas or oil-heated:

Carbonization of the binder in inert atmosphere at 800 to 1300 °C

- Ring furnaces (highest energy efficiency)
- Pit furnaces
- Rotary hearth kilns

The molded article is usually protected against air entry by coke poured over the surface which is partly burnt and also supports the unsintered molded article as the binder softens during heating.

At low temperatures it is heated up as slowly as possible, to ensure a high degree of carbonization and to give the gaseous pyrolysis products (ca. 30% by weight

The complete firing can take several weeks. The pyrolysis gases make up ca. 30% by weight of the binder

of the binder) time to escape through the fairly impermeable structure. In addition this avoids cracking as a result of high temperature differences between the surface and the interior of the unsintered molded article.

5.7.4.4 Graphitization of Synthetic Carbon

5.7.4.4.1 General Information

Graphitization processes divided into:

- Direct processes (Acheson and Castner processes)
- Indirect processes

Graphitization processes are carried out at 2600 to 3000 °C in inert atmospheres and can be divided into direct and indirect processes. In direct processes (e.g. the Acheson and Castner processes) the carbon articles are placed between electrodes, either directly or indirectly with resistive materials in between, and the required temperature for graphite formation attained by resistive heating. In indirect processes there is no physical contact between the energy source and the carbon article.

During graphitization the carbon articles are heated in inert atmospheres at 2600 to a max. 3000 °C

Upon graphitization the carbon article acquires a graphite-like microstructure. The degree of graphitization is dependent upon the final temperature (not the firing time) and the degree of orientation in the starting material

The high temperatures effect an increase in the ordering in the crystalline regions. The randomly oriented carbon layer planes become ordered and large regions acquire an almost ideal graphite lattice. In addition defects inside the crystallites are annealed. The achievable degree of structural ideality is limited by the degree of orientation in the starting material (coke or carbon solids).

The heating rate during graphitization must be chosen to avoid high temperature gradients in the carbon articles. The heating times are, however, shorter than for carbonization but the whole process still lasts ca. 1 to 3 weeks. The heating rate is also limited by the sudden escape of sulfur (puffing) at ca. 1600 °C during which the article undergoes an expansion in volume. This can be reduced by using inhibitors.

5.7.4.4.2 Acheson Process

Acheson process, most important graphitization process:

heating up of the carbon article by a resistive bed

The Acheson process is still the most important graphitization process. The carbon articles are placed with their long axes at right angles to the direction of the current in the furnace bed and are surrounded with a resistive bed of granular coke in which most of the Joule heat is produced. A constant furnace resistance is essential for uniform product quality (avoidance of local temperature peaks). The way in which the furnace is loaded is therefore critical. In Acheson furnaces articles with very different shapes can be graphitized, making it very versatile. The capacities of currently operated furnaces range up to net loads of 100 t and more.

Advantages of Acheson furnaces:

- Robust construction
- Virtually trouble-free operation
- Suitable for almost any type of carbon article

In the Acheson process only ca. 30% of the energy of ca. 3 to 10 kWh/kg supplied is utilized for graphitization. Other disadvantages are the high cost of the charge, the poor time and space utilization yield and the high emission of noxious gases (mainly SO_2), whose controlled disposal is exceptionally difficult due to the size of the older furnaces and the high operating temperatures. New plants have thus been designed to operate continuously or semicontinuously (transportable individual furnaces are placed between stationary mounted electrodes, the electrodes being run down from above in coupled U-shaped units which move in unison). This results in more rational production and easier disposal of noxious gases.

Most important disadvantages of the Acheson furnaces:

- Low energy efficiency
- Poor time and space utilization yield

5.7.4.4.3 Castner Process

The Castner process operates without intermediate resistive materials. The heat is thus exclusively produced in the carbon article itself, which is directly spanned between two electrodes at least one of which is movable to allow for changes in dimensions.

Castner process:

direct heating of the carbon article without intermediate resistive materials

The furnace construction is more complex than for the Acheson process and hence more liable to failure. However, the proportion of Castner plants is increasing due to easier control resulting in more uniform graphite products, a 50% lower energy consumption and a trend to continuously operating furnaces. The loading and cooling times can be considerably reduced by switching over from insulating packing to heat shields of carbon fiber and graphite foil.

Disadvantages:

- More complex
- More liable to failure

5.7.4.4.4 Other Graphitization Processes

The other graphitization processes transfer the heat indirectly to the carbon article which makes continuous operation easier.

Heat transfer by indirect processes:

- By induction
- By radiation

Heating can be carried out by induction in which the furnace is ringed by an induction coil and the dimensions of the article and the coil have to be precisely correlated with one another. Alternatively heating takes place by radiation, which is more flexible.

Indirect graphitization processes are the processes of the future due to:

- Considerably better energy utilization
- Higher time and space yields

Indirect processes enable a much better energy utilization and time and space yield to be achieved than with direct processes. These processes thus have good future prospects.

5.7.4.4.5 Purification Graphitization

Purification Graphitization:

- Purely thermal
- Chemothermal

Very pure sorts are available by modifying the processes described above. Purification can be accomplished purely thermally or thermochemically. Purely thermal processes entail increasing the firing temperature (to ca. 3000 °C) and firing time and using carbon black for the adsorption of escaping volatile impurities. In thermochemical processes impurities are converted into volatile halogen compounds by reaction with:

- Elemental chlorine or fluorine
- Halocarbon compounds
- Solid halides

5.7.4.5 Impregnation and Processing of Carbon and Graphite Articles

20 to 25% of the volume of carbon and graphite articles are open pores

Impregnation reduces the porosity and improves certain properties

The carbon and graphite articles obtained using the above-described processes contain up to 20 to 25% by volume of open pores formed by the escape of gaseous pyrolysis products upon binder carbonization. For many applications it is essential to fill up at least some of the pores by posttreatment or seal the surface to gas and liquids. This is accomplished by impregnation with:

- Pitch
- Resins
- Pyrolytic carbon
- Metals and alloys by the formation of surface layers of carbides, silicides, borides and nitrides

which may require further heat treatment at high temperatures. Impregnation can also improve certain properties e.g. density, strength and corrosion resistance.

The molded article may have to be pretreated to enable effective impregnation to be carried out. Carbon articles have to be worked with diamond or cemented carbide tools due to their hardness, whereas graphite articles are predictably much softer and are therefore easier to work.

Graphite articles can be much more easily and more accurately worked than carbon articles

5.7.4.6 Properties and Applications

Molded carbon articles are comparable in their hardness, strength and brittleness to ceramic articles. They exhibit a certain electrical and thermal conductivity and are chemically very resistant. Molded graphite articles on the other hand exhibit much higher thermal and electrical conductivities. Their strength and elasticity moduli at room temperature are lower, but increase strongly with increasing temperature. The other properties of artificial graphite increasingly resemble those of natural graphite as the degrees of crystallinity and order increase. Synthetic graphite is generally purer but less crystalline than natural graphite.

Properties of synthetic carbon molded articles:

- Hardness similar to ceramics
- Strength similar to ceramics
- Brittleness similar to ceramics
- Low electrical conductivity
- Chemical resistance

Graphite molded articles:

- High thermal and electrical conductivities
- Other properties increasingly resemble those of natural graphite as the degree of crystallinity increases

Synthetic carbon and graphite are utilized:

- In carbon and graphite electrodes for
  - melt electrolysis processes
  - aqueous electrolysis and electrodialysis (in chloralkali electrolysis graphite electrodes have already been largely replaced by titanium anodes and this is also happening in other electrolysis and electrodialysis processes)
  - arc furnaces for the production of electrosteel
  - smelting furnaces for alloy manufacture
  - carbide furnaces
  - reduction furnaces (e.g. for phosphorus and silicon manufacture)

  (Carbon electrodes are frequently not prefabricated, but are produced in situ, so-called "Söderberg electrodes", particularly for the manufacture of phosphorus)
- As blocks, bricks, plates and tamping compounds for the lining of furnaces, tanks and cast iron molds

Application sectors for mass-produced carbon and graphite:

- Aluminum production
- Electrosteel production
- Hard materials manufacture
- Metallurgy
- Foundry technology
- Chemical industry
- Manufacture of chemical plant
- Glass industry
- Ceramics industry
- Semiconductor industry
- Electrical industry
- Nuclear technology
- Machine tool industry

- In casting molds, sintering molds, cast iron molds and crucibles in foundry technology, metallurgical processes, the glass manufacturing and processing industries, the ceramic industry and semiconductor technology (zone melting, single crystal growth etc.)
- As a material in the manufacture of chemical plant e.g. heat exchangers, towers, containers, pumps, piping, rupture disks for pressure relief etc.
- As a material for machine tool components: bearings, seals etc.
- As very pure graphite in nuclear technology for moderators, reflectors and fuel elements (particularly for high temperature reactors) and as a construction material in reactors and reprocessing plants
- As carbon or graphite brushes, sliding contacts and sliding rings in electrical machines or for electrical conduction
- As small electrodes in the spark erosion machining of metals, as welding electrodes, arc lamp electrodes, spectral carbonization (an analytical technique).

5.7.5 Special Types of Carbon and Graphite

The types of carbon and graphite described in this section have been developed over the last 25 years. Their manufacture is in some cases much more complex than the mass-produced types and as a result their industrial importance is small. Their specific properties are however so exceptional that a bright future is forecast in special application sectors.

5.7.5.1 Pyrolytic Carbon and Pyrolytic Graphite

Pyrolytic carbon and pyrolytic graphite are produced by thermally decomposing gaseous or vaporized carbon-containing compounds

Pyrolytic carbon and pyrolytic graphite are deposited on substrates by the gas phase pyrolysis (chemical vapor deposition, CVD) of

- Methane
- Ethane
- Propene
- Benzene
- Carbon tetrachloride

etc. The pyrolysis reaction takes place in both the gas phase and on the substrate surface at temperatures between 800 and 3000 °C. The mix of properties can be varied within wide limits by suitable choice of process parameters. The production of an isotropic structure is just as possible as the obtaining of spiral, columnar or extremely anisotropic laminar structures. Graphite whiskers can also be produced.

Pyrolytic carbon is also utilized for impregnating porous electrographite by first filling the pores (low deposition temperature, low partial pressure or forced flow of the gases through the graphite article). The surface is then sealed gas- and liquid-tight at high temperatures. Such high grade graphite components are utilized for applications in which the highest resistance to erosion and oxidation is required e.g. guiding tubes for fuel rods in nuclear reactors. Nuclear fuel components are first rendered isotropic in a fluidized bed (to relax thermal and mechanical strain and absorb gaseous decomposition products) then lamellarly or granularly (gas tightness, strength) coated with pyrolytic carbon to give coated particles. As a result of their excellent biocompatibility isotropic pyrolytic carbon articles are being used more and more in medicine as cardiac flaps and valves and in prosthetic devices for which alloys with silicon carbide are also used. Pyrolytic graphite can also be utilized as the matrix material in single, two or three dimensional carbon fiber-reinforced carbon/carbon composite materials, which are also used in medicine for e.g. bone plates. Still further applications are as temperature resistant, highly abrasion resistant brake and coupling plates with residual slip properties for aircraft and ablation shields and cladding for rockets. Carbon fibers can even be strengthened by coating with pyrolytic carbon.

Application sectors for pyrolytic carbon and pyrolytic graphite:

- Nuclear technology
- Medicine
- Aerospace

Anisotropic pyrolytic graphite was formerly used on rocket noses, rocket jets, the leading edges of wings of supersonic aircraft etc. for the directional dissipation of thermal energy, due to its high thermal conductivity in the direction of the carbon layer plane. However, its anisotropic thermal expansion perpendicular to the layer plane results in very rapid scaling of this insulation which led to its discontinued use. Extremely anisotropic pyrolytic graphite articles obtained by additional heat treatment at 3000 °C under pressure applied perpendicular to the layer plane are used for monochromatizing X-rays and neutron radiation and as a starting material for the manufacture of intercalated compounds (Sec. 5.7.5.3).

5.7.5.2 Glassy Carbon and Foamed Carbon

Glassy carbon and foamed carbon from solid state pyrolysis

Glassy carbon and foamed carbon are also formed by the pyrolysis of carbon-rich materials, but in this case solid state pyrolysis at up to ca. 1000 °C rather than liquid or gas phase pyrolysis.

Thermosetting plastics i.e. nonmelting unfoamed or foamed organic polymers or polymers rendered nonmeltable by dehydration, crosslinking, cyclization or aromatization are used as starting materials e.g.:

Starting materials for glassy carbon and foamed carbon:

- Thermosetting polymers
- Polymers rendered nonmelting by chemical treatment

- Phenol formaldehyde resins
- Polyfurfuryl alcohols
- Poly(acenaphthalene) rendered nonmelting by treatment with sulfur
- Poly(acrylonitrile) rendered nonmelting by oxidation
- Phenol resin-impregnated poly(urethane) foam

The principle is the same as that used in the manufacture of carbon fibers (Sec. 5.2.3.3) only here molded solid or very well-defined uniform foam polymer articles are used instead of carbon fibers. The solid tailor-made molded components should not be too large and have wall thicknesses no greater than 5 mm. Otherwise even at extremely slow heating rates homogeneous escape of the pyrolysis products can not be guaranteed and cracks are formed. Foamed carbon on the other hand can be manufactured in plates or blocks with larger dimensions.

Glassy carbon is nongraphitizable

Glassy carbon can only be machined with diamond tools or with ultrasonic techniques. The shape must therefore be defined as exactly as possible during the molding of the polymer

Glassy carbon is a nongraphitizable modification of carbon with extremely small crystallites (max. 10 nm) with a ribbon structure. The strong intertwining of the carbon ribbons is reminiscent of a polymer coil. Heating under the usual graphitizing conditions does not lead to parallel ordering. Glassy carbon is therefore isotropic and also extremely hard and brittle. The slit-shaped micropores appear to shut with increasing temperatures making tempered glassy carbon almost impermeable to gases and very resistant to oxidation.

Small glassy carbon articles such as crucibles, boats or tubes are utilized under conditions in which high purity and temperature stability are required. Glassy carbon can, as with pyrolytic carbon, be used in prosthetic devices in human medicine. Furthermore the matrix in "all carbon composites" can consist of glassy carbon rather than pyrolytic graphite. The application sectors are the same as those mentioned in Sec. 5.7.5.1.

Application sectors for glassy carbon:

- Laboratories
- Chemical analysis
- Metallurgy
- Semiconductor industry
- Medicine
- Aerospace

Foamed carbon is also nongraphitizable. The cell structure remains intact upon careful carbonization and densities lower than 0.1 g/cm^3 are obtainable. The thermal conductivity is just above that of plastic foams, but foamed carbon can be used at much higher temperatures. Its relatively low compressive strength can be increased by impregnation with pyrolytic carbon, although the thermal conductivity increases at the same time. By comparison with glassy carbon, foamed carbon is easy to work, so that the shape does not have to be established at the start of the process. Foamed carbon is also corrosion resistant as are all carbon modifications.

Foamed carbon is nongraphitizable

Shaped foamed carbon components can be produced after carbonization by machining plates or blocks

Foamed carbon has 600 times the thermal insulation capacity of the same weight of firebrick and is therefore used as a self-supporting thermal insulation material in furnaces or in fire-retarding sandwich constructions. Foamed carbon cores are used in foundry technology, because they are easily produced and can be easily removed by combustion from the inaccessible cavities. Furthermore foamed carbon is utilized as a filter material for the filtration of corrosive materials, as sintered plugs in gas inlets, as an electrode material and as a catalyst carrier.

Application sectors for foamed carbon:

- Metallurgy
- Foundry technology
- Fire protection
- Catalyst production
- Laboratories

5.7.5.3 Graphite Foils and Membranes

High purity natural flake graphite or strongly anisotropic pyrolytic graphite are used as starting materials in the manufacture of graphite foil. In this process they are oxidized with fuming sulfuric acid or a mixture of sulfuric acid and nitric acid, hydrolyzed, washed and dried to graphitic oxide which is decomposed to expanded graphite by rapid heating to ca. 1000 °C. This is pressed without added binder into a foil.

Process steps in the production of graphite foil:

- Oxidation of flake graphite or pyrolytic graphite to graphitic oxide
- Thermal decomposition of the graphitic oxide
- Pressing the expanded graphite to a foil

Graphite membranes are obtained by drying a graphitic oxide gel on a polished surface, reducing with

Process steps in the manufacture of graphite membranes:

- Reduction of a graphitic oxide film with hydrogen
- Graphitization

hydrogen at temperatures above 500 °C and graphitizing the carbon membrane formed at temperatures above 2500 °C.

Application sectors for graphite foil and graphite membranes:

- Machine tool manufacture
- Chemical plant construction
- Metallurgy
- Foundry technology
- Electrical industry
- Seawater desalination

Graphite foils and graphite membranes are flexible, impervious to gases and liquids, exhibit high anisotropy and are more chemically resistant than most of the electrographite types. Graphite foils are preferably used in the manufacture of crimpable flat seals, self-lubricating sealing rings and packing seals. They are also utilized for lining pipes and containers in which corrosive substances are handled, for lining press and casting molds to obtain smooth surfaces and as efficient heat-reflecting shields in high temperature furnaces. Flat thermal conductors and rupture disks for pressure relief are also manufactured from graphite foil. Graphite membranes are utilized as dialysis membranes in seawater desalination.

5.7.6 Carbon Black

Carbon black has already been produced for several thousand years and is one of the oldest chemical products

Carbon black is the oldest synthetically produced carbon modification and is in fact one of the oldest chemical products. Carbon black was even used in the cave paintings of the Upper Paleolithic Age as a black dye. Its manufacture in the form of lamp black was, according to very early documents, known in old Egyptian and old Chinese cultures.

Structure of spherical primary carbon black particles: shell-shaped oriented individual graphite layers of different sizes, which in small regions are oriented parallel to one another and thus constitute quasicrystallites. Carbon black is therefore not amorphous

The 5 to 500 nm primary carbon black particles consist of more or less strongly deformed or twisted graphite layers, which are spherically oriented round a growth center. The size of the individual layers varies strongly but particularly near the surface can amount to several hundred Ångstrom in both directions. The layers can also extend from one growth center to another, often giving the aggregates formed a chain or coil structure. In small regions the fragments of several graphite layers may lie parallel and the same distance from one another. These separations are somewhat larger than in the case of single crystal graphite and the layer sequence is completely random.

5.7.6.1 Economic Importance

Table 5.7-4 gives a survey of the carbon black production statistics in the various regions (all the types together):

Table 5.7-4. Total Carbon Black Production in Various Regions in 10^3 t.

	1980	1981	1982	1983
Africa	45	45	48	49
North America (incl. USA)	1426	1541	1581	1584
USA	1156	1238	–	–
South America	227	199	204	202
Asia (excl. USSR)	726	693	649	743
Japan	575	557	504	568
Europe (excl. USSR)	1269	1235	1263	1306
West Germany	354	354	348	362
Oceania	74	74	74	74
Worldwide (excl. USSR)	3767	3787	3819	3957

The carbon black producing industry is dependent in two respects upon the prices of crude oil and natural gas because they are both important raw materials for carbon black manufacture and raw materials costs account for up to 75% of the manufacturing costs and because ca. 80% of the total production is utilized in the car industry, mainly as a reinforcing filler in tires, and the demand for cars is influenced by the cost of fuel.

The carbon black producing industry is very strongly dependent upon developments in the crude oil and natural gas sectors

Ca. 94% of the carbon black produced is consumed in the rubber industry, the remainder being mainly used in the printing ink, paint and plastics industries.

Demand is expected to increase at an annual rate of 1 to 2% in the next few years.

5.7.6.2 Manufacture

5.7.6.2.1 General Information

Carbon blacks are manufactured by the pyrolysis of gaseous (or materials converted into a gaseous state) or liquid (or molten) raw materials from the natural gas, crude oil or coal industries. There are two groups of processes: those in which part of the required thermal process energy is obtained by burning part of the raw material by intentional addition of air, the so-called "incomplete combustion processes", and those in which the required energy is completely supplied

Carbon black from carbon-rich raw materials by:

- Pyrolysis in the presence of oxygen
- Pyrolysis in the absence of oxygen

externally, the so-called "thermal decomposition processes".

The reaction conditions in the latter group of processes resemble those for the manufacture of pyrolytic carbon and pyrolytic graphite (Sec. 5.7.5.1). Carbon black formation, however, takes place at much higher partial pressures of the substances being pyrolyzed.

5.7.6.2.2 Pyrolysis Processes in the Presence of Oxygen

Furnace Black Processes

The furnace black process is the most commercially viable process. 95% of the carbon black is manufactured using this process

Whereas at the end of World War II almost all carbon black was produced using channel black processes, more than 95% is now manufactured using the furnace black processes. There are two basic reasons for this turn around: the higher carbon black yield and the range of qualities produced with furnace black processes.

Main raw materials for furnace black processes

- In the USA: residues from catalytic synthesis of petrol
- In the rest of the World: residues from ethylene production and coal tar oils

The first process of this type utilized gases as raw materials and produced carbon blacks for rubber with only moderate reinforcement properties at a maximum yield of 30%. The changeover to liquid and melting raw materials increased the yield of carbon black for rubber to 40 to 70% and enabled a much broader range of products to be produced, so that this type of raw material is now almost exclusively used e.g. aromatic compound-rich viscous residues from: oil refineries, coking plants, catalytic crackers and steam crackers for the production of ethene and from the catalytic synthesis of petrol.

Quality criteria for carbon black raw materials:

- High C/H-ratio
- Pour point
- Viscosity
- Density
- Low asphalt content
- Low sulfur content
- Very low alkali metal content

The required high carbon/hydrogen ratio in the feedstock is obtained by the highest possible proportion of condensed aromatic compounds conducive to their being pumpable at high temperatures. Furthermore, the content of asphalt (very high molecular weight polyaromatics) should not exceed a certain maximum, otherwise nonreinforcing microcarbon is formed. The sulfur content should also be low, since ever stronger emission regulations mean that sulfur emission must be severely restricted. The residual ash content should also be as low as possible to prevent the formation of hard abrasive and possibly catalytically active particles. Finally the alkali metal content should also be low since it hinders the formation of structured i.e. aggregated carbon black. Conversely the degree of aggregation can be controlled by dosing with small amounts of alkali metals.

The preheated feedstock is atomized radially as the smallest possible droplets into a tubular reactor, the so-called "carbon black furnace". There is a zone of high turbulence immediately in front of the injector in which the atomized feedstock intimately mixes with hot gases from the combustion of the fuel (preferably natural gas, but also coking gas or liquid gas) in an excess of preheated air. The air excess is so adjusted that up to ca. 50% of the carbon black-feedstock is also comsumed. The temperature obtained in the pyrolysis zone is 1200 to ca. 1800 °C. Carbon black formation is weakly endothermic.

The carbon black properties can be influenced by:

- Temperature
- Residence time
- Quantity of additives (particularly KOH, KCl)
- Type of feedstock atomization
- Turbulence in the mixing zone

The reaction mixture is quenched with water sprays to 800 to 500 °C and then passed through heat exchangers which cool it further to 350 to 250 °C. High pyrolysis temperatures, which can be controlled by the quantity and temperature of the combustion air, and short residence times, which can be achieved by adjusting the gas speed and the quenching position, lead to very finely divided carbon blacks with high specific surface areas. In modern plants the carbon black is separated in agglomeration cyclones with bag filters. The so-called tail gas has a given heat content due to its carbon monoxide, hydrogen and hydrocarbon contents and can be utilized for energy production.

High pyrolysis temperatures and short residence times result in very finely divided carbon black with large specific surface areas

After deposition the small proportion of coarse particles is separated by grinding and sieving and the impurities are removed. The very light carbon black powder is then compacted by "degassing" in which it is passed over rollers with a gas-permeable coating and a partial vacuum inside. Alternatively it is granulated dry without granulation aids in drums. This is only possible with very fine carbon blacks. Most of the carbon black produced is granulated wet with granulation aids in special pearling machines. The bulk density can be thereby increased to 500 g/L, which is about 10 times the bulk density of the noncompacted carbon black powder.

Compaction of fresh carbon black powder by:

- "Degassing"
- Dry pearling
- Wet pearling

The first two processes are mainly used for carbon blacks for coloring

Plants operating the furnace black process can produce up to 20×10^3 t/a in a single reactor. The reactor

The production capacity of a modern large reactor is up to 2500 kg carbon black/h

lifetime is almost two years despite the high thermal and mechanical stress.

Channel Black Process

Feedstock for the channel black process: natural gas

Disadvantages of the channel black process:

- High feedstock cost price
- Low carbon black yield
- Only fine particulate carbon black manufacturable

In the channel black process expensive natural gas is used as the feedstock and is burnt in many small glowing flames. These flames burn underneath and in contact with U-formed cooled iron rails, so-called "channels" hence the name of the process, onto which carbon black is deposited. The 3 to 6% yield is very low.

The channel black process is currently little used.

Gas Furnace Black Process

The gas furnace black process was developed by Degussa AG in the 1930's as an alternative to the channel black process, because natural gas was not available in sufficient quantities in Germany and is still currently operated.

Main feedstocks for gas furnace black process:

- Raw naphthalene
- Anthracene oil
- Anthacene oil residues

High boiling point aromatic compound-rich oils, preferably coal tar products, are used as the feedstock. This is evaporated at ca. 350 °C and is transported by a combustible carrier gas such as hydrogen, coking gas or methane to the burners. In the channel black process a whole battery of flames make contact with a cooled surface, whereas in this process it is a slowly rotating roller. The particle size of the carbon black produced is similar to that from the channel black process. The carbon black yield is between 10% for high quality coloring carbon blacks and 60% for carbon black for rubber applications.

Lamp Black Process

Feedstock for the lamp black process: oils rich in aromatic compounds

The lamp black process is just an industrialized version of the old lamp black production techniques. Oil in flat cast iron pans up to 2 m in diameter is burnt under an extraction hood with a nonstoichiometric quantity of air. The carbon black is separated from the combustion gases in cyclones and filters. The yield is ca. 50%.

The lamp black process is only sporadically operated today for the manufacture of coarse particulate "heavy" carbon black.

5.7.6.2.3 Pyrolysis Processes in the Absence of Oxygen

Thermal Black Process

The feedstock for the thermal black process is, as with the channel black process, expensive natural gas. A production unit consists of two refractory-lined cylindrical furnaces, or generators, alternating on about a 5-minute cycle between carbon black production and heating, making the overall production continuous from the two generators. The generators are lined with refractory bricks for heat storage and are mounted above one another with a large cavity in between. While the first furnace is being heated to the reaction temperature (ca. 1300 to 1500 °C) with a stoichiometric mixture of air and combustible gas (mainly natural gas), pure natural gas is pyrolyzed in the second furnace. Since the reaction is endothermic, the furnace cools and the gas flows are switched when a critical minimum temperature is reached.

Feedstock for the thermal black process:

natural gas

The reaction products are quenched by water sprays in a cooling zone and the carbon black separated in filter units. The yield is ca. 35% of the carbon content of the feedstock and combusted gas. Only inactive coarse particulate carbon blacks are produced which can be partly replaced by cheaper fillers. The tail gas, consisting almost exclusively of hydrogen, can be mixed with the natural gas used for heating up the furnace or used for diluting the natural gas being pyrolyzed to reduce the size of carbon black particle being produced.

Disadvatages of the thermal black process:

- High cost prices of feedstock and fuel
- Only coarse particulate inactive carbon black produced

Acetylene Black Process

In the acetylene black process the tubular reactor only has to be heated to 800 °C by the combustion of an acetylene/air mixture. Then the pyrolysis reaction can be carried out with pure acetylene, since the reaction is strongly exothermic and therefore self-sustaining. The reactor must in fact be cooled to maintain a constant temperature. The carbon black yield is 95 to 99%!

Feedstock for the acetylene black process:

acetylene occasionally mixed with hydrocarbon vapors

Pyrolysis of acetylene is exothermic!

Acetylene black consists of platelet-shaped primary particles in contrast with the normal carbon blacks and is used for special applications.

Electric Arc Process

Arc black produced as a byproduct in the production of acetylene by the Hüls electric arc process

In the electric arc process developed by Chemische Werke Hüls, acetylene is produced by pyrolyzing hydrocarbons with boiling points up to 200 °C in an electric arc. Fine particulate carbon black is produced as a byproduct in such quantities that working up is worthwhile.

Arc black consists of platelet-shaped primary particles.

5.7.6.2.4 Posttreatment

Species on the carbon black surface:

- Physically adsorbed organic substances
- Chemically bonded acid and alkaline groups

The chemical nature of the surface of carbon black is crucial to its applications-related behavior and in the first instance is a function of the manufacturing process. In addition to physically adsorbed organic substances, chemically combined surface oxygen is present on the surface, being formed in varying quantities in the presence of oxygen during the pyrolysis. Channel and gas furnace blacks have acid groups on the surface and furnace blacks have alkaline groups. The content of physically and chemically bonded species is known together as the volatile content, since it can be removed by heating to 950 °C in the absence of air.

Oxidative posttreatment increases the surface concentration of acidic groups and makes the carbon black more hydrophilic

In many cases it is desirable to augment the content of volatile components. This is accomplished by oxidative posttreatment in which acidic surface groups are formed. This oxidation can be carried out at high temperatures both in the liquid phase and in the gas phase with e.g.:

- Air
- NO_2/air mixtures
- Ozone
- Nitric acid
- Sodium hypochlorite

This posttreatment is mainly carried out on coloring carbon blacks. Strongly oxidized coloring carbon blacks can have volatile contents up to 20% by weight.

As the surface concentration of oxygen-containing groups increases the initially hydrophobic carbon black surface becomes increasingly hydrophilic. Such posttreated carbon blacks are preferred for use in aqueous and polar organic media.

5.7.6.3 Properties and Applications

The following characteristics are of primary importance for the applications of carbon blacks:

- Primary particle size, controlled by the choice of manufacturing process
- Carbon black structure
- Specific surface area, variable within wide limits
- Adsorption capacities for water, solvents, binders and polymers
- High absorption of visible, IR and UV light

These and other properties are quantified by a multitude of normalized measurement procedures and collated in the form of characteristic values.

Table 5.7-5 gives a survey of the characteristics of the carbon blacks produced by the different processes.

The physical properties important for the utilization of carbon black are:

- Primary particle size
- Structure (aggregate size and shape)
- Specific surface area
- Adsorption capacity
- Light absorption capacity

Table 5.7-5. Production Spectrum of the Different Carbon Black Manufacturing Process (including Posttreated Products).

Carbon black type	Average particle size[1] (in nm)	Specific surface area[2] (in m^2/g)	Application
furnace black	10 to 80	20 to 550	carbon blacks for rubbers and coloring
channel black	5 to 30	100 to 950	mainly coloring carbon blacks and special carbon blacks for rubbers
gas furnace black	10 to 30	90 to 800	
lamp black	50 to 120	20 to 50	mainly carbon blacks for rubbers and special carbon blacks for coloring
thermal black	120 to 500	6 to 15	carbon blacks for rubbers
acetylene black	35 to 42	65	special carbon blacks
arc black	35	80	coloring carbon blacks

[1] determined by electron microscope
[2] BET values

The largest consumers of carbon black are, as already mentioned in Sec. 5.7.6.1, the rubber and plastics industries. It is utilized as a filler and more particularly as a reinforcing filler, since the chance observation in 1912 that carbon black-filled rubber articles exhibit higher abrasion resistance than the then current zinc

Main application sector: rubber industry as a filler

oxide-filled rubbers. In the meantime a range of carbon blacks for rubber has been developed with graded reinforcing properties as well as inactive types. Table 5.7-6 gives a survey of the most important characteristics. This table shows that the reinforcing activity increases with decreasing particle size and increasing specific surface area. The DBP(dibutylphthalate)-adsorption gives an indication of the carbon black structure: carbon blacks with little structure adsorb little DBP, highly structured carbon blacks absorb large quantities of DBP.

Table 5.7-6. Survey of the Carbon Blacks Used in the Rubber Industry.

	Average particle size[1] (in nm)	Specific surface area[2] (in m^2/g)	DBP-adsorption (in mL/100g)
active carbon black	11 to 30	70 to 155	65 to 140
semi-active carbon black	31 to 100	20 to 70	65 to 130
inactive carbon black	101 to 500	20	40

[1] determined by electron microscope
[2] BET values

Since the beginning of the 1970's so-called "improved" carbon blacks have been developed, which in structure and specific surface area hardly differ from the corresponding conventional types, but have a higher surface energy and thereby improved reinforcing behavior.

Active carbon blacks are used in tire treads, conveyer belts and other highly stressed industrial rubber articles. Semi-active carbon blacks are utilized in applications in which little heat is generated under high dynamic stress e.g. casing mixtures, tire sidewalls or tire inner tubes. Inactive carbon blacks, having no effect on the elastic properties, are used in applications in which the elastic properties of rubber are important e.g. gloves, seals or cable insulation.

The applications spectrum for coloring carbon blacks is even broader than that for carbon blacks for rubber applications. Thus a rough classification on the basis of physical characteristics appropriate to particular applications is impossible. Instead the important characteristics of four typical coloring carbon blacks including one which has been oxidatively posttreated are given in Table 5.7-7. The oil numbers of coloring carbon blacks are, by analogy with the DBP-adsorption of carbon blacks for rubbers, a measure of the structure of the carbon black. The influence of compaction on the structure is evident from the two oil numbers in Table 5.7-7, the first relating to the pearlized form and the second to the powder. This Table also shows that oxidative posttreatment increases the specific surface area due to pore formation and imparts a deeper black tone to the carbon black together with a strong blue tinge, which is not evident from the M-value.

Table 5.7-7. Characteristics of Typical Coloring Carbon Blacks [the cited carbon blacks are trade marks of Degussa from whose publication "Schriftreihe Pigmente" (Series of publications about pigments) these figures have been taken].

	Average particle size[3)] (in nm)	Specific surface area[4)] (in m^2/g)	Color tone M-value[5)]	Oil number (flow point) (in % by weight)	Volatiles content (in % by weight)	pH value
Farbruß (coloring carbon black) FW 1[1)]	13	320	193	700/950	6.0	4.0
Farbruß (coloring carbon black) FW 200[1)] (oxidatively posttreated)	13	460	197	450/620	21.0	2.5
Printex 300[2)]	27	80	143	320/360	1.0	9.5
Flammruß (flame black) 101	95	20	112	225/280	1.0	7.0

1) very strongly colored gas furnace black
2) standard furnace black
3) determined by electron microscope
4) BET values
5) color-independent black scale, measured according to DIN 6174 (1979)
6) see Sec. 5.7.6.2.4.

Coloring carbon blacks are used in paints, printing inks, plastics and for coloring fibers, paper, cardboard and building materials.

A small proportion of the carbon black production is utilized in the manufacture of carbon and graphite articles, hard materials, foam glass and refractory bricks. Some carbon blacks are also used as UV-stabilizers, antioxidants and to increase the conductivity of plastics.

Further application fields for carbon blacks:

- Paint industry
- Manufacture of printing inks
- Plastics industry
- Fiber industry
- Paper industry
- Construction materials industry
- Electrical industry
- Manufacture of hard materials
- Refractories industry

5.7.7 Activated Carbon

5.7.7.1 Economic Importance

1985 activated carbon capacity in USA: ca. 158×10^3 t/a

Capacity utilization: ca. 60%

The production capacities and production of activated carbon in several countries and regions in 1983 are given in Table 5.7.7-1:

Table 5.7.7-1. Production Capacities and Production of Activated Carbon in 1983, in 10^3 t/a.

	USA	W. Europe	Asia	Japan
Capacity	158*	105*	85.3	64
Production	95	77		55.5

* 1985

5.7.7.2 Manufacture

5.7.7.2.1 General Information

Two activated carbon types:

- Powdered activated carbon
- Granulated activated carbon

There are two large groups of activated carbons:

- Powdered activated carbon
- Granulated activated carbon, available as irregularly shaped particles by grinding coarser material, or as cylindrical particles, produced by extruding a mixture of finely ground raw material (e.g. lignite coke) and a binding agent (e.g. an organic resin or pitch) and then activating.

Raw materials for the manufacture of activated carbon:

- Resinous, such as wood, sawdust, peat, coconut shells
- Carbonized, such as coal, low temperature lignite coke

A wide spectrum of organic materials is suitable as raw materials in the manufacture of activated carbons. Of the resinous fuels utilized, wood, sawdust, peat, coconut shells and even olive stones are preferred. Of the carbonized raw materials, coal, low temperature lignite coke, charcoal and coke from acid sludges (e.g. from the manufacture of lubricants) are utilized. The properties of the activated carbon are very considerably influenced by the type of raw material utilized.

Activation processes:

- "Chemical activation", generally with resinous raw materials
- "Gas activation", generally with carbonized raw materials

There are two fundamentally different processes for converting these raw materials into activated carbon:

- Chemical activation
- Gas activation

Chemical activation is generally carried out with resinous raw materials, gas activation generally with carbonized raw materials. The aim of both processes is to convert the given raw material into a material with a high specific surface area (BET values between 400 and 2500 m^2/g) and the optimum pore size distribution for the required application. There are three types of pores:

- Micropores, with diameters of 0.4 to 2 nm
- Mesopores, with diameters of 2 to 50 nm
- Macropores, with diameters greater than 50 nm.

5.7.7.2.2 Chemical Activation Processes

The preferred raw material in chemical activation processes (also Sec. 5.7.7.1) is wood in the form of sawdust. These processes are based on the dehydration of the starting material by reaction with dehydration agents (such as zinc chloride and in particular phosphoric acid with potassium sulfide being used in special cases) and are particularly effective with cellulose-containing materials. Sawdust is mixed with phosphoric acid or zinc chloride (e.g. 1 to 3 parts of zinc chloride to 1 part of sawdust), optionally molded and then heated to 400 to 600 °C when phosphoric acid is utilized or to 600 to 700 °C when zinc chloride is utilized.

Chemical activation:

dehydration of the raw material by reaction with phosphoric acid or zinc chloride at 400 to 700 °C with subsequent removal of the activation agent by repeated washing

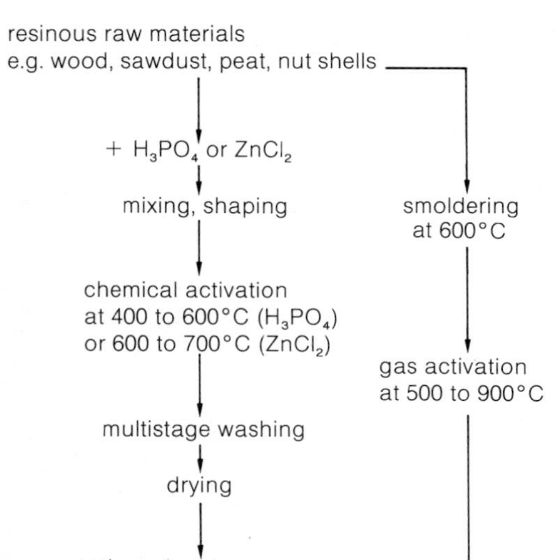

Fig. 5.7.7-1. Flow chart of the manufacture of activated carbon from resinous materials by chemical and gas-phase activation.

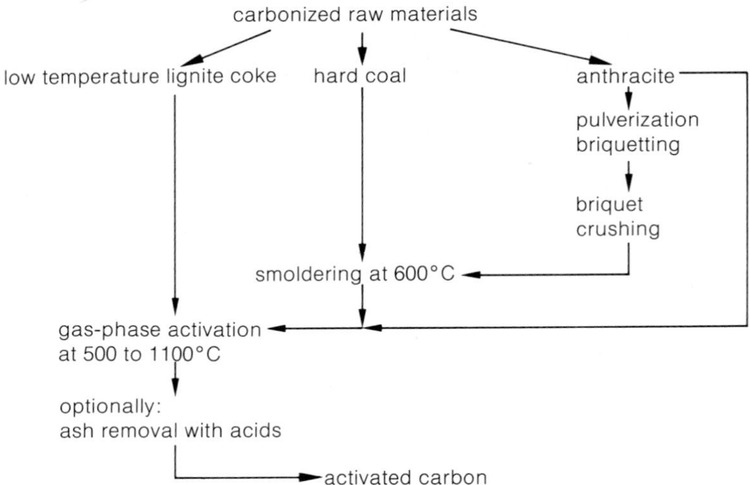

Fig. 5.7.7-2. Flow chart of the manufacture of activated carbon from carbonized raw materials by gas-phase activation.

The further processing of materials activated using this process is relatively difficult and complex, since on economic and ecological grounds most of the chemicals have to be recovered. They have to be repeatedly washed to obtain an activated carbon which is largely free of activation agent.

Powdered activated carbon is usually produced using these processes, but it is also possible to produce granulated activated carbon by molding (granulating) mixtures of sawdust or peat and activation agent.

5.7.7.2.3 Gas Activation Processes

Gas activation:

Production of pores in a carbon matrix by reaction of the carbon with

- Oxygen
- Steam
- Carbon dioxide

or their mixtures at 500 to 1100°C

Under the term "gas activation" is understood the formation of pores in a carbon matrix by the removal of carbon by chemical reaction with oxygen, steam or carbon dioxide or their mixtures. Whereas the reaction with oxygen proceeds rapidly and exothermically, those with steam 1) and 2) and particularly that with carbon dioxide 3) proceed endothermically and much more slowly.

1) $C + H_2O \longrightarrow CO + H_2, \quad \Delta H = 117\ kJ$

2) $C + 2H_2O \longrightarrow 2H_2 + CO_2, \quad \Delta H = 57\ kJ$

3) $C + CO_2 \longrightarrow 2CO, \quad \Delta H = 159\ kJ$

Thus the activation with oxygen is carried out at 500 to 600 °C, that with water vapor at 900 to 1100 °C, and that with carbon dioxide at still higher temperatures. The most important activation process, industrially, is the steam activation process, whose rate is, at the above-mentioned temperatures, determined by the reaction between carbon and steam. In this process carbon is removed from the pore walls thereby increasing the pore size. At temperatures > 1100 °C this reaction is diffusion controlled, the carbon only being eroded at the surface of the particles.

In gas activation of noncarbonized resinous raw materials, prior low temperature carbonization (pyrolysis in the absence of air) has to be carried out. This process has a considerable effect upon the pore size distribution in the final product.

Coal, which contains a high fraction of volatile components, has to be preoxidized and then carbonized at low temperatures, whereas anthracite can in principle be activated directly. In practice, however, pulverization, briquetting with the aid of a binder, briquette comminution and low temperature carbonization are usually carried out. This makes subsequent activation much easier. Low temperature lignite coke can be directly activated.

Coal has generally to be preoxidized and carbonized at low temperatures before gas activation

However, if molded activated carbon cylinders are manufactured from low temperature lignite coke, low temperature carbonization of the binding agent (pitch, organic resin)-bonded ground coke has to be carried out before activation. In this way the binder is also carbonized (see also Sec. 5.7.7.3).

The ash content of materials such as coal or lignite can be reduced by e.g. acid treatment (hydrochloric acid, nitric acid) both before and after activation. In this way the ash content of the powdered activated carbon produced from peat can be reduced from 6 to 10% by weight (mainly silica with 0.5% of calcium and magnesium as their salts) before washing to 4 to 6% by weight after acid-washing.

Ash removal by acid treatment before or after activation

The methods used by different companies for the manufacture of activated carbon are generally regarded as trade secrets, so that nothing is known of the actual units utilized for the manufacture of particular

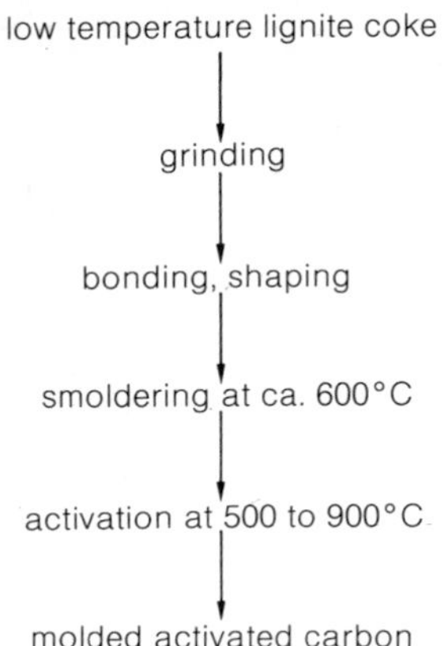

Fig. 5.7.7-3. Flow chart of the manufacture of molded activated carbon from low temperature lignite coke.

Units utilized for gas activation:

- Rotary kilns
- Multiple hearth furnaces
- Fluidized bed reactors
- Shaft furnaces

activated carbon types. In practice, units typically used for high temperature reactions, such as rotary kilns, multiple hearth furnaces, fluidized bed reactors, shaft furnaces etc., are employed. Activation gas and raw material can be directed in cocurrent or countercurrent. Heating can be direct or indirect. In direct heating, the gas formed upon activation (hydrogen, carbon monoxide) is immediately burnt in the reactor. A process is also known in which the gas formed is utilized for the indirect heating of a fluidized bed reactor.

The residence time in the reactors depends on the particular process conditions employed and has a considerable influence upon yield, pore size distribution and BET-surface area of the activated carbon produced.

5.7.7.3 Reactivation and Regeneration of Used Activated Carbons

Reactivation of used activated carbon (especially granular and molded activated carbon) mainly accomplished by steam treatment analogously to activation

Reactivation is effected using the methods of gas activation, for the most part using steam as the activating gas, particularly in multiple hearth furnaces, but also in rotary kilns and fluidized bed furnaces. The utilization of infrared systems for reactivation of activated carbon (also suitable for powdered activated carbon) is fairly recent. Depending upon the quantity, reactivation is carried out by the manufacturer or by the user. The reactivation of molded cylinders and granules of activated carbon is currently more important. The regeneration costs, including the replacement of the activated carbon losses, have been estimated to be ca. 10 to 15% of those of the activated carbon utilized. The reactivation of powdered activated carbon has latterly become more widely practiced despite its greater technical difficulty. It can be carried out by e.g. atmospheric oxidation of 5 to 10% powdered activated carbon slurries at 230 to 269 °C and pressures of ca. 60 bar (Zimpro process). In Du Pont's PACT-process for biological effluent purification in the presence of activated carbon, the powdered activated carbon is thermally regenerated. The gaseous, in some cases toxic, byproducts liberated upon reactivation have to be rendered harmless by subsequent combustion.

Activated carbon utilized for gas purification, e.g. for the adsorption of solvent vapors, can be regenerated by desorption with steam at 120 to 140 °C. The activated carbon then has to be dried. In special cases other types of regeneration are possible.

Solvent vapor-saturated gas purification activated carbon is regenerated with steam

5.7.7.4 Applications for Activated Carbon

Statistics regarding the utilization of activated carbon are summarized in Table 5.7.7-2.

Further application fields for activated carbon in the liquid phase are:

- Edible oil purification
- Decoloration of chemical products and natural products
- Treatment of drinks such as wine, fruit juice or spirits.

Applications of activated carbon:

in the liquid phase

- Decoloration of natural products such as sugars, edible oils, drinks
- Water and effluent treatment
- Purification of chemicals

In the gas phase activated carbon can also be utilized for:

- Air purification in buildings
- Gas mask filters
- Removal of sulfur compounds (sulfur dioxide, hydrogen sulfide, carbon disulfide) from flue gases and residual gases
- Gas separation, e.g. of nitrogen and oxygen ("activated carbon molecular sieves").

in the gas phase

- Solvent recovery
- Exhaust gas purification in cars
- Gas mask filters
- Gas separation

Table 5.7.7-2. USA Consumption of Activated Carbon in the Most Important Application Sectors in 10^3 t.

	1983 granular	1983 powdered	1983 total	1988 total (est.)
Applications in the liquid phase including:	28.5	40.5	69	74.5
• Sugar decoloration	5	13.5	18.5	17
• Potable water production	3	11.5	14.5	16
• Ground water purification	2	–	2	2.5
• Effluent purification	8	2	10	12.5
Applications in the gas phase including:	16.5	2.5	19	24
• In cars	3.5	–	3.5	4
• Solvent recovery	4.5	–	4.5	6.5

References for Chapter 5.7: Carbon Modifications

Diamonds:

Commercial Information:

World Mineral Statistics 1980–1984: Production, exports, Imports. British Geological Survey, Keyworth/Notts., 1986.

Reviews:

Kirk-Othmer: *Encyclopedia of Chemical Technology,* 3rd Ed., Vol. 4, pp 66–688, John Wiley & Sons, Inc., New York-Chichester-Brisbane-Toronto, 1978.
Ullmann's Encyclopedia of Industrial Chemistry, 5th Ed., Vol. A 5, pp 96–98, VCH Verlagsgesellschaft mbH, Weinheim-Deerfield Beach-Basel, 1986.
S. Toon, *Industrial Minerals,* pp 65–73 (December 1986).
Industrial Minerals No. 163, pp 47–59 (April 1981).

Natural Graphite:

Commercial Information and Applications:

World Mineral Statistics 1980–1984: Production, exports, Imports. British Geological Survey, Keyworth/Notts., 1986.
J. Robbins, *Industrial Minerals,* pp 37–55 (July 1984).

Reviews:

Kirk-Othmer: *Encyclopedia of Chemical Technology,* 3rd Ed., Vol. 4, pp 570–622, John Wiley & Sons, Inc., New York-Chichester-Brisbane-Toronto, 1978.
Ullmann's Encyclopedia of Industrial Chemistry, 5th Ed., Vol. A 5, pp 98–101, VCH Verlagsgesellschaft mbH, Weinheim-Deerfield Beach-Basel, 1986.

Synthetic Carbon and Synthetic Graphite:

Technical Information:

E. Wege, *High Temperatures – High Pressures* **8,** pp 293–305 (1976).
E. Fitzer and M. Heym, *High Temperatures – High Pressures* **10,** pp 29–66 (1978).
B. Findeisen, *Freiberger Forschungshefte* **A 618,** pp 19–56 (1980).

Reviews:

Kirk-Othmer: *Encyclopedia of Chemical Technology,* 3rd Ed., Vol. 4, pp 570–622, John Wiley & Sons, Inc., New York-Chichester-Brisbane-Toronto, 1978.
Ullmann's Encyclopedia of Industrial Chemistry, 5th Ed., Vol. A 5, pp 101–120, VCH Verlagsgesellschaft mbH, Weinheim-Deerfield Beach-Basel, 1986.
O. J. Vohler, *Erdöl u. Kohle Erdgas Petrochem.* **39,** pp 561–567 (1986) [in English].
E. Wege, *Sprechsaal* **113,** pp 432–440 (1980).

Special Types of Carbon and Graphites:

Technical Information:

E. Fitzer and M. Heym, *High Temperatures – High Pressures* **10,** pp 29–66 (1978).
E. Fitzer und D. Kehr, *Haus der Technik – Vortragsveröffentlichung* **338,** pp 29–39 (1974).
B. Findeisen, *Freiberger Forschungshefte* **A 618,** pp 19–56 (1980).

Reviews:

Kirk-Othmer: *Encyclopedia of Chemical Technology,* 3rd Ed., Vol. 4, pp 628–631, John Wiley & Sons, Inc., New York-Chichester-Brisbane-Toronto, 1978.
Ullmann's Encyclopedia of Industrial Chemistry, 5th Ed., Vol. A 5, pp 120–124, VCH Verlagsgesellschaft mbH, Weinheim-Deerfield Beach-Basel, 1986.
E. Wege, *Sprechsaal* **113,** pp 432–440 (1980).
E. Fitzer, *Angew. Chem.* **92,** pp 375–386 (1980).
O. Vohler, P. L. Reiser, R. Martina und D. Overhoff, *Angew. Chem.* **82,** pp 401–412 (1970).

Carbon Black:

Commercial Information:

Industrial Statistics Yearbook. 1983 Edition, Vol. II, Commodity Production Data 1974–1983, United Nations, New York, 1985.

Technical Information:

B. R. Corry, *Industrial Minerals,* Pigments and extenders supplement, pp 26–32 (May 1985).
A. Smith, *Paint and Resin,* pp 35–38 (May/June 1982).
A. Smith, *Paint and Resin,* pp 28–34 (September/October 1982).
E. J. Claassen, *Encycl. Chem. Process Des.* **6,** pp 262–279 (1978).
H. Ferch, *Polm. Paint Colour J.* **168,** pp 481–489 (1978).

Reviews:

Kirk-Othmer: *Encyclopedia of Chemical Technology,* 3rd Ed., Vol. 4, pp 631–666, John Wiley & Sons, Inc., New York-Chichester-Brisbane-Toronto, 1978.
Ullmann's Encyclopedia of Industrial Chemistry, 5th Ed., Vol. A 5, pp 140–158, VCH Verlagsgesellschaft mbH, Weinheim-Deerfield Beach-Basel, 1986.
"Was ist Ruß?" (What is carbon black?) *Degussa brochure.*

Microstructure and Morphology:

L. L. Ban and W. M. Hess, *Prep. Div. Pet. Chem. Am. Chem. Soc.* **20** (2), 405–419 (1975).

Activated Carbon:

Commercial Information:

"Activated Carbon", *Chemical Economics Handbook,* Stanford Research Institute, Menlo Park, California, USA, July 1985.
"Norit throughout the world", Prospectus of the Norit Company.

Technical Information:

J. Wilson, "Active Carbons from Coals", *Fuel* **60,** pp 823–831 (1981).

Applications:

H. Jüntgen, "New Applications for Carbonaceous Adsorbents", *Carbon* **15,** pp 273–283 (1977).
Active carbon in water purification: *Am. Water Works Assoc.,* **71**, Vol. 11 in several papers.

Regeneration:

R. H. Zanitsch and R. T. Lynch, "Selecting a Thermal Regeneration System for Activated Carbon", *Chem. Eng.* **85,** pp 95–100 (1978).

Reviews:

Ullmann's Encyclopedia of Industrial Chemistry, 5th Ed., Vol. A5, pp 124–140, VCH Verlagsgesellschaft, Weinheim-Deerfield Beach-Basel, 1987.
H. von Kienle und E. Bäder: *Aktivkohle und ihre industrielle Anwendung.* (Activated carbon and its industrial applications.) Ferdinand Enke Verlag, Stuttgart, 1980.

5.8 Fillers

5.8.1 General Information

Inorganic fillers are fine particulate solids, which differ from inorganic pigments in their refractive indices: fillers generally have refractive indices < 1.7 (vs. vacuum) and pigments have larger refractive indices. An additional distinguishing criterium is given by the term filler itself implying that these materials should fill and hence be a major component. Pigments on the other hand are mainly utilized in small quantities. Both distinguishing criteria are unsatisfactory in as much as a filler with n < 1.7 in a suitably chosen matrix can exhibit a certain covering power thus exhibiting a pigment-specific property, which is in fact used in-

Distinguishing criteria of inorganic fillers from inorganic pigments:

- Refractive indices < 1.7 (vs. vacuum)
- Mostly a major component of the particular formulation

dustrially. On the other hand certain fillers are only used in small quantities. The borderline is thus somewhat blurred.

Fillers should:

- Make the product cheaper
- Improve the properties of the product

Fillers have two functions: to make the product cheaper (classic definition) and to improve an existing property or add a new property compared with the unfilled product. The two functions cannot always be combined. Furthermore the utilization of a filler often makes the manufacture of the product easier.

Fillers often make the manufacture of the product easier

Classification of inorganic fillers in:

- Natural fillers
- Synthetic fillers

Inorganic fillers can be divided into natural and synthetic fillers. The former are obtained by the mechanical or chemical working up of naturally occurring minerals and the latter by chemical synthesis involving not only inorganic but also organic raw materials.

In this chapter the most important inorganic fillers will be discussed with the exception of fibers (Chapter 5.2) and carbon blacks (Sec. 5.7.6).

5.8.2 Economic Importance

In the filler sector there are hardly any reliable market surveys. This is probably due to the variety of types and the multiplicity of products.

As regards quantity the natural fillers dominate with ca. 70 to 80% of the total consumption, carbonates being the single largest group of products.

The future prospects are estimated to be promising particularly in the plastics and rubber sectors.

Table 5.8-1. USA Consumption Spectrum for Fillers for Plastics (without Fibers) in 10^3 t.

	1983	1984	1985
carbonates	1495	1645	1725
SiO_2 (quartz, novaculite etc.)	54	58	61
kaolin (clays)	92	100	104
talc	82	90	97
mica	7	9	11
microspheres	10	15	18
other inorganic fillers	140	155	170

5.8.3 Natural Fillers

5.8.3.1 Silicon-Based Fillers

The natural silicon-based fillers are divided into hydrated silicas and silicon dioxide (silicas), and "silicates". There is no sharp dividing line between these two groups, since the former do not always occur in a pure form and may contain silicates together with other impurities. Since their main component is SiO_2, this classification will be used here.

Silicon-based natural fillers classified into:

- Silicas and silicon dioxide: "silicas"
- "Silicates"

Silicon dioxide occurs in nature mainly as quartz in some cases in very large deposits (quartz gravel, quartz sand). Cristobalite is seldom found, but is produced synthetically (Sec. 5.8.4.1.5). Of lesser importance are tripolite, a special quartz type resulting from the weathering of sandstone, and novaculite, a lamellar-shaped microcrystalline quartz found in Arizona (USA). Novaculite has a large number of silanol groups on the surface and functions as an active reinforcing filler. The kieselguhrs, also known as diatomaceous or infusorial earths, contain between 70 and 90% by weight of amorphous silicon dioxide. They are formed from kiesel algae found in both freshwater and seawater, whose skeletons consist of silica. They are found all over the World in a variety of forms corresponding to the respective algae types. Aluminum silicate is mostly present in kieselguhr as a minor constituent. The kieselguhrs should not be confused with siliceous earth, a mineral of purely inorganic origin which contains up to 70 to 75% by weight of quartz and up to 25 to 30% by weight of kaolinite (see below) and is found only in the neighborhood of Neuburg on the river Danube (Neuburg siliceous earth). Quartz crystals and kaolinite platelets are fairly homogeneously distributed in the siliceous earth because they are formed together.

Natural fillers with silicon dioxide as the main component:

- Quartz
- Cristobalite
- Tripolite
- Novaculite
- Kieselguhr (diatomaceous or infusorial earths)
- Siliceous earth

Of the naturally occurring silicates, talc is a very important and versatile filler. It is a magnesium silicate with a lamellar structure and when pure has the composition $3\,MgO \cdot 4\,SiO_2 \cdot H_2O$. In Europe large deposits are found in the Fichtel Mountains, on the fringes of the Eastern and Southern Alps and in Scandinavia. Pyrophyllite is also a lamellar silicate but with aluminum as the cation: $Al_2O_3 \cdot 4\,SiO_2 \cdot H_2O$. It

Natural silicate fillers:

- Talc
- Pyrophyllite

is similar to talc in its properties and its applications. Another aluminum silicate with a lamellar structure, kaolinite $Al_2O_3 \cdot 2SiO_2 \cdot 2H_2O$, is the main constituent of kaolin, which in the English Speaking World is mostly known as "china clay" due to its original application in the manufacture of porcelain (Sec. 5.5.4.1). Kaolin is mainly formed by the weathering or hydrothermal decomposition of feldspars. The two largest deposits are in Cornwall (Great Britain) and Georgia (USA). Minerals in which aluminum is substituted on part of the silicon sites such as micas and particularly muscovite (with potassium as the compensating ion), phlogopite (potassium and magnesium) and vermiculite (magnesium) are also used as fillers. Apart from these laminate-structured silicates one silicate with a band structure is used as a filler: wollastonite $CaSiO_3$. The last natural silicate filler worthy of mention is perlite, which is manufactured from volcanic glassy stone by thermal expansion (Sec. 5.3.6.3) and is utilized both in unground and ground form.

- Kaolin (china clay)
- Mica
- Muscovite
- Phlogopite
- Vermiculite
- Wollastonite
- Perlite

5.8.3.2 Other Natural Fillers

Natural carbonate fillers:

- Calcite
- Dolomite
- Chalks

The most important natural fillers are calcium and magnesium carbonates. Of the numerous deposits, those suitable for processing are well-crystallized calcite ($CaCO_3$) and dolomite ($CaCO_3 \cdot MgCO_3$) as well as the chalks ($CaCO_3$), formed from the shells of maritime microorganisms and possessing a loose earthy structure.

Natural sulfate fillers:

- Heavy spar
- Light spar

Of the naturally occurring metal sulfates only heavy spar ($BaSO_4$) is of major importance for filler manufacture. Smaller quantities of gypsum ($CaSO_4 \cdot 2H_2O$, light spar) are also processed to fillers.

Micaceous iron oxide

Lamellar-shaped crystalline haematite (α-Fe_2O_3) can also be used as a filler. It is known as micaceous iron oxide, due to its morphological relationship to micas.

5.8.3.3 Beneficiation of Natural Fillers

Beneficiation:

- Washing
- Coarse crushing
- Coarse to medium crushing
- Classification
- Grinding

The route from mineral to filler is similar for all the above-mentioned materials. The important processing operations after mining are: washing, coarse crushing, coarse to medium crushing, classification, wet or dry grinding, filtration, drying and sieving. The order may not be adhered to and several of the steps may in fact be repeated. In the case of microcrystalline, loose raw materials such as siliceous earth, talc, kaolin (Sec. 5.5.4.2) or chalk, in particular, elutriation and flotation processes are used in the beneficiation. In the grinding of mica care must be taken to avoid destroying its lamellar structure. In some cases colored impurities are converted into colorless compounds with the help of additives or washed out (bleaching). Sometimes a calcination step is used. Some natural fillers are also posttreated with organic compounds.

- Filtration
- Drying
- Sieving

5.8.4 Synthetic Fillers

Some fillers have to be synthesized, because they do not occur in nature. Some fillers are synthesized, despite the fact that they occur as minerals in sufficient quantities and with sufficent purity, because even with the most careful processing it has not proved possible to realise the potentially available properties.

Reasons for the synthetic manufacture of fillers:

- No naturally available deposits
- Realization of better properties than those of the natural fillers

5.8.4.1 Silicas and Silicates

This group of fillers includes pyrogenic i.e. thermally or wet chemically produced silicas, wet chemically produced silicates, glasses and cristobalite.

5.8.4.1.1 Pyrogenic Silicas

The most important process for the manufacture of pyrogenic silicas is flame hydrolysis in which tetrachlorosilane or tetrafluorosilane is decomposed in an oxyhydrogen flame:

$$2\,H_2 + O_2 \longrightarrow 2\,H_2O$$

$$SiX_4 + 2\,H_2O \longrightarrow SiO_2 + 4\,HX$$

$$2\,H_2 + O_2 + SiX_4 \xrightarrow{\text{ca. } 1000\,^\circ C} SiO_2 + 4\,HX$$

$$X - Cl, F$$

Pyrogenic silicas (aerosols) are manufactured by three processes:

- Flame hydrolysis
- Electric arc process (currently insignificant)
- Plasma process (not yet introduced on a large scale)

The silica formed is like all synthetic silicas amorphous according to X-ray diffraction spectra. Their most important properties are given in Table 5.8-2.

The oldest, but in the meantime least significant, process for the production of fine particulate silica is the oxidation of gaseous silicon monoxide with air or water vapor. Quartz and fine particulate coke can be used as starting materials and are reacted with one another in an electric arc. Variants in which a plasma is used as an energy source instead of the electric arc have not yet been introduced on a large scale.

5.8.4.1.2 Wet Chemically Manufactured Silicas and Silicates

Silicas and silicates are wet chemically manufactured using:

- Precipitation processes
- Gel processes
- Hydrothermal processes

The wet chemical manufacturing processes can be divided into precipitation processes, gel processes and hydrothermal processes of which the precipitation processes are the most important.

Precipitation process:

water glass + acid

In the precipitation processes acid is added to a water glass solution in a large agitator vessel, as follows:

$$Na_2SiO_3 + 2\ HX \longrightarrow SiO_2 + 2\ NaX + H_2O$$

Colloidal primary particles are formed which agglomerate as the reaction proceeds and finally coalesce to aggregates. Metallic ions (Ca, Mg) are added to accelerate the often delayed precipitation. These ions are subsequently extracted with acid. The process can be very precisely controlled with respect to the properties of the required product. Further control can be achieved during working up (filtration, washing, drying and grinding). Table 5.8-2 gives the important properties of silicas manufactured by precipitation.

Silica properties can be influenced by:

- Speed of acid addition
- Intensity of stirring
- Working up procedure

If the water glass solution is mixed with alkaline earth or aluminum salts, silicates are obtained instead of silicas.

Gel processes:

Acid + water glass (mostly)

In contrast with the precipitation processes the formation of flocculent precipitates is specifically avoided in gel processes. Normally the sequence of addition is simply reversed, the water glass solution being added to the acid with the pH of the solution remaining acidic. Initially an unstable hydrosol is formed, which after a time becomes a hydrogel. This hydrogel is carefully washed, then dried and finally ground. A special drying process using dewatering under critical conditions leads to so-called aerogels. The properties of silica gels and aerogels are given in Table 5.8-2.

Silica gels dried under supercritical conditions are called aerogels

Table 5.8-2. Properties of Differently Manufactured Silicas [from H. Ferch, *Progr. Org. Coat.* **9,** 139 (1981)].

	Primary particle size [1] (in nm)	Specific surface area [2] (in m^2/g)	Tap volumes [3] (in mL/100 g)	Porosity
aerosols (from flame hydrolysis)	5 to 50	50 to 600	1000 to 2000	nonporous (up to ca. 300 m^2/g)
precipitated silicas	5 to 100	30 to 800	200 to 2000	low
silica gels	3 to 20	250 to 1000	100 to 200	very high
aerogels	3 to 20	250 to 400	800 to 2000	high

[1] determined by electron microscope
[2] BET values
[3] DIN 53 194

In the hydrothermal process aqueous suspensions of quartz sand and chalk are heated under pressure:

Hydrothermal process: fairly insignificant

$$CaO + SiO_2 \xrightarrow{H_2O} CaSiO_3\ aq$$

The proportions can be varied within wide limits. The calcium can be largely washed out with acid. Working up proceeds in the usual above-described manner.

5.8.4.1.3 Posttreatment of Silicas

The purely thermal posttreatment of dried silicas, which depending upon the temperature leads to more or less strong splitting off of water, is only of major importance for silica gels and is termed "activation".

Physical posttreatment of silicas: heating

Chemical posttreatment is already widely utilized and will probably be even more widely applied in the future. It renders hydrophilic silica organophilic or hydrophobic by treating preferably pyrogenic silicas, but also precipitated silicas, with organic or metallo-organic compounds which react with the silanol groups forming chemical bonds:

Chemical posttreatment of silicas with:

- Chlorosilanes
- Aminosilanes
- Silicones
- Alcohols
- Amines and quaternary ammonium compounds
- Isocyanates
- Waxes

and other compounds.

$$-Si-OH + X-R \longrightarrow -Si-OR + HX$$

or are strongly adsorbed on the surface. The reactions with chlorosilanes such as trichloromethylsilane and dichlorodimethylsilane are the most industrially important.

It is possible to endow the surface of silicas with functional groups e.g. by treatment with silane adhesion promoters $X_3Si-R-Y$, where Y is the functional group

5.8.4.1.4 Glass

Glass as a filler in the form of:

- Solid microspheres
- Hollow microspheres

Whereas glass in the form of powder or small flakes has attracted little interest as a filler, the demand for solid and hollow micro glass spheres is increasing. This is due to their geometric shape i.e. their surface to volume ratio, the absence of the sharp edges of the other fillers and in the case of hollow spheres their low density. Microspheres up to 1 μm in diameter have recently become available and lead to products with strongly isotropic properties. Ceramic spheres are also being marketed in increasing quantities. They are frequently posttreated with adhesion promoters to increase the interaction between the microspheres and the matrix (Sec. 5.8.4.1.3). For special applications e.g. attainment of electrical conductivity, metal-coated microspheres are obtainable.

5.8.4.1.5 Cristobalite

Cristobalite from quartz by heating in the presence of catalysts

Cristobalite (Sec. 5.8.3.1) is manufactured in large quantities by the thermal treatment of quartz sand. Addition of catalysts can reduce the transition temperature of ca. 1100 °C. The important advantage of cristobalite in comparison with quartz is its much higher whiteness.

5.8.4.2 Aluminum Hydroxide

Primary raw material for aluminum hydroxide manufacture is bauxite

The raw material for the manufacture of aluminum hydroxide is bauxite, a mineral with ca. 45 to 60% Al_2O_3 in the form of hydrargillite (gibbsite), böhmite and diaspore and with iron oxides or hydroxides, silica (as kaolinite) and titanium dioxide as the main minor constituents. The largest deposits are found in Africa and Australia.

Continuous digestion of bauxite with sodium hydroxide in autoclaves or tubular reactors at 140 to 250 °C under pressure using the Bayer process

Bauxite is digested with sodium hydroxide under pressure and at high temperatures using the Bayer process, whereupon soluble sodium aluminate is formed:

$$Al(OH)_3 + NaOH \longrightarrow NaAl(OH)_4$$

in the form of aquo-complexes. After pressure release the aluminate liquor is cooled, diluted and freed of red mud by decantation in gravity concentrators. Subsequent fine filtration removes the small residue of insoluble constituents. The clarified liquor, still at ca. 95 °C, is an important starting point for aluminum hydroxide manufacture.

Upon cooling, the liquor becomes supersaturated and aluminum hydroxide precipitates out in a reversal of the formation reaction. The solution is seeded with solid aluminum hydroxide to increase the precipitation rate. The hydrargillite formed is filtered off and washed. A large part of it is mashed and returned as seed for the following precipitation and the rest is carefully dried. Coarse particulate aluminum hydroxide is obtained, which can be processed by grinding and sieving to particular particle size fractions between 10 and 100 µm.

In the "normal" Bayer process coarse particulate aluminum hydroxide is formed, which is processed by grinding and sieving to particle sizes between 10 and 100 µm

Fine particulate aluminum hydroxide is obtained either by introducing special seeding processes into the Bayer process or by precipitation processes:

- Precipitation from cooled and strongly diluted aluminate liquor
- Precipitation by neutralization of aluminate liquor with e.g. CO_2, $NaHCO_3$, H_2SO_4
- Precipitation from aluminum salt solutions with bases such as e.g. NH_3, amines, NaOH
- Precipitation from aluminum salt solutions with sodium aluminate solutions

Fine particulate aluminum hydroxide (particle size < 1 µm) obtained by:

- Modified Bayer process
- Precipitation processes

Properties can be influenced during the precipitation processes by:

- Temperature
- pH value
- Rate
- Concentration
- Ripening time

The average primary particle sizes of commercial products are mainly below 1 µm and the specific surface areas (BET) between 4 and ca. 60 m^2/g. The $Al(OH)_3$-content is > 99.5% by weight. Organically posttreated types are also available.

5.8.4.3 Carbonates

Calcium carbonate is virtually the only important filler in the processing industry. Barium and magnesium carbonates are only manufactured for special purposes and will thus only be briefly mentioned here.

Synthetic carbonate fillers:

- Calcium carbonate
- (Barium carbonate)
- (Magnesium carbonate)

Calcium carbonate fillers are synthesized by precipitation processes. This accounts for the usual term "ccp" being an abbreviation of: calcium carbonicum praecipitatum. The following three precipitation processes are used.

Manufacture of calcium carbonate (ccp) by precipitation processes:

- Neutralization of milk of lime with carbon dioxide:

$$Ca(OH)_2 + CO_2 \rightarrow CaCO_3 + H_2O \quad (1)$$

This process is directly coupled with the manufacture of quicklime. Milk of lime is obtained by slaking and then carbon dioxide liberated by the burning of limestone is passed into the suspension forming calcium carbonate (Sec. 5.3.2.3 and 5.3.2.4).

- Neutralization of milk of lime with carbon dioxide

product properties influenced by:

- Temperature
- Concentration
- Pressure

- Methathesis of calcium chloride with sodium carbonate:

$$CaCl_2 + Na_2CO_3 \rightarrow CaCO_3 + 2\,NaCl \quad (2)$$

- Reaction of calcium salts, formed as byproducts in other processes, with carbonates in aqueous solution

Calcium chloride is formed as a waste product in the manufacture of sodium carbonate by the Solvay process (Sec. 3.1.1.3.3).

- Reaction of calcium nitrate with carbon dioxide and ammonia:

$$Ca(NO_3)_2 + 2\,NH_3 + CO_2 + H_2O \rightarrow CaCO_3 + 2\,NH_4NO_3 \quad (3)$$

Calcium nitrate is produced as a byproduct in the manufacture of fertilizers (Sec. 2.1.2.4).

The first process supplies purer products and is easier to control than the other two processes.

The calcium carbonate can be precipitated in either a calcite or an aragonite structure and can be obtained with a pure white color. The average particle size is in the range 0.01 to 1 µm and the specific surface area (BET) is in the range 5 to 100 m^2/g. The surface treatment of calcium carbonate with organic compounds is also becoming increasingly important.

5.8.4.4 Sulfates

Only synthetic sulfate filler:

barium sulfate (blanc fixe)

Only barium sulfate is synthetically manufactured as a filler and is known as "blanc fixe". As with calcium carbonate it is produced by precipitation reactions:

Blanc fixe *powder:* by reacting barium sulfide with sodium sulfate in aqueous solution and working up the precipitated products

$$BaS + Na_2SO_4 \longrightarrow BaSO_4 + Na_2S \quad (1)$$

The barium sulfide is obtained by reducing low silica heavy spar with coke. The reaction product is filtered washed, dried and ground. It is pure white and has particle sizes between ca. 0.3 and 10 µm.

Commercially available blanc fixe qualities as 25 to 30% aqueous pastes (blanc fixe paste) are manufactured by the reaction of barium chloride solution with a solution of Glauber's salts.

$$BaCl_2 + Na_2SO_4 \longrightarrow BaSO_4 + 2\ NaCl \qquad (2)$$

Blanc fixe *pastes:* by reacting barium chloride with sodium sulfate in aqueous solution and working up the suspension

The barium chloride solution is produced in the manufacture of lithopone or is produced by the reaction of barium sulfide with hydrochloric acid.

5.8.4.5 Other Synthetic Fillers

For some time titanium dioxide and aluminum(III) oxide manufactured by flame hydrolysis (Sec. 5.8.4.1.1) have been marketed as fillers.

5.8.5 Properties and Applications

The multitude of fillers mentioned above make it impossible to describe the properties of each one. Their application is not only dependent upon their chemical composition but also to a major degree upon their physical properties and their posttreatment, if any. Purity is an additional criterium particularly in the case of natural fillers.

Application possibilities of fillers substantially determined by:

- Particle size
- Particle shape
- Particle structure
- Specific surface area
- Density
- Color
- Refractive index
- Interaction with the matrix
- Chemical composition
- Type of surface treatment

A distinction between inactive and active fillers is at present hardly relevant, since the properties of the final product depend more or less strongly upon all the fillers utilized and their use has for a long time not been primarily determined in terms of cost reduction. The efforts of the filler producers in the direction of improved processing methods and dedicated manufacturing processes take this development into account. The surface treatment of natural and synthetic fillers has also acquired great importance. For years there have been products which in the classic sense are not fillers, but are in fact active substances (e.g. silica aerosols). Apart from any cost reduction considerations, fillers have essentially the following functions:

Trend: away from purely cost reducing fillers to high value multifunctional fillers

Fillers can influence:

- Rheological behavior
- Optical properties
- Mechanical properties
- Resistance to weathering and chemicals

Further:

- Improve the dispersion behavior of pigments
- Increase fire resistance (aluminum hydroxide)
- Improve the surface of paper
- Change the adhesion behavior between foils
- Reduce weight (perlite, hollow microspheres)

- Rheological:

  thickening and, if necessary, thixotropy by controlled adjustment of the time dependent structural viscosity (e.g. talc, wollastonite, kaolin, silica aerosols), in general: change the flow behavior of lacquers, coatings, printing inks, liquid polymers, sealing compounds

- Optical:
  - improvement of the whiteness and hiding power as (partial) replacement for real white pigments e.g. by synthetic silicates, blanc fixe, finest particulate calcium carbonate (homogeneous pigment distribution, supercritical pigmentation) in emulsion paints
  - as white pigments in paper coating paints: kaolin, natural ground calcium carbonate
  - gloss reduction of lacquers and printing inks: wet chemically manufactured silicas, kieselguhrs
  - opacification of plastics: barium sulfate
  - as a mineral UV-absorber: mica
- Mechanical:
  - reinforcing agent for rubbers: e.g. calcium carbonate, synthetic silicas. For the industrial utilization of reinforcing carbon black types see Sec. 5.7.6.3.
  - improvement of the mechanical properties of plastics: e.g. quartz, novaculite, wollastonite, microspheres
- Improvement in the resistance of paints and coatings to weathering and chemicals (corrosion protection) with e.g. mica, micaceous iron oxide, barium sulfate

Other functions of fillers are listed in the marginal notes.

The main utilization areas of the above discussed fillers are summarized in Table 5.8-3.

Table 5.8-3. Main Application Sectors of Fillers.

Filler	Plastics	Rubber	Lacquers, paints, coatings	Printing inks	Paper
natural SiO_2:					
quartz	×		×		
tripolite	×				
novaculite	×				
kieselguhr	×	(×)	×		(×)
siliceous earth		×	×		
synthetic SiO_2:					
cristobalite			×		
pyrogenic silica	(×)		×		
wet chemical silica	×	×	×	×	
talc, pyrophyllite	×	×	×		(×)
kaolin	×	×	×		×
mica	×		×		
wollastonite	×		×		
perlite	×		(×)		
synthetic silicates:					
wet chemical silicates	(×)	×	×		×
microspheres	×		×		
micaceous iron oxide			×		
$CaCO_3$:					
calcite, dolomite	×	×			×
chalk	×	×	×		×
ccp	×	×	×	×	
aluminum hydroxide	×	(×)	×	(×)	(×)
$BaSO_4$:					
heavy spar	×		×		
blanc fixe	×	(×)	×	(×)	(×)
$CaSO_4$:					
light spar		(×)		×	

References for Chapter 5.8: Fillers

Ullmanns Encyklopädie der technischen Chemie, 4th Ed., Vol. 18, pp 647–660, Verlag Chemie, Weinheim-Deerfield Beach-Basel, 1979.

H. S. Katz and J. W. Milewski: *Handbook of Fillers and Reinforcement for Plastics*. Van Nostrand Reinhold Co., New York, 1978.

H. Kittel: *Lehrbuch der Lacke und Beschichtigungen*. (Textbook on lacquers and coatings.) Vol. II, pp 284–466, Verlag W. A. Colomb, Berlin, 1974.

R. Juran, *Modern Plastics International*, pp 55–57 (September 1985).

PLASTverarbeiter **36**, pp 52–63 (1985).

R. C. Freas, *Industrial Minerals*, Pigments and Extenders Supplement, pp 5–9 (May 1985).

5.9 Inorganic Pigments

5.9.1 General Information and Economic Importance

Pigments: fine particle colorants insoluble in binders

The term colorant, as defined by DIN, includes both dyes and pigments. In contrast to dyes, pigments are practically insoluble in the binder. They consist of solid separate particles mostly of uniform chemical composition and for the most part of transition metal oxides, oxide hydrates, sulfides, silicates, sulfates or carbonates (see Table 5.9-1).

Carbon black and magnetic pigments can also be included, although in the latter case the optical effect is unimportant and the magnetic properties all

Table 5.9-1. Inorganic Pigments: Survey.

	Oxides	Sulfides	Chromates	Various
white	titanium dioxide zinc oxide	zinc sulfide lithopone ($ZnS/BaSO_4$)		white lead (formerly)
red	red iron oxide red lead	cadmium red cadmium orange	molybdate orange	
yellow	yellow iron oxide nickel titanate Cr-Sb-titanium buff	cadmium yellow	chrome yellow zinc chromate	
green	chromium oxide green spinel green		chrome green	
blue	cobalt blue	ultramarine		manganese blue prussian blue
brown	brown iron oxide ferrites			iron oxides with carbon black
black	black iron oxide spinel black			carbon black

Table 5.9-2. Classification of Inorganic Pigments.

Type	Function	Examples
natural inorganic pigments		ochre, sienna, graphite, green earth
synthetic inorganic pigments		
1 Uncolored pigments		
1.1 White pigments	mainly nonselective light scattering	titanium dioxide, zinc sulfide, lithopone, zinc oxide
1.2 Black pigments	mainly nonselective light absorption	carbon black, black iron oxide
2 Colored pigments	color by selective light absorption mostly with light scattering	iron oxide, cadmium sulfide, ultramarine, chrome yellow, cobalt blue
3 Speciality pigments		
3.1 Stains and oxides	as for colored pigments	cobalt blue, Ni-Sb-Ti yellow, nickel titanate, cadmium red
3.2 Corrosion-inhibiting pigments	electrochemical, chemical or physical protection of metallic surfaces	red lead, zinc yellow, zinc phosphate, zinc dust
3.3 Magnetic pigments	information storage by magnetization	magnetic iron oxides, chromium dioxide, metallic iron pigments
3.4 Luster pigments	luster effect by directed reflection of light from orientated flake-shaped pigment particles	
3.4.1 Metallics		gold bronze, aluminum bronze, copper bronze, micaceous iron oxide
3.4.2 Pearlescent pigments	pearlescence is due to multiple reflection on parallelly orientated lamellae. Lamellae of a particular thickness exhibit additional iridescent effects by interference	TiO_2-coated mica, basic lead carbonate, mother of pearl pigments
3.5 Luminescent pigments		
3.5.1 Fluorescent pigments	selective light absorption and undelayed luminescence, excited by high energy radiation (e.g. X-rays, cathode rays, U.V. and short wavelength visible light)	pigments for cathode ray and X-ray screens, fluorescent lamps, "radioactive" luminous pigments
3.5.2 Phosphorescent pigments	selective light absorption and scattering, superimposed on delayed luminescence excited by high energy radiation	zinc and alkaline earth sulfides doped with heavy metal ions (e.g. Cu, Ag, Au, Mn etc.)

important. Table 5.9-2 gives the DIN 55944 classification of inorganic pigments together with their functions and several examples.

Important pigment properties:

- Hiding power
- Tinting strength (colored pigments)
- Lightening power (white pigments)
- Brightness

Important pigment properties are their hiding power, their lightening power in the case of white pigments and their tinting strength in the case of colored and black pigments. In pigmented systems the color perceived, as for all nonluminescent body colors, is based on subtractive color mixing. Colored pigments absorb a particular part of the spectrum of the illuminating light, so that the eye only sees the residual spectrum.

White pigments, on the other hand, are characterized by their nonselective light scattering and black pigments by their nonselective absorption in the whole of the visible spectrum.

The selective absorption of colored pigments is due to the interaction of visible light with the valence electrons of the solid. The electronic structure and energy levels of a solid are determined by the elements present (cations and anions), their oxidation state and their electronic configuration as well as the crystal lattice geometry. The tinting power of a pigment increases with the number of light absorption-induced electron transitions. The steeper the absorption edge, the brighter the color will be.

High refractive index and optimum particle size are prerequesites for high light scattering power

The scattering power of pigments, and therefore the hiding power of a pigment coating, mainly depend upon the refractive index of the pigment and its particle size. For high scattering power the refractive index should be as high as possible above that of the binder. For a particular refractive index there is an optimum particle size to achieve maximum scattering power. For white and colored pigments this optimum particle size is between 0.1 and 1 µm. In large scale manufacture a particle size distribution is required which is as narrow as possible.

The increasingly colored nature of our surroundings has meant a strong increase in the use of colorants. Since many colored materials contain at least some inorganic pigment, their demand has steadily increased.

Production of naturally occurring pigments:

0.1×10^6 t (1986)

For a long time, the natural deposits have not been able to satisfy the demand partly due to the application-related steady tightening of qualitative specifications. Nevertheless, the production of natural pigments, mainly iron oxides, amounted to ca. 0.1×10^6 t in 1986.

By comparison, the 1985 worldwide capacity for synthetic pigments and carbon black was ca. 3.8×10^6 t/a. A survey of the production capacities of the most important synthetic inorganic pigments is given in Table 5.9-3. TiO_2 white pigments and iron oxide colored pigments together account for more than 80% of the world capacity. The main producers are in the USA and Western Europe. West Germany has a strong position in the market.

Worldwide production capacity for synthetic inorganic pigments:

3.8×10^6 t/a (1985) of which: 67% is for TiO_2 and 16% for iron oxides

Table 5.9-3. Production Capacity for Synthetic Inorganic Pigments in the Western World in 1985.

	10^3 t/a	%
titanium dioxide	2580	67.3
iron oxide	600	15.7
carbon black	380	9.9
lithopone	50	1.3
chromate yellow	75	2.0
ferrocyanide pigments	35	0.9
chromium oxide	53	1.4
ultramarine	20	0.5
molybdate red	20	0.5
mixed metal oxide pigments	12	0.3
cadmium pigments	8	0.2
Total	3833	100.0

Table 5.9-4. End Uses of Inorganic Pigments in Western Europe in %.

	Titanium dioxide	Iron oxides	Chromium oxide	Cadmium pigments	Mixed metal oxide pigments
paints and lacquers	63	24	50	10	65
construction materials	5	64	19	–	5
plastics	16	3	8	80	20
paper	9	2	–	–	–
enamel and ceramics	5	1	10	10	10
miscellaneous	2	6	14	–	–
	100	100	100	100	100

Whereas mass production pigments such as TiO_2 and iron oxide sell for a few US $/kg, the prices for individual special pigments may be 55 US $/kg or more. Assuming an average price of 2.40 US $/kg leads to an annual worldwide turnover of 6 to 7×10^9 US $.

Application fields for inorganic pigments:

- Paints and coating materials
- Construction materials
- Plastics
- Enamel and ceramics
- Printing inks
- Miscellaneous

The application fields of the individual pigment groups are very diverse. The main applications in Western Europe are: TiO_2 and mixed metal oxide pigments in the paint and lacquer sector, iron oxide pigments in construction materials and cadmium pigments in the plastics sector.

5.9.2 White Pigments

5.9.2.1 General Information

The production of white coatings or bulk white solids and the lightening of formulations with colored pigments is currently only technically possible by pigmentation with white pigments. All white pigments are produced synthetically.

Assuming an average worldwide price of ca. 2.35 US $/kg white pigment leads to a worldwide turnover of ca. 6.25×10^9 US $.

Largest group of inorganic pigments are the white pigments

1986 production:

2.6×10^6 t/a

Table 5.9-5. Production of the More Important White Pigments in West Germany and the World in 1986.

	West Germany 10^3 t	World 10^3 t
Titanium dioxide	320	2500
Lithopone	50	100
Zinc oxide,white lead	5	60

Zinc sulfide, zinc oxide and white lead are produced in much larger quantities as chemical reagents. The quantities in the above Table only refer to their pigment production.

TiO_2: white pigment with excellent properties

A high refractive index and an optimum particle size, upon which it is dependent, are prerequisites for achieving a maximum hiding power or lightening power. Titanium dioxide is more effective in this than any other white pigment (see Table 5.9-6).

Table 5.9-6. Refractive Index and Optimum Particle Sizes of Several White Pigments.

Compound	Average refractive index versus vacuum	Optimum particle size when using organic binders D_{opt}, in μm	Relative scattering power
TiO_2 rutile	2.80	0.19	up to 800
TiO_2, anatase	2.55	0.24	600
ZnS, zinc blende	2.37	0.29	350
ZnO, zinkite	2.01	0.48	100
($BaSO_4$, baryte)*	1.64	1.60	25

* for comparison

5.9.2.2 Titanium Dioxide Pigments

5.9.2.2.1 Economic Importance

TiO_2: by far the most important white pigment

TiO_2 pigments have a more than 90% share of the worldwide white pigment market, due to their excellent properties (see Table 5.9-6). White lead, lithopone and zinc oxide are being increasingly supplanted from the market (see Table 5.9-5). The worldwide capacity has risen in the last 20 years from 1.4×10^6 t/a (1965) to 2.6×10^6 t/a (1986). Over this period, the chloride process has been increasingly preferred and in the last few years almost exclusively used in new plants (see Table 5.9-7).

Table 5.9-7. Worldwide Capacities for TiO_2 Production (10^3 t/a).

Year	Sulfate process	Chloride process	Total
1965	1254	135	1389
1970	1499	437	1936
1972	1632	441	2073
1974	1810	538	2348
1976	1922	657	2579
1978	1819	729	2548
1980	1757	964	2721
1985	1635	945	2580
1986	–	–	2887

5.9.2.2.2 Raw Materials for TiO_2 Pigments

The starting materials for TiO_2 production are ilmenite and titaniferous slag in the case of the sulfate process (Sec. 5.9.2.2.3) and leucoxene, rutile, synthetic

Raw materials for TiO_2 pigments

natural:
- Ilmenite
- Leucoxene
- Anatase
- Rutile

synthetic:
- Titaniferous slag
- Synthetic rutile

rutile, ilmenite, titaniferous slag and in the future possibly also anatase for the chloride process (Sec. 5.9.2.2.3).

Ilmenite (Fe, Mg, Mn) TiO_3 is found in nature in primary or secondary deposits. The TiO_2 content is 43 to 61% and the iron oxide content 34 to 49%. The impurities are mainly Mn, Mg, Al, Si, Cr, V, Nb, and Sb. Primary ore deposits of ilmenite are found in Norway, the USSR, Finland, Canada and the USA. Secondary ore deposits are more important raw materials in view of their easier processing (sands). They are found in South Africa, Australia, India, Brazil, Malaysia and Egypt.

Rutile is extracted from river, coastal and dune sands yielding concentrates with 90 to 98% TiO_2, which may be contaminated with Fe, Zr, V, Cr, Al, Si, Mg etc. Australia is the main supplier with 0.27×10^6 t in 1985.

Anatase deposits with a TiO_2 content of 30 to 40% have been found in Brazil. They contain 220×10^6 t TiO_2 which is more titanium dioxide than in all the currently known rutile deposits. It can be possibly used as a starting material for $TiCl_4$ in the production of TiO_2 pigments or in the production of metallic titanium. Plants for upgrading ores with a combined capacity of 0.4×10^6 t/a are currently under construction.

Production of titaniferous slag:

by reduction of ilmenite at 1200 °C and separation of the liquid iron

Titaniferous slags from ilmenite are because of their high TiO_2 content important raw materials for the sulfate and chloride processes. The iron contained in ilmenite is for the most part reduced with carbon to the metal in an arc furnace at 1200 to 1600 °C. Slags are formed with a TiO_2 content of 70 to 85%, depending upon the ilmenite used. The main producers are the Quebec Iron and Titanium Corp. (QIT) in Canada and Richards Bay Minerals Ltd. (RBM) in South Africa. The capacity of QIT is 1.3×10^6 t/a, that of RBM 0.44×10^6 t/a.

Manufacture of synthetic rutile from ilmenite:

by removal of iron using various processes while retaining the granular form of the starting material

Synthetic rutile. In recent years the prices for natural rutile have increased due to the increasing scarcity of exploitable deposits. To provide a cheap raw material for the chloride process, processes for the removal of iron from ilmenite ore have been developed in which the particle form is retained, as this is particularly

Table 5.9-8. Production Plants for the Manufacture of Synthetic Rutile (1986/1987).

Developer of process	Process steps	Byproducts	Operating company/site/ Capacity in 10^3 t/a
Benilite Corp. of America (BCA)	– partial reduction to Fe(II) – hydrochloric acid leaching – calcination	20 to 23% $FeCl_2$ solution, pyrolysis into Fe_2O_3 and HCl	Kerr McGee/Mobile, USA/ 100; Kerala Minerals and Metals Ltd./Chavara, Kerala/ 25; Indian Rare Earths/Orissa, India/100
Western Titanium	– oxidation to Fe(III) – reduction to iron – leaching with $FeCl_2$ while oxidizing with air	iron hydroxide	Associated Minerals Consolidated/Capel, Australia/60; Associated Minerals Consolidated/Narngulu, Australia/ 112 (1987)
Lurgi	– reduction to iron (rotary tube furnace) – leaching while aerating – separation in hydrocyclone – calcination	iron hydroxide	Westralian Sands Ltd/Capel, Australia/100 (1987)
Ishihara Sangyo Kaisha	– reduction to Fe(II) – sulfuric acid leaching – calcination	$FeSO_4$ solution (converted with NH_3 to ammonium sulfate and iron hydroxides)	Ishihara/Yokkaichi, Japan/ 48
Dhrangadhra Chemical Works	– reduction – hydrochloric acid leaching – calcination	$FeCl_2$ solution	Dhrangadhra Chemical Works Ltd./Suhupuram, Tamil Nadu, India/25
			World total: 570

favorable for fluidized bed chlorination (see Table 5.9-8).

Synthetic rutile has a TiO_2 content of 85 to 90%. In all these processes iron is either removed in the form of a salt solution which has to be worked up or as valueless oxide. In the process for the production of titaniferous slag, on the other hand, metallic iron is obtained, which makes this process more interesting both ecologically and economically.

5.9.2.2.3 Manufacturing Processes for TiO_2 Pigments

TiO_2 pigments are either manufactured using the older sulfate or newer chloride processes. The economics of the two processes are very much dependent upon the raw materials available.

Production processes for TiO_2 pigments:

- Sulfate process
- Chloride process

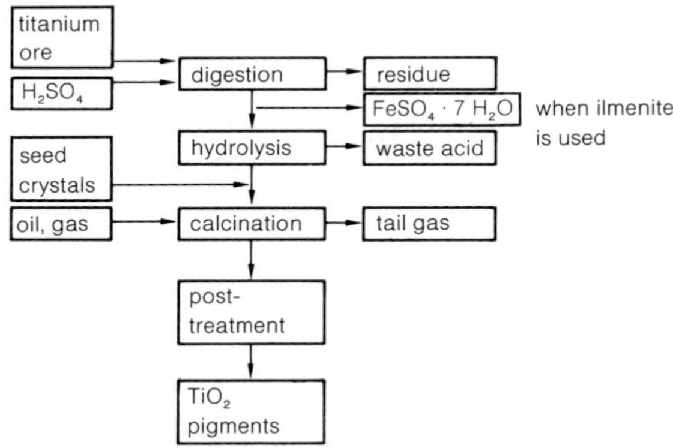

Fig. 5.9-1. TiO_2 pigment manufacture using the sulfate process.

Sulfate Process

Sulfate process:

- Digestion of ilmenite or titanium slag with H_2SO_4
- Reduction of Fe(III) to Fe(II)
- Separation of $FeSO_4 \cdot 7H_2O$
- Hydrolysis of $TiOSO_4$
- Washing and "bleaching"
- Calcination

The sulfate process (Fig. 5.9-1) utilizes ilmenite or titaniferous slag.

The finely ground raw materials are digested in an exothermic reaction with concentrated sulfuric acid, the digested cake dissolved in cold water and the residue separated off. To prevent their precipitation during the subsequent hydrolysis the iron(III) ions are reduced to iron(II) by adding a Ti(III) solution or scrap-iron. Upon evaporating the solution, the large quantities of iron(II) sulfate heptahydrate produced when ilmenite is used, crystallize out. The titanium sulfate is then hydrolyzed by heating the clarified solution with steam at 95 to 110 °C. TiO_2 seed crystals are added or are formed before hydrolysis to ensure yields of 93 to 96% TiO_2 and to obtain a hydrolysis product which yields the optimum particle size of ca. 0.2 μm upon firing. Dilute sulfuric acid remains as "waste acid".

The hydrolysis product is washed, treated with a Ti(III) solution to remove adsorbed heavy metal ions (Fe, Cr, Mn, V) and then calcined at temperatures between 800 and 1000 °C. Anatase or rutile pigments can be produced in the calcination process depending upon the choice of additives, which determine the characteristics of the product. Rutilization is promoted by the addition of "rutilization nuclei" prior to calcination.

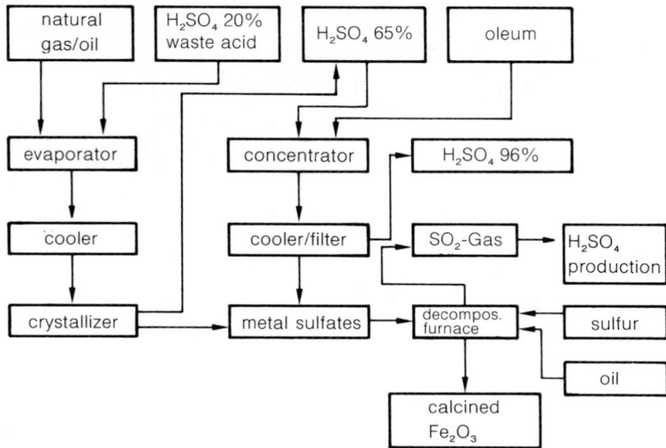

Fig. 5.9-2. Upgrading of waste acid and thermal decomposition of metal sulfate.

The byproducts, dilute sulfuric acid (waste acid) and, if ilmenite is used, ferrous sulfate heptahydrate, are to an increasing extent being processed further. The process used by Bayer AG since 1958 (Fig. 5.9-2) has been optimized to reduce energy consumption. Waste acid is concentrated to 65% in multistage vacuum evaporators, whereupon all the sulfates precipitate out. After removing the sulfate, the sulfuric acid is upgraded from 65% to 96% by adding oleum. It can then be reused for the decomposition of ilmenite or titaniferous slag.

Disposal of the waste products from the sulfate process:

- Upgrading of waste acid and recycling
- Thermal decomposition of $FeSO_4$ to iron oxide and SO_2 ($\rightarrow H_2SO_4$)

The iron sulfates are thermally decomposed to sulfur dioxide and iron oxide in a fluidized bed reactor at 800 to 1000 °C. The energy is provided by the combustion of sulfur or other fuels, the sulfur dioxide formed being, after purification, processed to sulfuric acid. The iron oxide residue can be employed in the cement industry.

Chloride Process

In the chloride process natural or synthetic rutile ore with a TiO_2 content of ca. 96% is chlorinated in a fluidized bed with oil coke and chlorine. The raw $TiCl_4$ produced is mixed with reducing agents to convert impurities such as vanadium oxychloride to lower

Chloride process:

- Fluidized bed chlorination of rutile ore
- Removal of vanadium
- $TiCl_4$ distillation
- $TiCl_4$ combustion to TiO_2 and Cl_2

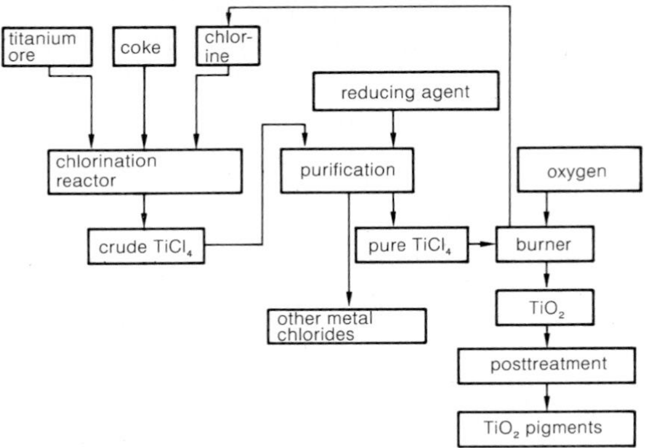

Fig. 5.9-3. TiO_2-manufacture by the chloride process.

valency state compounds. It is then distilled yielding titanium tetrachloride in almost any required purity. Finally it is combusted with pure oxygen to TiO_2 and chlorine, which is reused in the chlorination (Fig. 5.9-3).

The formation of metal chlorides other than $TiCl_4$ is dependent upon the raw material. If there were sufficient rutile available, this process would be particularly ecologically favorable. However, in view of its scarcity and hence its high price, raw materials with low titanium contents, such as 85% RBM slag (Sec. 5.9.2.2.2), are currently being used in the chloride process. Allowance has then to be made for the resulting increased metal chloride formation. TiO_2 pigment manufacture without byproducts is impossible. These are disposed of in similar ways to those already described for the sulfate process.

Posttreatment of TiO_2

Posttreatment of TiO_2 pigments to:

- Improve their dispersibility
- Increase their weather resistance

Most TiO_2 pigments undergo an inorganic and organic posttreatment, to increase their weather resistance and improve their dispersibility in paints and plastics.

Untreated anatase pigments, in particular, exhibit poor weather resistance due to the absorption of near UV light not only by TiO_2 itself but also by paint binders and plastics. This leads to UV-induced degradation of the organic media and to the formation of OH and HO_2 radicals on the anatase surface. These accelerate the photochemical degradation of the binder resulting in the exposure of pigment particles (chalking).

These effects can be suppressed or even reversed by using the more stable rutile pigments, by doping the hydrolysis product with Zn^{2+}, Al^{3+}, Zr^{4+} or Si^{4+} prior to calcination and by precipitation of poorly soluble colorless inorganic compounds such as hydrated SiO_2, $Al(OH)_3$, hydrated ZrO_2, hydrated aluminum silicates, aluminum phosphates, etc., onto the TiO_2-surface. The so treated stabilized rutile pigment still absorbs UV light, but without secondary photochemical reactions and hence protects the organic binder by screening it from UV light.

5.9.2.2.4 Applications for TiO_2 Pigments

The hiding and lightening powers of TiO_2 pigments, their chemical stability, nontoxicity and favorable price/effect ratio has gained them entry into many application fields and supplanted the other white pigments. TiO_2 pigments are thus used in the manufacture of paints and coating materials, plastics, printing inks, fibers (delustering), papers, laminate papers (laminates based on melamine or urea-formaldehyde resins), construction materials, enamel and ceramics, cosmetic powders, ointments, toothpaste and even cigar wrappers. There is hardly a white-colored or tinted object in our environment, which does not contain TiO_2 pigments.

Application fields for TiO_2 pigments:

- Paints and coating materials
- Plastics
- Printing inks
- Fibers
- Papers
- Laminates
- Construction materials
- Enamel and ceramics
- Cosmetics

Table 5.9-9. Application Fields for TiO_2 Pigments in Western Europe and the USA.

	Western Europe %	USA %
paints and other coating materials	63	50
plastics	16	17
paper	9	23
construction materials	5	—
enamel and ceramics	5	2
miscellaneous	2	8

TiO_2 slurries: highly concentrated, flow-, pump- and volumetrically dosable TiO_2 suspensions

Recently the use of TiO_2 pigments as highly concentrated, easy flowing, pumpable and volumetrically dosable suspensions (so called slurries) has increased. These slurries with up to 75% by weight of TiO_2 are rheologically designed so that no sediment is formed even after long storage times. Such TiO_2 pigments are dust free.

There is an increasing demand for TiO_2 in the ceramic sector e.g. for TiO_2 substrates in the SCR (Selective Catalytic Reduction) technology for the removal of NO_x from power station flue gases. The activity of these substrates, which may be honeycombed, can be increased by doping with other metal oxides such as WO_3 and V_2O_5.

5.9.2.3 Lithopone and Zinc Sulfide Pigments

Lithopones:

Mixtures of ZnS and $BaSO_4$

The lithopones comprise a group of white pigments, which are obtained by the calcination of coprecipitated zinc sulfide and barium sulfate. The quality of the pigment is determined by the zinc sulfide content which varies between 29 and 60%. Special pigments may consist of pure zinc sulfide. ZnS-containing white pigments are used for special paints, printing inks for wallpaper and in the plastics and rubber industries. The use of zinc sulfide-containing pigments in recent decades has steadily declined and with 100 t/a is currently 4.2% of the TiO_2 production (see Table 5.9-5).

Manufacture of lithopones:

- Precipitation
  $ZnSO_4 + BaS \rightarrow ZnS + BaSO_4$
- Firing of the mixture

The lithopones are manufactured by mixing zinc sulfate and barium sulfide solutions:

$$ZnSO_4 + BaS \longrightarrow ZnS + BaSO_4$$

The zinc salt solutions currently used come mainly from zinc and copper smelters and contain impurities which have to be completely removed prior to use, because all heavy metals form colored sulfides. In the first purification step the solution is oxidized with chlorine (pH 4), which precipitates most of the iron and manganese as the oxide-hydrate and part of the cobalt, nickel and cadmium present as their hydroxides. In the second step, the elements nobler than zinc

(Ni, Co, Cd, In, Tl, Pb, Cu, Ag) are precipitated as the metals by adding zinc dust and the metals returned to the copper smelters for noble metal extraction. After purification the solutions are adjusted to a particular zinc content. Mixtures of zinc sulfate and zinc chloride solutions are used for manufacturing lithopone types with more than 30% ZnS.

Purification of the $ZnSO_4/ZnCl_2$ solution by removal of foreign metals:

- Oxidation with chlorine
- Precipitation of nobler metals with zinc dust

Barium sulfide solutions are obtained by the reduction of low silica natural baryte with coke or coal in rotary furnaces at ca. 1250 °C and leaching with water. Since all the heavy metal sulfides remain in the residue, the filtered solution can be directly used for the precipitation of the zinc salt.

Preparation of the BaS solution:

$$BaSO_4 + 2C \rightarrow BaS + 2CO_2$$

The precipitated raw pigment does not exhibit pigment properties due to its small particle size and has to be calcined at a particular firing temperature (dependent upon the residual salt content) to yield the optimum particle size of ca. 0.35 μm. Lithopones with satisfactory light stability are obtained by incorporating 0.02 to 0.5% of cobalt. Chemical or mechanical posttreatment (e.g. steam-jet grinding with organic grinding and dispersion aids if required) yields optimum lithopone types for particular application fields.

Light stability of lithopones increased by doping with 0.02 to 0.5% Co

5.9.2.4 Zinc Oxide White Pigments

5.9.2.4.1 Manufacture

Zinc oxide white pigments are designated "zinc white" or zinc oxide depending upon the manufacturing process.

"Zinc White" by the French Process

In the manufacture of zinc white by the "French" process, metallic zinc is melted, evaporated and oxidized in the vapor state to zinc oxide with air. If the starting material is insufficiently pure, the zinc has to be purified by evaporation and condensation prior to the oxidation step. The particle size and the shape of the zinc oxide crystals can be controlled by the oxidation conditions. After combustion the ZnO is precipitated from the ZnO/air mixture in settling chambers, in which fractionation of the zinc oxide particles takes place according to their size.

Manufacture of zinc white (ZnO) by the "French" process:

- Melting of metallic Zn
- Evaporation of the Zn
- Vapor phase oxidation with air to ZnO

Zinc Oxide by the American Process

Manufacture of zinc oxide by the "American" process:

- Reduction of oxidic zinc compounds
- Subsequent direct oxidation to ZnO

The raw materials for the manufacture of zinc oxide by the "American" process are roasted ores, concentrates of zinc ores or more recently zinc hydroxide or zinc carbonate from the processing of scrap. These are reduced with coal and directly reoxidized to ZnO in the same reaction chamber. The purity of the zinc oxide is normally somewhat less than that from the "French" process. Grades of various purities can be produced by fractionation.

Zinc Oxide by the Wet-Chemical Process

Manufacture of zinc oxide wet-chemically:

- Precipitation of zinc hydroxide or carbonate
- Calcination

Zinc is precipitated from purified zinc salt solutions as the hydroxide, basic carbonate or carbonate, which is washed, filtered and dried. The pigment is obtained by subsequent calcination. Pigments suitable for particular applications can be obtained by varying the precipitation and calcination conditions. At low calcination temperatures particularly fine particles are produced, so-called transparent zinc oxide, which is really basic carbonate. Wet-chemically produced zinc oxides are particularly pure, since very pure zinc salt solutions can be produced (Sec. 5.9.2.3).

5.9.2.4.2 Applications

Applications for ZnO:

- Rubber industry
- Coloration of plastics
- Paints
- Photocopier paper
- Pharmaceutical and cosmetic products

Zinc oxide is mainly used in the rubber industry in which it serves as an activator in the vulcanization process. Its low hardness and high compatibility with polymers are important for its utilization in vulcanization products. In its utilization in paints, its tendency to form zinc soaps is important. These hinder the attack of coatings by fungus and improve the aging resistance (UV absorption). Further application fields for zinc oxide are copier papers, glues, pharmaceutical and cosmetic products and dental cements.

5.9.3 Colored Pigments

5.9.3.1 Iron Oxide Pigments

Iron oxide pigments:

group of colored pigments with largest turnover in quantity

Iron oxide pigments with an annual world production of ca. 0.6×10^6 t of synthetic pigments and ca. 0.1×10^6 t of natural products, represent by far the most important group of colored pigments (Sec. 5.9.1). This group consists of chemically very different coloring substances e.g. α-FeOOH (yellow), γ-FeOOH (yellow orange), Fe_3O_4 (black), γ-Fe_2O_3 (brown) and α-Fe_2O_3 (red), from which a wide range of brown hues can be produced by blending. The production of mixed oxides and the control of hue by adjustment of particle size are also important. Iron oxide pigments are the cheapest inorganic pigments and therefore have found widespread use, particularly in construction products and coating materials.

1985 production: ca. 0.7×10^6 t

5.9.3.1.1 Natural Iron Oxide Pigments

The naturally occurring mineral pigments contain various impurities depending upon their mining locality, but owe their color to iron compounds. Their poor tinting strength and less saturated color shade are due to their low purity and broader particle size distributions compared with synthetic pigments. The natural iron pigments are therefore, with a few exceptions, only used for less demanding applications. We have its fastness, which it has in common with synthetic products, to thank for our being able to admire the colored cave paintings from 15 000 B.C.

Natural reds such as Persian red (found at Hormuz in the Persian Gulf), Spanish red and burnt sienna contain ca. 50% iron oxide and are the most important group of natural mineral pigments. Their tinting strength increases with increasing α-Fe_2O_3 content, which can amount to 95% in the purest sorts. Their utilization is mostly limited to primers, cheap ships' paints and house paints (USA, Scandinavia).

Natural iron oxides:

- Natural reds such as Spanish red or Persian red
- Ochre
- Sienna
- Umber
- Micaceous iron oxide

Ochre, whose color is almost exclusively due to limonite, (α-FeOOH), is still mined in appreciable quantities in South Africa and France. It is being supplanted more and more by synthetic pigments, because of its low α-FeOOH content (20% Fe_2O_3 in French ochre, up to 55% in South African ochre and ca. 87% Fe_2O_3 in synthetic α-FeOOH) and the poorer constancy of its coloristic properties.

Sienna, consisting mainly of a mixture of finely crystalline goethite and colloidal silica (Fe_2O_3 content ca. 50%), has translucent (semi-transparent) properties

Table 5.9-10. Natural and Synthetic Iron Oxide Pigments.

Color	Natural	Synthetic	Hue shift with increasing particle size
yellow	limonite yellow ochre sienna	goethite, α-FeOOH lepidokrokite, γ-FeOOH	green-yellow → brown-yellow yellow → orange
red	hematite persian red, spanish red, red ochre, siderite (burnt) sienna (burnt)	hematite, α-Fe_2O_3	yellow-red → red-violet
brown	umber (untreated and burnt) siderite (burnt) sienna (burnt)	maghemite, γ-Fe_2O_3 mixtures of hematite, goethite and magnetite	bright brown → dark brown
black	magnetite	magnetite	green-brown → black blue-tinged black

and as a result is particularly important in artists' paints and other special fields. The manganese oxide (up to 1%) in sienna gives it its brown-yellow hue, which changes to bright red-brown upon burning.

Umber (Fe_2O_3 content between 45 and 70%, MnO_2 content between 5 and 20%), whose largest deposits are on Cyprus, is mainly exported from there in a calcined form.

Micaceous iron oxide, a flake-shaped α-Fe_2O_3 with a metallic gray color, is mainly mined in Carinthia (Austria). It is used in the top coat of corrosion protection coatings. Application fields are steel superstructures (stations, Eifel tower), petrochemical plant and ship building. The annual capacity is ca. 10×10^3 t/a.

5.9.3.1.2 Synthetic Iron Oxide Pigments

General Information

The stringent requirements of the paint and construction material industries, can almost only be satisfied by synthetic iron oxide pigments. Their chemical purity and their availability in various particle sizes enable variation in both hiding power and hue to be achieved. Iron oxide pigments attain their maximum scattering power at a particle size of ca. 0.2 μm, the hiding power being lower for both larger and smaller particles. Particles smaller than ca. 0.01 μm lose their scattering power completely. These iron oxide pigments produce transparent varnishes.

Hue adjustment with synthetic iron oxide pigments:

- Chemically through choice of materials
  - α-FeOOH yellow
  - α-Fe_2O_3 red
  - Fe_3O_4 black
  - mixtures brown
- Physically through particle size variation:
  - ca. 0.2 μm for optimum hiding power
  - < 0.01 μm for transparent iron oxide pigments

Table 5.9-11. Particle Size, Hue and Hiding Power of Iron Oxide Pigments.

	Particle diameter (μm) 0.001 → 0.01	0.1 → 1.0	10.0 → 100.0
	Transparent iron oxides	Hiding iron oxides	Micaceous iron oxides
hue:			
red iron oxide (α-Fe_2O_3)	yellow-red	yellow-red to violet	gray-brown (metallic effect)
yellow iron oxide (α-FeOOH)	lemon yellow	green-yellow to orange-yellow	–
black iron oxide (Fe_3O_4)	–	brown-black to blue-black	–
hiding power:	low	optimum	low

The production sites for synthetic iron oxide pigments are mainly in West Germany, the USA, Japan and Great Britain. The largest manufacturer is Bayer AG (FRG). The different production processes can, in principle, all be used to manufacture all the different iron oxides and thus all the different colors. The economics of the processes is the deciding factor.

The purity of raw materials, such as ferrous sulfate and ferrous chloride from steel industry pickling plants and scrap iron from the metal-working industries, determines their suitability for the manufacture of pigments.

Manufacture

Thermal Decomposition of Iron Compounds

Red iron oxide pigments are mainly obtained by roasting and calcining processes. α-Fe_2O_3 is obtained by oxidative calcination of all decomposable iron compounds. The decomposition of iron sulfate and α-

Manufacture of synthetic iron oxide pigments:

- Thermal decomposition of iron compounds
- Oxidative processes in aqueous media

FeOOH and the oxidation of Fe_3O_4 are industrially important.

Thermal decomposition of iron compounds to α-Fe_2O_3:

- Roasting of $FeSO_4$
- Calcination of α-FeOOH
- Oxidative calcination of Fe_3O_4
- Oxidative high temperature decomposition of $FeCl_2$

After prior dehydration to the monohydrate, ferrous sulfate heptahydrate is oxidatively roasted at temperatures above 650 °C, producing clear red-colored α-Fe_2O_3 pigments (Copperas Reds):

$$6\,FeSO_4 \cdot H_2O + 3/2\,O_2 \longrightarrow Fe_2O_3 + 2\,Fe_2(SO_4)_3 + 6\,H_2O$$

$$2\,Fe_2(SO_4)_3 \longrightarrow 2\,Fe_2O_3 + 6\,SO_3\,.$$

The sulfur dioxide formed can be processed to sulfuric acid. A disadvantage of this process is the contamination of the effluent with the soluble unroasted sulfate when the roasted charge is washed, due to the incomplete roasting of the sulfate.

Ferrous chloride, the salt available in the largest quantities from pickling plants,can also be thermally decomposed. The iron oxide formed does not, however, possess the usual pigment qualities and is normally pelletized and reused in steel manufacture. The $FeCl_2$ solutions from the Ruthner process are an exception. These are utilized in the hard- and soft-ferrite industries.

Oxidative processes used for the manufacture of:

- Yellow α-FeOOH pigments
- Black Fe_3O_4 pigments
- Red α-Fe_2O_3 pigments

In principle, all the α-FeOOH or Fe_3O_4 pigments produced by the precipitation process, the Penniman-Zoph process and the aniline process can be calcined to red α-Fe_2O_3 pigments. The oxidative calcination of Fe_3O_4 from the aniline process is the most important. A complete range of red tones can be obtained by rotary kiln calcination by varying the particle size through the choice of material and temperature conditions. Working up only requires a grinding step, a washing step being unnecessary due to the purity of the starting materials. The calcination of yellow pigments also leads to needle-shaped red pigments. Transparent α-Fe_2O_3 pigments result from the calcination of very fine α-FeOOH particles. These can also be obtained by oxidative decomposition of $Fe(CO)_5$.

Oxidation Processes in Aqueous Media

Oxidation processes in aqueous media take place in two steps:

FeOOH-yellow, Fe_3O_4-black and Fe_2O_3-red pigments are obtainable from ferrous sulfate with the correct choice of reaction conditions and the use of certain seed crystals. The manufacture of α-FeOOH-yellow pigments is the most important. The first reaction step is the production of the seed crystals so important to the properties of the final pigment: sodium hydroxide is added to a $FeSO_4$ solution precipitating basic ferrous sulfate, which is converted into a α-FeOOH seed crystal suspension by air oxidation.

- Seed crystal formation
- Pigment particle growth

$$2\,FeSO_4 + 4\,NaOH + 1/2\,O_2 \longrightarrow 2\,FeOOH + 2\,Na_2SO_4 + H_2O\,.$$

The crystalline α-FeOOH seed crystals are very small and can be used as a transparent pigment. Pigments with hiding properties are obtained by crystal growth on the α-FeOOH seed crystals, by feeding them with additional iron sulfate solution, sodium hydroxide and air, to the particle size corresponding to the hue required (precipitation process).

Pigment particle growth:

- By the precipitation process through the addition of $FeSO_4$
- By the Penniman-Zoph Process through the dissolution of scrap iron

Particles can also be grown by adding scrap iron and oxidizing it with air without the consumption of any other chemicals (Penniman-Zoph Process).

$$2\,FeSO_4 + 3\,H_2O + 1/2\,O_2 \longrightarrow 2\,FeOOH + 2\,H_2SO_4$$

$$Fe + H_2SO_4 \longrightarrow FeSO_4 + H_2\,.$$

The effluent from this process does not contain any additional salt, because the sulfuric acid formed reacts with the scrap to $FeSO_4$. The reaction is accelerated by maintaining the temperature at 70 to 90 °C with steam.

In the manufacture of black pigments by the precipitation process, iron(II) salt solutions are neutralized and oxidized with air at 90 to 100 °C to a Fe(II)/Fe(III) ratio of 0.5 (Fe_3O_4). Using a similar process and appropriate seed crystal formation conditions, small α-Fe_2O_3 seed crystals are formed upon complete oxidation, which after growth yield easily dispersible red α-Fe_2O_3 pigments of high tinting strength.

The long known reduction of nitrobenzene with metallic iron, previously used exclusively for the manufacture of aniline, has been so adapted by Laux (Bayer AG) that high tinting strength Fe_3O_4 pigment is

Aniline process:

- Reduction of nitrobenzene with scrap iron to α-FeOOH or Fe_3O_4 pigments and aniline
- Distillative separation of the aniline

formed instead of the previously unusable iron oxide-containing residue. This pigment can be directly used or calcined to strongly tinting red α-Fe_2O_3 pigments. The addition of aluminum chloride to the reaction mixture yields yellow α-FeOOH pigments.

The raw materials for this "aniline process" are ground and sieved largely degreased cast iron or wrought iron chips. Particle size, metallurgical state and addition rate determine the quality of the resulting pigment. The reaction is started by passing in steam and is exothermic:

$$C_6H_5NO_2 + 2\,Fe + 2\,H_2O \xrightarrow{AlCl_3} C_6H_5NH_2 + 2\,FeOOH$$

or

$$4\,C_6H_5NO_2 + 9\,Fe + 4\,H_2O \xrightarrow{FeCl_2} 4\,C_6H_5NH_2 + 3\,Fe_3O_4.$$

Pigment is formed in the absence of alkali and nitrobenzene is used as an oxidation agent. At the end of the reaction most of the aniline is separated off, the residue being distilled off by steam distillation. The pigment suspension is freed from excess iron chips by sieving and then washed free of salt in a series of sludge thickeners and vacuum rotary filters. The pigment paste is dried in belt driers, drum driers, spray driers or fluidized-bed driers and is then ground or micronized (pulverization in a jet mill).

Applications for Iron Oxide Pigments

Applications for iron oxide pigments:

- Construction materials coloring
- Paints and lacquers
- Coloring of plastics
- As a raw material for hard and soft ferrites.

There are considerable regional differences in the applications of iron oxide pigments. In Europe the construction materials sector is dominant, whereas in the USA it is the paint and coating materials sector.

Table 5.9-12. Applications of Iron Oxide Pigments in %.

	Western Europe	USA
construction materials colorant	64	33
paints and lacquers	24	46
coloring of plastics	3	13
miscellaneous	9	8

In the construction materials coloring sector chemical stability, light fastness and weathering stability are important. Concrete bricks, paving stones, roofing tiles, asbestos cement, mortar, bitumen and facade plaster are all currently colored with iron oxide pigments.

Natural and synthetic iron oxides not possessing pigment properties are used as raw materials in the production of hard or soft ferrites (Sec. 5.5.5.2 Electronic and Magnetic Ceramics) for radio, television and telephone technology, for adhesive magnets, for rotors in dynamos, for low-loss magnetic layers, for DC-motors, for transformer cores, electronic calculators and high-frequency furnaces. This amounts to an annual world production of more than 300×10^3 t.

α-FeOOH or α-Fe_2O_3 pigments with particles sizes less than 0.01 μm (transparent pigments) are mainly used in metallic paints and in transparent wood-protection lacquers (UV protection). The annual production is of minor importance compared with that of iron oxide pigments.

5.9.3.2 Chromium(III) Oxide Pigments

5.9.3.2.1 Manufacture

All the technically important production processes for chromium(III) oxide pigments start from chromates or dichromates, which are obtained by alkaline oxidative digestion of chrome ores.

Manufacturing Processes in Aqueous Media

Chromium hydroxide or chromium oxide hydrate are formed when chromates are used as oxidizing agents in organic chemistry. These can be calcined to α-Cr_2O_3 or directly used as a pigment. Such processes have declined in importance, due to the increasing use of catalytic oxidation in organic chemistry.

Molasses, saw-dust and sulfur are particularly cheap reducing agents for chromates or dichromates. Polysulfide-containing sulfur suspensions are most widely used producing chromium hydroxide which is difficult to filter together with sodium thiosulfate, which can be used for reducing further chromate. This is carried out by mixing the thiosulfate-containing sus-

Manufacture of Cr_2O_3 pigments:

- In aqueous media:

  reduction of chromate or dichromate solutions with organic waste products or polysulfides and subsequent calcination to α-Cr_2O_3

pension with additional sodium dichromate in a kneader. The resulting granulate is then calcined to α-Cr_2O_3 at 900 to 1100 °C.

Manufacturing Processes in the Solid State or Melt

- In the solid state or melt:
  reduction of dichromate with sulfur or ammonium salts

Sodium dichromate is reduced with sulfur by intensively mixing the components and then calcining in rotary kilns or rotary-plate furnaces at 800 to 1000 °C.

$$Na_2Cr_2O_7 + S \longrightarrow Cr_2O_3 + Na_2SO_4$$

The calcination product, which contains sodium sulfate in addition to chromium oxide and possibly unreacted chromate, is purified by mashing with water, filtering and washing.

The presence of fluxes such as NaCl or Na_2SO_4 during the decomposition of ammonium dichromate ensures that Cr_2O_3 pigments are formed in addition to nitrogen and water instead of extremely fine powders. The isolation of ammonium dichromate can be dispensed with by calcining mixtures of sodium dichromate and ammonium chloride or ammonium sulfate at 800 to 1000 °C.

$$Na_2Cr_2O_7 + 2NH_4Cl \longrightarrow Cr_2O_3 + 2NaCl + N_2 + 4H_2O$$

$$Na_2Cr_2O_7 + (NH_4)_2SO_4 \longrightarrow Cr_2O_3 + Na_2SO_4 + N_2 + 4H_2O$$

It is important that the intermediate formation of ammonium dichromate be as complete as possible to ensure a good yield and high quality pigments.

The use of stoichiometric quantities of ammonium chloride or slightly understoichiometric quantities of ammonium sulfate in the reaction mixture yields low sulfur chromium(III) oxide for the manufacture of chromium metal.

The Cr_2O_3 pigments produced by the different processes are further processed by washing, filtering, drying and grinding.

5.9.3.2.2 Properties and Applications of Chromium(III) Oxide

Cr_2O_3 was already in use in 1802 as a green stain in the manufacture of Sèvres porcelain, due to its extreme thermal and chemical stability. Some hue variation is possible by changing the pigment particle size (0.3 to 0.5 μm). As the particles become coarser the color changes from green to green-blue and the hiding power decreases. A yellow-green hue can be obtained by precipitating aluminum and titanium hydroxide onto the pigment particles and then calcining at 1200 °C.

A permanent green coloring of cement or lime-bonded construction materials can only be achieved using Cr_2O_3. Particularly high purity chromium oxide is used in the aluminothermic manufacture of metallic chromium.

Applications for Cr_2O_3:

- As a pigment for paints, plastics, construction materials, enamels
- For the aluminothermic manufacture of metallic chromium
- In refractory ceramics

The importance of green chromium oxide hydrate pigments (CrOOH, Guignet's green) has decreased considerably since the introduction of phthalocyanine pigments.

The worldwide production capacity for Cr_2O_3 is ca. 53×10^3 t/a. In Western Europe ca. 50% is used in paints and coatings, 19% in construction materials, 8% in plastics, 10% in enamels and ceramics and 13% for miscellaneous applications such as e.g. refractory ceramics and the manufacture of chromium metal.

5.9.3.3 Chromate and Molybdate Pigments

Chrome yellow, molybdate orange, molybdate red, chrome orange and chrome red belong to this pigment group and are characterized by their brilliant hues, high hiding power, high tinting strength, light fastness and solvent, migration and weathering stability.

Chromate and molybdate pigments:

- Chrome yellow
- Chrome orange
- Chrome red
- Molybdate orange
- Molybdate red

By far the most important products are chrome yellow and molybdate red. The chrome yellow pigments are $Pb(Cr,S)O_4$ mixed phases or pure $PbCrO_4$ and are produced by precipitation from the aqueous phase. The different hues from green-tinged yellow to bright yellow to golden yellow can be controlled by the crystal modification (orthorhombic or monoclinic) or by varying the lead sulfate content.

Most important properties:

- Brilliant hue
- High tinting strength
- High hiding power

Table 5.9-13. Properties of Chromate and Molybdate Pigments.

Pigment type	Formula	Color	Particle size in µm
chrome yellow			
orthorhombic	$Pb(Cr, S)O_4$	green-tinged yellow	0.2 to 0.6
monoclinic (needle-shaped)	$Pb(Cr, S)O_4$ $PbCrO_4$	bright yellow golden yellow	0.1 to 0.8 long length to width 3 to 1
molybdate orange (tetragonal)	$Pb(Cr, Mo, S)O_4$	orange to red	0.15 to 0.25
molybdate red (tetragonal)			0.5 to 0.75
chrome orange (monoclinic)	$PbCrO_4 \cdot PbO$	orange to red	0.1 to 1.0
chrome red			1 to 12

Most important pigments:

- Chrome yellow: $Pb(Cr,S)O_4$-mixed phase, or $PbCrO_4$
- Molybdate orange or red: $Pb(Cr,Mo,S)O_4$-mixed phases

The orthorhombic modification can be stabilized by mixed crystal formation with lead sulfate ($PbSO_4$ content ca. $35 \pm 15\%$), appropriate precipitation conditions and precipitation of protective colloids on it e.g. aluminum hydroxide. The stable monoclinic modification of the golden yellow $PbCrO_4$ is formed at precipitation temperatures above $50\,^\circ C$. Monoclinic lead chromate with bright to mid-yellow hues can be produced by mixed-phase formation with lead sulfate ($PbCr_xS_{1-x}O_4$ with up to 50% $PbSO_4$). The tetragonal $Pb(Cr,Mo,S)O_4$ pigments can be produced by incorporating tetragonal white $PbMoO_4$ resulting in the composition: 75 to 80% $PbCrO_4$, 10 to 15% $PbMoO_4$ and 3 to 10% $PbSO_4$ for molybdate orange or molybdate red pigments.

The chrome orange and red pigments are manufactured by precipitating the starting materials in the presence of alkali and then boiling. Raw materials for the precipitation reaction are sodium dichromate, sodium sulfate or sulfuric acid, lead salts such as lead nitrate, PbO suspensions and litharge. The latter is converted into lead salt solutions by dissolution in nitric or acetic acid. Their application-related properties are improved by posttreatment with phosphates, colorless oxides and hydroxides and amorphous SiO_2. This considerably improves their light fastness, temperature stability and chemical inertness.

The high tinting strength, light fastness, hiding power, brilliant hue and good price/performance ratio of these pigments has led to their widespread application in paints, printing inks and plastics. Chrome green or fast chrome green is formed by mixing chrome yellow with iron blue or phthalocyanine blue. Since the mixing ratio of the two components can be varied within a wide range, a wide range of green hues can be obtained.

Applications in:

- Paints
- Printing inks
- Plastics

The adoption of appropriate safety measures during manufacture, such as closed units and dust removal plants, ensures emission values for chromate and lead below their respective MAK values (0.1 mg/m^3 as CrO_3 and 0.1 mg/m^3 as Pb).

5.9.3.4 Mixed Metal Oxide Pigments, Stains and Oxides

Colored mixed metal oxide pigments result from the incorporation of color-giving transition metal ions into an oxidic host-lattice (see Table 5.9-14). Depending upon the particle size and properties of the chosen host, pigments (0.2 to 2 µm) or ceramic stains (up to ca. 10 µm) result, which in many cases are characterized by high thermal and chemical stability and thus are suitable for the coloring of vitreous enamels and ceramics.

Mixed metal oxide pigments, stains and oxides:

Obtained by the incorporation of color-giving transition metal ions into an oxidic host-lattice

Particle size:

- Pigments : 0.2 to 2 µm
- Stains and oxides: average size up to ca. 10 µm

The starting materials in the manufacture of mixed metal oxide pigments (as a rule carbonates, hydroxides, oxides and oxide-hydrates) are intimately mixed and heated, if necessary with added flux, at temperatures of 1000 to 1300 °C. The particle size can be controlled, and hence the coloristic properties influenced, by varying the calcination temperature and added flux. The 0.2 to 2 µm particles required for the pigment sector are obtained by intensive grinding of the calcined clinker in ball or sand mills.

Manufacture by solid state reactions at 1000 to 1400 °C

Stains and oxides, which have an average particle size up to ca. 10 µm and a broad particle size distribution, are in some cases manufactured at still higher temperatures, up to 1400 °C. The higher primary particle sizes are necessary because a more or less strong solubilization of the particles takes place during the firing of the enamel or ceramic frits. The component to be

Applications in the pigmentation of:

- Paints
- Plastics
- Vitreous enamels and ceramics

enameled or glazed is coated with a mixture of intensively mixed frits and colorants then fired at temperatures between 500 and 1200 °C.

The thermal stability of stains and oxides is particularly important in view of the high processing temperatures, so that oxidic host-lattices with spinel-, corundum-, rutile- and silicate-phases as well as zircon, phenacite, garnet and sphene structures are favored.

The stability of cadmium sulfide or sulfoselenide pigments for use as ceramic colorants, can be increased by coating with zirconium silicate.

Table 5.9-14. Mixed Metal Oxide Pigments.

Pigment group	Chemical composition	Crystal structure	Color
cobalt aluminate blue	$CoAl_2O_4$ $Co(Al, Cr)_2O_4$	spinel	red-tinged blue to green-tinged blue
spinel green	$(Co, Ni, Zn)_2O_4$	spinel	green
zinc ferrite	$ZnFe_2O_4$	spinel	bright to mid-brown
spinel brown	$Zn(Cr, Fe)_2O_4$	spinel	red-brown
spinel black	$Cu(Fe, Cr)_2O_4$	spinel	black
chrome antimony titanium buff	$(Ti, Cr, Sb)O_2$	rutile	orange-yellow
nickel antimony titanium yellow	$(Ti, Ni, Sb)O_2$	rutile	lemon yellow
manganese rutile brown	$(Ti, Mn, Sb)O_2$	rutile	bright to dark brown
pseudobrookite yellow	$Fe_2TiO_5 \cdot xTiO_2$	pseudobrookite	yellow-brown
iron manganese black	$(Fe, Mn)_2O_3$	bixbyite	black
iron manganese brown	$(Fe, Mn)_2O_3$	hematite	bright to red-brown
iron chrome brown	$(Fe, Cr)_2O_3$	corundum	black-brown
manganese blue	$BaSO_4 \cdot Ba_3(MnO_4)_2$	barite	green-tinged blue

The color range can be considerably widened by incorporating several color- giving or lattice-modifying ions (Table 5.9-15). The pure red-tinged blue pigment cobalt aluminum spinel can thus be changed into a green-tinged blue pigment by the additional incorporation of chromium. The further incorporation of nickel and titanium leads to an inverse titanium spinel with a brilliant green color. Brown zinc-iron-brown spinel is converted into a black pigment by replacing the zinc by copper.

The rutile lattice is particularly suitable for incorporating color-giving ions. Almost all of the transition metal ions can be accommodated in the rutile lattice.

The most important rutile mixed metal oxide pigments contain nickel and chromium (lemon yellow and ochre colors respectively). By incorporating manganese a brown pigment is obtained. Niobium or antimony is incorporated to compensate for charges lower than 4+. This incorporation principle is used to a considerable extent in stains and oxides for enamels and ceramics. In this case, the number of possible element combinations is further expanded, since host-lattices such as zircon, phenacite and sphene, which have lower refractive indices and hence are less interesting as pigments but have very high thermal stability, can be used. Ca. 10×10^3 t/a of ceramic stains are currently produced in the USA and Europe.

Table 5.9-15. Stains and Oxides (Selection).

Colorant	Chemical composition	Crystal structure	Temperature up to which colorant is stable, °C
blue:			
Co-blue	$CoAl_2O_4$	spinel	1450
Co-phenacite	$(Co, Zn)_2SiO_4$	phenacite	1450
zircon blue	$(Zr, V)SiO_4$	zircon	1350
green:			
chromium(III) oxide	Cr_2O_3	corundum	1350
Co-titanate	$(Co, Ni, Zn)_2TiO_4$	spinel	1350
Co-chromite	$Co(Cr, Al)_2O_4$	spinel	1400
yellow:			
zircon yellow	$(Zr, Pr)SiO_4$	zircon	1300
zirconium vanadium yellow	$(Zr, V)O_2$	baddeleyite	1400
Cd-yellow	$(Cd,Zn)S$ coated with $ZrSiO_4$	wurzite	1250
orange and red:			
pink red	$Ca(Sn, Cr)SiO_5$	sphene	1250
zircon iron pink	$(Zr,Fe)SiO_4$	zircon	1250
manganese pink	$(Al, Mn)_2O_3$	corundum	
Cd-red	$Cd(S,Se)$ coated with $ZrSiO_4$	wurzite	1250
violet:			
Co-olivine	Co_2SiO_4	olivine	1250
pink violet	$(Sn, Cr)O_2$	rutile	1250
brown:			
oxidic mixed phases	$Zn(Cr, Fe, Al)_2O_4$	spinel	1300
	$Fe(Fe, Cr)_2O_4$	spinel	1300
	$(Zn, Fe)Fe_2O_4$	spinel	1300
	$(Fe, Mn)_2O_3$	corundum	
black:			
oxidic mixed phases	$(Cu,Co,Ni)(Cr,Fe,Mn)_2O_4$	spinel	1300
	$(Fe, Co)(Fe, Cr)_2O_4$	spinel	1300
gray:			
tin antimony gray	$(Sn, Sb)O_2$	rutile	1250
zircon gray	$(Zr, Co, Ni)SiO_4$	zircon	1250

5.9.3.5 Cadmium Pigments

Most important properties of cadmium pigments:

- High tinting strength
- High brightness
- Thermal stability during the pigmentation of plastics

Cadmium pigments based on yellow CdS or its mixed phases are among the most brilliant inorganic colored pigments. As a result of their high thermal stability and their not bleeding in plastics, they are mainly incorporated into plastics with high processing temperatures such as styrene-polymers, poly(ethylene), poly(propylene) and poly(carbonate), in which organic pigments with comparable brilliance suffer from thermal degradation. They are also used in especially brilliant paints. Their UV absorption protects the organic matrix (the binder) from UV degradation.

Variation of the hue of yellow cadmium sulfide pigments by mixed-crystal formation:

- With ZnS, greenish yellow
- With HgS or CdSe, orange to bordeaux

In yellow CdS, which is the basis of the cadmium pigments, both the cadmium and the anion can within certain limits be replaced by ions with similar radii. The incorporation of zinc has practical importance in the manufacture of the greenish-yellow pigments (Cd,Zn)S and that of mercury in (Hg,Cd)S and selenium in Cd(S,Se) for the production of a color range from orange to bordeaux red.

Manufacture of cadmium pigments by:

- Precipitation with $Na_2S(Se)$ and calcination
- Calcination of $CdCO_3$, a zinc salt and S (yellow pigments) or calcination of $CdCO_3$, S and Se (red pigments)

Cadmium yellow pigments (CdS) are manufactured by adding sodium sulfide to dissolved metallic cadmium or a cadmium salt solution to precipitate the fine particle raw colorant, which does not have pigment properties. Calcination at temperatures between 600 and 700 °C induces particle growth to ca. 0.2 μm at which optimum tinting strength and hiding power are attained.

An alternative frequently used process is the calcination of a mixture of $CdCO_3$, zinc salt and sulfur (for the manufacture of yellow pigments), or with selenium in the place of zinc salt in the case of red pigments.

Optimum particle size:

0.2 to 0.4 μm

For the manufacture of orange to deep red cadmium sulfoselenide using the precipitation process, selenium is dissolved in the sodium sulfide solution in an amount consistent with the color required. Particularly brilliant hues are obtained, if the precipitated raw colorant undergoes brief heat treatment in molten salt at ca. 800 °C. The conditions can be controlled to yield the optimum particle size of ca. 0.2 to 0.4 μm.

LD_{50} of cadmium pigments:

> 10 g/kg rat

Cadmium pigments are among the least soluble cadmium compounds and therefore their toxicity should not be compared with that of soluble cadmium compounds. The LD_{50}-values are greater than 10 g/kg rat for oral ingestion. The cadmium solubility of commercial pigments in simulated stomach acid (0.1% hydrochloric acid) is less than 0.1% and that of cadmium pigments incorporated in paints and plastics orders of magnitude lower, so that cadmium pigment-colored plastics can be used in the manufacture of household apparatus and packaging materials without reservation.

5.9.3.6 Ferrocyanate Pigments

Ferrocyanate pigments (formerly known as Prussian blue, Milori blue or Turnbull's blue) have the composition

$$Me^{+}[Fe^{2+}Fe^{3+}(CN)_6] \cdot xH_2O \text{ with } Me(Na^{+},K^{+},NH_4^{+})$$

They are precipitated as a white intermediate when ferrous salts react with complex iron(II)-cyanides. This is then converted to iron blue by oxidation with chlorates or dichromates.

Manufacture of ferrocyanate pigments $Me^{+}[Fe^{2+}Fe^{3+}(CN)_6]$:

- Precipitation of Fe(II)-salts with hexacyanoferrate(II)
- Oxidation with chlorates or dichromates

$$Fe^{2+} + Me_4^{+}[Fe^{2+}(CN)_6] \cdot (x+y)H_2O \longrightarrow$$
$$Me_2^{+}[Fe^{2+}Fe^{2+}(CN)_6] \cdot xH_2O + 2Me^{+} + yH_2O$$

$$Me_2^{+}[Fe^{2+}Fe^{2+}(CN)_6 \cdot xH_2O \xrightarrow{\text{Oxidation}}$$
$$Me^{+}[Fe^{2+}Fe^{3+}(CN)_6] \cdot xH_2O + Me^{+}$$

The white intermediate is either precipitated by adding the hexacyanoferrate(II) solution to the iron(II)-solution or the simultaneous addition of the dissolved components to an agitator vessel at pH values between 2 and 6. The particle size can be influenced by the choice of temperature (20 to 60 °C) and the concentrations of the initial solutions. The white intermediate is aged by boiling and then oxidized to the ferrocyanate pigment which is filtered off, gently dried and ground.

Dark ferrocyanate types have particle sizes from 0.01 to 0.05 µm and have jet mass tones, whereas the bright-blue types have particle sizes between 0.05 and 0.2 µm. Ferrocyanate pigments have extremely high tinting strengths but are difficult to disperse, due to their

Applications for ferrocyanate pigments:

- Car paint
- Printing ink
- Colored paper

tendency to agglomerate. They are stable over short periods at temperatures up to 180 °C and hence are usable in stoving varnishes. They are mainly used in printing inks (particularly gravure), in the coloring of fungicides (ferrocyanate is used e.g. in vineyards as a leaf fertilizer), in paints (car paints) and in colored paper manufacture.

Mixtures of ferrocyanate pigments with chrome yellow and zinc yellow are known as chrome green and zinc green respectively and are used in paints and printing inks.

LD_{50} of ferrocyanate pigments: 8 to 10 g/kg rat

The DL_{50} of ferrocyanate pigments is 8 to 10 g/kg rat.

5.9.3.7 Ultramarine Pigments

Ultramarine pigments: blue, green, red, violet

Ultramarine pigments are sodium aluminum silicates with the composition $Na_8[Al_6Si_6O_{24}] \cdot S_x$ (Na-rich) or $Na_{8-y}[Al_{6-y}Si_{6+y}O_{24}] \cdot S_x$ (Si-rich) which can be blue, green, red or violet, depending upon the composition of the chromogenic S_x-group.

Manufacture of ultramarine pigments:

- Partial calcination of kaolinite to metakaolinite
- Formation of a zeolite structure by calcination with Na_2CO_3 in a SO_2 atmosphere
- Reaction with Na_2S_x to white ultramarine
- Slow oxidation to ultramarine

The multistage manufacture of ultramarine pigments begins with the careful partial calcination of the mineral kaolinite (China Clay) to metakaolinite:

$$\underset{\text{(kaolinite)}}{Al_4(OH)_8Si_4O_{10}} \xrightarrow[-xH_2O]{500 \text{ to } 600\,^\circ C} \underset{\text{(metakaolinite)}}{Al_4(OH)_{8-2y}O_ySi_4O_{10}}$$

A wide mesh three-dimensional zeolite structure is built up out of metakaolinite and the other raw materials (Na_2SO_3, S, reducing agent) in a very complex firing process in a reducing atmosphere (SO_2):

$$1.5\,Al_4(OH)_{8-2y}OySi_4O_{10} + 3\,Na_2CO_3 \xrightarrow[48 \text{ to } 80\text{ h}]{750 \text{ to } 800\,^\circ C} \underset{\text{(silica deficient)}}{Na_6[Al_6Si_6O_{24}]} + 3\,CO_2 + (6-1.5y)\,H_2O$$

The reaction of this with the intermediately formed Na_2S_x yields white prae-ultramarine:

$$Na_6[Al_6Si_6O_{24}] + Na_2S_x \rightarrow \underset{\text{(white prae-ultramarine)}}{Na_6[Al_6Si_6O_{24}] \cdot Na_2S_x}$$

Blue ultramarine (with the chromogenic group S_3^-, herein generally represented as S_x) is formed by slow (up to 20 days!) oxidation of the sulfur in the cavities and channels in the zeolite cage:

$$2\ Na_6[Al_6Si_6O_{24}] \cdot Na_2S_x + \xrightarrow[-SO_4^{2-}]{O_2,\ SO_2} 2\ Na_8[Al_6Si_6O_{24}] \cdot S_x \quad \text{(Na-rich)}$$

Ultramarine blue pigments rich in silica are obtained with the same reaction sequence. Green ultramarine requires 1/7 to 1/8 of the sulfur required for blue pigment, but 2.5 times the amount of reducing agent. The firing process occurs at temperatures between 900 and 1000 °C in a reducing atmosphere and the reaction and cooling times (20 to 25 h) are much shorter than for ultramarine blue rich in sodium. The color-giving species are S_2^- and S_3^- in cavities in the cage-structure. Oxidation of ultramarine blue or green with air at 130 to 280 °C in the presence of ammonium chloride results in ultramarine violet with the chromophore S_4^-. Ultramarine red is obtained from ultramarine blue or green by air oxidation at 100 to 150 °C in the presence of HCl and Cl_2.

Variation in manufacturing conditions produces:

- Blue
- Green
- Violet
- Red

} ultramarine

5.9.4 Corrosion Protection Pigments

Paints pigmented with active corrosion inhibiting pigments are applied as coatings or primers to metallic surfaces. Active corrosion protection pigments should inhibit the corrosion processes. This can occur by a multiplicity of mechanisms, which explains the wide range of corrosion protection pigments (Tables 5.9-16 and 5.9-17).

Mode of operation of active corrosion inhibiting pigments:

Inhibition of corrosion processes by

- Passivation
- Cathodic protection
- Formation of protective layers
- Formation of metal soaps
- pH change
- Neutralization of corrosion promoters

As a rule, corrosion protection pigments possess several of the cited inhibition properties, so that the corrosion protection of a particular pigment is a combination of mechanisms. Optimum efficiency requires combination with a suitable binder. For example the utilization of a corrosion protection pigment which forms metal soaps only makes sense, if the binder is able to form soaps.

Table 5.9-16. Active Corrosion Protection Pigments.

Classification	Material(s)	Formula	Manufacture	World production quantities t/a
lead-containing	red lead	Pb_3O_4	oxidation of PbO	50 000
	lead cyanamide/lead phosphate/lead phosphite	$PbCN_2/Pb_3(PO_4)_2 \cdot 3\,H_2O$ $2\,PbO \cdot PbHPO_3 \cdot 0.5H_2O$	precipitation	3 000
	basic lead silicochromate	$PbSiO_3 \cdot 3\,PbO/SiO_2$ $PbCrO_4 \cdot PbO/SiO_2$	precipitation on SiO_2	6 000
	calcium plumbate	Ca_2PbO_4	PbO + CaO	3 000
chromate-containing	zinc yellow	$3ZnCrO_4 \cdot K_2CrO_4 \cdot Zn(OH)_2 \cdot 2H_2O$	precipitation with ZnO	10 000
	basic zinc chromate	$ZnCrO_4 \cdot 4Zn(OH)_2$	$ZnO + CrO_3$	2 000
	strontium chromate	$SrCrO_4$	precipitation	2 000
phosphate-containing, borate-containing	zinc phosphate	$Zn_3(PO_4)_2 \cdot 2$ to $4\,H_2O$	precipitation	15 000
	chromium phosphate	$CrPO_4 \cdot 3H_2O$	precipitation	<1 000
	barium borate	$BaO \cdot B_2O_3 \cdot H_2O$	precipitation	2 000
metallic	zinc dust	Zn	sublimation	70 000
	lead powder	Pb	melt spraying	1 000
oxide, mixed oxide	zinc oxide	ZnO	oxidation of Zn	7 000
	calcium ferrite/ zinc ferrite	$CaO \cdot xFe_2O_3/$ $ZnO \cdot xFe_2O_3$	calcination	non applicable
inhibitor-containing	organic compound	Zn-salt of nitroisophthalic acid	precipitation	100

Table 5.9-17. Corrosion Protection Processes with Different Pigment Types.

Type of process	Mode of protection	Typical examples
electrochemical	passivation by development of an appropriate potential	chromate
	cathodic protection	zinc dust
chemical	formation of protective layers on metal surfaces	chromate, phosphate
	formation of metal soaps with binder fatty acids	ZnO, Pb_3O_4
	pH shift in the passivity range of iron (no Fe-corrosion at pH 11 to 12)	Ca_2PbO_4, ferrites such as $CaO \cdot xFe_2O_3$, $ZnO \cdot xFe_2O_3$
	neutralization of corrosion promoters (Cl^-,SO_4^{2-})	Pb_3O_4, alkaline reacting corrosion protection pigments
physical	coverage of undercoat	micaceous iron oxide
	improvement in physical properties of the coating layer (improved flexibility and adhesion, reduced water permeability)	Pb_3O_4, $Pb(CN)_2$ ferrites $CaO \cdot xFe_2O_3$, $ZnO \cdot xFe_2O_3$

The use of the optimum pigment concentration, which can vary considerably, is equally important. Thus zinc salts of nitroisophthalic acid are used at a concentration of ca. 0.5 wt% in paints, whereas ca. 90% is necessary in the case of zinc-dust pigments, to guarantee the zinc-iron contact necessary for cathodic protection.

Corrosion protection pigments normally also contain inactive corrosion protection pigments as additives (TiO_2, Fe_2O_3, fillers, $CaCO_3$ etc.), which support the operation of the active pigments.

Toxicological reservations in connection with several corrosion protection pigments has led to the increased use of nontoxic pigments and to new developments such as Ca- or Zn-ferrite pigments.

5.9.5 Lustrous Pigments

The term lustrous pigments includes metallic, nacreous and interference pigments. The luster effect is due to directed reflection on planar-shaped and ordered pigment particles.

The term lustrous pigments includes:

- Metallic pigments
- Nacreous pigments
- Interference pigments

5.9.5.1 Metallic Pigments

Metallic pigments:

flakes or lamella-shaped metal particles of

- Aluminum
- Copper
- Bronze

Metallic pigments consist of high luster flakes or lamella-shaped particles of soft ductile metals such as aluminum, copper or bronze. They are produced by cold forming of granules or pieces of sheet, foil or wire in specially constructed ball mills. Cold welding is hindered by the addition of fatty acids or alkylamines as lubricants which coat the freshly formed surfaces. Particles from a few milli- to micrometers in diameter with aspect (thickness : diameter) ratios of 1 : 50 to 1 : 250 can be produced, depending upon the grinding technique and grinding time. The metallic pigments are used in paints, printing inks and plastics, either alone (mass tone) or colored with transparent colorants.

5.9.5.2 Nacreous Pigments

Manufacture of synthetic nacreous luster pigments by precipitating TiO_2 on mica platelets

Pearlescence comes about by multiple partial reflection of every incident ray of light on several platelet-shaped pigment particles at different depths in the coating.

The naturally occurring nacreous pigment fish scales (guanine) is no longer available in sufficient quantities. Basic lead carbonate serves as a synthetic substitute, but is suspect on toxicological grounds. Bismuth oxychloride nacreous pigments are expensive and are being supplanted by cheaper products. Synthetic titanium dioxide nacreous pigments produced by the precipitation of hydrous titanium dioxide onto colorless mica fractions of uniform platelet size and subsequent calcination are very important. The uncoated mica platelets should be at most 200 to 500 μm thick and the double-sided titanium dioxide coating is ca. 50 nm thick. Normally the particle diameter is 10 to 30 μm.

5.9.5.3 Interference Pigments

Interference pigments are nacreous pigments, which fulfill the interference criteria and hence exhibit color effects

If the interference condition is fulfilled by the nacreous pigments produced by the precipitation process i.e. the product of oxide coating thickness and refractive index lies in the 200 to 500 nm range, then interference colors are observed. Precise maintenance of coating thickness is important for a uniform color effect. The color effects possible with interference pigments (complementary colors in transmission and reflection) are utilized in the manufacture of cosmetics, artificial mother of pearl and costume jewelry. An example of an industrial application is the manufacture of infrared reflecting light-cupolas from acrylic glass. The worldwide annual consumption of nacreous and interference pigments is ca. 3.5 to 4.0×10^3 t.

5.9.6 Luminescent Pigments

Luminescent pigments are solid fine particulate (1 to 5 µm) luminescent materials, which reemit absorbed energy as light at lower energy than the energy absorbed (mainly in the visible spectral region) either during (fluorescence), or subsequent to (phosphorescence) excitation. Depending upon the structure and composition of the luminescent material, the stored energy can be reemitted within fractions of a second or up to several hours after excitation.

Luminescent pigments:

fine particulate activated and sensitized solids, producing luminescence upon excitation

Activators (e.g. transition metal and rare earth ions) which act as luminescent centers optionally with sensitizers (e.g. Sb^{3+}, Pb^{2+}, Ce^{3+}, Eu^{3+}, Tb^{3+}) are incorporated in concentrations of 10^{-2} to 10^{-4} g/mol into a crystal lattice generally consisting of colorless oxides, oxysulfides, sulfides, silicates, phosphates, borates or halides of zinc, alkaline earth or rare earth metals. The impurity level must be well below the activator concentration, since impurities act as quenching centres and considerably reduce the emission yield. Emission decay time and color are mainly dependent upon the choice of activators and the crystal field influences of the matrix.

Luminescent pigments are manufactured by the repeated calcination/sintering of homogeneously mixed raw material at 1000 to 1600 °C under reducing or oxidizing conditions,depending upon the activator(s) and the crystal lattice, interspersed with gentle grinding. They are coated in thin layers from suspensions which contain an adhesive by e.g. precipitation. Table 5.9-18 gives several selected examples.

Applications in:

- Cathode ray tubes
- Fluorescent lights
- Television screens
- Radar screens
- Flying spot scanners
- Image intensifiers
- X-ray screens

Table 5.9-18. Selected Examples of Luminescent Pigments and their Fields of Application.

Activator	Phosphor composition		Main emission maxima, nm	Color	Fields of application
Mn^{2+}	zinc orthosilicate	Zn_2SiO_4	525	green	oscilloscopes
Mn^{2+}/Sb^{3+}	Ca-halogenphosphates	$Ca_5(PO_4)_3(Cl,F)$	480 and 580	blue and yellow-orange	fluorescent lights
Mn^{4+}	Mg-fluorogermanate	$Mg_2GeO_4 \cdot 1.5\ MgO \cdot 0.5\ MgF_2$	710	red	high pressure mercury lamps
Sn^{2+}	(Sr,Mg)-orthophosphate	$(Sr,Mg)_3(PO_4)_2$	630	rose-red	fluorescent lights & high pressure mercury lamps
Ce^{3+}	yttrium aluminum garnet	$Y_3Al_5O_{12}$	550	yellow	flying spot scanners
Eu^{2+}	barium fluorobromochloride	$BaF(Br,Cl)$	440	blue	X-ray screens
Eu^{3+}	yttrium oxide	Y_2O_3	625	red	fluorescent lights and television tubes
Tb^{3+}	yttrium oxysulfide	Y_2O_2S	540	green	television tubes
			440	blue	television tubes, radar tubes & X-ray screens
Tb^{3+}	gadolinium oxysulfide	Gd_2O_2S	540	green	X-ray screens
Ag^+/Cl^-, Cu^+/Cl^-	zinc sulfide	ZnS	525	green	radar tubes
Zn^{2+}	zinc oxide	ZnO	505	green	flying spot scanners
none	Ca-wolframate	$CaWO_4$	415	blue-violet	fluorescent lights & X-ray screens

5.9.7 Magnetic Pigments

5.9.7.1 General Information and Properties

Magnetic information storage on tapes, drums and rigid and floppy discs is based on the magnetization of miniscule solids, the so-called magnetic pigments, dispersed in organic binders. These are needle-shaped particles 0.03 to 0.1 μm in diameter with aspect (length/width) ratios of 5:1 to 10:1. The products used are based on the ferrimagnetic compounds γ-Fe_2O_3 and Fe_3O_4 or ferromagnetic substances such as CrO_2 and metallic iron. The composition, shape and size of the needles are crucial for obtaining the desired magnetic properties in the tape. Platelet-shaped barium ferrite is a recent development in this field (see Table 5.9-19).

Important properties of magnetic pigments:

- Ferri- of ferromagnetic
- Needle-shaped
- Aspect ratios of 5:1 to 10:1
- Coercive forces between 300 and 1500 Oe
- Tape remanences between 1200 and 3200 Gauß

The coercive force I_{HC} and remanence I_R achievable in the storage medium are important properties*).

The coercive force I_{HC} represents the resistance of the tape to re- and demagnetization. An oppositely directed field strength of the size of I_{HC} is necessary to demagnetize an already magnetized object.

The coercive forces of magnetic pigments lie between those of weak and hard magnetic materials. Values between 300 and 1500 Oe are desirable. These are sufficiently high to hinder the interference of a recorded signal by small stray fields, but sufficiently low to enable easy erasure and rerecording. Pigments for magnetic identification systems may have coercive forces well above 1500 Oe to prevent an involuntary erasure of the stored information.

The higher the residual magnetization (I_R, remanence) in the tape after switching off the magnetizing field, the higher will be the intensity of the signal reproduced. The remanence of a tape depends upon the degree of filling as well as the choice of pigment. Depending on the tape quality, remanences of 1500 to 2100 Gauß are attained with oxidic magnetic materials and 2600 to 3200 Gauß with metallic iron pigments.

Composition of widely used magnetic pigments:

- γ-Fe_2O_3
- Fe_3O_4/γ-Fe_2O_3-mixed phases
- γ-Fe_2O_3 Co-modified
- Fe_3O_4/γ-Fe_2O_3-mixed phases Co-modified
- CrO_2
- Fe metallic pigments, optionally alloyed
- Barium ferrite

*) The internationally used units Oersted, Gauß and emu/g (electromagnetic units) are converted to SI-units as follows:
$1\ Oe = 10^3/(4\nu)\ [Am^{-1}]$; $1\ Gauß = 10^{-4}\ T$ (Tesla);
$1\ emu/g = 12.56 \times 10^4\ [Tcm^3g^{-1}]$

Operating points for tape recorders are set on the basis of the coercivity of the tapes:

I. normal (γ-Fe_2O_3)
II. CrO_2
III. double-layer tapes CrO_2/γ-Fe_2O_3
IV. metal

To obtain optimum reproduction, the apparatus has to be adapted to the different coercive forces of these materials. This is accomplished by setting operating points (OP):

OP I, normal (γ-Fe_2O_3);
OP II, chromium dioxide (and higher coercivity iron oxides);
OP III, ferrochrome (doubly coated tapes with CrO_2/iron oxide; largely historical);
OP IV, metal (iron metal or alloys).

5.9.7.2 Manufacture of Magnetic Pigments

Manufacture of CrO_2:

1. $CrO_3 \xrightarrow[200 \text{ to } 300 \text{ bar}]{250\,^\circ C} CrO_2 + 0.5O_2$
2. $2CrOOH + CrO_3 \xrightarrow[50 \text{ to } 800 \text{ bar}]{300 \text{ to } 450\,^\circ C} 3CrO_2 + H_2O$

Very complicated processes are currently necessary for the production of needle-shaped magnetic pigment particles. In relation to needle-shaped pigments, chromium dioxide with a rutile structure is favored, since it crystallizes as needles under hydrothermal production conditions (decomposition of CrO_3 in an oxygen atmosphere or reaction of CrOOH with CrO_3).

Manufacture of the iron-based magnetic pigments γ-Fe_2O_3, Fe_3O_4 and Fe from needle-shaped α-FeOOH or γ-FeOOH

The manufacture of iron oxide-based needle-shaped magnetic pigments is much more complicated as γ-Fe_2O_3, Fe_3O_4 and metallic iron (all cubic) do not crystallize in a needle shape. It is therefore necessary to start from needle-shaped precursors. Nonmagnetic α-FeOOH (goethite) or γ-FeOOH (lepidokrokite) produced by precipitation and oxidation from Fe(II)-salt solutions are suitable precursors which can be converted into ferrimagnetic magnetite by dehydration and reduction and into ferromagnetic metallic iron by complete reduction.

A careful oxidation of magnetite to the ferrimagnetic γ-Fe_2O_3, under well defined conditions, is necessary to synthesize γ-Fe_2O_3 in a spinel structure (maghemite):

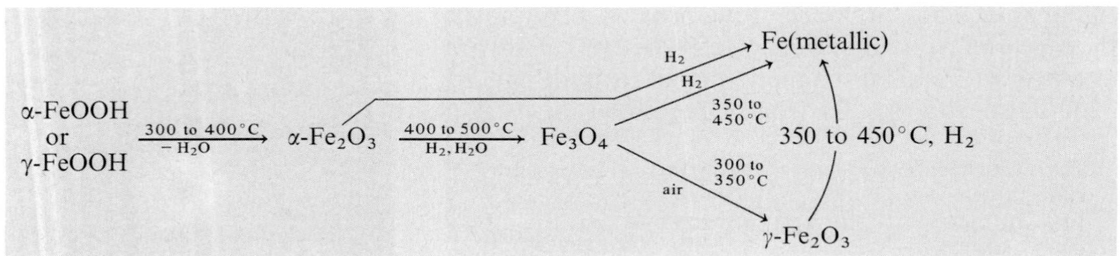

It is important that the initial needle shape be retained as completely as possible during the synthesis by, for example, using protective layers in the sequence of processes in which it is transformed both chemically and structurally several times.

Coercivity increase for iron oxide magnetic pigments by Co-doping or Co-ferrite coating

Recently more and more use has been made of the coercivity-increasing effect of cobalt in the magnetic iron oxides. The cobalt is either incorporated homogeneously into the iron oxide lattice (bulk doping) or the needles of the finished γ-Fe_2O_3 or Fe_3O_4 magnetic pigment are epitaxially coated with a layer of cobalt ferrite. Coercive forces of 380 to 750 Oe require ca. 1 to 4% Co. Such products have comparable properties to CrO_2-pigments (see Table 5.9-19).

Table 5.9-19. Properties of Magnetic Pigments.

Magnetic pigment	Coercivity I_{HC} in Oersted	Saturation magnetization π_s in emu/g theoretical	practical	Typical application field
Fe_3O_4 (isometric)	60 to 320	84	82	magnetic toner for photocopiers, magnetic printing inks
γ-Fe_2O_3	330 to 370	74	65 to 73	audio cassettes OP I, amateur & professional tape recorders, computer tapes, floppy discs
Co-modified γ-Fe_2O_3-pigments	370 to 450	74	76	audio cassettes, OP I
	550 to 650			audio cassettes, OP II
	650 to 750			video tapes (amateur recorders)
berthollide Fe_3O_4/γ-Fe_2O_3	400 to 440	74 to 84	80	audio cassettes, OP I
Co-modified Fe_3O_4/γ-Fe_2O_3)	550 to 650		80	audio cassettes, OP II
	650 to 750		80	video tapes (amateur recorder)
	900 to 1600		54	magnetic identification systems
CrO_2	450 to 650	75	75	audio cassette, OP II computer tapes
	600 to 700	75	75	video tapes (amateur recorder)
Fe (metal)	1100 to 1200	178	120 to 140	audio cassettes, OP IV
	1400 to 1500	178	120 to 140	micro cassettes, 8 mm video tapes, integrated video camera recorder, digital audio
Ba-ferrite	600 to 1400	68	55	floppy discs
	2700 to 6000		60	magnetic identification systems

The world production capacity for magnetic pigments was ca. 65×10^3 t in 1985, magnetic iron oxide pigments accounting for over 90%. Metallic iron pigments are still being introduced into the market.

References for Chapter 5.9: Inorganic Pigments

Pigments:

Colored pigments:

Winnacker-Küchler: *Chemische Technologie*. 4th Ed., Vol. 3, p. 349 (1983).

Ullmanns Encyklopädie der technischen Chemie. 4th Ed., Vol. 18, pp 545–645.

G. Winter; *Fortschr. Miner.* **57(2),** pp 172–202 (1979).

F. Hund; *Angew. Chem.* **93**, pp 763–771 (1981).

P. M. Koxholt, *Industrial Minerals,* pp 22–25 (June 1986).

White pigments:

J. Barksdale: *Titanium*. 2nd Ed., New York, Ronald Press Co. (1966).

M. Cremer, *Industrial Minerals,* Pigments & Extenders Supplement pp 16–25 (1985).

P. Woditsch, *defazet* **33**, Nr. 6/7, pp 191–194 (1979).

G. Clarke: *Industrial Minerals*. pp 47–55 (June 1986).

Special pigments:

H. Wienand und W. Ostertag; *Modern Paint and Coatings* pp 38–48 (Nov. 1984).

R. Fischer and P. Kleinschmitt, *Keramische Zeitschrift* **38(7),** pp 365–367 (1986).

Magnetic pigments:

G. Bates in *Ferromagnetic Materials,* E. P. Wohlfahrt (Ed.), Vol. II, Chapter 7, Amsterdam, North Holland Pub. Co. (1980).

6 Nuclear Fuel Cycle

6.1 Economic Importance of Nuclear Energy

The worldwide consumption of energy has increased strongly in recent decades, particularly in the 1960's, and will increase steadily in the future (see Table 6-1), but with considerable regional variations.

Table 6-1. Primary Energy Consumption 1960–2000 (in 10^9 t SKE*).

	1960	1970	1980	1985	2000 (est.)
World	4.0	6.4	8.5	9.4	11.2
North America	1.6	2.4	2.6	2.7	2.5
Western Europe	0.8	1.3	1.5	1.5	1.3

* 1 t SKE $\hat{=}$ energy content of 1 t of hard coal $\hat{=}$ 29.3 GJ $\hat{=}$ 27.8×10^6 BThU

1985 primary energy consumption:

- Worldwide 9.4×10^9 t SKE
- North America 2.7×10^9 t SKE
- Western Europe 1.5×10^9 t SKE

Above average growth is predicted for Communist Countries and the Third World

The energy consumption in the industrialized countries in North America and Western Europe is decreasing slightly at the moment due to intensive energy saving efforts. On the other hand the annual growth of energy consumption in the Communist Countries and the Third World is expected to be between 10 and 80% (with strong regional differences). This is due both to the considerable shortfall in supply in these countries and the steadily increasing world population (1985: 4.5×10^9 people, 2000: ca. 6.4×10^9).

At the moment, fossil sources (crude oil, coal and natural gas) are mainly used to supply the world's primary energy needs. The proportion of nonfossil energy in 1983 was only 4.2% (see marginal notes). Conservation of these resources, particularly crude oil and natural gas, is, however, urgently required since these are only available in limited quantities and are better used as raw materials in the chemical industry (ca. 90% of organic chemicals are manufactured from crude oil or natural gas, 10% from coal).

1983 primary energy source breakdown in the World, the USA and West Germany (in %):

	World	West Germany	USA
crude oil	41.5	43.1	39.4
coal	32.2	32.3	24.0
natural gas	22.1	15.3	23.1
nuclear energy	1.2	5.9	5.2
water } various }	3.0	3.4	3.4 / 4.9

Combustion of fossil fuels leads to a further increase in the CO_2-content of the atmosphere and thereby to global climatic changes ("greenhouse effect")

The World coal reserves will last considerably longer (300 years or more) than those of crude oil and natural gas (ca. 40 to 70 years) and are also more uniformly distributed in the Earth's crust. However, its more intensive utilization is hardly ecologically acceptable due to the resulting increased release of carbon dioxide, sulfur dioxide and nitrogen oxides into the environment. In particular, it is feared that the release of carbon dioxide into the atmosphere (ca. 23×10^9 t worldwide in 1980) will lead to major long term climatic changes.

Alternative energy sources such as solar energy, wind power, wave power and nuclear energy are therefore becoming increasingly important. The first three are in principle available in unlimited quantities but their cost is still too high. Nuclear energy on the other hand with a higher energy density can be utilized worldwide at competitive prices.

The use of nuclear energy can not be dispensed within the next decades

Nuclear energy is thus being increasingly utilized particularly for electricity production (ca. 20 to 25% of the primary energy is utilized for electricity generation). Table 6-2 gives a survey of the contribution of nuclear energy to electricity production in selected industrialized countries.

Table 6-2. Contribution of Nuclear Energy to Electricity Production in Selected Industrialized Countries in 1986.

	France	Belgium	Sweden	West Germany	Japan	Great Britain	USA	Canada
percentage of electricity produced from nuclear energy	69.8	67.0	50.3	36.0	24.7	18.4	16.6	14.7
total electricity production from nuclear energy (10^9 kWh)	224	35	59	126	153	61	405	64

Worldwide, nuclear energy accounts for ca. 15% of total electricity production (7000×10^9 kWh in 1986). This percentage will increase further in the future despite the accidents at Harrisburg (USA) and Chernobyl (USSR). The acceptance of nuclear energy has hardly been changed by the accident in 1986 at Chernobyl, the most serious accident in the history of civil atomic energy (Sec. 6.4.3.2). This accident has,

Accidents at Harrisburg and Chernobyl have not led globally to a reduction in the utilization of nuclear energy for electricity production

however, led to a critical examination, also in the USSR, of the minimum standards of nuclear safety under the guidance of the International Atomic Energy Organization (IAEO).

Safety measures are being strengthened

The relative cheapness of the electricity produced using nuclear energy compared with that from fossil fuels also favors the increased utilization of nuclear energy, in addition to the conservation of resources and reduction in crude oil dependence. This is the case for all the important industrialized countries, with the exception of the USA, and takes into account very different national considerations e.g. for coal-fired power stations: desulfurization and nitrogen oxide removal from flue gases and in nuclear power stations: uranium enrichment, reprocessing of exhausted fuel elements and conditioning and disposal of radioactive waste.

Increased utilization of nuclear energy:

- Reduces dependence upon imported fossil fuels
- Makes cheaper electricity available

A comparison of electricity production costs for coal- and nuclear power stations (see Table 6-3) shows that fuel costs are significantly higher for coal combustion, whereas plant costs (including waste disposal) are considerably higher for nuclear power. The cost advantage of nuclear power is highest in France, which has the most extensive nuclear power program and currently exports large quantities of electricity (ca. 30×10^9 kWh/a) to neighboring countries. If a free market for electricity is declared within the EEC in 1992, this will have wide ranging consequences for the West European economies.

Fossil fuel-fired power stations:

- High fuel costs
- Low plant costs

Nuclear power stations:

- High plant costs
- Low fuel costs

Table 6-3. Survey of the Average Electricity Production Costs for Coal and Nuclear Power Stations in Several Countries on 1.1.1981 (in 10^{-2} ECU/kWh).

Country	**Nuclear energy**				**Coal**			
	Plant costs	Maintenance & repairs	Fuel	Total	Plant costs	Maintenance & repairs	Fuel	Total
France	1.02	0.36	0.69	2.07	0.83	0.29	2.50	3.62
West Germany	1.58	0.47	0.82	2.87	0.79	0.60	3.32	4.71
Italy	0.99	0.22	0.78	1.99	0.56	0.19	2.38	3.13
Japan	1.34	0.47	0.76	2.57	0.95	0.42	2.51	3.88
Great Britain	2.85	0.34	0.93	4.12	1.73	0.35	3.82	5.90
Sweden	1.75	0.45	0.85	3.05	0.84	0.49	2.74	4.07
USA	1.85	0.37	0.67	**2.89**	1.03	0.37	1.52	**2.92**

6.2 General Information about the Nuclear Fuel Cycle

Nuclear fuel cycle

- Fuel element manufacture
- Nuclear fission (energy production)
- Reprocessing, waste disposal

The utilization of energy produced by nuclear fission requires highly sophisticated nuclear technology in which the nuclear reactor is the key component. A functioning reactor requires the provision of fissile nuclear fuel (uranium, plutonium) and the disposal of spent strongly irradiated fuel elements.

A nuclear fuel cycle is not absolutely necessary, but represents the optimum form of nuclear fuel utilization

The individual nuclear physical and chemical process steps of nuclear fuel management are mainly determined by the reactor type and are thereby closely connected with the specific fuel and fission product properties. They can be divided into:

Nuclear fuel cycle:

1. Supply of fuel elements to nuclear power stations:

- Mining and dressing of uranium ores
- Production of UF_6
- Enrichment of the ^{235}U isotope
- Production of nuclear fuel from UF_6
- Manufacture of fuel elements

1. Fuel element manufacture:

- Mining and upgrading of uranium ores
- Conversion of uranium concentrate into uranium(VI) fluoride
- Enrichment of the fissile uranium isotope ^{235}U from 0.7% (natural uranium) to ca. 2 to 4%
- Conversion of uranium(VI) fluoride into uranium(IV) oxide or metallic uranium (nuclear fuel)
- Fuel element manufacture

2. Nuclear fission e.g.:

$$^{235}_{92}U + ^{1}_{0}n \rightarrow \underbrace{^{89}_{36}Kr + ^{144}_{36}Ba + 3\,^{1}_{0}n}_{\text{and other fission products}} + \text{energy}$$

2. Fuel element removal from nuclear power stations by

- Reprocessing of spent fuel elements i.e. separation of U and Pu from one another and from the radioactive fission products,
- Recycling of the fissile materials to the fuel element production and
- Treatment and permanent storage of radioactive waste

or by direct final storage of spent nuclear fuel elements

3. Reprocessing and waste disposal:

- Separation and recycling of unchanged (fissile) uranium and plutonium produced in the reactor
- Conditioning and secure final storage of the radioactive fission products

The reprocessing of spent fuel elements and the associated recycling of uranium and plutonium completes the fuel cycle and represents an optimal form of energy utilization (see Fig. 6-1). It is, however, not absolutely necessary and is therefore under intense discussion in a number of countries in connection with the danger of proliferation, environmental consequences, safety etc. Moreover, it is uneconomic for particular reactor

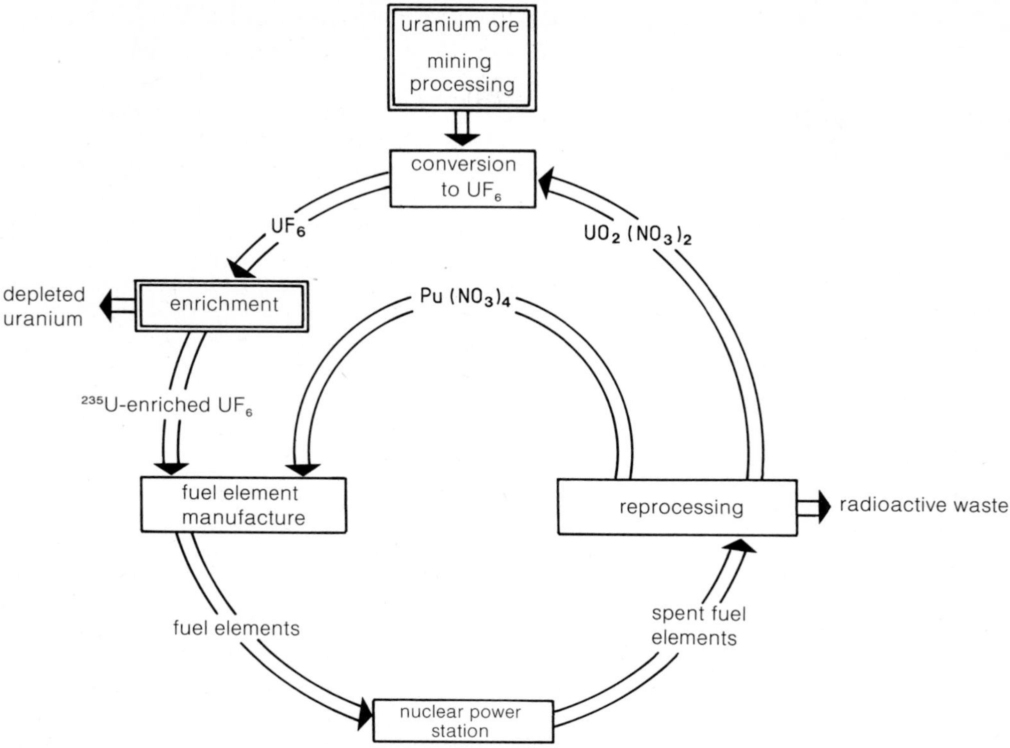

Fig. 6-1. Fuel cycle for light water reactors.

types (Candu reactors) and currently technically infeasible for others (high temperature reactors).

The direct final storage of spent nuclear fuel elements in e.g. granite or salt formations is an alternative. This is the preferred disposal option in the USA, Canada and Sweden, but requires extensive development work before it can become a practical option and is unlikely to be in use before the turn of the century.

Direct final storage preferred in the USA, Canada and Sweden, but not yet feasible

6.3 Availability of Uranium and Thorium

The only widely occurring natural fissile nuclear fuels are uranium and thorium. Plutonium is exclusively produced by breeding processes in nuclear reactors, ca. 300 kg of plutonium accruing annually in a 1000 MWe light water reactor.

Uranium occurrence:

- Earth's crust 4 ppm
- Seawater 3 ppb

Uranium reserves outside Communist Countries (excluding seawater and sedimentary phosphate ores):

- Reasonably secure: 3.5×10^6 t U

  of which 32.8% is in Africa
  24.4% is in Australia
  24.1% is in North America
  8.5% is in Europe
  7.9% is in Latin America
  2.1% is in Asia

Speculative deposits of 6.6 to 16×10^6 t U in the Western World include:

- Uranium in sedimentary phosphate ores: ca. 6.3×10^6 t
- Uranium in seawater: ca. 4.5×10^9 t

The Earth's crust contains ca. 4 ppm uranium and seawater 3 ppb uranium. The largest deposits are in Australia, Africa (South Africa, Niger, Namibia) and North America (USA, Canada). In 1985 the economically exploitable reserves were estimated to be 1.6×10^6 t with an extra 0.9×10^6 t of probable reserves (as uranium) on the basis of a cost of < 80 $/kg uranium. An estimated additional 0.63×10^6 t and 0.4×10^6 t respectively can be added for uranium prices in the range 80 to 130 $/kg. The speculative reserves are much higher (up to 16×10^6 t). These include, for example, the uranium deposits in sedimentary phosphate ores (mainly apatite) amounting to 6.3×10^6 t, but do not include the practically inexhaustable (but difficult to extract) uranium reserves in seawater (ca. 4.5×10^9 t).

1985 uranium production in the Western World:

35×10^3 t

Main producer countries:

Canada, South Africa, USA, Namibia (in the future also Australia and P.R. China)

The uranium demand in the Western World is currently ca. 42×10^3 t/a and is expected to rise to 52.5×10^3 t/a by the year 2000. The worldwide production of uranium in 1985 was 35×10^3 t, the shortfall between production and consumption being made up from national reserves. The main producing countries are Canada (currently providing 30% of the World production), South Africa, the USA, Namibia and Niger with the production capacities in Australia and the Peoples Republic of China being strongly expanded at the moment. The total uranium production from 1938 to the end of 1985 was ca. 800×10^3 t of which 25% was used for military purposes.

Thorium reserves:

Worldwide 3.3×10^6 t
Western World 1.3×10^6 t
In 1986, 377 nuclear reactor blocks in operation

The main natural source of thorium is monazite sand. The worldwide reserves are ca. 3.3×10^6 t of which 1.3×10^6 t are to be found in the Western World. The 1985 production of monazite concentrate in the Western World excluding the USA was ca. 31.5×10^3 t. The main producer countries are Australia, Brazil, India and Malaysia.

Thorium is mainly utilized in the manufacture of refractory materials for the metallurgical industry. Only a small part is used in the manufacture of fuel elements (e.g. only 5 t in the USA in 1985). This can change rapidly in the future should thorium (high temperature) reactors be widely installed.

In terms of satisfaction of demand and price stability, the supply of uranium and thorium to the end of the century is much more secure than all the fossilized energy sources. The demand for fissile materials can be fairly reliably calculated from the capacities of the operating and planned nuclear power stations. It can be strongly reduced in future by intensified reprocessing of spent fuel elements and the use of fast breeder reactors as can be seen by comparing the fuel required to produce 8.5×10^9 kWh of electricity:

In terms of price stability and satisfaction of demand, the supply of uranium and thorium is more stable than that for fossilized energy sources

Quantity of fuel consumed in the production of 8.5×10^9 kWh:

- Ca. 33 t uranium in light water reactors without reprocessing
- Ca. 5 t uranium and plutonium in fast breeder reactors

- Light water reactor (1300 MWe) without reprocessing: 33 t uranium
- Light water reactor (1300 MWe) with reprocessing: 22 t uranium
- Fast breeder reactor (1300 MWe): 5.1 t uranium and plutonium

A closed fuel cycle in light water reactors and the more widespread use of fast breeder reactors with fuel cycles makes nuclear energy a long term energy source

compared with:

- Oil-fired power stations: 1.6×10^6 t fuel oil
- Coal-fired power stations: 2.7×10^6 t coal

These figures show that there is sufficient nuclear fuel for the long term operation of nuclear power stations.

6.4 Nuclear Reactor Types

6.4.1 General Information

At the end of 1986, 26 countries had a total of 377 nuclear power station blocks with a total capacity of 286 GWe. 73 of the blocks with a capacity of 42.5 GWe are in Comecon countries (see Table 6-4).

In 1986, 377 nuclear reactor blocks in operation

If these are analyzed on the basis of reactor type, the pressurized water reactor is the most widely built type (see marginal notes). Light water reactors, gas-cooled graphite-moderated reactors and heavy water reactors are also often built, the latter particularly in Canada. High temperature reactors (HTR) and fast breeder reactors are not yet widely used, but will make an increasing contribution to electricity production in the long term.

Analysis of nuclear power stations installed worldwide by 1987 according to reactor type:

Reactor type	In use	Under construction
light water		
– boiling water	79	19
– pressurized water	206	100
graphite		
– gas/graphite	27	7
– water/graphite	27	7
– high temperature	3	–
heavy water	28	14
fast breeder	7	2

Light waterreactors are currently the most widespread

Table 6-4. Nuclear Power Stations and Nuclear Power Station Capacities at the End of 1986.

	In operation		Under construction	
	Number	Net capacity in GWe	Number	Net capacity in GWe
World	377	285.7	149	142.7
USA	98	87.6	27	32.0
France	49	47.2	15	20.0
USSR	50	29.3	37	36.9
Japan	34	25.8	11	9.9
West Germany	21	19.9	4	4.3
Great Britain	22	12.6	4	2.6
Canada	18	11.5	5	4.6
Sweden	12	9.8	–	–

These reactor types differ in their nuclear fuels (uranium(IV) oxide, thorium/uranium oxide or carbide, uranium/plutonium-mixed oxides), moderators (light water, heavy water, graphite), coolant (water, carbon dioxide, helium, graphite) and particularly in reactor construction and hence operational safety.

6.4.2 Light Water Reactors

Two types of light water reactor:

- Boiling water reactors
- Pressurized water reactors

Water is here used simultaneously as a neutron absorber and as a coolant

There are two types of light water reactors: boiling water and pressurized water reactors. Both reactor types use water both as a coolant and as a moderator for slowing down the fast neutrons produced during nuclear fission to the thermal energies required for the fission of further ^{235}U-nuclei.

6.4.2.1 Boiling Water Reactors

Boiling water reactors:

- Coolant water is evaporated at ca. 70 bar in the reactor core
- Steam goes directly to turbines

- Fuel:

  ^{235}U-enriched UO_2

In this reactor type water is brought to boiling point in the reactor core at a pressure of 70 bar. The resulting steam is directly fed fom the pressurized reactor vessel to turbines. Fuel element bundles, each consisting of e.g. a lattice-like 8×8 fuel rod array, are to be found in the reactor core. The nuclear fuel consists of enriched uranium(IV) oxide (with e.g. 3.2% ^{235}U). The reactor is controlled by lowerable cross-shaped control rod assemblies consisting of neutron-absorbing materials such as boron carbide.

6.4.2.2 Pressurized Water Reactors

In pressurized water reactors the heat produced in the reactor core is transferred by a primary loop (containing liquid water at 155 bar and ca. 320 °C) to a nonradioactive secondary loop in which steam is produced. The fuel rods in the reactor core are ca. 3 to 5 m long and 10 mm in diameter and are arranged in square bundles very close to one another. A typical fuel element consists of 236 fuel rods and 20 control rods in a 16 x 16 arrangement. 193 such fuel elements totalling 117 t of ^{235}U-enriched UO_2 nuclear fuel make up the heat producing core of a 1300 MW nuclear power station. The chain reaction is regulated (moderated) with soluble neutron-absorbing materials in the coolant (e.g. boric acid) with the assistance of the control rods. The components of the primary loop are enclosed in a hermetically sealed cylindrical or spherical pressure containment vessel which is designed to withstand the pressure build up in the event of damage and to protect the environment from the release of any of the radioactive materials formed inside.

Pressure water reactors:

- Heat produced in the reactor core is absorbed by a primary coolant loop at 155 bar and ca. 320 °C in which no steam is produced and is transferred to a nonradioactive secondary loop
- Fuel:
  ^{235}U-enriched UO_2, fuel complement for a 1300 MW nuclear power station: 117 t UO_2
- Primary coolant loop enclosed in a hermetically sealed safety containment vessel

6.4.3 Graphite-Moderated Reactors

Graphite-moderated reactors have played an important rôle in nuclear technology since the early days of nuclear energy utilization. Thus the first self-sustaining nuclear reactor at Chicago which came on stream in 1938, was moderated with graphite as are the modern advanced gas-cooled reactors and high temperature reactors. The reactor involved in the accident at Chernobyl was also a graphite-moderated (pressure tube boiling water) reactor.

6.4.3.1 Gas-Cooled Reactors

Gas cooled, graphite-moderated reactors:

Magnox reactors:

- Coolant: CO_2
- Fuel: natural uranium
- Temperature of exiting gas: 420 °C

This category includes the few Magnox reactors still operated in Great Britain and France and the *advanced gas-cooled reactors* (AGR) much used in Great Britain. They are cooled with carbon dioxide and require weakly enriched uranium(IV) oxide fuels. The exit temperature of the gas is ca. 690 °C.

AGR-reactors are further developed Magnox reactors:

- Coolant: CO_2
- Fuel: lightly ^{235}U-enriched UO_2
- Temperature of exiting gas: ca. 690 °C

In the *high temperature reactors* (HTR) currently under development helium is used as a coolant gas thereby avoiding the corrosion of graphite with carbon dioxide which is a problem in AGR-reactors:

$$C + CO_2 \rightleftharpoons 2CO$$

High temperature reactors (HTR):

- Coolant: He
- Fuel: ^{235}U-enriched uranium(IV) with ^{232}Th oxide or carbide mixture in beads
- Temperature of exiting gas: 700 to 1000 °C

The fuel consists of highly enriched uranium and thorium oxide or carbide. It can be utilized in the form of fuel rods (such as in experimental reactors in the USA and Great Britain) or as fuel beads (such as in the prototype reactor at Hamm-Uentrop in West Germany). The latter consist of fuel particles containing ca. 1 g ^{235}U and 10 g ^{232}Th coated with SiC or pyrolytic carbon and embedded in a graphite matrix to prevent the release of fission products.

Fuel beads each consist of 35 000 SiC- or pyrolytic carbon-coated fuel particles embedded in a graphite matrix. These fuel beads are continuously circulated

360 000 such fuel beads together with 280 000 graphite moderator beads and 35 000 boron-containing absorber beads were required just for the first filling of the THTR-300 reactor (with a capacity of 300 MWe or 750MW$_{therm}$), which came on stream in West Germany in 1985. The 6 cm diameter beads are present as a loose layer and have to be continuously circulated in the reactor core while the reactor is in use (pebble bed reactor).

Hot helium is suitable for

- Steam generation
- Chemical reactions, such as coal gasification

The power density of such helium-cooled reactors is significantly higher than that of CO_2-cooled reactors and the temperature of the exit gas can exceed 1000 °C (750 °C in the THTR-300). The heat abstracted from the helium can be used for steam generation or as process heat for chemical reactions at high temperatuires (e.g. high temperature steam electrolysis see Sec. 1.2.2.2 and coal gasification see Sec. 1.4.1.2.2).

Advantage: breeding of fissile ^{233}U from nonfissile ^{232}Th

The advantage of thorium-HTR's is that the neutrons produced during uranium fission are utilized to convert nonfissile ^{232}Th into fissile ^{233}U.

6.4.3.2 Light Water-Cooled Reactors

This is the reactor type used in Chernobyl and is the type preferred in the USSR. It essentially consists of a graphite block ca. 14 m in diameter and 8 m high, which is traversed by 2500 vertical pressure pipes in which a total of ca. 1700 fuel elements with lightly enriched uranium and 250 control and moderating rods are placed. In normal operation the fuel elements are cooled with boiling water at a pressure of 100 to 150 bar and steam is produced at a temperature of ca. 550 °C.

H_2O-cooled graphite-moderated reactors:

- Fuel: lightly enriched uranium compounds
- Coolant: H_2O at high pressure, > 500 °C steam being produced
- Moderator: graphite

An analysis of the Chernobyl accident has shown that the absence of a rapid shut-down system, the complexity of the control and coolant systems, the reactivity of the coolant water with graphite and zirconium and the absence of a pressure-resistant reactor containment vessel, which is statutory in Western Countries, were particularly critical.

The Chernobyl accident has shown that this type of reactor is relatively poorly regulable and was not installed according to Western safety standards

It is estimated that ca. 2×10^9 GBq (1 GBq = 10^9 decays/s) of radioactive materials excluding rare gases were released during the reactor explosion and its subsequent melt-down.

6.4.4 Heavy Water Reactors

Heavy water reactors utilize heavy water (D_2O) as a moderator. They can be operated with natural uranium due to the low capture cross section for thermal neutrons of D_2O compared with H_2O. The high price of heavy water (only present as 0.015% in natural water) is, however, a disadvantage. The resulting higher investment costs compared with light water reactors and the lower power density have hindered a wider use of this reactor type. Only in Canada has this reactor type been installed in large numbers.

Heavy water reactors:

- Fuel: natural uranium or enriched uranium
- Moderator: D_2O

whose production is very expensive. As a result this reactor type is not widely used

To achieve the required moderator/fuel ratio of 20:1 (light water reactor: 2:1), bundles of 19 to 37 tightly packed fuel rods are placed 25 cm apart. The coolant flows through the fuel rod bundle in cooling channels and the region between the fuel rod bundles is filled with the D_2O moderator. Coolants with low neutron absorption coefficients, such as CO_2, D_2O or He, are used in natural uranium reactors. This restriction does

- Coolant:
  - for natural uranium reactors: CO_2, D_2O, He
  - for enriched uranium reactors: H_2O

not apply to enriched uranium reactors and thus water can be used in such reactors.

Most widely used type:
Candu reactor (Canada)

The most widely used heavy water reactor type is the D_2O-cooled pressure tube (Candu) reactor developed in Canada. It uses natural uranium fuel, a pressurized D_2O coolant and low pressure D_2O as a moderator. In Canada, Candu reactors have been utilized for years in the extraction of oil from oil tar sands.

6.4.5 Fast Breeder Reactors

Fast breeder reactors:

- Operated with "fast" neutrons
- These unmoderated neutrons are used to breed new fissile materials (^{239}Pu, ^{233}U)
- Fuel: highly enriched UO_2 or PuO_2
- More fissile material can be produced than is consumed
- Uranium up to ca. 60 times better utilized than e.g. in light water reactors

Fast breeder reactors are operated with unmoderated "fast" neutrons, which can convert nonfissile ^{232}Th and ^{238}U (breeding materials) into fissile ^{233}U and ^{238}Pu respectively (fuel). With suitable arrangement of breeder and fuel materials in the core (centre: ^{233}U, ^{235}U, ^{239}Pu; shell: ^{232}Th, ^{238}U) it is, in principle, possible to produce more fissile material than is consumed in the nuclear fission (breeding rate > 1). In this way the uranium used can be up to 60 times more efficiently employed than in comparable light water reactors.

- Liquid Na is the heat transfer medium in both the primary and secondary loops; He-cooled fast breeder reactors are under development

Na-cooled fast breeder reactors are technically realizable

A mixture of highly enriched UO_2/PuO_2 with ca. 10 to 20% ^{239}Pu (or ^{235}U) is generally used as the fuel and ^{235}U-deficient uranium from isotope-separation plants and spent fuel reprocessing is utilized as the breeder material. A moderator is not necessary. The preferred heat transfer medium is liquid sodium (primary loop), which by way of a ca. 400 °C secondary loop (also liquid sodium) is coupled to a conventional water/steam loop.

USSR and France in the forefront of fast breeder technology development

There are currently in addition to a number of small experimental plants with capacities between 100 and 300 MWe (in the USA, the USSR, France, Great Britain), two large capacity reactors in operation (USSR: 600 MWe, France: 1200 MWe). France has with its Superphénix-breeder, which came on stream in 1985, by far the most advanced breeder power station in the World. Other fast breeder reactors are under construction in West Germany, India, Japan and the USSR.

6.5 Nuclear Fuel Production

The most widely used nuclear fuel material for nuclear power stations (particularly for light water reactors) is uranium(IV) oxide. Also of importance are uranium-plutonium oxide mixtures, various carbides (UC, UC_2, $U_{1-x}Pu_xC$, ThC_2) and metallic uranium (for the technically outmoded Magnox reactors). Nitridic fuels have to date only been used in experimental reactors.

Fuel production comprises: mining and concentration of uranium ore, conversion into UF_6, enrichment of ^{235}U and reconversion of UF_6 into U_3O_8. This is finally cold pressed into cylindrical pellets and heat-processed into sintered UO_2 pellets.

Nuclear fuel manufacture (for light water reactors):

- Production of uranium ore concentrate
- Conversion to UF_6
- Enrichment of ^{235}U
- Reconversion of UF_6 into UO_2

The starting materials for uranium nuclear fuels are uranium compounds produced from natural uranium ore deposits and fissile material separated during the reprocessing of spent uranium fuel.

Nuclear fuels from:

- Natural uranium ore
- Spent uranium fuel rods

6.5.1 Production of Uranium Concentrates

Uranium is found in nature mainly as pitchblende (UO_2) and uraninite ($UO_2 \cdot 2\,UO_3$) in pegmatites (Namibia, Canada, USSR), conglomerates (Canada, South Africa), oxygen-deficient black slate (Sweden), phosphates (USA, Morocco) and gangue and sandstone deposits (Australia, USA, Niger) from which they are extracted by open pit or underground mining (for U_3O_8 contents $> 0.03\%$). Uranium extraction from phosphoric acid and the "in situ leaching" process are gaining in importance, currently accounting for 5% of the total production. By far the largest, almost untapped, uranium reserves are to be found in seawater.

Uranium ore deposits:

- Pegmatites
- Gangue, conglomerate and sandstone deposits
- Phosphates, coal, bituminous slates

Uranium ores are mined when

- U_3O_8 content $> 0.03\%$
- Deposits contain economically extractable quantities

6.5.1.1 Uranium from Uranium Ores

The ore extracted from open pit or underground mines is sorted on the basis of radioactivity intensity, then coarsely ground in crushers to particle sizes < 10 mm and finally ground during which the radon liberated is sucked off as a radiation prevention measure. The

Uranium-containing ore is crushed, ground and finally leached uranium is extracted with the aid of hydrometallurgical (leaching) processes and then concentrated by ion exchange or extraction processes and finally precipitated.

6.5.1.1.1 Leaching Processes

Leaching of uranium ores preferably carried out in acidic media

The leaching of uranium ores is mainly carried out with dilute sulfuric acid, uranium(VI) oxide being digested to uranyl(VI) sulfate at atmospheric pressure:

$$UO_3 + H_2SO_4 \longrightarrow UO_2SO_4 + H_2O$$

UO_2 in primary pitchblende must first be oxidized with trivalent iron

If uranium is present as uranium(IV) oxide (e.g. in pitchblende), it must first be oxidized with iron(III) salts or nitric acid to hexavalent uranium. In some cases hydrochloric acid or phosphoric acid are used in the digestion of the ore.

Alkaline leaching with:

- Na_2CO_3
- $NaHCO_3$
- $(NH_4)_2CO_3$

Ores with alkaline gangue (e.g. carbonates) are preferably leached with sodium carbonate, sodium hydrogen carbonate or ammonium carbonate under pressure (5 to 6 bar, 95 to 120 °C) or at atmospheric pressure (75 to 80 °C), if necessary with added atmospheric oxygen, forming the uranyl(VI) triscarbonato complex:

A uranyl(VI) triscarbonato complex is formed with the hexavalent uranium

$$UO_3 + Na_2CO_3 + 2\,NaHCO_3 \longrightarrow Na_4[UO_2(CO_3)_3] + H_2O$$

Uranium is extracted from low grade uranium ores by trickling the leaching medium through heaps without preliminary grinding or into the seam itself (in situ leaching).

In situ leaching from seams proceeds alkalinely

Uranium extraction 60 to 85%; importance increasing

In "in situ leaching" the alkaline leaching agent is introduced into the rock seam (in particular sandstone) by way of injection pipes and conveyed by a central Archimedian screw to ground level. The extraction efficiency is 60 to 85% (relative to the uranium present). The advantage of this method is the considerably lower investment costs (20 to 30 \$/lb U_3O_8 compared with 20 to 75 \$/lb for mining).

6.5.1.1.2 Separation of Uranium from the Leaching Solutions

After separating off the undigested solids by multistep filtration or countercurrent decantation, the uranium present in the solution is further enriched and freed of impurities (particularly Cd, Cr, Mo, W, V) by ion exchange processes, countercurrent extraction or a combination of these processes.

Separation of uranium from the leaching solution by:

- Ion exchange
- Extraction
- Combination of these processes

Separation by ion exchange: Strongly basic or weakly basic ion exchange resins are used to separate selectively the uranium from the weakly acidic or alkaline solutions from the leaching step. The uranium is eluted from the ion exchange resins with the aid of nitrate or chloride solutions as anionic (carbonato- or sulfato-) complexes.

Uranium separation by ion exchange:

- Strongly basic or weakly alkaline ion exchange resins used
- Elution with nitrate or chloride solutions

The ion exchange is carried out either by the (conventional) fixed bed process, continuous ion exchange or (for unfiltered leaching solutions) with ion exchange resin suspensions. The advantage of the ion exchange process is that uranium can be extracted from very dilute solutions as well as from turbid unclarified solutions.

Ion exchange carried out:

- In fixed beds
- In suspension
- Continuously

Separation by solvent extraction: Extractive uranium separation from mainly acidic digestion solutions is currently much more important than ion exchange processes. It essentially consists of two steps namely:

Solvent extraction preferred worldwide

- Extraction of the uranium in an organic solvent with the aid of an extraction agent
- Stripping of the uranium present into the aqueous phase as a uranyl(VI)-complex e.g.

$$[UO_2(SO_4)_3]^{4-}_{org.} + 5Na_2CO_3 \rightarrow [UO_2(CO_3)_3]^{4-}_{aq.} + 2H_2O + 2CO_2 + 3Na_2SO_4$$

Both steps are carried out in multistage mixer-settlers operated countercurrently.

Extractors used: mixer-settlers

Extraction agents: phosphorus compounds and alkyl ammonium salts in kerosene

The extraction agents used are phosphorus compounds such as tri-n-butylphosphate (TBP), di-(2-ethylhexyl)-phosphate (D2EHPA), tri-n-octyl-phosphinoxide (TOPO) and various quaternary alkyl ammonium salts. The choice of extraction agent is dependent upon the digestion process utilized, it being employed as ca. 4 to 10% by volume solutions in organic solvents such as kerosene. In some cases 3 to 5% of a long chain alcohol such as isooctanol is added to increase the solubility of the alkyl ammonium salts in the organic phase.

Stripping of the U-compounds from the organic phase with aqueous salt solutions

The stripping of the uranium compounds into the aqueous phase is carried out with hydrochloric acid, ammonium sulfate or sodium carbonate solutions.

6.5.1.1.3 Manufacture of Marketable Uranium Compounds ("Yellow Cake")

Uranium compound concentrates "yellow cake" have to be very pure

To obtain marketable uranium compounds, which due to their yellow color are known as "yellow cake", the aqueous uranium salt solutions produced by ion exchange or solvent extraction have to be processed further and freed of interfering impurities (e.g. Mo, V, P). After the removal of iron, phosphate and excess sulfate ions, the uranium concentrate is precipitated as sodium or ammonium diuranate. This is filtered off, washed and then dried.

Precipitation of uranium from alkaline solution as diuranate with:

- NaOH
- NH_3 or MgO after prior acidification of the solution

The precipitation proceeds in either alkaline solution at pH > 12 and a temperature of 80 °C by adding sodium hydroxide

$$2\,Na_4[UO_2(CO_3)_3] + 6\,NaOH \rightarrow Na_2U_2O_7 + 6\,Na_2CO_3 + 3\,H_2O$$

or in acidic solution at pH 5 by passing in ammonia

$$2UO_2SO_4 + 6\,NH_3 + 3\,H_2O \rightarrow (NH_4)_2U_2O_7 + 2\,(NH_4)_2SO_4$$

in some cases also with the addition of magnesium or calcium hydroxide.

The coprecipitated vanadium can in the first process be driven out by roasting the yellow cake at 850 °C with added sodium carbonate. In the second process this is not possible. Thus it can only be used for uranium ores with low molybdenum and vanadium contents.

Diuranate precipitates are washed and dried:

- Ammonium diuranate at ca. 750 °C during which it is converted into U_3O_8
- Na-diuranate at 120 to 175 °C

Sodium diuranate is dried at 120 to 175 °C, whereas ammonium diuranate has to be heated in multiple hearth roasters to 750 °C to complete its decomposition and remove the impurities still present (sulfite, chloride):

$$9(NH_4)_2U_2O_7 \rightarrow 6U_3O_8 + 14NH_3 + 15H_2O + 2N_2$$

6.5.1.2 Uranium from Phosphates

Much of the uranium reserves in the Earth's crust are found in apatites and can be extracted by wet digestion with sulfuric acid (Sec. 1.5.1.2.1). However, the average U_3O_8 concentration of 100 to 200 ppm is very low and enrichment is only economic for large phosphoric acid plants and not too low uranium prices.

Apatites contain 100 to 200 ppm U_3O_8. Processing requires high enrichment factors

Separation from phosphoric acid is achieved either by extraction of hexavalent uranium with a mixture of TOPO and D2EHPA in kerosene (ORNL-process) or by extraction of tetravalent uranium with a mixture of mono and dioctylphenyl esters of orthophosphoric acid in kerosene. Subsequent processing is either by the reduction strip or oxidation strip processes.

U_3O_8 separation from raw phosphoric acid by extraction with:

- TOPO/D2EHPA in kerosene for U^{6+}
- A mixture of mono and dioctylphenyl esters of ortho phosphoric acid in kerosene for U^{4+}

In the reduction strip process tetravalent uranium in the raw phosphoric acid is oxidized to hexavalent uranium by treatment with sodium chlorate, hydrogen peroxide or air at 60 to 70 °C, extracted with TOPO/D2EHPA in an organic phase (kerosene) and finally stripped by multistage reduction with aqueous iron(II) solutions. The resulting uranium solution enriched by a factor of 40 is further concentrated by renewed oxidation and extraction. Hexavalent uranium is finally stripped with ammonium carbonate and precipitated as the ammonium uranyl(VI) triscarbonato-complex. This is filtered off, washed and calcined to U_3O_8.

Reduction strip process:

- Oxidation of U^{4+} to U^{6+}
- Extraction of U^{6+} with TOPO/D2EHPA in kerosene
- Reduction stripping of the organic phase with Fe^{2+} in water
- Reoxidation of U^{4+} to U^{6+}
- Renewed extraction of U^{6+}
- Stripping of the organic phase with aqueous $(NH_4)_2CO_3$ with $(NH_4)_4[UO_2(CO_3)_3]$ precipitating out

In the oxidation strip process the hexavalent uranium in the raw phosphoric acid is initially reduced to tetravalent uranium, which is extracted with a mixture of the mono and dioctylphenyl esters of ortho phosphoric acid in kerosene. Oxidation with sodium chlorate in phosphoric acid transfers the uranium to the aqueous phase and it is then extracted with TOPO/D2EHPA, as in the reduction strip process.

Oxidation strip process:

- Extraction with mono/dioctylphenyl-esters of phosphoric acid in kerosene
- Oxidation of U^{4+} to U^{6+} in aqueous phosphoric acid
- Extraction with TOPO/D2EHPA in kerosene

The uranium content in the apatite deposits in the USA alone is estimated to be ca. 600×10^3 t (as U_3O_8). The 1981 capacity for the extraction of uranium from raw phosphoric acid was ca. 2×10^3 t/a worldwide.

Capacity for the extraction of uranium from raw phosphoric acid:
worldwide: ca. 2000 t/a

6.5.1.3 Uranium from Seawater

In order to be able to satisfy the uranium demand for light water reactors well into the next century, ways are already being explored of exploiting the practically inexhaustible uranium reserves in seawater.

Uranium content in seawater very low compared with other deposits:

- Seawater 0.003 ppm
- Phosphates 100 to 200 ppm
- Conventional U ores 350 to 5000 ppm

Ca. 1 km^3 seawater required for 1 t uranium

The problem is the extremely low uranium content in seawater: 0.003 ppm compared with 350 to 5000 ppm in mined uranium ores and 100 to 200 ppm in phosphates. This means that ca. 1 km^3 ($= 10^{12}$ L) of seawater have to be processed to produce 1 t of uranium at a uranium yield of 30%. This is only possible by utilizing extremely selective absorption agents.

Development work aimed at U extraction from seawater by adsorption or ion exchange processes

The aim of current development work in the USA, Western Europe and particularly Japan is to extract uranium from seawater with the aid of adsorption agents (such as titanium(IV) oxide hydrate) or ion exchangers [e.g. poly(acrylamidoxime)] by exploiting the tides or major ocean currents.

Enrichment factors of 10^5 to 10^6 have thus far been achieved at a production cost of several hundred $/lb U_3O_8. This could change rapidly with further development. The first operational production plant came on stream in Japan in 1986 (capacity: 10 kg uranium/a).

6.5.2 Conversion of Uranium Concentrates to Uranium Hexafluoride

6.5.2.1 General Information

Conversion of uranium concentrates into UF_6 used for:

- Purification of uranium
- ^{235}U enrichment

The basic principles for the conversion of uranium concentrates into gaseous uranium compounds required for the enrichment of ^{235}U were established in the early 1940's in the USA during the Manhattan Project, in which they were used for the production of the first atomic bomb. They still form the basis of all military and civil uranium enrichment plants and are essentially based on the conversion of prepurified yellow cake into uranium(VI) fluoride, by far the most thermally stable volatile uranium compound (sublimation temperature: 56.5 °C). UF_6 has also the advantage that the molecular weight important for isotope enrichment is only dependent upon the uranium isotope ratio, there being only one naturally occurring fluorine isotope.

Two processes are employed for the production of UF_6 namely the wet and dry processes. In both processes UO_2 and UF_4 are formed as intermediates. In the wet process the UO_2 is produced from the uranium concentrate by way of uranyl(VI) nitrate, whereas in the dry process the uranium concentrate is directly reduced to UO_2.

Conversion to UF_6 by two processes:

- Wet process: uranium concentrate → $UO_2(NO_3)_2$ (purification) → UO_2 → UF_4 → UF_6
- Dry process: uranium concentrate → UO_2 → UF_4 → UF_6 (purification)

6.5.2.2 Wet Process for UF_6 Manufacture

The purification of uranium concentrates in the wet process is almost exclusively carried out by solvent extraction. Uranium concentrate is dissolved in 0.5 to 1 M nitric acid and extracted with 30% tri-n-butylphosphate in kerosene as $UO_2(NO_3)_2 \cdot 2TBP$ into the organic phase in mixer-settlers or in pulsed plate columns. After removing the coextracted impurities (Th, Mo) in a scrubber, the uranium salt is stripped into the aqueous phase with 60 °C water. The resulting dilute uranyl(VI) nitrate solution is evaporated to uranyl(VI) nitrate hexahydrate $UO_2(NO_3)_2 \cdot 6H_2O$ (UNH) and finally converted into orange-colored UO_3 in fluidized bed kilns at 350 to 450 °C:

$$UO_2(NO_3)_2 \cdot 6H_2O \longrightarrow UO_3 + NO + NO_2 + 6H_2O + O_2$$

Finally UO_3 is reduced with hydrogen to UO_2 at 500 to 600 °C in fluidized bed kilns.

Wet process consists of the following steps:

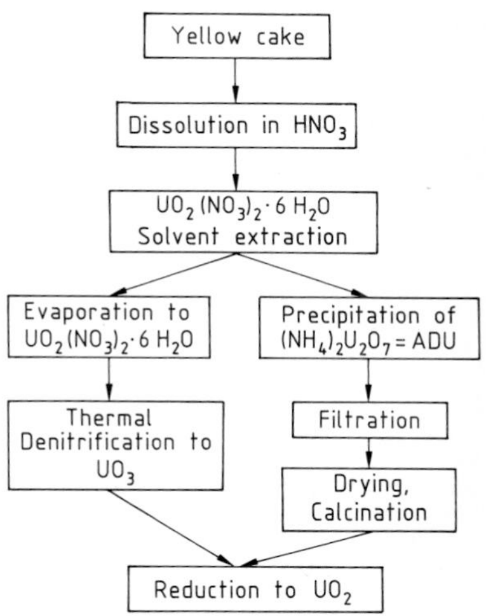

Alternatively, ammonium diuranate (ADU) can be precipitated by adding ammonia to the purified uranyl(VI) nitrate solution, filtered off, washed, dried, calcined and reduced with hydrogen to UO_2 (French Comurhex-process). Typical plant sizes are of the order of 10^4 t/a uranium.

The highly reactive UO_2 produced by the reduction of UO_3 is directly converted to UF_4 (green salt) by hydrofluorination with hydrogen fluoride gas. This highly exothermic reaction proceeds with air cooling at 550 to 650 °C in fluidized bed, moving bed or screw reactors e.g. 17 m long and 1.2 m in diameter. To achieve as complete a reaction as possible the hydrofluorination is carried out in several zones in a single reactor or in a series of connected reactors. In the moving bed reactor UO_3 reduction and UO_2 hydro-

Conversion of UO_2 into UF_4 proceeds with anhydrous hydrogen fluoride in fluidized-bed, moving bed and screw reactors

Moving bed reactors consist of:

- An upper reduction zone for reacting UO_3 with H_2 and
- A lower hydrofluorination zone for UF_4 manufacture

fluorination are arranged above one another in a single unit.

The conversion of UF_4 to UF_6 proceeds exclusively by reaction with elemental fluorine:

$$UF_{4(solid)} + F_{2(gas)} \rightarrow UF_{6(gas)} \quad \Delta H = -247 \text{ kJ/mol}$$

Manufacture of UF_6 from $UF_4 + F_2$ in flame or fluidized bed reactors

and is either carried out in water-cooled flame reactors whose walls are maintained below 540 °C or in air-cooled fluidized bed reactors of a new design at ca. 500 °C. The latter contain calcium fluoride as a heat transfer agent and fluidized bed stabilizer. It also prevents the caking of UF_4.

In both reactor types, the UF_6 gas formed is freed of UF_4 dust by fitration through Monel metal filters and solidifies upon cooling to ca. −40 °C.

The fluorine required for fluorination (e.g. 1100 t/a for a 10 000 t/a UF_6 plant) is produced in a nearby dedicated fluorine plant.

6.5.2.3 Dry Process for UF_6 Manufacture

Allied Chemical Corp. dry process for UF_6 manufacture:

uranium concentrate
↓
reduction to UO_2
↓
hydrofluorination to UF_4
↓
fluorination to UF_6
↓
UF_6 purification by fractional distillation

In the dry process, introduced by Allied Chemical Corp., the uranium concentrate is pelletized and directly reduced with hydrogen to UO_2 at temperatures between 540 and 650 °C in a fluidized bed reactor. Hydrofluorination to UF_4 proceeds in two fluidized bed reactors connected in series. After fluorination of the UF_4 formed in a unit consisting of a flame reactor and a fluidized bed reactor, the UF_6 produced is purified in a two stage pressure distillation process. This distillative purification process is necessary, because, unlike the wet process, no purification step is included in the preceding process steps.

1986 capacity for uranium conversion to UF_6 in 10^3 t uranium:

Western World	52.3
produced by:	
• Allied Chemical, USA	12.7
• BNFL, GB	9.5
• Comurhex, France	12.0
• Eldorado, Canada	9.0
• Sequoyah Fuels, USA	9.1

The nominal capacity for uranium conversion to UF_6 in the five major companies in the Western World was ca. 52×10^3 t uranium/a in 1986 of which 75% was for production using the wet process. This compares with a current demand of 42×10^3 t/a.

6.5.3 ^{235}U-Enrichment

In the manufacture of fissile nuclear fuel the 0.7% ^{235}U present in natural uranium has to be enriched to ca. 2 to 4% (in the case of fast breeder reactors and high temperature reactors up to 93%). This is accomplished by exploiting the minimal differences in physical properties between gaseous $^{235}UF_6$ and gaseous $^{238}UF_6$ in huge cascade-like isotope separation plants. The industrially important processes are:

Enrichment of the ^{235}U-isotope from 0.7% in natural uranium to ca. 3% is accomplished by:

- The classical *gas diffusion process* which is mainly operated in the USA, the USSR and France. It is based on the difference in diffusion rate of UF_6 molecules with different molecular weights through a very fine pore Ag/Zn alloy membrane. Aluminum, sintered nickel, Al_2O_3 or PTFE foils (with e.g. 10^9 pores/cm^2) can also be used. The separation factor is ca. 1.002/stage.

- Gas diffusion

- In the *gas centrifuge process* developed by URENCO and operated in a number of plants in Great Britain, the Netherlands and West Germany the heavier $^{238}UF_6$ is enriched at the walls of the long overcritical rotating pipes with a separation factor of > 1.2 per stage. The energy requirements are only ca. 5% of those of the gas diffusion process.

- Gas centrifuge

- The *nozzle separation process* has up to now only been used as preliminary cascades (with 24 of the envisaged total of 300 nozzle separation stages) in Brazil. It is based on the propulsion of gaseous mixtures of UF_6 and hydrogen or helium at high speed along a curved wall and through a 20μm thin stripping metal plate in which the isotopic molecules are separated.

- Nozzle separation

There are also a number of other separation processes based on a wide range of separation principles. The most promising is the laser separation process, which is being particularly strongly pursued in the USA but will not be ready for production before 1995. It is based on the selective isotope excitation of $^{235}UF_6$ or ^{235}U vapor by laser beams and the subsequent separation of the excited molecules or atoms respectively from the unexcited matrix (e.g. $UF_6 \rightarrow UF_6^* \rightarrow UF_5$).

- Laser isotope separation, in the future

In all cases the wall material of the separation plants has to be selected to resist attack by the extremely corrosive UF_6 (e.g. nickel-coated steel).

1985 capacity of uranium SWU's (in 10^3 t/a):

Western World	33.3
comprising:	
• Urenco	1.5
• Eurodif	10.8
• DOE-Diffusion	19.0
• USSR (for export)	2.0

1985 demand: 23×10^3 t uranium SWU

The enrichment capacity in the Western World currently amounts to ca. 33×10^3 t uranium separation work units (SWU) per year, of which 95% uses the gas diffusion process (see marginal notes). The annual demand is currently 23×10^3 t uranium SWU. A 1300 MWe light water reactor requires ca. 150 t uranium SWU/a.

6.5.4 Reconversion of UF_6 into Nuclear Fuel

UF_6 has to be converted into a solid, chemically stable compound (UO_2)

The direct utilization of UF_6 for nuclear energy production is limited to exceptional cases such as the homogeneous gas phase reactor. As a rule the enriched UF_6 produced in the isotope separation plants has to be converted into a chemically more stable solid compound (UO_2, metallic uranium, UO_2/PuO_2), which fulfills the high specifications regarding fuel density, thermal conductivity, corrosion resistance, thermal stability and irradiation resistance.

In the manufacture of enriched UO_2 criticality must be avoided by special safety precautions i.e. by:

- Limiting the quantity of fissile material
- Minimizing the container dimensions
- Incorporating heterogeneous neutron poisons into the containers

During the reconversion of UF_6 to UO_2 and the reprocessing of spent fuel elements, criticality (i.e. the critical mass of the nuclear fuel) must be avoided at all costs. This can be achieved by limiting the quantity of fissile material, minimizing the container dimensions and incorporating heterogeneous neutron poisons e.g. boron carbide or cadmium. In addition the container walls have to be shielded against irradiation due to the increased γ-radiation from ^{235}U.

6.5.4.1 Conversion into Uranium(IV) Oxide

Conversion of UF_6 into UO_2 by:

two wet and one dry process

ADU-process is currently the most important, but is being superceded by the AUC process and the (dry) IDC process

Uranium(IV) oxide is by far the most important nuclear fuel. It is generally manufactured using either one of two wet processes or a dry process. The currently most widely used process is the ammonium diuranate process. The two more recent variants: the ammonium uranyl(VI) carbonate process and the dry process result in purer (in particular lower fluoride content) uranium(IV) oxide and hence are preferred in new reconversion plants.

6.5.4.1.1 Wet Processes

In the *ammonium diuranate (ADU) process* developed in the USA in the 1950's, the UF_6 produced by the enrichment plant is first hydrolyzed:

$$UF_6 + 2H_2O \rightarrow UO_2F_2 + 4HF$$

and then treated with ammonia in a precipitation container, whereupon ammonium diuranate (with a nonstoichiometric composition) is precipitated:

$$2UO_2F_2 + 8HF + 14NH_3 + 3H_2O \rightarrow (NH_4)_2U_2O_7 + 12NH_4F$$

The precipitate is filtered off or centrifuged, the residual HF and NH_4F removed by extraction or reprecipitation and dried at 200 °C. Finally the ammonium diuranate is converted into U_3O_8 by reaction with a H_2O/H_2 mixture at ca. 500 °C and the U_3O_8 reduced with hydrogen at 800 °C to UO_2 (resulting in a fluoride content of ca. 2% by weight).

The reductive decomposition to U_3O_8 and its reduction can also be achieved in a single step e.g. in a rotary kiln. Since the UO_2 formed can be pyrophoric, it is then weakly reoxidized.

ADU-process:

- UF_6 hydrolysis (UO_2F_2)
- Precipitation of $(NH_4)_2U_2O_7$
- Filtration, purification (reduction of the fluorine content)
- Drying, reductive calcination
- Weak reoxidation of UO_2

In the *ammonium uranyl(VI) carbonate (AUC) process* developed by NUKEM, the UF_6 is introduced with carbon dioxide and ammonia into an aqueous ammonium carbonate solution at 70 °C, whereupon ammonium uranyl(VI) carbonate is precipitated:

$$UF_6 + 5H_2O + 3CO_2 + 10NH_3 \rightarrow (NH_4)_4[UO_2(CO_3)_3] + 6NH_4F$$

The product properties and the fluoride content (generally < 0.05% by weight) are strongly influenced by the precipitation conditions.

After the precipitate is filtered off and washed with ammonium carbonate and methanol, it is converted into UO_2 in a fluidized bed kiln at 500 to 700 °C with water as the fluidizing medium and hydrogen as the reducing medium. The resulting finely divided UO_2

AUC-process (provides UO_2 with a low F content):

- UF_6 reaction with CO_2, NH_3 and water to $(NH_4)_4[UO_2(CO_3)_3]$
- Filtration, purification
- Drying, reductive calcination
- Weak reoxidation

powder (with a fluoride content of < 100 ppm) can be pyrophoric and therefore has to be weakly reoxidized with air at < 100 °C after cooling.

6.5.4.1.2 Dry Process

IDR process:

reaction of UF_6 gas with superheated H_2O vapor to UO_2F_2 (solid). Reduction to UO_2

The *integrated dry route (IDR) process* developed by British Nuclear Fuels Ltd. particularly for the manufacture of AGR-fuel is based on the two stage reaction of gaseous UF_6 with steam and hydrogen with the formation of UO_2F_2 (intermediate) and UO_2:

$$UF_6 + H_2 + 2H_2O \rightarrow UO_2 + 6HF \text{ (overall reaction)}$$

The two partial reactions take place in separate zones in a rotary kiln and produce a very pure UO_2 powder. 4000 t of enriched UO_2 had already been produced using this process by the end of 1986.

The fluoride-containing process water and gases produced in these processes are worked up to inorganic fluorine compounds or are used directly (hydrogen fluoride).

6.5.4.1.3 Manufacture of UO_2 Pellets

UO_2-sintered articles are manufactured from UO_2 by compression and sintering in the presence of H_2 at ca. 1700 °C

The uranium(IV) oxide produced by the above-described processes is exclusively used for the manufacture of UO_2-sintered pellets. The UO_2 powder is ground, mixed with an organic binder, compressed to pellets with e.g. hydraulic presses and finally sintered at 1550 to 1700 °C in the presence of e.g. hydrogen during which it shrinks to the required density (90 to 98% of the theoretical density can be attained depending upon the starting material and the process). The resulting pellets with a diameter of 10 mm are finished by grinding with cylindrical grinders, washed and dried.

The dry stored pellets have to fulfill stringent quality requirements and are constantly examined with extensive and expensive quality assurance tests.

6.5.4.2 Other Uranium Nuclear Fuels

U-metal fuel is unimportant compared with UO_2

Uranium metal: Metallic uranium is only utilized as a fuel in Magnox reactors and a few research reactors and is of minor importance compared with UO_2. It is manufactured by reducing UF_4 with metallic magnesium or calcium in batches of up to 150 kg:

Manufactured by reduction of UF_4 with Mg or Ca

$$UF_4 + 2Mg \longrightarrow U + 2MgF_2$$

The starting material for the manufacture of ^{235}U-enriched uranium metal, UF_4, is produced by reducing enriched UF_6 with hydrogen or chlorohydrocarbons.

Uranium-plutonium mixed oxides: Uranium-plutonium mixed oxides are becoming increasingly more important, since plutonium is separated during the reprocessing of spent fuel elements and irradiated breeder material. In these mixed oxide fuel elements a mixture of UO_2 and PuO_2 with a Pu content of 3 to 5% is used instead of ca. 3% ^{235}U-enriched UO_2. Such fuel elements have similar nuclear physical properties to the standard elements with ^{235}U and can therefore be used in their place.

U-Pu mixed oxides utilize the Pu isolated during the reprocessing of spent fuel elements and irradiated breeder material

Uranium-plutonium mixed oxides are also utilized in the manufacture of fuel elements for fast breeder reactors (10 to 20% PuO_2).

In their manufacture UO_2 is mixed with the corresponding quantity of PuO_2, the mixture pressed into pellets and then sintered (termed coprocessing in the USA). UO_2 is produced by one of the processes described above and PuO_2 from the aqueous nitrate solution produced during reprocessing by precipitating it as plutonium oxalate and calcining the oxalate (Sec. 6.6.2.3).

Manufacture of U-Pu mixed oxides:

1. By mixing UO_2 and PuO_2

These mixed oxides can also be manufactured by mixing the uranium and plutonium nitrate solutions produced during the reprocessing of spent nuclear fuels and converting these metal nitrate mixtures into a mixed oxide (coprecipitation). In this process the plutonium is first reoxidized, then gaseous ammonia and carbon dioxide are introduced into the aqueous nitrate mixture, whereupon ammonium uranyl(VI) plutonyl(VI) carbonate is precipitated. This can be calcined to U-Pu mixed oxides at temperatures above 500 °C.

2. By mixing the nitrate solutions of U and Pu, precipitation of ammonium uranyl plutonyl carbonate with NH_3 and CO_2, calcining to U-Pu mixed oxide

6.5.5 Fuel Element Manufacture

Fuel elements consist of nuclear fuel in cladding tubes

The nuclear fuel pellets are generally placed in tubular metal cladding to avoid leaching by coolant in the reactor core and to prevent the release of fission products into the coolant circuit. In light water reactors, for example, zirconium alloy (zirkaloy) cladding can be used.

The cladding has to be able to retain the radioactive fuel and the fission gas formed

The fuel rods for boiling and pressurized water reactors are constructed similarly. They are filled with helium to improve the heat transfer from the pellets to the casing tube and to withstand better the external pressure in the reactor and contain no fuel at the top end of the fuel rods to achieve improved fission gas retention. The latter can be ensured by holding the fuel in place with the aid of a plenum spring.

The actual fuel elements in pressurized water reactors consist of individual fuel rods and control rod tubes mounted in a self-supporting construction of spacers fitted with a top and feet. Fuel elements for boiling water reactors have in contrast no control rod tubes, the fuel element zirkaloy casings being used to guide the control rods and the coolant.

6.6 Disposal of Waste from Nuclear Power Stations

6.6.1 General Information

Transmutation of nuclear fuel in a reactor:

original fuel	after burn-up (33×10^3 MWd/t U)
	94.3% ^{238}U
3.2% ^{235}U	0.8% ^{235}U
	0.4% ^{236}U
96.8% ^{238}U	0.9% Pu
	0.1% Np,Am,Cm
	3.5% other fission products

After a residence time of on average three years and an average burn-up of 35×10^3 MW days/t uranium, the irradiated fuel elements have to be removed from the reactor, interimly stored in decay containers and then disposed of. In the case of a 1000 MWe light water reactor, ca. 30 t of spent fuel elements have to be disposed of annually.

The main task of waste disposal is the conditioning and safe-keeping of the radioactive fission products contained in the irradiated fuel elements. The following disposal procedures can be used depending on the way in which all the uranium remaining and plutonium bred (see marginal notes) is disposed of:

- Reprocessing and recycling of the fissile nuclear fuel (closed fuel cycle)
- Direct final storage of completely or partially dismantled fuel elements in geological seams

Disposal of spent nuclear fuel by:

- Reprocessing and recycling of nuclear fuels (uranium, plutonium)
- Direct final storage of fuel elements (fuel + fission products)

The reprocessing of nuclear fuel represents the only disposal route practiced on a large scale since the beginning of the peaceful use of nuclear energy. The demand for natural uranium can be reduced by up to 35% and the capacity of isotope separation plants by ca. 25% by the recovery and recycling of fissile uranium and plutonium from spent fuel elements. Reprocessing also enables entry into fast breeder technology and thereby guarantees the long term supply of electricity produced from nuclear energy.

Reprocessing represents the only currently feasible disposal route

Advantages:

- Considerable saving of natural uranium & uranium separation work
- Entry into fast breeder technology (long term securing of electricity from nuclear energy)

Disadvantages:

- High potential risk
- Uneconomic at low uranium prices

As a result of the current very low world prices for uranium concentrate and the supposed high potential risk from reprocessing plants, the direct final storage of spent fuel elements is gaining noticeably in importance in a number of countries (Sec. 6.2). Direct storage is, however, not yet technically mature and thus merits only a brief description here.

Direct final storage still being developed

6.6.2 Stages in Nuclear Waste Disposal

6.6.2.1 Interim Storage of Irradiated Fuel Elements

The spent fuel elements have to be temporarily stored for a period of at least 150 days (but generally 2 to 7 years) to allow for the decay of short-lived radioactive fission products emitting very high radioactivity and considerable heat, before reprocessing can begin or they can be transported to final storage sites. The interim storage mostly takes place in large water basins at nuclear power stations, but also sometimes in external fuel element interim storage sites. The fuel elements are thereby water cooled and at the same time surrounded by a radiation absorbing medium. Underwater storage has been practiced for decades, individual fuel elements having been "temporarily" stored in this way for up to 30 years.

Interim storage of spent fuel elements takes place initially in water basins at nuclear power station

Weakening of the cladding tube or structural components of a fuel element has never been observed. In the future, spent fuel elements will be packed together more closely to enable them to be kept longer in interim storage (compact storing site). Experience in the USA, Japan and West Germany has shown that the fuel elements do not become unduly warm and that despite the close packing of the spent fuel no self-sustaining chain reaction occurs.

Compact storage sites will be built in nuclear power stations in the future to increase their storage capacities

After two years wet interim storage in water basins, the fuel elements can be temporarily stored dry with passive air cooling. In West Germany, for example, thick-walled cast iron containers 6 m in length and 2 m in diameter are used, which at the same time function as radiation shielding and as transport containers.

In West Germany interim storage also carried out in air-cooled cast iron containers (also used as transport containers)

The interim storage of spent fuel elements in geological formations has also been recently discussed ("retrievable final storage"). The idea is that the fuel elements can stay underground if a decision is made against reprocessing.

"Retrievable final storage" in geological formations under discussion

6.6.2.2 Reprocessing of Spent Fuel Elements

The reprocessing of spent and temporarily stored fuel elements forms the link between nuclear waste disposal and fuel manufacture from used uranium and in particular from used uranium-plutonium mixed oxides, thereby closing the fuel cycle. This process also enables the utilization of more advanced reactor types such as the high temperature reactor and the fast breeder reactor.

Reprocessing of fuel elements closes the fuel cycle

The essential function of reprocessing is the separation of uranium and plutonium from one another and both of them from the highly radioactive fission products. The PUREX process (**P**lutonium and **U**ranium **R**ecovery by **EX**traction) based on extractive separation has become accepted worldwide for this purpose. Many hundred thousand tons of spent nuclear fuel have been reprocessed in the USA, Great Britain, France and several Eastern European Countries since 1954 using this USA-developed process. Most of this (ca. 800×10^3 t uranium) was from fuel elements used in the military sector (with burn-up of a few 100 MWd/t U), ca. 25×10^3 t uranium pertained to Magnox and

PUREX process used for reprocessing

Years of experience with the PUREX process

Reprocessing plants currently in operation in the Western World in:

- France
- Great Britain
- USA (only military use)
- India
- West Germany (experimental plant)
- Japan (experimental plant)

AGR fuel elements (in Great Britain) and at least 2000 t to fuel elements from light water reactors (mainly in France). Other civil reprocessing plants are currently under construction in West Germany, in France and in Great Britain and will be operational by 1995. In the USA civil reprocessing has come to a standstill.

New plants under construction in:
- France
- West Germany
- Great Britain

The available civil reprocessing capacity outside the Communist Countries (ca. 1200 t/a) is totally inadequate for the processing of the irradiated fuel elements currently being stored and those accruing in the near future. Most of the spent fuel elements accruing at the present time (ca. 3300 t/a) will therefore have to be "temporarily stored" long term. This is particularly the case in the USA, Canada and Sweden.

1986 reprocessing plant capacity outside the Communist Countries was ca. 1200 t/a

The PUREX process is in principle suitable for the reprocessing of fuel with low, medium and high burn-up originating in military, gas-graphite and light water reactors respectively as well as for the processing of fuel from fast breeder reactors with very high burn-up (up to 100×10^3 MWd/t). Thus in Western Europe (mainly France and Great Britain) 38 t of fuel from fast breeder reactors had already been reprocessed by the end of 1986.

PUREX process suitable for fuels with:

- Low burn-up (weapon plutonium)
- Medium burn-up (gas-graphite reactors)
- High burn-up (light water reactors)
- Very high burn-up (fast breeder reactors)

Fuel elements from fast breeder reactors differ from those from light water reactors in their higher plutonium and fission product content, their much shorter cooling time and their (fourteen times) higher afterheat output and therefore require a modified PUREX process.

PUREX process: Reprocessing is normally divided into three process steps:

PUREX process consists of the following steps:

- Head-End
- Extraction
- Tail-End

In the *Head-End* step the fuel elements are disassembled and cut with rod or bundle shears into ca. 5 cm long pieces in the entry basin of the reprocessing plant. The nuclear fuel is then leached out of the zircalloy-clad rods with hot concentrated nitric acid, the cladding remaining behind being collected in so-

- Cutting of the fuel elements with
  - rod shears
  - bundle shears

- Dissolution of fuel in boiling nitric acid

- Waste gas purification (separation of ^{85}Kr, ^{129}I, tritium)

called dissolver baskets. The gases released during the chop/leach step (containing radioactive ^{85}Kr, ^{131}Xe, ^{129}I, $^{14}CO_2$, tritium) are separately processed (Sec. 6.6.2.4) and then vented into the atmosphere using a 200 m high chimney.

- Removal of solids from the fuel solution

The fuel solution emerging from the dissolver is clarified in a centrifuge or filtered through a poly(propene) membrane and an accurate analysis made of the radioactive materials present. The remaining cladding pieces, the structural components of the fuel elements, the clarification sludge and the filter from waste gas purification are added to the medium active waste (MAW) for treatment. The fuel solution thus obtained contains uranium, plutonium and radioactive fission products (see Fig. 6-2).

- 1st extraction cycle:
  - separation of U and Pu from the fission products by extracting with tributylphosphate in dodecane

The clarified fuel solution is adjusted to fulfill the requirements for chemical separation (3 M nitric acid, 240 to 300 g/L uranium) and is *extracted* with a 30% solution of tributylphosphate in dodecane, kerosene or hydrogenated propene tetramers (HPT). In the first cycle the valuable materials (uranium and plutonium) are extracted into the organic phase using pulsed sieve plate columns or mixer-settlers. Most of the fission materials are left behind in the aqueous phase and after concentration and acid recovery are added to the high activity waste (HAW) for disposal.

  - separation of U and Pu through
    - reduction of Pu^{IV} to Pu^{III}
    - extraction of Pu^{III} into the aqueous phase
    - extraction of U with dilute nitric acid into the aqueous phase

The uranium and plutonium-containing organic phase is then reacted with a reducing agent e.g. hydrazine or a uranium(IV) nitrate solution, or electrochemically, whereupon Pu^{IV} is reduced to Pu^{III}. Pu^{III} is insoluble in the organic phase and can therefore be stripped into the aqueous phase. The uranium remaining in the organic phase is finally stripped with 0.01 M nitric acid. The first extraction cycle thus separates the fuel solution into three aqueous solutions containing uranium, plutonium and the fission products respectively (see Fig. 6-2).

- 2nd and 3rd extraction cycles:

  separation of entrained fission products (Np, Tc) and plutonium (from the uranium cycle); acceptable residual plutonium content: 10 ppb

The residual fission products (e.g. neptunium, technetium) are removed from both uranyl(VI) and plutonium nitrate solutions in two further extraction cycles in which Pu^{III} has to be reoxidized to Pu^{IV} either with the aid of nitrogen oxides or anodically.

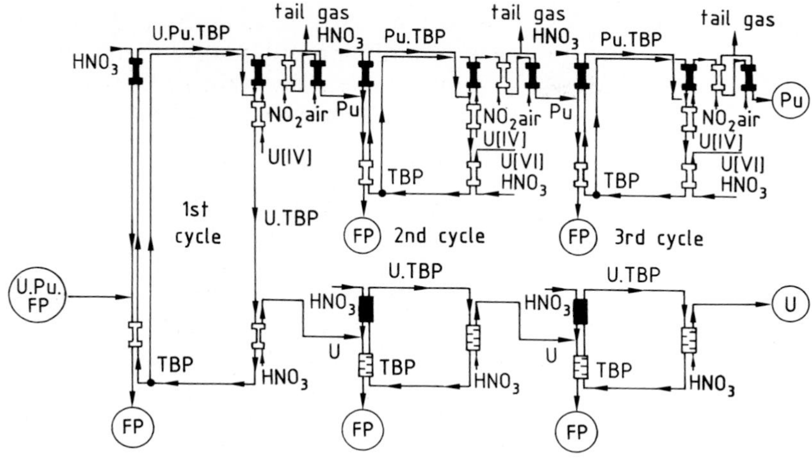

Fig. 6-2. Flow sheet of the mass transport in the PUREX process.

Through this multicyclic extraction process, uranium and plutonium are separated from the fission products with decontamination factors of 10^6 to 10^7. The residual plutonium content in the uranium recovered may not exceed 10 ppb, since uranium can then be further processed with reduced protective measures. The recovery efficiency of uranium and plutonium is 98 to 99%.

Decontamination factor for U and Pu: 10^6 to 10^7

Recovery efficiency of U and Pu: 98 to 99%

The radiolysis products formed by the radiation of tributylphosphate have to be continuously removed by washing the extraction agent with sodium carbonate solution. Entrained or dissolved organic compounds can form difficultly decomposable products with metal nitrates and therefore have to be removed by steam distillation or scrubbing with kerosene before the aqueous solution is concentrated.

Tributylphosphate is recycled after interfering impurities have been removed

In the *Tail-End* the end-products of the PUREX process are finally produced. In the case of uranium, this is a highly concentrated uranyl(VI) nitrate solution, the concentration of the plutonium nitrate solution being somewhat lower: ca. 250 g/L.

The PUREX process is carried out at temperatures not exceeding 130 °C at or slightly below atmospheric pressure and utilizes aqueous solutions and extraction processes tried and tested in the chemical industry. The entire operation is carried out by remote control in bunker-like rooms with meter-thick concrete walls (hot cells) for radiation protection reasons.

6.6.2.3 Further Processing of Uranium and Plutonium Solutions

Further processing of U directed to further U-enrichment:

- For low ^{235}U contents: conversion to UO_3
- For sufficient ^{235}U contents: conversion to UF_4 or UF_6

The further processing of the uranyl(VI) nitrate solution, which in some plants is postpurified with silica gel, is directed towards further enrichment of the uranium. If this is not worthwhile due to a too low ^{235}U-content, the product is converted into uranium(VI) oxide, a storable compound. This can serve as a starting material suitable for possible later utilization in a fast breeder reactor. The uranium(VI) oxide is either produced indirectly by way of ammonium diuranate or by direct calcination. If further enrichment is foreseen, uranium(VI) fluoride or uranium(IV) fluoride is produced, the latter being fluorinated to UF_6.

Further processing of Pu: production of PuO_2 by way of Pu-oxalate and conversion to U-Pu mixed oxide fuels

Many years experience in the manufacture of mixed oxide fuel elements

Plutonium is usually precipitated as the oxalate, which is converted into plutonium(IV) oxide from which mixed oxide fuel elements for light water reactors or fast breeder reactors can be manufactured. In the planned West German reprocessing plant, uranium-plutonium oxides will be produced by coprecipitation. Countries such as the USA, France, Great Britain, Japan, Belgium and West Germany possess mixed oxide fuel element manufacturing knowhow. Thus in West Germany in 1986 alone, 27 000 fuel rods were produced from reprocessed fuel for light water reactors and 26 000 fuel rods for fast breeder reactors.

6.6.2.4 Conditioning of Radioactive Waste

Quantities of waste produced during the reprocessing of fuel elements (30 t, burn-up 33×10^3 MWd/t):

- LAW 120 t
- MAW 60 t
- HAW 1 t
- Cladding scrap 12 t
- Gaseous 6 Nm^3

The solid, liquid and gaseous radioactive waste (see marginal notes) produced during the reprocessing of irradiated fuel elements has to be securely stored over a period of up to 100 000 years (and longer) to prevent its entry into the biosphere. This holds particularly for the high activity aqueous refined product (HAW) from the first extraction cycle. Medium and low activity waste (MAW, LAW) must also be converted into a chemically stable (mostly solid) and final storable form (conditioned). These include the washing liquors from the scrubbing step, cladding and structural components from fuel elements, contaminated solid waste of all sorts (paper, plastics, protective clothing etc.), dissolver sludge and radioactive gases.

The treatment of liquid radioactive waste is directed to concentrating the activity into the smallest possible volume. The high activity fission product solutions are thus concentrated to ca. 1/10 of their initial volume by distilling off the nitric acid. The resulting waste is a self-heating liquid, which can be stored in cooled stainless steel containers. Vitrification has proved to be particularly useful in converting these HAW-solutions into a permanently storable form. Borosilicate glass is the preferred glass matrix for the embedding of up to 30% by weight of concentrated, denitrated and dried HAW.

Conditioning of high activity solutions (HAW):

- Concentration to ca. 1/10 of the initial volume by distilling off the nitric acid
- Vitrification by embedding in a glass matrix

These glasses are stable to irradiation and are only very slightly leached by water or aqueous salt solutions, which is important in the envisaged permanent storage of the radioactive fission products, and dissipate the decay-heat efficiently to the environment. The vitrification of highly active waste also enables above ground long term interim storage e.g. in air-cooled dry stores.

Glasses are stable to irradiation, resistant to leaching and dissipate the decay-heat efficiently to the environment

The vitrified end-product resulting from the disposal of HAW solutions takes up a very small volume e.g. 3 m^3 for 30 t fuel with a burn-up of 30×10^3 MWd/t uranium.

Vitrification is operated on a large scale in France (Marcoule). Since 1978 ca. 1000 m^3 of highly concentrated fission product solutions have been processed there using the AVM(Atelier-de-Vitrification Marcoule)-process. The concentrate is first evaporated, denitrated and dried, then melted in an electrically heated rotary furnace with ground glass frit and finally processed to glass cylinders 50 cm in diameter and 1 m long.

Vitrification using the AVM-process carried out on a large scale in France

Vitrification process consists of:

- Concentration
- Drying
- Calcination
- Production of the glass melts
- Solidification of the glass

A pilot plant at Mol (Belgium) based on an alternative process developed in West Germany, has already processed 50 m^3 of HAW concentrate since coming on stream in 1986. In this process the liquid fission product concentrate is fed without prior solidification into a ceramic furnace and melted at 1200 °C with separately

End products of vitrification:

- Glass cylinders (0.5 m in diameter, 1 m long)
- Glass beads embedded in lead

1 t fuel results in 0.06 m^3 glass

added glass frit. The melt can then be either cast to glass cylinders (as with the AVM-process) or processed with a slowly rotating steel disc into glass beads. The latter are put into cylindrical steel containers and lead poured over them.

Similar plants exist in the USA, Great Britain and India. Other solidification possibilities are embedding the fission products in ceramic masses, in sintered oxide or in graphite.

Conditioning of medium activity liquid waste (MAW):

- Concentration to 1/10 of the initial volume
- Solidification by stirring into bitumen or concrete

Medium activity waste from extraction agent scrubbing and the decontamination operations particularly important in connection with repairs and maintenance work is first treated by evaporation to ca. 1/10 of the initial volume. The concentrate obtained can then be solidified by stirring into hot bitumen or concreting resulting in ca. 1.6 m^3 bituminized mass/t fuel. Utilization of hydrazine carbonate as the scrubbing extraction agent (instead of sodium carbonate) and electro-oxidation of the used hydrazine scrubbing solution to nitrogen and carbon dioxide, reduces the waste volume still further.

Conditioning of medium activity solids (fuel element cladding etc.):

- Encased in concrete
- Dissolver sludge must be previously solidified and cooled

The fuel element cladding and assembly parts (still containing ca. 0.1% nuclear fuel and tritium) left behind after leaching the nuclear fuels are also encased in concrete barrels. Dissolver sludge is a problem due to its high ruthenium content, which means that it can only be encased in concrete after long cooling times.

Conditioning low activity waste (LAW):

- Distillative concentration and solidification of the concentrate
- Burning and solidification of the ash
- Tritium-containing water are encased in concrete or injected into porous reservoirs

The weak activity waste materials (aqueous and organic solutions, filter membranes) resulting from different process steps can be concentrated by burning and solidified by encasing in concrete in long-term storable barrels. The decontaminated distillate produced during distillative concentration can be returned to the reprocessing process.

Tritium-containing water (HTO) from acid recovery can be disposed of by encasing in concrete in casks, by injection into pore-containing geological formations or, after prior enrichment, by permanent fixing in metallic reservoirs (as HT), zeolites etc.

Conditioning of gaseous radioactive products:

- Aerosols:
  retained by scrubbers and filters

The gaseous radioactive products (^{129}Iodine, radioactive aerosols, ^{85}Krypton, $^{14}CO_2$, tritium) released by the dissolution of nuclear fuel are handled differently. Radioactive aerosols are removed by scrubbing and with electrostatic and air filters. ^{129}Iodine can be absorbed on silver nitrate-impregnated silica-based solid bed filters. The decontamination factors achievable in this process are very high ($> 10^3$). ^{85}Krypton is currently vented into the atmosphere from all reprocessing plants. In a process developed in West Germany, it can be separated from the prepurified residual gas (mainly nitrogen) by low temperature rectification and stored in gas cylinders. ^{133}Xenon and ^{3}Hydrogen (tritium) are also vented into the atmosphere in small quantities under strictly controlled conditions, the environmental contamination being negligible.

- ^{85}Krypton:
  currently vented into the atmosphere. Separation by low temperature rectification possible
- ^{129}Iodine:
  absorbed on $AgNO_3$-impregnated silica
- Tritium:
  60% bound in fuel element cladding. Gaseous tritium vented into the atmosphere

The contamination of the population by the small amounts of radioactive materials released from reprocessing plants in the tail gases and effluents, despite the measures described above, is radio-ecologically innocuous and well below the legal limits.

Release of small quantities of radioactive materials with the tail gases and effluent of reprocessing plants is radio-ecologically innocuous

6.6.2.5 Permanent Storage of Radioactive Waste

The last stage in the disposal of nuclear power station waste is a long-term secure final storage of the solidified high, medium and low activity waste. The fission products ^{90}Strontium, ^{137}Cesium and the long-lived α-emitting actinides (neptunium, plutonium) represent a particular risk, so that safe-keeping of the waste in a permanent store can only be achieved if its stability can be guaranteed over geological periods. Nowhere in the world is there long-term experience of the permanent storage of self-heating HAW. The first high activity permanent storage barrel will probably be stored in a salt cavern at Gorleben (West Germany) at the end of this century. The lowering of HAW encased in cast iron into 300 m deep bore holes is envisaged, with reusable shielding containers being utilized for the transport.

Permanent storage site for radioactive waste must be stable over geological periods. Under discussion are:

- Rock salt formations deep underground
- Granite
- Clay
- Anhydrite

The first permanent storage site for HAW will probably be operational in the year 2000 in West Germany in salt formations

Other countries are looking at the long-term interim storage of radioactive waste (e.g. in France) or at direct final storage (Sec. 6.6.2).

Medium and low activity wastes are, due to their much larger volume, less suitable for above-ground interim storage. Storage in underground repositories (mines etc.) is therefore practiced in a number of countries.

Medium and low activity wastes are currently in repositories near the surface

6.6.3 Direct Permanent Storage of Irradiated Fuel Elements

Direct permanent storage:

development began in the mid 1970's particularly in the USA, Canada and Sweden

The direct permanent storage of unchopped fuel elements first began to be considered in the mid 1970's. However, a number of countries (the USA, Canada, Sweden, Finland and Spain) have already opted for this form of waste disposal. At present it is unlikely that permanent storage by deposition of burnt-up (light water reactor) fuel elements in geological formations will be realized before the turn of the century (this is most advanced in the USA).

Process steps:

- Interim storage (decay of fission product radioactivity)
- Conditioning (packing of unchopped fuel elements in closed final storage containers)
- Transport to the permanent storage site in thick shielded containers
- Deposition in up to 800 m deep bore holes in geologically stable formations

The concepts for the direct permanent storage in the different countries are technically different. They have in common the encapsulation of several previously interimly stored fuel elements in corrosion resistant cylindrical containers of structural steel, copper or titanium alloys and the deposition of these containers in geologically stable formations (rock salt, granite and other hard rocks) in up to 800 m deep holes. The handling of the very large and very heavy waste barrels (up to 60 t) required for shielding during transportation is a problem.

The basic advantage of direct permanent storage over the permanent storage of reprocessed fuel elements is the comparatively easy process management (only a single waste product is formed). Disadvantageous is the higher heat developed in nonreprocessed plutonium-containing fuel elements. The question of long-term monitoring of nuclear materials remains open.

References for Chapter 6: Nuclear Fuel Cycle

General Information:

Encyclopedia of Physical Science and Technology, Vol. 9, pp 321 – 388, Academic Press Inc., Orlando/Florida (1987).

Jahrbuch der Atomwirtschaft **18,** Handelsblatt GmbH, Verlag für Wirtschaftsinformationen, Düsseldorf, Frankfurt (1987).

Ullmanns Encyklopädie der technischen Chemie, 4th Ed., Vol. 14, pp 13 – 170, Verlag Chemie, Weinheim-Deerfield Beach-Basel (1977).

E. Gärtner, "Uran, Produktion und Gewinnung, Brennstoff-Kreislauf und möglicher Beitrag zur Energieversorgung" (Uranium, production and extraction, nuclear fuel cycle and possible contributions to energy supply), *Jahrbuch für Bergbau, Energie, Mineralöl und Chemie (1977/78).*

Specific Information:

Economic Importance of Nuclear Energy:

International Nuclear Fuel Cycle Evaluation (INFCE), *International Energy Agency, IAEA*, Vol. 1 to 9, Vienna (1980).

"The costs of generating electricity in nuclear and coal fired power stations", OECD/NEA study (1982).

"Energistik" (Energistics), Study of ESSO AG (1986).

K. P. Messer: "Wirtschaftlichkeitsfragen des Kernbrennstoffkreislaufs" (Economic questions concerning the nuclear fuel cycle), *Atomwirtschaft* **29,** pp 422–425 (1984).

K. Heinloth: "All unsere Energie-Quellen" (All our energy sources), *Physik i. u. Zeit* **18,** pp 47–51 (1987).

K. Knizia: "Energie für die Welt von morgen" (Energy for the World of tomorrow), *Atomwirtschaft* **32,** pp 74–79 (1987).

J. Haggin and J. Krieger: "Biomass becoming more important in the US energy mix", *Chem. Eng. News,* pp 28–29 (March 1983).

Monthly Energy Review, Energy Information Administration, pp 1–38, Washington (March 1987).

Statistik der Energiewirtschaft 1985/86 (Statistics of energy economics 1985/86), Vereinigung Industrielle Kraftwirtschaft, Verlag Energieberatung GmbH, Essen (1987).

H. Magerl: "Welt-Elektrizitätsversorgung" (World electricity supply), *Atomwirtschaft* **31,** pp 27–30 (1986).

"Spannung aus Paris" (Electricity from Paris), *Wirtschaftswoche* Nr. 34 (1987).

G. H. Kohlmaier: "Energiestrategien und CO_2-Risiko" (Energy strategies and the CO_2 risk), *Umsch. Wiss. Tech.* **81,** pp 648–651 (1981).

A. Schlüter: "Kernfusion und das Energieproblem" (Nuclear fusion and the energy problem), *Naturwissenschaften* **69,** pp 226–235 (1982).

Nuclear Reactors:

Kirk-Othmer: Encyclopedia of Chemical Technology, 3rd Ed., Vol. 16, pp 138–247, John Wiley & Sons, Inc., New York-Chichester-Brisbane-Toronto (1981).

W. Braun: "Advanced light water reactors – even more reliable and cost effective", *AtomKernenergie-Kerntechnik* **48,** pp 2–6 (1986).

E. Treuchert et al.: "Physics Features of the HTR for Process Heat", *Nuclear Eng. and Design* **78,** pp 147–154 (1984).

I. R. Cameron: *Nuclear Fission Reactors,* Plenum Press, London (1982).

"Chernobyl Report, Coping with the human factor, *Nature* **323,** pp 25–28 (Sept. 1986).

J. G. Collier and G. F. Hewitt: *Introduction to Nuclear Power,* Springer Verlag, Heidelberg (1987).

H. Eckey, K. G. Emmert and P. Kilian: "Kernphysikalische Schwerpunkte der Reaktortechnik" (Important nuclear physical aspects of nuclear reactor technology), *Physik in unserer Zeit* **11,** pp 18–21, pp 47–59 (1980).

Supply:

Gmelin: *Handbuch der anorg. Chemie,* 8th Ed., "Uran, Ergänzungsband A 1 Uranlagerstätten" (Uranium, Supplementary Volume A 1 Uranium Deposits), Springer-Verlag (1979).

Gmelin: *Handbuch der anorg. Chemie,* 8th Ed., "Uran, Ergänzungsband A 3 Technologie, Verwendung" (Uranium, Supplementary Volume A 3 Technology, Applications), Springer-Verlag (1981).

A. Naylor, J. F. Ellis and R. H. Watson: "Chemical Aspects of Nuclear Fuel Fabrication Processes", *J. Chem. Tech. Biotechnol.* **36,** pp 162–168 (1986).

H. D. Fuchs and D. Wentzlau: "Uran – Exploration, Bergbau und Technik der Erzaufarbeitung" (Uranium – exploration, mining and technology of ore processing), *Chemiker-Zeitung* **109,** pp 383–389 (1985).

K. Schwochau et al.: "Probleme und Ergebnisse der Uran-Gewinnung aus Meerwasser" (Problems and results of uranium production from seawater), *Chemiker-Zeitung* **107,** pp 177–189 (1983).

D. A. O'Sullivan: "Strong Rise Forecast in Global Uranium Demand", *Chem. Eng. News* pp 26–27 (March 1987).

W. Bacher and E. Jacob: "Uranhexafluorid – Chemie und Technologie eines Grundstoffs des nuklearen Brennstoff-Kreislaufs" (Uranium hexafluoride – chemistry and technology of a raw material of the nuclear fuel cycle), *Chem. Z.* **106,** pp 117–136 (1982).

W. de Ruiter: "Laser-Trennung von Isotopen des Urans und Plutoniums", *Endeavour* **8,** 128 (1984).

B. R. T. Frost: *Nuclear Fuel Elements,* Pergamon Press, New York (1982).

The Uranium Market 1986–2000, Uranium Institute (Nov. 1986).

J. Pellisier Tanon and L. F. Durret: *Trends in the Conversion Market,* Uranium Institute 11th Annual Symposium, London (Sept. 1986).

W. Ehrfeld and U. Ehrfeld, "Uranisotopentrennung" (Uranium isotope separation), *Chem. Z.* **101,** pp 53–63 (1977).

Nuclear Waste Disposal:

J. D. Bredehoeft and T. Maini: "Strategy for Radioactive Waste Disposal in Crystalline Rocks", *Science* **213,** pp 293–296 (July 1981).

H. Eccles and A. Naylor: "Solvent extraction in the nuclear fuel cycle", *Chemistry and Industry* **6,** pp 174–179 (March 1987).

G. Baumgärtel and R. P. Randl: "Zur Wiederaufarbeitung abgebrannter Kernbrennstoffe aus Schnellen Brutreaktoren" (Concerning reprocessing spent nuclear fuel from fast breeder reactors), *Chemiker-Zeitung* **111,** 95 (1987).

K. D. Closs: "Direkte Endlagerung ausgedienter Kernbrennelemente" (Direct final storage of spent nuclear fuel elements), *Chemiker-Zeitung* **110,** 251 (1986).

U.S. Department of Energy, Office of Civilian Radioactive Waste Management: "Mission Plan for the Civilian Radioactive Waste Program", DOE/RW–0005, Vol I (1985).

P. E. Ahlström et al.: "A Possible Strategy Based on Final Disposal of Unreprocessed Spent Nuclear Fuel, The Swedish KBS–3 Concept". *Proc. Int. Meeting on Fuel Processing and Waste Management,* Aug. 26–29, 1984, Jackson, WY, pp 1–276.

W. T. Hancox et al.: "Radioactive Waste Management in Canada." Paper presented at the *Int. Symposium on the Siting, Design and Construction of Underground*

Repositories for Radioactive Wastes, Hanover, West Germany, 3 – 7 March 1986, IAEA-SM 289/2.

J. Mischke: "Wiederaufarbeitung von Kernstoffen" (Reprocessing of nuclear materials), *Chem.-Ing.-Tech.* **57,** 102 (1985).

F. Baumgartner: *"Chemie der Nuklearen Entsorgung"* (Chemistry of nuclear waste disposal). Verlag K. Thiemig, München (1978).

R. Kroebel: "Die Entsorgung von Kernkraftwerken" (Waste disposal from nuclear power stations), *Nachr. Chem. Tech. Lab.* **30,** pp 372 – 377 (1982).

W.-J. Schmidt-Küster: "Das Entsorgungskonzept der Bundesregierung" (The nuclear waste disposal concept of the West German Government), *Atomwirtschaft* **25,** pp 294 – 299 (1980).

K. Müller-Christiansen and M. Wollessen: *Plutonium.* Gesellschaft für Reaktorsicherheit mbH, Reihe Stellungnahmen zu Kernenergiefragen (GRS – S – 27) Köln (1979).

W. Koelzer: "Auswirkungen radioaktiver Emissionen von Kernkraftwerken und Wiederaufarbeitungsanlagen auf die Umgebung" (Effects of radioactive emission from nuclear power stations upon the environment). *Reihe "Kernthemen",* Deutsches Atomforum e.V., Bonn (1979).

Company Abbreviations Index

Abbreviation	Official Name and Headquarters
Albright and Wilson	Albright and Wilson PLC, London, UK; subsidiary of Tenneco Inc., Houston, USA
Alcoa	Aluminum Company of America, Pittsburgh, USA
Allied Chemical	Allied Chemical Corp., New York, USA
American Chrome	American Chrome & Chemicals Inc., Corpus Christi, USA; subsidiary of Harrisons & Crossfield PLC, London, UK
Asahi Chemical	Asahi Chemical Industry Co., Ltd., Tokyo, Japan
Asahi Glass	Asahi Glass Co., Ltd., Tokyo, Japan
Associated Minerals	Associated Minerals Consolidated Ltd. (AMA), Newcastle and Southport, Australia
Atelier-de-Vitrification Marcoule	Atelier-de-Vitrification, Marcoule, France
BASF	BASF AG, Ludwigshafen, West Germany
Bayer	Bayer AG, Leverkusen, West Germany
Bertrams	Bertrams AG, Basle, Switzerland; subsidiary of Davy Int. Ltd., London, UK
de Beers	de Beers Consolidated Mines Ltd., Kimberly, South Africa
Benilite Corp. of America	Benelite Corp. of America, New York, USA
Braun	John Braun, Engineering & Constructors Ltd., London, UK
Breyer Heurty	Heurty S.A., Paris, France
British Chrome Chemicals	British Chrome & Chemicals, Eagles Cliff, UK
British Nuclear Fuel	British Nuclear Fuel Ltd., subsidiary of the United Kingdom Atomic Energy Authority, London, UK
Business Communications Co.	Business Communications Co., Stamford, USA
Buss	Buss Ltd., Pratteln, Switzerland
CdF-Chimie	Charbonnage de France, Paris, France

Company Abbreviations Index (cont.)

Abbreviation	Official Name and Headquarters
Chemie Linz	Chemie Linz AG, Linz, Austria
Chiyoda	Chiyoda Chem. Eng. & Construction Co. Ltd., Tokyo, Japan
CIBA-Geigy	CIBA-Geigy AG, Basle, Switzerland
Comurhex	Comurhex, Paris, France
Davy McKee	Davy McKee AG, Frankfurt/M.; subsidiary of Davy Int. Ltd., London, UK
Degussa	Degussa AG, Frankfurt/M., West Germany
Dhrangadhra Chemicals	Dhrangadhra Chemical Works, Bombay, India
Diamond Shamrock	Diamond Shamrock Corp., Cleveland, USA
Dorr	Dorr Co. Inc., Stamford, USA
Dow Chemical	Dow Chemical Comp., Midland, USA
DSM	N.V. DSM, Nederlands Staatsmijnen, Heerlen, The Netherlands
DuPont	E. I. DuPont de Nemours and Comp. Inc., Wilmington, USA
Dynamit Nobel	Dynamit Nobel AG, Troisdorf, West Germany
Eldorado Nuclear	Eldorado Nuclear, Ltd., Ontario, Canada
Erco	Erco Industries Ltd., Islington, Canada
Fairmont	Fairmont Chemical, Newark, USA
Fisons	Fisons Ltd., Felixstowe, UK
FMC	Food Machinery & Chemicals Corp., New York, USA
Ford	Ford Motor Comp., Dearborn, USA
Franklin Key	Franklin Key Inc., Valley Forge, USA
General Electric	General Electric Comp., New York, USA
Gesellschaft f. Elektrometallurgie	GfE, Gesellschaft für Elektrometallurgie mbH, Düsseldorf, West Germany; subsidiary of Metallurg Inc., New York, USA
Giulini	Gebr. Giulini GmbH, Chemische Werke, Ludwigshafen, West Germany
Gulf	Gulf Oil Corporation, Pittsburgh, USA

Company Abbreviations Index (cont.)

Abbreviation	Official Name and Headquarters
Haldor-Topsoe	Haldor-Topsoe A/S, Lyngby, Denmark
Hargreaves	Hargreaves Fertilizers Ltd., York, UK; subsidiary of ICI PLC, London, UK
Hoechst	Hoechst AG, Frankfurt/M., West Germany
Hooker	Hooker Electrochemical Co., New York, USA
Hüls	Chemische Werke Hüls AG, Marl, West Germany
ICI	Imperial Chemical Industries PLC, London, UK
Interox	Interox Chemicals Ltd., Widnes, Cheshire, UK; subsidiary of Laporte Industries (Holdings) Ltd., London,UK and Solvay & Cie SA, Brussels, Belgium
Inventa	Inventa AG für Forschung und Patentverwertung, Zürich, Switzerland; subsidiary of EMS-Chemie Holding AG, Donat, Ems, Switzerland
Ishihara Sangyo Kaisha	Ishihara Sangyo Kaisha Ltd., Osaka, Japan
Kali-Chemie	Kali-Chemie AG, Hannover, West Germany; subsidiary of Hoechst AG, Frankfurt/M., West Germany
Kellog	Kellog Int. Corp., Houston; subsidiary of Pullmann Inc., Chicago, USA
Knauf	Gebr. Knauf, Westdeutsche Gipswerke Iphofen, West Germany
Koppers	Krupp-Koppers GmbH, Essen, West Germany
Krupp	Fried. Krupp GmbH, Essen, West Germany
Lonza	Lonza AG, Basle; subsidiary of Alusuisse, Zürich, Switzerland
Lurgi	Lurgi Gesellschaften, Frankfurt; subsidiary of Metallgesellschaft AG, Frankfurt/M., West Germany
Malaysian Titanium Corp.	Malaysian Titanium Corp., (SDM Berhad), Ipoh Perank, Malaysia
Merck	Merck AG, Darmstadt, West Germany
Metallgesellschaft	Metallgesellschaft AG, Frankfurt/M, West Germany
MIAG	Bühler-Miag GmbH, Braunschweig, West Germany (Plant Constructor)
Mitsubishi	Mitsubishi Chemical Industries Ltd., Tokyo, Japan

Company Abbreviations Index (cont.)

Abbreviation	Official Name and Headquarters
Mitsui-Toatsu	Mitsui-Toatsu Chemicals Inc., Yokohama, Japan
Mobay	Mobay Chemical Corp., Pittsburgh, USA; subsidiary of Bayer AG, Leverkusen, West Germany
Monsanto	Monsanto Comp., St. Louis, USA
Montedison	Montedison S.p.A., Milan, Italy
MPLC	Minerals Processing Licensing Corp., Monaco
Nippon Carbide	Nippon Carbide Kogyo K. K., Tokyo, Japan
Nippon Carbon	Nippon Carbon Co. Ltd., Tokyo, Japan
Nippon Chemical	Nippon Chemical Ind., Tokyo, Japan
Nippon Denko	Nippon Denko K. K., Tokyo, Japan
Nippon Hydrazine	Nippon Hydrazine Co., Inc., Tokyo, Japan
Nissan	Nissan Chemical Ind., Ltd., Tokyo, Japan
Norsk Hydro	Norsk Hydro, Porsgrunn, Norway
Nukem	Nukem, Hanau, West Germany
Nynäs Petroleum	Nynäs Petroleum AB, Stockholm, Sweden
ORNL	Oak Ridge National Laboratory, Oak Ridge, USA
Occidental	Occidental Chemical Co., Houston, Texas, USA
Olin	Olin-Mathieson Chem. Corp., New York, USA
Otsuka Kagaku	Otsuka Kagaku Yakuhin K. K., Osaka, Japan
ÖSW	Österreichische Stickstoffwerke AG, Linz, Austria
Outokumpu	Outokumpu Oy, Helsinki, Finland
PCUK	Produits Chimiques Ugine Kuhlmann, Paris, France; subsidiary of Pechiney Ugine Kuhlmann (up to 1982)
Pennwalt	Pennwalt Corp., Buffalo, USA
Pilkington	Pilkington Brothers, St. Helens, UK
Plinke	Adolf Plinke Söhne, Chemisch-technisches Büro, Bad Homburg, West Germany
PPG	Pittsburgh Plate Glass, Pittsburgh, USA

Company Abbreviations Index (cont.)

Abbreviation	Official Name and Headquarters
Prayon	Société de Prayons S.A., Forêt-Trooz, Belgium
Rhône-Poulenc	Société des Usines Chimiques Rhône-Poulenc, Paris, France
Rhône-Progil	Rhône-Progil SA, since 1980 Rhône-Poulenc Agrochimie SA, Courbevoie, France; subsidiary of Rhône-Poulenc, Paris, France
Richards Bay Minerals (RBM)	Richards Bay Minerals,Iron & Titanium Ltd., Richards Bay, South Africa
Ruthner	Ruthner Industrieanlagen AG, Vienna, Austria
Sèvres	Cristal de Sèvres, Choisy-la-Roi, Hauts de Seine, France
Shieldalloy	Shieldalloy Corp., Newfield, USA
Siemens	Siemens AG, Berlin-Munich, West Germany
Shell	Royal Dutch Shell Konzern, The Hague, The Netherlands
Snam-Progetti	Snam-Progetti S.p.a., Milan, Italy
H. C. Starck	Hermann C. Starck Berlin GmbH & Co K.G., Düsseldorf, West Germany; subsidiary of Bayer AG, Leverkusen, West Germany
Swift	Swift & Comp., Chicago, USA
Stamicarbon	Stamicarbon B.V.; subsidiary of N.V. DSM, Heerlen, The Netherlands
St. Gobain	Compagnie de Saint Gobain SA, Paris, France
Stoppani	Stoppani, Milan, Italy
Taiwan Alkali	Taiwan Alkali Comp., Taipeh, Taiwan
Texaco	Texaco Inc., New York, USA
Thann	Fabriques des Produits Chimiques de Thann et de Mulhouse, Thann, France; subsidiary of Rhône-Poulenc, Paris, France
Titanium Corp. (Quebec Iron)	Quebec Iron and Titanium Corp. (QIT), Quebec, Canada
Tokuyama Soda	Tokuyama Soda Ltd., Tokuyama City, Japan
Union Carbide	Union Carbide Corp., New York, USA
Ugine Kuhlmann	Pechiney Ugine Kuhlmann SA (PUK), Paris, France; since 1983 Pechiney

Company Abbreviations Index (cont.)

Abbreviation	Official Name and Headquarters
Uhde	Friedrich Uhde GmbH, Dortmund, West Germany
UKF	Unie van Kunstmestfabrieken B.V., Utrecht, The Netherlands
VAW	Vereinigte Aluminiumwerke, Berlin-Bonn, West Germany
Wacker Chemie	Wacker Chemie GmbH, Munich, West Germany
Wedgewood	Wedgewood Potteries, Etruria, UK

Subject Index

activated carbon 498 ff.
– applications 6 ff., 123 ff., 503 ff.
– economic importance 498
– reactivation and regeneration 502
activated carbon manufacture 498 ff.
– chemical activation 499 ff.
– gas activation 500 ff.
– general information 498 ff.
active aluminas 252
active carbon black 496
ADU process 581
aerated concrete bricks 376
alkali bromates
– applications 183
alkali chromates
– manufacture 258 ff.
alkali compounds
– general information 215
alkali metals
– general information 215
alkaline earth compounds
– general information 231
alkaline earth metals
– general information 231
alkali peroxodisulfates
– applications 30
– economic importance 22
alkali peroxodisulfide
– production 27 ff.
alumina cement 384 ff.
aluminum 247 ff.
– applications 249 ff., 515
– economic importance 248
– manufacture 248 ff.
– occurrence 247 ff.
aluminum chloride 253 ff.
aluminum fluoride 140 ff., 249
aluminum fluoride manufacture
– Chemie Linz process 141
– from hexafluorosilic acid 141
– from hydrogen fluoride 141
– UKF process 141
aluminum hydroxide
– applications 252 ff., 512
– economic importance 250
aluminum hydroxide filler
– manufacture 512 ff.
aluminum hydroxide manufacture 251 ff.
– Bayer process 251 ff.
aluminum nitride products 455
aluminum oxide 250 ff., 440 ff., 445
– applications 252 ff., 515
aluminum oxide fibers 362 ff.
– applications 364
– manufacture 363 ff.
aluminum oxide manufacture 251
– Bayer process 251
aluminum silicates 310
aluminum sulfate 4, 253
aluminum titanate 444
ammonia 30 ff.
– applications 44 ff., 46, 48, 53, 56, 66, 101, 114, 195 ff., 201, 203 ff., 581
– economic importance 31
ammonia manufacture 31 ff.
– catalysts 39 ff.
– synthesis gas production 33 ff.
ammonium 144
ammonium carbonate 572, 581
ammonium dichromate
– applications 265
– manufacture 262
ammonium diuranate 581
ammonium hydrogen carbonate 53
ammonium hydrogen fluoride 143 ff.
ammonium nitrate
– economic importance 200 ff.
– manufacture 203 ff.
ammonium perchlorate
– applications 177
– economic importance 169
– manufacture 175
ammonium phosphate
– economic importance 80 ff.
ammonium phosphates 80 ff., 191
– applications 81
– economic importance 191
– liquid fertilizer manufacture 196 ff.
– manufacture 84
– solid fertilizer manufacture 195 ff.
ammonium sulfate
– economic importance 200
– manufacture 201 ff.
ammonium thiosulfate 126 ff.
– applications 127
– manufacture 127
ammonium uranyl(VI) carbonate 581
anhydrite, natural 393
anhydrite, synthetic 139, 393 ff.
Anhydrite II 390 ff.
Anhydrite III 390 ff.
apatite 68 ff., 192, 197, 575
asbestos cement 385
asbestos fibers 333 ff.
– deposits 333
– economic importance 333

– extraction 335
– properties and structure 335ff., 338
– toxic properties 337
– types and compositions 334
asbestos fibers applications 337
– general information 337
– in asbestos textiles and filter materials 341
– in composite materials 340ff.
AVC process 581
AVM process 591

barium
– economic importance 244
– natural deposits 244
barium carbonate 244ff., 513
– applications 245
– economic importance 244
– manufacture 244ff.
barium sulfate 245, 514ff.
– applications 508
barium sulfate filler
– manufacture 514ff.
barium sulfide 245
barium titanate 445ff.
basic chromium sulfate
– application 266
basic chromium sulfates 264ff.
beryllium
– applications 232
– economic importance 232
– manufacture 232
– oxide 232, 443
bismuth oxychloride 552
bleaching powder 171, 177
bonding agents 371
bone china 430
boron carbide products 453ff.
boron fibers 364ff.
– applications 365
– manufacture 364ff.
– properties 365
boron nitride products 454
boron trifluoride 145
brackish water 10
bromates, alkali 182
bromides 181ff.
– applications 183
bromine 178ff.
– applications 182ff.
– economic importance 179
– natural deposits 178
bromine manufacture
– from bromide-enriched starting materials 180ff.
– from seawater 179ff.

cadmium pigments 546ff.
– toxicity 546ff.
cadmium yellow pigments
– manufacture 546
caesium 215
calcium 239ff.
– natural deposits 239
calcium carbide 242ff.
– applications 243
– economic importance 242
– manufacture 242ff.
calcium carbonate 240, 513ff.
– applications 126, 199, 240, 377, 508
– economic importance 240
– mining and manufacture 240
calcium carbonate filler
– manufacture 513
calcium chloride 241ff.
– applications 242
– economic importance 241ff.
– manufacture 242
calcium cyanamide 243
calcium fluoride 133ff.
calcium hydride 243
calcium hydrogen sulfite 126
calcium hydroxide 240ff., 374ff.
– applications 6, 241, 377
– by dry slaking 375
– by wet slaking 375
– from calcium carbide 376
– manufacture 241, 374ff.
calcium hypochlorite
– applications 177
– economic importance 169
– manufacture 169
calcium hypochlorite manufacture
– Griesheim Elektron process 171
– Olin process 171
– PPG process 172
calcium oxide 240ff., 373ff.
– applications 239, 241, 376ff.
– economic importance 240ff.
– manufacture 241, 373ff.
calcium phosphates
– applications 81ff.
– economic importance 81
– manufacture 85
calcium sulfate 389ff.
– applications 508
calcium sulfate manufacture 72, 75
calcium sulfate modifications 389ff.
– formation 390ff.
– properties 391
carbon, synthetic 476ff.
– applications 483ff.
– article impregnation and processing 482ff.
– economic importance 476

– properties 483 ff.
carbon black 488 ff.
– applications 495 ff.
– economic importance 488 ff.
– posttreatment 494
– properties 495 ff.
carbon black manufacture 489 ff.
– acetylene black process 493
– channel black process 492
– electric arc process 494
– furnace black processes 490 ff.
– gas furnace black process 492
– general information 489 ff.
– lamp black process 492
– pearling 491
– thermal black process 493
carbon disulfide 131
– applications 131
– manufacture 131
carbon felt 358 ff.
carbon fibers 356 ff.
– applications 357
– economic importance 362
– manufacture 356 ff.
– properties 356 ff.
– silicon carbide-coated 365 ff.
carbon manufacture 476 ff.
– carbonization 479 ff.
– densification and molding 479
– processing 478
– raw materials 477 ff.
carbon wool 358 ff.
cement
– compositions 378
– economic importance 378
cemented carbides 463
ceramic fibers 351
ceramics 414 ff.
– general information 414 ff.
– general manufacturing steps 417
– product classification 415 ff.
ceramic stains 543 ff.
CERPHOS process 396
chloralkali electrolysis 149, 152 ff.
– diaphragm process 155 ff.
– evaluation of the processes 161 ff.
– membrane process 157 ff.
– mercury process 152 ff.
– starting materials 150 ff.
chlorinated trisodium phosphate 172, 177
chlorine 149 ff.
– applications 3 ff., 7, 162, 280
– economic importance 149
– manufacture 152 ff., 166 ff.
chlorine dioxide
– applications 7, 172, 177
– economic importance 169
– manufacture 174 ff., 177
(chloro)methylphenysilanes
– manufacture 292 ff.
(chloro)methylsilanes
– manufacture 291 ff., 292
chlorosulfonic acid 125
chromate pigments 541 ff.
chrome-tanning agents
– manufacture 264 ff.
chromic acid *see chromium(VI) oxide*
chromite 257 ff.
chromium 266 ff.
– economic importance 266
chromium borides 467
chromium carbide 461 ff.
chromium compounds
– economic importance 255 ff.
– manufacture 258 ff.
– raw material 257 ff.
chromium manufacture 266
– chemical reduction 266 ff.
– electrochemical reduction of chromium(VI) oxide 268
– electrochemical reduction of chrome alum 267 ff.
chromium(III) oxide 263 ff.
– applications 266 ff.
chromium(III) oxide pigments
– applications 541
– manufacture in aqueous media 539 ff.
– manufacture in the solid state or melt 540
– properties 541
chromium(VI) oxide
– applications 266
– manufacture 263
clay ceramics *see silicate ceramics*
CNA 62
coarse ceramic products 398 ff.
coloring carbon blacks 495 ff.
COMURHEX process 577
concrete 371
corrosion protection pigments 549 ff.
– active 550
– chemical 551
– electrochemical 551
– physical 551
cristobalite 507, 512
cryolite *see sodium aluminum hexafluoride*

dental ceramics 418, 430
dialkyl phosphites 100
diamond 468 ff.
– applications 472
– economic importance 469
– manufacture 470 ff.
– mining 470

– properties 472
(dichloro)dimethylsilane
– hydrolysis 294
– methanolysis 295
Dipterex
– manufacture 104
disulfur dichloride 123
dolomite bricks 436ff.
DSN processes 62ff.

electro-ceramics 445ff.
electrochemical fluorination of organic compounds 146ff.
– product applications 148
electronic grade (EG) silicon 269, 271ff.
– manufacture 271ff.
enamel 406ff.
– applications 413
– firing 413
– general information 406
enamel application on sheet steel 412ff.
– dry processes 412ff.
– wet processes 412
enamel classification 406ff.
– coloring and opacifying systems 408ff.
– layer arrangement 406
enameling procedure 409
expandable clays 401
expanded products 399ff.
– applications 404ff.
– general information 399
expanded products (foam glass)
– from glasses 404
expanded products from clays and shales 399ff.
– gas-forming reactions 401ff.
– manufacture 402ff.
expanding cement 385

feldspar 419
ferrites 446
– applications 448ff.
– cubic 446ff.
– general information 446
– hard 448ff.
– hexagonal 448ff.
– manufacture 447ff., 449
– soft 446ff.
ferrochrome 267
ferrocyanate pigments 547ff.
ferrophosphorus 88
ferrosilicon 269ff.
– applications 273
– manufacture 270ff.
fillers 505ff.
– applications 516ff.
– economic importance 506
– general information 505ff.
– properties 515ff.
fillers, natural 507ff.
– beneficiation 508ff.
– silicas and silicon dioxide 507ff.
– silicates 507ff.
fillers, synthetic 509ff.
fine earthenware 428ff.
fireclay brick products 434ff.
fluorapatite 135
fluorine 133ff.
– applications 133
– economic importance 133
fluorine manufacture 135ff.
– raw materials 133ff.
fluorosulfonic acid 125
fluorspar 134
foamed carbon 486ff.
forsterite bricks 436
FP fiber 363
frit manufacture 410
– raw materials 410
– smelting 410ff.
furnace acid 70
furnace acid manufacture 78ff.

glass 307ff.
– applications 316, 512
– economic importance 307
– general information 307ff.
– properties 316ff.
glass-ceramics 309ff.
glass compositions 308
glass fiber
– textiles 346
– yarns and twists 346
glass fillers 512
glass manufacture 310ff.
– chemical decoloration 311
– container glass 315ff.
– flat glass 315
– melting furnaces 312ff.
– melting process 311ff.
– molding 314ff.
– raw materials 310ff.
– tank furnaces 313ff.
glass wool 350ff.
glassy carbon 486ff.
graphite 468, 472ff., 567ff.
graphite, natural 472ff.
– applications 475
– deposits and mining 473ff.
– economic importance 472ff.
– properties and applications 474ff.
graphite, synthetic 476ff.
– applications 483ff., 568

– article impregnation and processing 482ff.
– economic importance 476
– properties 483ff.
– purification 482
graphite foils and membranes 487ff.
graphite manufacture 476ff.
graphitization of synthetic carbon 480ff.
– Acheson process 480ff.
– Castner process 481
– general information 480
Guillini process 397
gypsum
– economic importance 389
gypsum, natural 392
gypsum, synthetic
– from flue gas desulfurization 394ff.
– from organic acid manufacture 394
– from phosphoric acid production 395ff.

hafnium carbide 460
hafnium nitride 465
heavy water 569ff.
heraklith 386
hexafluorosilicates 144
hexafluorosilicic acid 72, 133, 135, 141
HM carbon fibers
– applications 357
– manufacture 357ff., 358ff.
– properties 358
HST carbon fibers 360
HT carbon fibers
– applications 357
– manufacture 357ff., 358ff.
– properties 358
hydrazine 45ff.
– applications 51
– economic importance 45
hydrazine manufacture 45ff.
– Bayer process 48
– H_2O_2 process 50
– Raschig process 46
– urea process 48
hydrochloric acid 164ff., 280
– economic importance 165
– electrolysis 166
hydrogen 14ff.
– applications 19ff.
– economic importance 14ff.
– transport 20
hydrogen bromide 181
– applications 183
hydrogen chloride 164ff.
– applications 125, 165ff., 280
– economic importance 165
– manufacture 164
hydrogen fluoride 137ff.
– applications 125, 140
– economic importance 137
hydrogen fluoride manufacture 137ff.
– Bayer process 138ff.
– Buss process 139.
– Du Pont process 139
hydrogen iodide 186ff.
hydrogen manufacture 15ff., 152ff.
– as a byproduct 18
– petrochemical processes and coal gasification 15
– water electrolysis 16
hydrogen peroxide 21ff., 50
– applications 29
– economic importance 21ff.
– production 23ff.
hydrogen sulfide 129ff.
– applications 129
– manufacture 129
hydroxylamine 52ff.
– applications 52ff.
– economic importance 52ff.
hydroxylamine manufacture 53ff.
– nitrate reduction process 54
– nitric oxide reduction process 53
– Raschig process 53
hypophosphites 93ff.
hypophosphorous acid 93ff.

IDR process 582
inorganic fibers
– general information 330
inorganic fibers applications 331ff.
– insulation materials 331
– reinforcement sector 332ff.
inorganic peroxo compounds 21ff.
interference pigments 552ff.
iodates, alkali 187
iodides
– applications 187ff.
iodine
– applications 187ff.
– economic importance 184ff.
– natural deposits 184
iodine manufacture
– from brines 185
– from niter residual solutions 186
iron(II) sulfate 4, 119ff., 526
iron oxide pigments 532ff.
– aniline process 538
– by oxidation processes in aqueous media 536ff.
– by thermal decomposition 535ff.
– natural 533ff.
– synthetic 534ff.
iron oxide pigments, synthetic
– applications 538ff.
– general information 534ff.

kaolin 418 ff., 508
Knauf process 396

lead glass 309
LECA process 403
light bricks 398
lime 372 ff.
– applications 310, 377
– economic importance 372 ff.
– raw materials 372 ff.
lime hydrate *see calcium hydroxide*
lithium 215 ff.
– applications 216
– economic importance 215
– occurrence 215
lithium carbonate 217
lithium chloride 218
lithium hydride 217 ff.
lithium hydroxide 217
lithium hypochlorite 172
lithopone pigments 530 ff.
luminescent pigments 553 ff.
– applications 553 ff.
– properties 554
lustrous pigments 551

magnesia bricks 436
magnesia cement 386
magnesium
– applications 234 ff.
– economic importance 233
– natural deposits 233
magnesium carbonate 235 ff., 513
– applications 236, 508
– economic importance 235
– manufacture 235 ff.
– natural deposits 235
magnesium chloride 237 ff.
– applications 238
– economic importance 237
– manufacture 237 ff.
magnesium manufacture 233 ff.
– melt electrolysis 233 ff.
– silicothermal 234
magnesium nitrate 64
magnesium oxide 236 ff., 436, 444
– applications 237
– economic importance 236
– manufacture 236
magnesium sulfate 238 ff.
– applications 238 ff.
– economic importance 238
– manufacture 238
magnetic pigments 555 ff.
– applications 557 ff.
– general information 555
– manufacture 556
– properties 555 ff.
magneto-ceramics 445 ff.
Malathion
– manufacture 103
manganese
– electrochemical manufacture 287
– importance and applications 287
manganese compounds
– economic importance 277
– manufacture 279 ff.
– raw materials 277 ff.
manganese(II) carbonate
– applications 286
– manufacture 281
– applications 286
– manufacture 280
manganese(II) oxide
– applications 286
– manufacture 279
manganese(II) sulfate
– applications 283, 286 ff.
– economic importance 277
– manufacture 279 ff.
manganese(II, III) oxide
– applications 286
– manufacture 281
manganese(III) oxide
– applications 286
– manufacture 281
manganese(IV) oxide
– applications 284, 286
– economic importance 277
– manufacture 281 ff.
manganese(IV) oxide manufacture
– aftertreatment of manganese(IV) oxide minerals 282 ff.
– electrolytic manganese(IV) oxide (EMD) 283 ff.
– oxidation of manganese(II) carbonate 283
melt phosphates 192
metal borides 466 ff.
metal fibers 367 ff.
– applications 369
– manufacture 367 ff.
– properties 368 ff.
metallic hard materials 457 ff.
– general information 457 ff.
– general manufacturing processes 458 ff.
– properties 459
metallic pigments 552
metallurgical grade (MG) silicon 269 ff.
– manufacture 270
metal nitrides 464 ff.
metal silicides 467
mineral fiber compositions 351
mineral fiber insulating materials

– applications 354 ff.
– economic importance 349 ff.
– fiber manufacture 352 ff.
– general information 349
– manufacture 350
– melt manufacture 351
– processing of fibers into insulating materials 353
– raw materials 350
mineral fiber insulating materials manufacture
– blowing process 352
– centrifugal processes 352
– two-step centrifugal jet process 353
mixed metal oxide pigments 543 ff.
molybdate pigments 541 ff.
molybdenum carbide 462
molybdenum disilicide 467
mortar 371
Munich process 174

nacreous pigments 552
niobium carbide 461
niobium nitride 465
nitric acid, highly concentrated 62 ff.
– direct strong nitric processes 62 ff.
– indirect extractive distillation process 64 ff.
nitric acid 55 ff.
– applications 67 ff., 197 ff., 213, 577, 587
– economic importance 55 ff.
nitric acid manufacture 56 ff.
– catalytic combustion of ammonia 57 ff.
– conversion of nitrogen(IV) oxide into nitric acid 58 ff.
– oxidation of nitrogen(II) oxide 58
– plant types 59 ff.
– process description 60 ff.
– waste gases 65 ff.
nitric oxide 53
nitrogen-containing fertilizers 199 ff.
– economic importance 199
nitrophosphates
– economic importance 191
nitrophosphates manufacture 197
– carbonitric process 198
– sulfonitric process 198
nonoxide ceramics 449 ff.
– carbides 450 ff.
– economic importance 450
nuclear energy
– economic importance 559
nuclear fuel
– manufacture 571 ff.
– manufacture from uranium ore concentrates 571 ff.
nuclear fuel cycle 559 ff.
– general information 562
nuclear fuel elements 584
nuclear fuel manufacture
– fuel elements 584
– uranium metal 582 ff.
– uranium-plutonium mixed oxides 583
nuclear fuel reprocessing
– separation of uranium and plutonium 588 ff.
– working up of uranium and plutonium 590
nuclear reactor types
– advanced gas-cooled reactors (AGR) 568
– boiling water reactors 567
– Candu 570
– fast breeder reactors 570 ff.
– gas-cooled reactors 568
– general information 565
– graphite-moderated reactors 567
– heavy water reactors 569 ff.
– high temperature reactors (HTR) 568
– light water-cooled reactors 569
– light water reactors 566
– Magnox 568, 586 ff.
– pressurized water reactors 567
nuclear waste disposal 584 ff.
– concrete encasement 592
– conditioning of radioactive waste 590 ff.
– deposition in bore holes 594
– direct permanent storage of irradiated fuel elements 594
– gaseous radioactive products 592 ff.
– general information 584 ff.
– interim storage of irradiated fuel elements 585 ff.
– liquid radioactive waste treatment 591
– permanent storage 593 ff.
– reprocessing of spent fuel elements 586 ff.
– vitrification 591

optical glass 309
optical (glass) fibers 348 ff.
– applications 349
– manufacture 348 ff.
oxide ceramics 439 ff.
– aluminum oxide products 440 ff.
– beryllium oxide products 443
– general information 439 ff.
– other products 444 ff.
– thorium oxide products 444
– uranium oxide products 443 ff.
– zirconium(IV) oxide products 442
oxide pigments 543 ff.

Parathion
– manufacture 103
perchloric acid
– economic importance 169
– from Merck process 175
perlite 508
Phosdrin

– manufacture 104
phosphonic acids 100ff.
– applications 105
phosphoric acid 70ff.
– applications 70ff.
– economic importance 70ff.
phosphoric acid manufacture 72ff.
– hemihydrate processes 72ff.
– impurity removal 72, 77ff.
– wet-process acid by dihydrate process 74ff.
phosphoric acid triesters
– applications 102 ff.
– economic importance 96
– manufacture 96ff.
phosphorous acid 94
phosphorous acid esters
– applications 104
phosphorous(V) ester acids 98
– applications 103
phosphorus 85ff.
– applications 86
– economic importance 85
phosphorus compounds 68ff.
phosphorus-containing fertilizers 189ff.
– economic importance 189
– manufacture 192
phosphorus oxychloride 92ff.
phosphorus pentachloride 92
phosphorus pentasulfide 90ff.
phosphorus pentoxide 90
phosphorus raw materials manufacture 68ff.
phosphorus sulfochloride 93
phosphorus trichloride 91ff.
pigments 518ff.
– applications 521ff.
– classification 519
– economic importance 518ff.
– general information 518ff.
plasters 371
plaster setting processes 397ff.
plutonium 570, 586
plutonium carbide 571
plutonium(IV) oxide 570ff., 583
poly(organosiloxanes), branched
– manufacture 297
poly(organosiloxanes), linear 294ff.
– acidic ring opening polymerization 296
– alkaline ring opening polymerization 296
– nonequilibrating polycondensation 297
– synthesis of oligomeric precursors 294f.
polyphosphates 11, 83
porcelain 429ff.
– soft 418, 429
– hard 418, 430
Portland cement 378ff.
– applications 383
– clinker compositions 379
– raw materials 378
Portland cement manufacture 379, 381ff.
– dry processes 381ff.
– half-dry process 381
– half-wet process 381
– Müller-Kühne process 119
– wet process 381
Portland cement solidification processes 386ff.
posttreatment of silicas 511
potassium 215, 228ff.
potassium carbonate
– applications 37, 230
– manufacture 229ff.
potassium chlorate
– applications 177
– economic importance 169
– manufacture 173
potassium chloride 210ff., 212ff.
– extraction 210ff.
potassium-containing fertilizers 208ff.
– economic importance 209ff.
– manufacture 210ff.
potassium dichromate
– applications 265
– manufacture 262
potassium hydrogen fluoride 143ff.
potassium hydroxide
– applications 229
– economic importance 229
– manufacture 229
potassium nitrate 213
potassium perchlorate
– manufacture 175
potassium permanganate
– applications 286
– economic importance 277
potassium permanganate manufacture 284
– single-step liquid phase processes 285ff.
– two-step roasting processes 284ff.
potassium salts
– extraction 210ff.
– occurrence 208ff.
potassium silicates
– applications 321
– economic importance 319
– general information 319
– manufacture 319ff.
potassium sulfate
– economic importance 212
– manufacture 212
potassium titanate fibers 369ff.
pozzolan cements 384
PUREX process 586ff.
pyrogenic silicas 509ff.

pyrolytic carbon 484 ff., 568
pyrolytic graphite 484 ff.

quartz glass 308
quicklime *see calcium oxide*

rapidly fired porcelain 430 ff.
red phosphorus
– manufacture 89 ff.
refractory ceramics 431 ff.
– alumina-rich 435
– classification 433
– definition 431 ff.
– properties of important constituents 432 ff.
– silica products 434
– silicon carbide products 437 ff.
– speciality products 437
reinforcing carbon fibers *see HT, HM and HST carbon fibers*
reverse osmosis *see water desalination*
rock wool 351
roofing materials 371
rubidium 215

Saffil fiber 364
sand-line bricks 376
seawater 10
semi-active carbon black 496
silanes
– manufacture 291 ff.
silicate ceramics 417 ff.
– composition 417
– economic importance 431
– extraction and treatment of kaolin 419
– glazes 427 ff.
– properties and applications 428
– raw materials 417 ff.
silicate ceramics firing 425 ff.
– conditions 426 ff.
– physical-chemical processes 425 ff.
silicate ceramics manufacture
– casting processes 422 ff.
– drying processes 425
– forming processes 421
– of clay ceramic batches 420
– plastic forming 423
– powder pressing 423 ff.
silicic acid esters 276
silicon 269 ff.
– applications 273 ff.
– economic importance 269 ff.
– general information 269
– inorganic compounds 274 ff.
silicon carbide 437 ff.
silicon carbide fibers 365 ff.
– applications 366
– manufacture 365 ff.
– properties 366
silicon carbide products 450 ff.
– economic importance 438 ff.
– fine ceramics 451 ff.
– manufacture 450 ff.
silicone
– manufacture 291 ff.
silicone block copolymers 305
silicone copolymers 305
silicone graft copolymers 305 ff.
silicone oils
– emulsions 299
– pastes and greases 299
– properties and applications 298
silicone-polyester resins 304 ff.
silicone resins 304 ff.
silicone rubbers 299 ff.
– hot vulcanizable addition crosslinkable 302 ff.
– hot vulcanizable liquid 303
– hot vulcanizable peroxide crosslinked 302
– properties 303 ff.
– room temperature vulcanizable 299 ff.
– single component 299 ff.
– two component room temperature vulcanizable 301 ff.
silicones
– economic importance 290
– structure and properties 289 ff.
silicon nitride products 452 ff.
silicon tetrachloride 275 ff.
siloxanes
– organo-functional 293
single crystal fibers calcium sulfate 370
sinter phosphates 192
slag cement 383 ff.
slag fibers 351
slaked lime *see calcium hydroxide*
sodium 215
– applications 219, 570
– economic importance 218
– manufacture 218
sodium aluminate 254
sodium aluminum hexafluoride 133, 142 ff., 248 ff.
sodium borates 226 ff.
– applications 227
– extraction 226 ff.
– natural deposits and economic importance 226
sodium carbonate 215, 219 ff.
– applications 151, 222, 258, 310, 572
– economic importance 220
– general information 219 ff.
– manufacture of synthetic 221 ff.
– production from natural deposits 221
sodium carbonate perhydrate
– applications 29
– economic importance 22

– production 27
sodium chlorate
– applications 176 ff.
– economic importance 169
– manufacture 173
sodium chloride 151 ff., 215, 224 ff.
– applications 176 ff.
– economic importance 151 f., 169
– manufacture 172
sodium dichromate
– applications 265
sodium dichromate dihydrate manufacture
– carbon dioxide process 262
– sulfuric acid process 261 ff.
sodium disulfite 126
sodium dithionite 127 ff.
– applications 129
– production 127 ff.
sodium fluoride 143
sodium hydrogen carbonate 222 ff.
– applications 223, 572
– economic importance 223
– manufacture 223
– natural deposits 222 ff.
sodium hydrogen sulfate 225 ff.
sodium hydrogen sulfide 126, 130 ff.
sodium hydroxide 48, 149 ff., 215
– applications 126, 162 ff., 251, 254
– economic importance 149
– manufacture 152 ff., 166 ff.
sodium hydroxymethanesulfinate 129
– applications 129
– manufacture 129
sodium hypochlorite 46, 48, 68 ff., 177
– applications 177
– economic importance 168
– manufacture 169
sodium metasilicate 320
sodium perborate
– applications 29
– economic importance 22
– production 26 ff.
sodium percarbonate *see sodium carbonate perhydrate*
sodium perchlorate
– manufacture 175
sodium peroxide
– applications 30
– exonomic importance 22
– production 28 ff.
sodium phosphates 79 ff.
– applications 79 ff.
– economic importance 79 ff.
– manufacture 82
sodium silicates
– applications 321
– economic importance 319
– general information 319
– manufacture 319 ff.
sodium sulfate 215, 223 fff.
– applications 225, 312
– economic importance 224
– from natural deposits 224
– manufacture 225
sodium sulfide 130
– applications 130
– manufacture 130
sodium sulfite 126
sodium tetrahydroborate 128
sodium thiosulfate 126 ff.
– applications 127
– manufacture 126
sol-gel process 317 f.
sorel cement 385
stoneware 429
strontium
– applications 243
– natural deposits 243
strontium carbonate 243 ff.
strontium nitrate 244
sulfur 105 ff.
– applications 108, 120
– economic importance 106
– occurrence 105 ff.
sulfur dichloride 123
sulfur dioxide 120 ff.
– applications 122 ff., 126, 164
– from flue gas desulfurization 121
– manufacture 110 ff., 120 ff.
sulfur hexafluoride 145 ff.
sulfuric acid 108 ff., 527
– applications 11, 64, 68, 120, 224 ff., 261, 263, 279
– economic importance 108 ff.
– recycling and metal sulfate disposal 115 ff.
sulfuric acid from metal sulfates
– iron(II) sulfate disposal 119
– Müller-Kühne process 119
sulfuric acid manufacture
– contact process 112
– double absorption process 113 ff.
– from sulfur dioxide 109
– moist gas catalysis process 115
– nitrous process 115
sulfuric acid recycling
– Bayer-Bertrams process 117
– oxidative cracking 118
– Pauling-Plinke process 117
– reductive cracking 118
sulfur production 106 ff.
– Clauss process 107
– COPE process 108
– Frasch process 107
sulfur trioxide 122

– applications 122 ff., 125
– manufacture 122
sulfuryl chloride 124 ff.
superphosphate
– economic importance 190
– manufacture 192 ff.

talc 507 ff.
tantalum carbide 461
tantalum nitride 465
tetrafluoroboric acid 145
tetrapotassium diphosphate 81
– manufacture 84
textile glass fibers
– applications 347
– classification 343
– economic importance 342 ff.
– general information 342
– raw materials 344
textile glass fibers manufacture 343 ff.
– bead melt process 345
– direct melt process 344
– rod drawing process 345
textile glass mats 346
thermal and basic slag phosphates
– economic importance and manufacture 191 ff.
thionyl chloride 123 ff.
– applications 124
– manufacture 123 ff.
thiophosphoric acid derivatives 99
Thomas phosphates 192
thorium 570
– economic importance 564 ff.
– occurrence 564 ff.
thorium carbide 464, 568, 571
thorium(IV) oxide 444, 568, 570
titanates 445 ff.
titanium carbide 460
titanium diboride 467
titanium dioxide 444 ff., 515, 523, 525 ff., 552
– applications 515
titanium dioxide pigments 523
– applications 529 ff.
– by chloride process 527 ff.
– by sulfate process 526 ff.
– economic importance 523
– manufacture 525 ff.
– posttreatment 528 ff.
– raw materials 523
titanium nitride 465
trialkyl phosphites 100
triaryl phosphites 99
trichlorosilane 275
triple superphosphate
– economic importance 190
– manufacture 194
tungsten carbide 462 ff.

ultramarine pigments 548 ff.
uranium 569 ff., 586
– economic importance 564 ff.
– occurrence 564 ff.
uranium 235 enrichment 579 ff.
– gas centrifuge process 579
– gas diffusion process 579
– laser isotope separation 579
– nozzle separation process 579
uranium carbide 464
uranium hexafluoride 144 ff., 571
– conversion into nuclear fuel 580
uranium hexafluoride from uranium concentrates 576
– dry process 578
– general information 576 ff.
– wet process 577 ff.
uranium(IV) oxide 443 ff., 566 ff., 580 ff., 583
– pellet manufacture 582
uranium(IV) oxide from uranium hexafluoride 580
– dry process 582
– wet processes 581 ff.
uranium production
– from phosphates 575
– from seawater 576
uranium production from ores 571 ff.
– leaching processes 572
– separation from leaching solutions 573
urea 48
– economic importance 201
urea manufacture 204 ff.
– Mitsui-Toatsu process 206
– Montedison process 206
– Snam Progetti process 207
– Stamicarbon process 207

vanadium carbide 461
vanadium(V) oxide 88, 111

water 1 ff.
– chlorination 3 ff.
– chlorine dioxide treatment 7
– chlorohydrocarbon removal 8
– deionized 8
– desalination 2, 9, 10 ff., 12 ff.
– economic importance 1 ff.
– nitrate removal 7 ff.
– ozonization 3ff.
– purification 2 ff.
– usage 1
water glass
– manufacture 320 ff.
wet chemically manufactured silicas 510 ff.
wet chemically manufactured silicates 510 ff.
wet process acid 70

whiskers 369 ff.
white cement 385
white phosphorus
– manufacture 86 ff.
white pigments 522 ff.
woodstone 386
woven carbon 358

yellow cake 574
yttrium oxide 444

zeolites 321 ff.
– characteristics 324
– dehydration 327
– economic importance 321 ff.
– natural types 325
– pelletization 327
– structure 323 ff.
– types 322 ff.
zeolites applications 328 ff.
– as adsorption agents 328
– as catalysts 329 ff.
– as ion exchangers 328
– for separation processes 329
zeolites manufacture 325
– by cation exchange 327
– from natural raw materials 325
– from synthetic raw materials 325 ff.
zinc oxide white pigments 531 ff.
– applications 532
– by American process 532
– by French process 531
– by wet chemical process 532
– manufacture 531 ff.
zinc sulfide pigments 530 ff.
zirconium carbide 460
zirconium nitride 465
zirconium(IV) oxide 442
zirkaloy 584